Nano-bioelectronics for Precision Health Monitoring

Kuldeep Mahato • Pranjal Chandra
Editors

Nano-bioelectronics for Precision Health Monitoring

Springer

Editors
Kuldeep Mahato
Department of Biotechnology
Motilal Nehru National Institute of Technology Allahabad
Prayagraj, Uttar Pradesh, India

Pranjal Chandra
School of Biochemical Engineering
Indian Institute of Technology (BHU) Varanasi
Varanasi, Uttar Pradesh, India

ISBN 978-981-95-2116-6 ISBN 978-981-95-2117-3 (eBook)
https://doi.org/10.1007/978-981-95-2117-3

This Springer imprint is published by the registered company Springer Nature Singapore Pte Ltd.
The registered company address is: 152 Beach Road, #21-01/04 Gateway East, Singapore 189721, Singapore

Preface

The twenty-first century has witnessed a paradigm shift in healthcare, from the generalized, population-wide treatment models of the past to an era of personalized, precision health. At the confluence of this transformation lies an exciting, interdisciplinary frontier: nanobioelectronics, a domain where nanotechnology, biosensing, and electronics converge to advance individualized health monitoring, disease prevention, and therapy optimization. This book, *Nanobioelectronics for Precision Health Monitoring*, seeks to elucidate the fundamentals, innovations, and future potential of this evolving field with a particular emphasis on the enabling role of sensors, biomarkers, and bioelectronic systems. Precision health represents a holistic, proactive approach that goes beyond disease management to emphasize early detection, continuous monitoring, and personalized interventions tailored to an individual's unique biological, environmental, and behavioral profiles. Achieving such granularity in healthcare necessitates advanced technological infrastructures that can capture multidimensional physiological data in real-time, with high sensitivity and specificity. This aspiration has driven the rapid advancement of wearable, implantable, and ingestible biosensors, technologies that promise to redefine our interaction with health information. The concept of bioelectronics, enhanced at the nanoscale, emerges as a cornerstone of this shift, offering unprecedented capabilities in interfacing biological systems with electronic platforms.

The first chapter, "Precision Health Monitoring and Wellness Tracking: Fundamentals and Principles," establishes the conceptual groundwork for understanding precision health. It explores the scientific underpinnings that differentiate precision health from conventional medical paradigms. Central to this discussion are the principles of predictive analytics, continuous monitoring, and the integration of data streams from diverse biological, physiological, and environmental sources. This foundational knowledge serves as the springboard for appreciating how nanobioelectronics can meet the intricate demands of personalized healthcare. In Chapter 2, "Introduction to Nano-Bioelectronics for Precision Health Monitoring," we discuss the interdisciplinary synthesis that defines nanobioelectronics. This chapter surveys the underlying technologies in nanofabrication, biointerface engineering, and signal transduction mechanisms that form the basis of modern bioelectronic devices. Special attention is given to the unique properties that emerge at the nanoscale, such as quantum effects and enhanced surface-to-volume ratios, which significantly improve sensor performance, miniaturization, and integration with biological systems. Chapter 3, "Fundamentals of Biomarkers and Biosensors for Precision Healthcare," addresses the crucial role biomarkers play in informing precision health strategies. Biomarkers, whether genomic, proteomic, metabolic, or physiological, offer invaluable insights into an individual's health status and disease risk. This chapter comprehensively covers the criteria for effective biomarkers, methods of detection, and their translational relevance in clinical and nonclinical settings. Building upon this, Chapter 4, "Biochemical, Biophysical, and Digital Biomarkers in Precision Healthcare," provides an in-depth exploration of specific biomarkers pertinent to various health domains, including metabolic health, cardiovascular function, neurological integrity, and immune status. This chapter not only categorizes these biomarkers but also examines the current and emerging biosensing technologies capable of detecting and quantifying them with clinical-grade accuracy.

The spectrum of biosensing modalities is broad, and Chapter 5, "Exploring Novel Healthcare Monitoring Modalities: Wearable, Ingestible, and Implantable Biosensors," provides a detailed examination of these diverse form factors. Each modality presents unique advantages and challenges in terms of biocompatibility, data fidelity, power consumption, and user compliance. The chapter also navigates through case studies and prototypes that illustrate how these biosensors are being deployed in real-world scenarios for continuous, unobtrusive health monitoring. Drug therapy remains a cornerstone of disease management, yet therapeutic efficacy is often limited by the lack of real-time drug monitoring. Chapter 6, "Electrochemical Biosensors for Therapeutic Drug Monitoring: Next Generation of Disease Management," addresses this gap by presenting electrochemical sensing technologies that enable precise monitoring of pharmacokinetics and pharmacodynamics at the point of care. The discussion extends to the materials science aspects and electrode functionalization strategies that drive sensitivity and specificity in these biosensors. The subsequent chapters apply these technologies to specific disease contexts. Chapter 7, "Biosensors for Cardiovascular Health Monitoring," reviews the latest biosensing solutions aimed at tracking critical cardiovascular biomarkers and physiological parameters such as heart rate variability, blood pressure, and circulating biomarkers of cardiac stress. Such devices are pivotal for early detection of cardiovascular events and longitudinal health management. Similarly, Chapter 8, "Bioelectronic Devices for Diabetes: Monitoring, Management, and Precision Health," highlights the innovations in glucose sensing, insulin delivery systems, and closed-loop devices that are redefining diabetes care. The chapter addresses both invasive and noninvasive approaches, exploring how nanomaterials and smart algorithms enabling more responsive and less burdensome management of diabetes.

Neurological diseases present unique challenges due to the complexity of brain function and the blood-brain barrier. Chapter 9, "Nanobioelectronics for Neurodegenerative Disease Monitoring and Management," presents the current state and future prospects of biosensing technologies designed for neurological applications. Topics include wearable EEG sensors, neurochemical biosensors, and implantable devices that can monitor neurodegeneration and potentially facilitate early interventions. Healthcare monitoring is inherently multidimensional. Chapter 10, "Multimodal Sensing Arrays for Comprehensive Health Monitoring and Disease Management," explores sensor systems that integrate multiple sensing modalities, chemical, physical, and electrophysiological, to capture a holistic picture of health. This integration not only enhances diagnostic accuracy but also enables predictive analytics through advanced data fusion techniques. Wearable technologies continue to evolve beyond conventional devices. Chapter 11, "Smart Skin and Beyond: Unlocking the Full Potential of Wearable Health Sensors," introduces cutting-edge developments in flexible, stretchable electronics that seamlessly integrate with the human body. Smart skin technologies, capable of conformal wear, offer high-fidelity data capture while maintaining user comfort and long-term wearability. Finally, Chapter 12, "Integrated Bioelectronics for Precision Health Monitoring," synthesizes the thematic threads explored throughout the book, proposing frameworks for fully integrated health monitoring systems. This chapter discusses the integration of biosensors with wireless communication, cloud computing, and machine learning for real-time data analytics and personalized health recommendations.

In assembling this book, our intent is not only to inform but also to inspire further research and innovation at the interface of nanotechnology, biology, and electronics. The audience for this text spans a broad spectrum from academic researchers and clinicians to engineers and innovators in the biomedical industry. We envision this volume as a reference and a guide that underscores both the current achievements and the boundless potential of nano-bioelectronics in shaping the future of precision health.

Prayagraj, Uttar Pradesh, India Kuldeep Mahato
Varanasi, Uttar Pradesh, India Pranjal Chandra

Contents

Editors and Contributors

About the Editors

Kuldeep Mahato is a faculty in the Department of Biotechnology at Motilal Nehru National Institute of Technology (MNNIT) Allahabad, Prayagraj, India. Prior to joining MNNIT, he was a post-doctoral fellow at Aiiso Yufeng Li Family Department of Chemical and NanoEngineering, University of California San Diego, USA. He earned his PhD from the Indian Institute of Technology Guwahati, Assam, India. His research bridges bioengineering, bioelectronics, biotechnology, biomedical devices, and material sciences through his multidisciplinary approach. Currently, his research focus is on the development of hybrid multimodal bioelectronics devices aimed at comprehensive health monitoring. His work has led to the innovation of several devices for biomedical diagnosis and onsite quality controls for food. He is an active member of several prestigious organizations, including the American Association for the Advancement of Science, the American Chemical Society, and the Institute of Electrical and Electronics Engineers. His expertise is further recognized through his role as a reviewer for top-tier journals in the domains of biosensors, bioelectronics, and analytical chemistry. With over 70 publications to his name, including research papers, review articles, invited book chapters, and edited books on various domains of biosensor development, he has significantly contributed to scientific literature. His research has earned him the prestigious Gandhian Young Technological Innovation Award. Additionally, he serves as a guest editor for several renowned journals in the domain of sensors, biosensors, bioelectronics, etc. Notably, he has also been honored by being listed among the world's top 2% of scientists by Stanford University, USA.

Pranjal Chandra is an Associate Professor at the School of Biochemical Engineering, Indian Institute of Technology (BHU) Varanasi, India. He earned his PhD from Pusan National University, South Korea, and did post-doctoral training at the Technion-Israel Institute of Technology, Israel. His research focus is highly interdisciplinary, spanning a wide range in biotechnology, nanobiosensors, material engineering, and biomedical devices. He has designed several commercially viable biosensing prototypes that can be operated for onsite analysis in the arena of biomedical diagnostics and environmental monitoring. He is an associate editor/ editorial board member for various international journals including *Scientific Reports*, *Microchemical Journal*, *Green Analytical Chemistry*, etc. He has been appointed as an Advisor for the Biomedical Sensors Domain and Sensor Networks Systems at the Institution of Engineering and Technology (IET), London. He is also a member of the International Centre for Sensor Science and Engineering, Chicago, Illinois, USA. He has authored more than 200 publications including research/ review papers and invited book chapters. He is an accomplished inventor of 14 patents; he has co-edited 14 books on various aspects of biosensors/medical diagnostics/material engineering. He has guided 5 PhD students, and currently, he is supervising 13 graduate students (PhD and MTech) and a Post-Doctoral Fellow. He is the recipient of many prestigious awards and coveted honors such as Fellow of the Indian Chemical Society; Shakuntala Amirchand Award by Indian Council of Medical Research; FSSAI: Eat Right Award; Prof. B K Bacchawat Young Scientist Award Medal from National Academy of Sciences, Member of INYAS_INSA, DST Ramanujan fellowship; Early Career Research Award; BK-21 and NRF fellowship, South Korea; Technion Post-Doctoral Fellowship, Israel; Young Scientist Award of Biotech Research Society India; and Young Engineers Award. He is also listed among the world's top 2% scientists from last 4 years by Stanford University, USA.

Contributors

Amal Abbas Aiiso Yufeng Li Family Department of Chemical and NanoEngineering, University of California San Diego, La Jolla, CA, USA

Adyant Balaji Shu Chien-Gene Lay Department of Bioengineering, University of California, San Diego, CA, USA

Laís Canniatti Brazaca Instituto de Química de São Carlos, Universidade de São Paulo, São Carlos, SP, Brazil

Instituto Nacional de Ciência e Tecnologia de Bioanalítica - Lauro Kubota (INCTBio-LK), Campinas, SP, Brazil

Kalyani Bukya Department of Chemistry, Indian Institute of Technology Palakkad, Palakkad, Kerala, India

Emanuel Carrilho Instituto de Química de São Carlos, Universidade de São Paulo, São Carlos, SP, Brazil

Instituto Nacional de Ciência e Tecnologia de Bioanalítica - Lauro Kubota (INCTBio-LK), Campinas, SP, Brazil

Gert Cauwenberghs Institute for Neural Computation, University of California, San Diego, CA, USA

Shu Chien-Gene Lay Department of Bioengineering, University of California, San Diego, CA, USA

Nuenghathai Chaiya Chemistry program, Faculty of Science, Maejo University, Chiang Mai, Thailand

Department of Chemical and Nano Engineering, University of California San Diego, La Jolla, CA, USA

Pranjal Chandra Laboratory of Bio-Physio Sensors and Nanobioengineering, School of Biochemical Engineering, Indian Institute of Technology (BHU), Varanasi, Uttar Pradesh, India

Chen Chen School of Life Science and Technology, Xidian University, Xi'an, China

José Bonilla Cruz Advanced Functional Materials and Nanotechnology Group, Centro de Investigación en Materiales Avanzados S. C. (CIMAV-Subsede Monterrey), Apodaca-Nuevo León, México

Omeed Djassemi Shu Chien-Gene Lay Department of Bioengineering, University of California, San Diego, CA, USA

Gairika Ghosh Department of Chemistry and Biochemistry, University of California San Diego, La Jolla, CA, USA

Xinyue Guo State Key Laboratory of Geomicrobiology and Environmental Changes, Engineering Research Center of Nano-Geomaterials of the Ministry of Education, Faculty of Materials Science and Chemistry, China University of Geosciences, Wuhan, China

Ga-Eun Kim Aiiso Yufeng Li Family Department of Chemical and NanoEngineering, University of California San Diego, La Jolla, CA, USA

Naveen Kotagiri Department of Chemistry, Indian Institute of Technology Palakkad, Palakkad, Kerala, India

Yugender Kotagiri Goud Department of Chemistry, Indian Institute of Technology Palakkad, Palakkad, Kerala, India

Shekher Kummari Department of Electrical and Computer Engineering, University of California San Diego, La Jolla, CA, USA

Min Suk Lee Shu Chien-Gene Lay Department of Bioengineering, University of California, San Diego, CA, USA

Hui Li State Key Laboratory of Geomicrobiology and Environmental Changes, Engineering Research Center of Nano-Geomaterials of the Ministry of Education, Faculty of Materials Science and Chemistry, China University of Geosciences, Wuhan, China

Linhan Li Faculty of Biotechnology, Universitat Autònoma Barcelona, Barcelona, Spain

Shaoguang Li State Key Laboratory of Geomicrobiology and Environmental Changes, Engineering Research Center of Nano-Geomaterials of the Ministry of Education, Faculty of Materials Science and Chemistry, China University of Geosciences, Wuhan, China

Renzhe Lu State Key Laboratory of Biogeology and Environmental Geology, Engineering Research Center of Nano-Geomaterials of the Ministry of Education, Faculty of Materials Science and Chemistry, China University of Geosciences, Wuhan, China

Kuldeep Mahato Department of Biotechnology, Motilal Nehru National Institute of Technology Allahabad, Prayagraj, Uttar Pradesh, India

Elsa Maria Materón Instituto de Física de São Carlos, Universidade de São Paulo, São Carlos, SP, Brazil

Ziyin Mei State Key Laboratory of Geomicrobiology and Environmental Changes, Engineering Research Center of Nano-Geomaterials of the Ministry of Education, Faculty of Materials Science and Chemistry, China University of Geosciences, Wuhan, China

Larissa Meneghin Michilini Instituto de Química de São Carlos, Universidade de São Paulo, São Carlos, SP, Brazil

Instituto Nacional de Ciência e Tecnologia de Bioanalítica – Lauro Kubota (INCTBio-LK), Campinas, SP, Brazil

Chochanon Moonla Department of Chemical and Nano Engineering, University of California San Diego, La Jolla, CA, USA

Marcos Lazaro Moreli Instituto de Química de São Carlos, Universidade de São Paulo, São Carlos, SP, Brazil

Instituto de Ciências da Saúde - Universadade Federal de Jataí, Jataí, GO, Brazil

Alok Pandya Department of Biotechnology and Bioengineering, Institute of Advanced Research, Gandhinagar, Gujarat, India

Lakshmi R. Panicker Department of Chemistry, Indian Institute of Technology Palakkad, Palakkad, Kerala, India

Nirmal Roy Electronics and Communication Engineering Department, Motilal Nehru National Institute of Technology Allahabad, Prayagraj, Uttar Pradesh, India

Desiree Tamara Scheidt Instituto de Química de São Carlos, Universidade de São Paulo, São Carlos, SP, Brazil

Instituto Nacional de Ciência e Tecnologia de Bioanalítica - Lauro Kubota (INCTBio-LK), Campinas, SP, Brazil

Ravish Sharma Department of Chemistry and Biochemistry, University of California San Diego, La Jolla, CA, USA

Tanin Tangkuaram Chemistry program, Faculty of Science, Maejo University, Chiang Mai, Thailand

Alok Tripathi Department of Chemical and Biochemical Engineering, School of Engineering, Indrashil University, Gujarat, India

Abhinav Uppal Shu Chien-Gene Lay Department of Bioengineering, University of California, San Diego, CA, USA

Fan Xia State Key Laboratory of Biogeology and Environmental Geology, Engineering Research Center of Nano-Geomaterials of the Ministry of Education, Faculty of Materials Science and Chemistry, China University of Geosciences, Wuhan, China

Yuchen Xu Institute for Neural Computation, University of California, San Diego, CA, USA

Aizip inc., Cupertino, CA, USA

Martins Lateef Yusuf State Key Laboratory of Biogeology and Environmental Geology, Engineering Research Center of Nano-Geomaterials of the Ministry of Education, Faculty of Materials Science and Chemistry, China University of Geosciences, Wuhan, China

Wanxue Zhang State Key Laboratory of Geomicrobiology and Environmental Changes, Engineering Research Center of Nano-Geomaterials of the Ministry of Education, Faculty of Materials Science and Chemistry, China University of Geosciences, Wuhan, China

Yuhong Zhang Shu Chien-Gene Lay Department of Bioengineering, University of California, San Diego, CA, USA

Hubin Zhao Division of Surgery and Interventional Science, University College London, London, UK

Haozhe Zhou State Key Laboratory of Biogeology and Environmental Geology, Engineering Research Center of Nano-Geomaterials of the Ministry of Education, Faculty of Materials Science and Chemistry, China University of Geosciences, Wuhan, China

1 Precision Health Monitoring and Wellness Tracking: Fundamentals and Principles

Kuldeep Mahato, Amal Abbas, Ga-Eun Kim, Alok Pandya, and Pranjal Chandra

Abstract

The integration of bioelectronic technologies into healthcare is transforming precision health monitoring and wellness tracking. This chapter explores the foundational principles, current advancements, and future directions of precision health, emphasizing the synergy between wearable devices, biosensing mechanisms, and digital health platforms. Wearable devices equipped with electrochemical, optical, and mechanical sensors now enable continuous, real-time monitoring of physiological parameters, ranging from glucose levels and cardiovascular metrics to stress markers and hydration status. These innovations are reshaping health management by shifting the focus from episodic care to proactive, personalized, and predictive interventions. Emerging technologies, including nanotechnology, flexible electronics, and artificial intelligence (AI), are accelerating the evolution of bioelectronic systems. Next-generation devices will offer multimodal sensing capabilities, enabling a comprehensive understanding of individual health dynamics. Real-time, closed-loop systems, integrating

K. Mahato (✉)
Department of Biotechnology, Motilal Nehru National Institute of Technology Allahabad, Prayagraj, Uttar Pradesh, India
e-mail: kuldeep.ucsd@gmail.com

A. Abbas · G.-E. Kim
Aiiso Yufeng Li Family Department of Chemical and Nano Engineering, University of California, San Diego, La Jolla, CA, USA

A. Pandya
Department of Biotechnology and Bioengineering, Institute of Advanced Research, Gandhinagar, Gujarat, India

P. Chandra
Laboratory of Bio-Physio Sensors and Nanobioengineering, School of Biochemical Engineering, Indian Institute of Technology (BHU), Varanasi, Uttar Pradesh, India

K. Mahato, P. Chandra (eds.), *Nano-bioelectronics for Precision Health Monitoring*, https://doi.org/10.1007/978-981-95-2117-3_1

advanced analytics and therapeutic responses, are poised to automate disease management for chronic conditions such as diabetes and hypertension. Additionally, the seamless integration of wearable devices with telemedicine platforms and electronic health records (EHRs) promises enhanced accessibility and personalization in healthcare delivery. Despite the transformative potential of these technologies, challenges related to accuracy, data privacy, regulatory compliance, and healthcare integration persist. Addressing these concerns is essential for realizing the full benefits of precision health. This chapter provides a comprehensive assessment of the technological, ethical, and regulatory dimensions of precision health monitoring, highlighting its role in fostering a data-driven, patient-centered, and preventive healthcare paradigm. The discussion underscores the importance of innovation in advancing global health outcomes.

Keywords

Precision health monitoring · Real-time health tracking · Wearable biosensors · Bioelectronics · Multimodal sensors

1.1 Introduction

Precision health monitoring refers to a tailored approach to healthcare, where real-time data from individual patients are collected, analyzed, and applied to optimize health and wellness strategies (Mahato et al. 2017, 2024a). Unlike traditional medical practices, which rely on general guidelines and reactive treatments, precision health seeks to customize prevention, diagnosis, and treatment plans based on a person's unique physiological and biochemical profile (Mahato and Chandra 2024). This approach aims to shift the healthcare paradigm from disease management to proactive wellness optimization, facilitating early detection, prevention, and timely intervention. The emergence of bioelectronic systems, which integrate biological and electronic elements, plays a pivotal role in enabling this transformation (Zhao et al. 2024). These systems harness the power of sensors to measure various physiological parameters, providing continuous, real-time feedback (Kim et al. 2018). These data points can reveal early signs of disease, monitor chronic conditions, and even offer insights into a person's lifestyle habits, such as physical activity, nutrition, and sleep patterns. The goal is to create a comprehensive health profile that allows healthcare providers to make data-driven decisions that are specific to an individual's health status and lifestyle. Bioelectronics bridges the gap between medicine, engineering, and information technology, providing tools to monitor various health parameters with high precision. From blood glucose monitoring in diabetics to wearable sensors that track heart rate and activity levels, bioelectronics are the backbone of the precision health movement. With the advancement of digital technologies such as AI and machine learning, bioelectronics-enabled precision health monitoring is rapidly evolving into a sophisticated system that empowers both patients and clinicians to take more informed actions (Shim et al. 2023; Chen et al. 2023).

Historically, healthcare has been centered around diagnosing and treating diseases after symptoms appear. Conventional health monitoring methods often rely on sporadic checkups, where patients provide subjective feedback, and clinicians make decisions based on a snapshot of their health status at a given moment (Mahato et al. 2022). This reactive approach, while effective in many cases, is limited by the lack of continuous and individualized data, which can lead to delayed diagnoses, ineffective treatments, and inefficient use of healthcare resources. In contrast, precision health monitoring offers a more proactive, personalized approach. With continuous data streams provided by wearable and implantable devices, healthcare providers can track a patient's health metrics in real time (Mahato and Wang 2021). This ongoing monitoring allows for the detection of subtle changes that may indicate the early onset of diseases such as cardiovascular conditions, metabolic disorders, or even mental health conditions (Jafleh et al. 2024; Ates et al. 2022). For instance, in the case of heart disease, bioelectronic sensors can continuously monitor heart rate variability, blood pressure, and oxygen levels, alerting both the patient and the physician if any abnormalities are detected (Scholte et al. 2024). This shift from traditional, episodic monitoring to precision-based health tracking is made possible by advancements in several key areas. First, the miniaturization of sensors has enabled the development of small, comfortable devices that can be worn or implanted without disrupting daily life (Purohit et al. 2022a). Second, advancements in data transmission and cloud computing make it possible to collect and analyze vast amounts of data from individuals (Gao et al. 2023). AI and machine learning algorithms can sift through this data, identifying trends and making predictions that would be impossible for humans to detect. These technologies, when combined with bioelectronic sensors, create a healthcare system that is predictive rather than reactive, aimed at preventing disease before it occurs (Mahato et al. 2024a; Shajari et al. 2023).

The power of precision health monitoring lies in its ability to detect deviations from a person's normal baseline long before symptoms become noticeable. Early detection is critical in a wide range of diseases, particularly chronic conditions like diabetes, hypertension, and cancer. For instance, continuous monitoring of blood pressure and heart rate can help in the early identification of cardiovascular issues, which is key in preventing severe outcomes like heart attacks or strokes (Wang et al. 2023). Wearable bioelectronic devices track metrics such as physical activity, sleep patterns, and even stress levels, providing insights into an individual's overall wellness. This data can encourage healthier behaviors, such as increased physical activity or better sleep hygiene, which have been proven to reduce the risk of various lifestyle-related diseases. Wellness tracking can also help individuals identify the impact of diet, exercise, and stress on their health, enabling them to make more informed decisions that support long-term well-being (Kang and Exworthy 2022). Such, preventative care not only benefits individuals but also has the potential to reduce the overall burden on healthcare systems. For instance, by preventing the progression of chronic diseases, precision health monitoring reduces the need for costly interventions, such as hospitalizations and surgeries (Walter et al. 2024). This shift toward a prevention-based healthcare model has significant implications for public health policy and healthcare delivery, as it could lead to improved patient outcomes and reduced healthcare costs.

The bioelectronic sensing systems are composed of sensors that detect specific biomarkers, transducers that convert biological signals into electrical signals, and processing units that analyze and interpret the data. These enable continuous, non-invasive monitoring of key health parameters, allowing for the development of personalized treatment plans. One of the most well-known applications of bioelectronics in personalized health is glucose monitoring for diabetic patients. Traditional methods require finger-prick blood tests multiple times a day, which are uncomfortable and provide only intermittent data. In contrast, bioelectronic continuous glucose monitoring (CGMs) provide real-time data, giving individuals and their healthcare providers a comprehensive view of glucose trends over time. This not only improves glycemic control but also empowers patients to take a more active role in managing their condition (Saha et al. 2023). In addition to diabetes, bioelectronic systems are also being developed to monitor a range of other health conditions, from cardiovascular diseases to neurological disorders. Wearable electrocardiograms (ECGs) can detect irregular heart rhythms, while electroencephalograms are used to monitor brain activity in patients with epilepsy or other neurological conditions (Mahato et al. 2024a). These devices are becoming increasingly sophisticated, offering wireless connectivity, long battery life, and advanced data analytics, making them indispensable tools for precision healthcare. Beyond disease monitoring, bioelectronic systems are also being used for wellness tracking. Devices like smartwatches and fitness trackers incorporate a variety of sensors that monitor activity levels, heart rate, and even blood oxygen saturation. These devices provide real-time feedback on health metrics, enabling individuals to make lifestyle changes that improve their overall well-being. This trend toward consumer-driven health monitoring is transforming the way we think about wellness, with bioelectronic devices playing a central role in this shift.

1.2 Bioelectronic Monitoring Technologies

The foundation of bioelectronics and precision healthcare is the development of devices that can continuously and accurately monitor physiological and biochemical parameters encompassing nearable, wearable, and implantable biosensing devices. Fundamentally, the biosensors are analytical devices that combine a biological sensing element with a transducer to convert a biological signal into an electrical signal (Mahato et al. 2018a). The biological element, which can be an enzyme, antibody, aptamer, or microorganism, interacts with the target analyte (e.g., glucose, lactate, or cortisol) (Mahato et al. 2019). This interaction produces a signal, often electrical, that the transducer converts into a measurable output, which is analyzed to provide real-time data about the individual's health (Mahato 2019). The versatility of biosensors allows them to detect a wide array of biomarkers, from metabolic products to proteins and hormones. Wearable devices have become the most visible manifestation of bioelectronic monitoring, used widely by both healthcare professionals and consumers. These devices, typically worn on the body as patches, wristbands, or smartwatches, contain sensors that continuously monitor various health parameters, such as heart rate, oxygen saturation, and physical activity (Kim et al. 2019). Wearable bioelectronics are designed to provide continuous feedback, allowing for real-time monitoring and early detection of abnormal trends that could indicate the onset of disease or health deterioration (Gao et al. 2023). Implantable devices, while less commonly used than wearables, offer another tier of precision health monitoring. These devices are surgically implanted into the body to continuously monitor more invasive biomarkers, such as glucose or intraocular pressure, over extended periods (Mahato 2024a). They are particularly useful for patients with chronic conditions requiring constant management. The potential of implantable bioelectronic devices is tremendous, as they provide a more reliable and stable data stream compared to external wearables, avoiding issues such as patient compliance or skin contact sensitivity. Together, these bioelectronic technologies form the basis of precision healthcare, enabling personalized monitoring that is tailored to the unique health profile of each individual.

At the heart of bioelectronic devices are sensing platforms that convert biological signals into measurable outputs. The three most common types of transducers used in bioelectronics with biosensors are electrochemical, optical, and mechanical transducers (Mahato 2024b). Among these, electrochemical bioelectronic sensors are the most widely used platform due to their better sensitivity, low cost, and compatibility with miniaturization. These sensors measure the current, voltage, or impedance changes that result from a biochemical reaction occurring at the sensor surface. Commonly, these sensors are categorized as amperometric sensors, potentiometric sensors, and impedimetric sensors (Purohit et al. 2022a). Another optical sensors, relying on the interaction of light with biological molecules to generate a analytical signal. These sensors can measure changes in light absorption, fluorescence, or refractive index as a result of biochemical interaction. Optical sensors are commonly used in bioelectronic systems for non-invasive monitoring of oxygen saturation, glucose levels, and other biomarkers (Rehg et al. 2017). For example, pulse oximeters use optical sensors to measure the absorption of light by oxygenated and deoxygenated hemoglobin in the blood, providing a real-time measure of oxygen saturation. Optical sensors offer several advantages, including high sensitivity, fast response times, and the ability to operate in real-time. However, they can be more complex and expensive to manufacture compared to electrochemical sensors. Additionally, they are more prone to interference from environmental factors such as ambient light or movement, which can affect their accuracy in wearable applications (Mahato and Chandra 2019). Mechanical sensors measure physical changes, such as pressure, strain, or deformation, that occur in response to a biological interaction. These sensors are commonly used in bioelectronics to monitor mechanical properties such as blood pressure, intraocular pressure, or skin deformation in response to physiological changes (Prasad et al. 2016). Mechanical sensors are essential in wearable devices that monitor physical activity, where accelerometers and gyroscopes are used to track motion and orientation (Yin et al. 2022). One innovative application of mechanical sensors is in smart textiles, where sensors are embedded into fabrics to monitor physiological parameters, such as the respiratory rate or body temperature, in real time (Leal-Junior et al. 2020). These sensors offer the potential for unobtrusive, continuous monitoring without the need for external devices. Each of these sensor platforms has unique strengths and weaknesses, making them suitable for different applications in precision healthcare. The choice of sensor platforms depends on factors such as the target analyte, the required sensitivity and selectivity, and the specific application environment.

The integration of bioelectronic devices with digital health technologies, such as smartphones, cloud computing, and AI, has dramatically enhanced the capabilities of precision health monitoring (Wu et al. 2023). This integration enables real-time data collection, analysis, and interpretation, providing actionable insights into a person's health status. Smartphones have become central hubs for bioelectronic monitoring, acting as interfaces for wearable devices (Purohit et al. 2020). With built-in sensors, high processing power, and wireless connectivity, smartphones can collect data from bioelectronic devices, analyze it using AI algorithms, and display the results in user-friendly formats. Mobile health (mHealth) applications enable individuals to track their health metrics, receive personalized feedback, and communicate with healthcare providers from anywhere (Rehg et al. 2017). These platforms are not only improving patient engagement but also enhancing clinical decision-making by providing healthcare professionals with continuous, real-time data from their patients. Cloud computing further expands the capabilities of bioelectronic devices by enabling the storage and analysis of vast amounts of health data. By uploading bioelectronic sensor data to cloud-based platforms, individuals and healthcare providers can access longitudinal data sets that reveal trends over time (Mahato et al. 2024a). This is particularly valuable in the management of chronic diseases, where small changes in physiological parameters over long periods can indicate the progression of a condition or the effectiveness of a treatment. The use of AI and machine learning (ML) in bioelectronic monitoring is revolutionizing the field of precision health. AI algorithms can process vast amounts of data from bioelectronic devices, identifying patterns that would be impossible for humans to detect (Sahu et al. 2022). For example, AI can analyze heart rate variability data to detect early signs of cardiovascular disease or use glucose data to predict the likelihood of hypoglycemic events in diabetics. Machine learning algorithms are also being used to personalize healthcare interventions by learning from an individual's unique health data, enabling more accurate and tailored treatment plans (Porumb et al. 2020). Moreover, AI-powered bioelectronic devices are beginning to offer predictive capabilities, alerting individuals to potential health issues before they become symptomatic (Wasilewski et al. 2024). This proactive approach to healthcare has the potential to prevent the onset of chronic diseases, reduce hospitalizations, and improve overall health outcomes. The integration of bioelectronic devices with digital health technologies represents a significant leap forward in precision healthcare, enabling more personalized, data-driven care that is accessible from anywhere, at any time (Shajari et al. 2023). The conceptualization of various AI-enabled real-time monitoring devices are emerging in the fields of sleep monitoring, respiratory monitoring, and hydration tracking. These devices use a variety of sensors to track physiological parameters such as oxygen saturation, respiratory rate, and hydration levels, providing individuals with a comprehensive view of their health status.

1.3 Biosensing Mechanisms for Monitoring Health Indicators

Biosensors are analytical devices that combine a biological recognition element with a transducer to detect specific biological molecules, or analytes, in a sample. At the core of every biosensor is a biological recognition element, a component that interacts selectively with the target analyte. This selective recognition is critical for the sensor's specificity and accuracy in detecting various health parameters, such as glucose levels, lactate concentrations, or stress hormones. The recognition element is responsible for the initial interaction with the analyte, which triggers a measurable response that is then processed by the sensor. These recognition elements can take many forms, and the choice depends on the analyte being targeted, the required sensitivity and specificity, and the environment in which the sensor will be deployed. The most common biological recognition elements used in biosensors include:

- *Enzymes*: Enzymes are proteins that catalyze specific biochemical reactions, making them ideal for detecting small molecules such as glucose or lactate. Enzyme-based biosensors have been widely used in medical diagnostics, particularly in glucose monitoring for diabetes management. For example, glucose oxidase (GOx) is an enzyme commonly used in glucose biosensors, where it catalyzes the oxidation of glucose to produce gluconic acid and hydrogen peroxide. The concentration of glucose is determined by measuring the products of this enzymatic reaction (Tehrani et al. 2022).
- *Antibodies*: Antibodies are highly specific proteins that bind to particular antigens, making them useful for detecting proteins, pathogens, or toxins in biological samples. Immunosensors, which rely on antibody-antigen interactions, are widely used in diagnostic applications. For instance, antibody-based biosensors can detect biomarkers of cancer, such as prostate-specific antigen, with high specificity (Mahato et al. 2020a).
- *Aptamers*: Aptamers are synthetic single-stranded deoxyribonucleic acid (DNA) or ribonucleic acid (RNA) molecules that fold into specific three-dimensional structures capable of binding to a target molecule. Aptamers offer several advantages over antibodies, including higher stability, ease of production, and the ability to bind a wide range of target molecules, from small organic compounds to proteins. Aptamer-based biosensors are increasingly being explored for the detection of drugs, toxins, and biomarkers in health monitoring (Mahato et al. 2018b).
- *Nucleic acids*: DNA and RNA sequences can be used as recognition elements in biosensors for the detection of genetic material, such as viral RNA or DNA

mutations associated with diseases. Nucleic acid biosensors, also known as genosensors, have become essential tools for early diagnosis of infectious diseases, genetic disorders, and cancer. For example, biosensors that detect specific gene mutations associated with cancer can help in early diagnosis and personalized treatment (Bhatnagar et al. 2018).

- *Whole cells or micro-organisms*: In some cases, entire cells or microorganisms can serve as the biological recognition element. These biosensors, often referred to as whole-cell biosensors, are particularly useful for detecting toxic compounds, environmental pollutants, or pathogens. For example, a bacterial biosensor might be engineered to produce a measurable signal, such as luminescence, in the presence of a toxic substance (Purohit et al. 2019).

The interaction between the biological recognition element and the analyte is fundamental to the function of a biosensor, as it determines the selectivity and sensitivity of the device. After the recognition event, a measurable change occurs, which is captured by the transducer and converted into an electrical, optical, or mechanical signal (Mahato et al. 2018a).

Once the recognition element interacts with the target analyte, the next critical step in the functioning of a biosensor is the conversion of this interaction into a measurable signal. This process is known as signal transduction, and it is performed by a transducer, which converts the biochemical interaction into an electrical, optical, or mechanical signal that can be quantified. The choice of transducer depends on the type of biological recognition element and the application for which the biosensor is designed. The most common types of signal transduction mechanisms are electrochemical, optical, and mechanical. Electrochemical biosensors are among the most widely used in healthcare applications, particularly because of their high sensitivity and compatibility with miniaturization for point-of-care testing and wearable devices (Mahato et al. 2018c). Electrochemical transducers convert the biochemical interaction between the recognition element and the analyte into an electrical signal, typically by measuring changes in current, voltage, or impedance (Mahato et al. 2020b, 2023; Kumar et al. 2020; Yadav et al. 2023; Mahato and Chandra 2022).

- *Amperometric sensors* measure the current generated by the oxidation or reduction of an electroactive species, such as glucose, at an electrode. In glucose biosensors, for instance, the oxidation of glucose catalyzed by glucose oxidase produces electrons, which generate a current proportional to the glucose concentration. These sensors are highly sensitive and are commonly used in CGMs for diabetes management (Mahato and Wang 2021).
- *Potentiometric sensors* measure the voltage difference between a reference electrode and a working electrode in response to the concentration of an analyte. These sensors are used for measuring pH levels or ion concentrations, such as potassium or sodium ions in blood (Laochai et al. 2024).
- *Impedimetric sensors* measure the change in impedance, or resistance to an electrical current, caused by the binding of the analyte to the recognition element.

These sensors are particularly useful for detecting large biomolecules such as proteins, cells, or pathogens (Mahato et al. 2020a).

Electrochemical transduction is widely used in bioelectronics due to its high sensitivity, low cost, and ease of integration into portable or wearable devices. The small size and low power requirements of electrochemical sensors make them ideal for continuous, real-time monitoring of health parameters.

Optical biosensors rely on the interaction of light with the biological recognition element to generate a measurable signal. These sensors measure changes in light absorption, fluorescence, or refractive index because of the interaction between the analyte and the recognition element. Optical biosensors are often used in diagnostic applications where non-invasive or label-free detection is desired (Priyadarshni et al. 2023).

- *Absorbance-based sensors* measure the amount of light absorbed by the analyte or by the products of a biochemical reaction. For example, pulse oximeters use absorbance-based sensors to measure oxygen saturation in the blood by detecting the absorption of light by oxygenated and deoxygenated hemoglobin at different wavelengths (Mahato et al. 2017).
- *Fluorescence-based sensors* use fluorescent molecules or probes that emit light when excited by a specific wavelength. These sensors are highly sensitive and are commonly used in assays to detect proteins, nucleic acids, or other biomolecules in clinical diagnostics. For example, fluorescence-based biosensors are used in ELISA (enzyme-linked immunosorbent assay)-based tests to detect biomarkers of diseases such as cancer or infectious diseases (Purohit et al. 2019).
- *Surface plasmon resonance (SPR)* is an optical technique that measures changes in the refractive index at the surface of a sensor when an analyte binds to the recognition element. SPR sensors are widely used for real-time, label-free detection of biomolecular interactions, making them valuable tools for studying protein-protein interactions or detecting pathogens (Mahato et al. 2019).

Optical biosensors offer several advantages, including high sensitivity, the ability to perform label-free detection, and the capacity for real-time monitoring. However, they can be more complex and expensive to manufacture than electrochemical sensors, and they are often more sensitive to environmental factors such as ambient light or temperature.

Mechanical biosensors detect changes in mechanical properties, such as mass, pressure, or force, that result from the interaction between the analyte and the recognition element. These sensors are commonly used to monitor physical parameters such as blood pressure, intraocular pressure, or tissue stiffness.

- *Quartz crystal microbalance (QCM)* is a type of mechanical sensor that measures changes in mass by detecting shifts in the resonance frequency of a quartz crystal. When an analyte binds to the recognition element on the surface of the crystal, the added mass causes a measurable change in the frequency of oscillation. QCM sensors are highly sensitive and are used in applications such as detecting pathogens or monitoring biomolecular interactions (Mahato et al. 2018a).

- *Microcantilever sensors* are another type of mechanical sensor that measures changes in the deflection or bending of a tiny cantilever beam when an analyte binds to the recognition element on its surface. These sensors are highly sensitive and can be used to detect changes in mass, force, or pressure. Microcantilever sensors are being explored for applications such as detecting cancer biomarkers, viruses, or environmental toxins (Zhang et al. 2020).
- *Piezoelectric sensors* convert mechanical stress into an electrical signal and are used in applications such as blood pressure monitoring or respiratory monitoring. Piezoelectric sensors are highly sensitive to changes in pressure or force, making them useful for detecting mechanical changes in tissues or fluids (Pohanka 2018).

Mechanical transducers are particularly useful for detecting physical changes in the body, such as pressure or mass, making them ideal for applications in wearable devices or implantable sensors.

The versatility of biosensors and their transduction mechanisms allows for the measurement of a wide range of health parameters. For instance, glucose monitoring is one of the most well-established applications of biosensors, particularly for the management of diabetes (Saha et al. 2023). CGMs use enzyme-based electrochemical sensors to measure interstitial glucose levels in real-time. These sensors typically use GOx, a glucose cleaving enzyme as the recognition element, which catalyzes the oxidation of glucose to produce hydrogen peroxide (Purohit et al. 2022b). The concentration of glucose is then determined by measuring the current produced by the electrochemical reaction. CGMs have transformed diabetes management by providing real-time data on glucose trends, enabling individuals to make informed decisions about insulin dosing, diet, and exercise. These devices are essential for maintaining tight glycemic control and reducing the risk of complications such as hypoglycemia, hyperglycemia, and long-term damage to organs (Banday et al. 2020). Lactate is a byproduct of anaerobic metabolism and is often used as a biomarker for physical exertion, muscle fatigue, and tissue hypoxia. Lactate biosensors, similar to glucose biosensors, use lactate oxidase as the recognition element to catalyze the oxidation of lactate, producing a measurable current (Li et al. 2021). Lactate monitoring is particularly useful for athletes, as it provides insights into their metabolic performance during exercise (Windmiller et al. 2011). In medical applications, lactate biosensors are used to monitor critically ill patients, as elevated lactate levels can indicate tissue hypoxia, sepsis, or organ failure (Andersen et al. 2013). Lactate monitoring is also valuable in neonatal care, where it can provide early indications of oxygen deprivation or metabolic stress. Cortisol, a hormone produced by the adrenal glands, is commonly referred to as the "stress hormone" because its levels increase in response to physical or emotional stress. Elevated cortisol levels over prolonged periods can lead to various health problems, including hypertension, immune dysfunction, and metabolic disorders (Sic et al. 2024). Biosensors for cortisol monitoring are being developed to provide non-invasive, real-time data on stress levels, enabling better management of stress-related conditions (Tang et al. 2021). Wearable cortisol sensors hold great potential for monitoring stress in real time, allowing individuals to take preventive measures before stress-related health problems arise. Similarly, monitoring the pH and ionic composition of biological fluids is critical for assessing various physiological processes, including acid-base balance, electrolyte levels, and metabolic function. Potentiometric biosensors are commonly used to measure pH and ion concentrations in samples such as blood, urine, or sweat. pH monitoring is particularly important in critical care settings, where rapid changes in acid-base balance can indicate respiratory or metabolic distress (Wang 2023; Li et al. 2022). These sensors use ion-selective electrodes that are sensitive to specific ions, such as hydrogen (H^+), sodium (Na^+), or potassium (K^+). Similarly, ion monitoring is essential for managing conditions such as electrolyte imbalances, dehydration, or renal dysfunction (Menon et al. 2016). Wearable ion sensors, integrated into patches or smart textiles, offer continuous monitoring of sweat electrolytes, providing valuable insights into hydration levels during physical activity or illness. Not only do sensors monitoring biochemical parameters contribute to the personalized health promotion, but simultaneous recording of biophysical parameters also leverages multitudinous data. In this context, monitoring of oxygen saturation offers the key health information of respiratory function and overall cardiovascular health (Zhou et al. 2024). Pulse oximeters, which use optical biosensors, measure the percentage of oxygenated hemoglobin in the blood by analyzing the absorption of light at different wavelengths. This non-invasive technology is widely used in hospitals, clinics, and home settings to monitor patients with respiratory conditions, such as chronic obstructive pulmonary disease (COPD), asthma, or COVID-19 (Luks and Swenson 2020). Pulse oximeters are also integrated into wearable devices, allowing for continuous monitoring of oxygen saturation during sleep or exercise. This real-time data can help detect early signs of hypoxemia, sleep apnea, or other conditions that affect oxygen delivery to tissues (Leppänen et al. 2022). As precision healthcare evolves, there is a growing demand for multimodal sensors that can simultaneously monitor multiple health parameters (Shahid et al. 2025). These multimodal systems integrate different types of biosensors—such as electrochemical, optical, and mechanical sensors—into a single device, providing a more comprehensive view of an individual's health status. Multimodal sensors offer several advantages over single-parameter sensors. By measuring multiple biomarkers or physiological parameters simultaneously, these devices provide a more holistic view of a patient's health, enabling earlier detection of diseases and more personalized interventions (Mahato et al. 2024a). For example, a multimodal wearable device could simultaneously monitor glucose levels, heart rate, and hydration status, providing valuable insights into the interaction between metabolism, cardiovascular function, and hydration. The development of multimodal sensors is particularly important for managing complex chronic diseases, such as diabetes, cardiovascular disease, or cancer, where multiple physiological systems are affected. By integrating multiple biosensors into a single platform, healthcare providers can gain a more complete understanding of a patient's condition and tailor treatment plans accordingly.

1.4 Wearable Devices for Wellness Tracking

Wearable devices have revolutionized wellness tracking by providing individuals with real-time access to data about their physiological and physical activity parameters (Kim et al. 2019). These devices are typically worn on the body in the form of wristbands, smartwatches, patches, or even integrated into clothing (Mahato and Wang 2021). Wearable sensors have become a cornerstone of precision health monitoring, with applications ranging from fitness tracking to chronic disease management (Mahato and Chandra 2024). The types of wearable sensors commonly used for wellness tracking can be classified into several categories based on the parameters they monitor (Fig. 1.1). Physical activity sensors, such as accelerometers and gyroscopes, are among the most widely used in wearable devices. These sensors measure movement, acceleration, and orientation, providing valuable data on physical activity levels, including walking, running, cycling, or even sedentary behavior (Dobkin 2013). The integration of these sensors into fitness trackers and smartwatches allows users to monitor daily step counts, calories burned, and active minutes, promoting a more active lifestyle. Accelerometers measure linear acceleration, typically along three axes (X, Y, and Z), allowing for the detection of movement and changes in velocity. These are commonly used to monitor activities like walking, running, or climbing stairs. By analyzing changes in acceleration patterns,

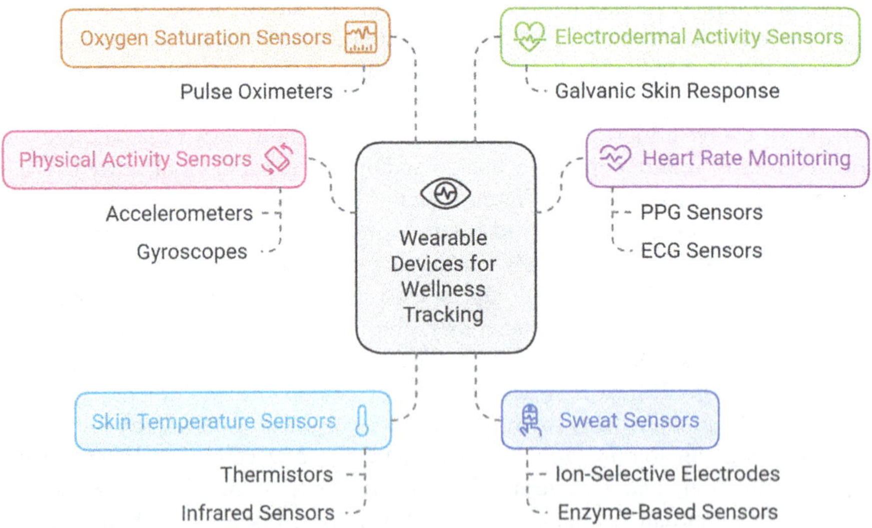

Fig. 1.1 Wearable sensors for health monitoring: (a) oxygen saturation, (b) electrodermal activity, (c) motion, (d) heart rate, (e) skin temperature, (f) sweat biomarkers

accelerometers can also detect periods of inactivity, which is important for assessing sedentary behavior and its associated health risks. Gyroscopes measure rotational motion or angular velocity. These sensors complement accelerometers by providing additional data on the orientation and rotation of the body. When used together, accelerometers and gyroscopes can offer a more comprehensive analysis of physical activity, such as detecting whether a person is standing up, sitting down, or lying down (Teymourian et al. 2022). Gyroscopes are particularly useful for monitoring activities that involve complex movements, such as sports or yoga. Pedometers are a type of wearable sensor specifically designed to count steps. Pedometers use accelerometer data to detect the rhythmic movement of walking or running. These are simple but effective tools for encouraging individuals to increase their daily step count, which is a key factor in promoting physical activity and improving overall health (Mudgal et al. 2020). Heart rate monitoring is a fundamental feature of most modern wearables, particularly fitness trackers and smartwatches. Heart rate sensors provide continuous or on-demand measurements of heart rate, allowing users to track their cardiovascular health during exercise, rest, and sleep (Li et al. 2023). The most common types of heart rate sensors in wearable devices are photoplethysmography (PPG) sensors and ECG sensors. PPG sensors use optical techniques to measure changes in blood volume by shining light (usually green or infrared) through the skin and detecting how much light is absorbed by blood vessels. The pulsatile nature of blood flow corresponds to the heart's activity, allowing the PPG sensor to estimate heart rate. PPG sensors are widely used in smartwatches and fitness bands because they are non-invasive, low-cost, and can provide continuous heart rate monitoring (Namvari et al. 2022). However, their accuracy can be affected by factors such as skin tone, movement, and ambient light interference. ECG sensors, in contrast, measure the electrical activity of the heart by detecting the small electrical signals produced by the heart's muscles as they contract and relax. ECG sensors provide a more accurate and detailed assessment of heart function compared to PPG sensors. Some advanced wearable devices, such as certain smartwatches, incorporate ECG sensors to detect irregular heart rhythms including atrial fibrillation (AFib) (Shahid et al. 2025), and to provide a more clinical-grade measurement of heart health. ECG sensors are particularly valuable for users with cardiovascular conditions or those at risk of heart disease. Wearable devices with skin temperature sensors monitor changes in skin temperature, which can provide insights into various health metrics, such as metabolic activity, stress, and even early signs of illness. For example, elevated skin temperature may indicate fever, while lower-than-usual temperature readings can suggest poor circulation or hypothermia. Skin temperature sensors use thermistors or infrared sensors to detect temperature changes. These sensors are commonly found in fitness trackers and smartwatches to track how body temperature fluctuates throughout the day, during physical activity, or sleep. Continuous monitoring of skin temperature can provide useful insights into circadian rhythms, sleep patterns, and overall well-being (Zhang et al. 2025a).

Sweat sensors are an emerging class of wearable sensors that provide real-time monitoring of sweat composition, offering insights into hydration levels, electrolyte balance, and even metabolic markers such as glucose or lactate (Gao et al. 2023). Sweat contains various biomarkers that can reflect an individual's physiological state, including sodium, potassium, chloride, and lactate concentrations, as well as certain hormones and metabolites. Sweat sensors typically use ion-selective electrodes or enzyme-based sensors to detect specific ions or molecules in sweat (Baker 2019). Wearable sweat sensors can be integrated into patches or wristbands, providing non-invasive, continuous monitoring of hydration levels during physical activity. This type of sensor is particularly useful for athletes, as it helps prevent dehydration, electrolyte imbalances, and heat-related illnesses (Baker et al. 2020). Furthermore, advancements in sweat sensing technologies are expanding their applications to include the monitoring of glucose for diabetic patients or cortisol levels for stress management (Samson and Koh 2020). Oxygen saturation sensors, also known as pulse oximeters, measure the percentage of oxygenated hemoglobin in the blood. These sensors use optical techniques, similar to PPG, to measure the absorption of light at different wavelengths by oxygenated and deoxygenated hemoglobin (Abay and Kyriacou 2018). Wearable devices with SpO_2 sensors can provide continuous monitoring of blood oxygen levels, which is particularly important for individuals with respiratory conditions, such as asthma or COPD. Wearable SpO_2 sensors are also valuable for tracking changes in oxygen saturation during exercise or sleep. For example, low oxygen saturation during sleep can indicate the presence of sleep apnea, a condition in which breathing temporarily stops or becomes shallow during sleep (Block et al. 1979). By providing real-time SpO_2 data, wearables can help individuals detect potential respiratory issues early and seek medical attention if needed. Electrodermal activity (EDA), also known as galvanic skin response (GSR), measures changes in the electrical conductivity of the skin, which can be influenced by sweat gland activity. EDA sensors are used to monitor emotional arousal or stress levels, as increased sweating due to stress or anxiety increases the skin's conductivity (Hickey et al. 2021). EDA sensors are often integrated into wellness wearables designed to track mental health and stress management. By measuring changes in skin conductivity in response to emotional or physical stressors, EDA sensors can provide real-time feedback on stress levels and help individuals practice relaxation techniques or mindfulness exercises to reduce stress.

Wearable devices provide a comprehensive range of wellness metrics by continuously monitoring physiological and physical activity parameters. These metrics offer valuable insights into an individual's overall health and well-being. Below are some of the key parameters commonly tracked by wearable devices: heart rate variability (HRV) refers to the variation in time between consecutive heartbeats, measured in milliseconds. It is a critical metric for assessing the autonomic nervous system (ANS), which controls the body's response to stress and recovery. A high HRV indicates a well-balanced and resilient ANS, capable of adapting to stress and environmental changes. Conversely, low HRV can be a sign of poor autonomic function, often associated with stress, fatigue, or illness. Wearable devices equipped with heart rate sensors, particularly ECG-based sensors, can accurately measure HRV in real time. HRV is commonly used as an indicator of physical and mental well-being and is particularly useful for stress monitoring, physical recovery, and sleep quality (Li et al. 2023). In addition, the physical activity is a cornerstone of wellness, and wearable devices have made it easier than ever to track and quantify movement throughout the day. By using accelerometers, gyroscopes, and heart rate sensors, wearables can monitor various aspects of physical activity, including step count, active minutes, and calorie burnt. By combining data on movement, heart rate, and personal characteristics such as age, weight, and gender, wearables can estimate the number of calories burned during physical activity (Shei et al. 2022).

1.5 Integration with Digital Health Platforms

The convergence of wearable biosensors and digital health platforms has created a transformative ecosystem in healthcare (Vo and Trinh 2024). This integration enables real-time data acquisition, advanced analytics, and personalized interventions, offering immense potential for improving health outcomes, reducing healthcare costs, and increasing patient engagement. Digital health platforms, encompassing cloud computing, mobile health (mHealth) applications, telemedicine, and AI, serve as the connective tissue between data generated by bioelectronics and actionable insights used by individuals, healthcare professionals, and researchers (Stoumpos et al. 2023). This section discusses the critical components of such digital health platforms, how they interact with bioelectronic devices in this evolving landscape.

1.5.1 Data Acquisition and Management: Real-Time Monitoring, Data Storage, and Analysis

One of the defining characteristics of wearable bioelectronic devices integrated with digital health platforms is the ability to continuously collect and analyze real-time physiological data. This continuous monitoring is a key differentiator from traditional episodic health assessments, providing a dynamic view of an individual's health over time. However, managing the vast amount of data generated by these devices requires robust infrastructure for data acquisition, storage, and analysis. The first step in the integration of bioelectronics with digital health platforms is data acquisition. Wearable devices collect diverse streams of health information, including heart rate, oxygen saturation, glucose levels, and physical activity metrics, using various sensors (Mahato et al. 2024a). This health information is then transmitted wirelessly to a connected device, such as a smartphone or tablet, which serves as a local hub for initial data processing. The communication between the wearable device and the smartphone is typically facilitated through bluetooth low energy (BLE) or Wi-Fi, providing real-time transmission of health data without requiring significant power consumption. BLE is the preferred communication protocol for most wearable devices due to its energy efficiency and wide compatibility with consumer electronics. In certain clinical applications, more sophisticated medical-grade devices may utilize cellular networks or satellite communication to transmit data, especially in remote patient monitoring scenarios. This is particularly important for patients with chronic conditions who need constant oversight, such as those with cardiovascular disease, diabetes, or respiratory conditions. Once data is acquired by a smartphone or other connected device, it is often processed locally for basic metrics (e.g., calculating steps, heart rate averages, or displaying current glucose levels). For more advanced analysis, such as predictive modeling or pattern recognition, the data is then uploaded to a cloud platform. The continuous flow of data from wearable devices requires scalable and secure data storage solutions. Cloud computing has emerged as the backbone of digital health platforms, offering the storage capacity and computational power needed to handle the massive influx of health data from millions of users. Cloud platforms also provide the flexibility and scalability needed to accommodate growing datasets as more individuals adopt bioelectronic devices (Gu et al. 2025).

Data storage in the cloud ensures that health information is available anytime and anywhere, enabling the providers to access a patient's health records in real-time. This is particularly important in emergencies or during telemedicine consultations, where immediate access to up-to-date information is critical. Security and privacy are major considerations in cloud-based data storage (Sachdeva et al. 2024). Given the sensitive nature of health data, it must be protected against unauthorized access, tampering, and breaches. Cloud service providers must comply with stringent regulations, such as the Health Insurance Portability and Accountability Act in the United States and the General Data Protection Regulation in the European Union, to ensure that personal health information is handled securely (Vashishth et al. 2025). Data encryption, both during transmission and while at rest, is a critical component of maintaining the integrity and confidentiality of health data. End-to-end encryption ensures that only authorized parties—such as patients, healthcare providers, or approved researchers—can access the data. Additionally, role-based access controls and two-factor authentication are used to limit access to sensitive data and enhance security (Jaime et al. 2023).

The true value of integrating bioelectronics with digital health platforms lies in data analysis. By processing vast amounts of real-time health data, digital health platforms can identify trends, detect anomalies, and make personalized recommendations that enhance individual health and wellness. The AI and ML are central to this data analysis, transforming raw sensor data into actionable insights. AI-powered algorithms can recognize complex patterns in physiological data that may indicate early signs of disease, deterioration in chronic conditions, or response to treatment. For example, predictive analytics can anticipate adverse health events, such as hypoglycemia in diabetes or arrhythmia in patients with cardiovascular disease, enabling preventive interventions before symptoms manifest. Anomaly detection models can identify unusual patterns in a patient's data, such as a sudden spike in heart rate or oxygen saturation, which may signal a medical emergency. Behavioral analytics can provide personalized wellness recommendations by correlating physical activity levels, sleep patterns, and nutritional intake with long-term health outcomes. Beyond individual health monitoring, data analysis plays a pivotal role in population health management. Aggregated, anonymized data from thousands or millions of wearable devices can reveal trends in public health, such as the spread of infectious diseases or the impact of lifestyle factors on chronic conditions. Public health authorities and researchers can use this information to develop targeted interventions and improve overall healthcare delivery (Shajari et al. 2023).

1.5.2 Integration with Telemedicine, Mobile Apps, and Cloud Platforms

The integration of bioelectronic devices with digital health platforms is most evident in the growing adoption of telemedicine and mHealth applications, which enable patients and healthcare providers to connect remotely. This has been particularly critical in the wake of the COVID-19 pandemic, which accelerated the shift toward remote healthcare solutions. Telemedicine refers to the delivery of healthcare services remotely through telecommunications technology. By integrating bioelectronic data into telemedicine platforms, healthcare providers can monitor a patient's condition in real-time, provide timely interventions, and conduct virtual consultations without the need for in-person visits. Wearable devices play a crucial role in telemedicine by providing continuous streams of physiological data that can be shared with healthcare providers in real-time. This data can be accessed during teleconsultations to guide decision-making and enable more personalized care (Clement David-Olawade et al. 2024). For example, a patient with hypertension may wear a continuous blood pressure monitor that transmits real-time data to their healthcare provider. During a teleconsultation, the provider can review the patient's recent blood pressure trends and adjust their treatment plan accordingly. Telemedicine also benefits from AI-powered analytics, which can automatically flag potential health issues based on real-time data. For instance, if a wearable detects an abnormal heart rhythm, the telemedicine platform can alert the healthcare provider to schedule an immediate consultation with the patient. The integration of telemedicine and bioelectronic devices is particularly valuable for managing chronic diseases, post-surgical care, and remote patient monitoring (Zhang et al. 2025b). It enables healthcare providers to track patients' progress, adjust treatments, and intervene early, when necessary, ultimately improving health outcomes and reducing hospital readmissions. Mobile health applications, or mHealth apps, are software applications designed to run on smartphones, tablets, or wearable devices. These apps serve as an interface between the individual and the bioelectronic device, providing real-time data visualization, health insights, and personalized recommendations. Most wearable devices are paired with dedicated mHealth apps that display data such as heart rate, step count, sleep patterns, and calorie expenditure.

In addition to providing real-time feedback, these apps often include features like goal setting, reminders, and health coaching to encourage individuals to meet their wellness objectives. Data visualization is one of the core functions of mHealth apps, providing users with easy-to-understand visualizations of their health data. Charts, graphs, and trend lines allow individuals to track their progress over time, helping them identify patterns and make informed decisions about their health. The mHealth apps often incorporate AI-driven analytics to provide personalized health insights based on the data collected by bioelectronic devices. For example, an app might analyze sleep data to suggest changes in bedtime routines or use heart rate variability to recommend stress-reducing activities like meditation. Many mHealth apps include features designed to promote healthy behaviors, such as goal setting, reminders, and rewards for meeting fitness or wellness goals. These behavior change strategies are crucial for helping individuals make long-term lifestyle changes, such as increasing physical activity, improving sleep quality, or managing stress. These apps increasingly allow individuals to share their health data directly with healthcare providers, either through secure messaging platforms or EHR systems. This integration enables more comprehensive and continuous care, as healthcare providers can monitor patients' health remotely and intervene when necessary (Deniz-Garcia et al. 2023).

1.5.3 Personalized Feedback and Actionable Insights for Preventive Healthcare

The integration of bioelectronics with digital health platforms enables personalized feedback and actionable insights that empower individuals to take a proactive role in their health. By continuously monitoring physiological data and analyzing trends, these platforms can provide recommendations tailored to the unique needs of each individual. Personalized feedback refers to the delivery of health insights and recommendations that are specific to an individual's unique health profile, lifestyle, and goals. Rather than relying on generalized health advice, personalized feedback is based on the continuous monitoring of real-time data from wearable devices, ensuring that recommendations are relevant to the individual's current state of health (Canali et al. 2022). For example, a wearable device might monitor an individual's HRV and provide personalized feedback on stress management techniques based on fluctuations in HRV for the day. Similarly, a device that tracks sleep patterns might recommend adjustments to bedtime routines or suggest relaxation exercises to improve sleep quality.

Personalized feedback is a key driver of behavior change, as it allows individuals to understand how their actions impact their health in real-time. By providing timely, actionable insights, digital health platforms can help users adopt healthier habits and make informed decisions about their wellness. Preventive healthcare focuses on identifying and addressing potential health risks before they become serious medical issues. The integration of bioelectronic devices with digital health platforms is instrumental in shifting the healthcare paradigm from reactive to preventive care (Stoumpos et al. 2023). Wearable devices provide continuous, real-time monitoring of physiological parameters, allowing for the early detection of health issues before symptoms arise. For example, CGMs can detect rising blood glucose levels, enabling individuals with diabetes to take corrective actions before reaching dangerous levels of hyperglycemia or hypoglycemia. For individuals with chronic conditions, such as hypertension or heart disease, bioelectronic devices integrated with digital health platforms can provide ongoing monitoring, ensuring that any deterioration in the patient's condition is detected early. Telemedicine platforms can then alert healthcare providers to intervene before the patient experiences serious complications. Personalized feedback based on wearable data can encourage individuals to make long-term lifestyle modifications, such as increasing physical activity, improving sleep hygiene, or reducing stress. Preventive healthcare platforms often include features like goal setting, health coaching, and social support networks to help individuals stay motivated and accountable in their health journey (Haleem et al. 2021).

1.5.4 Data Privacy and Security Concerns in Precision Health Tracking

As bioelectronic devices and digital health platforms become increasingly integrated, concerns about data privacy and security are growing. The collection, storage, and sharing of sensitive personal health data raises significant ethical and legal challenges that must be addressed to protect individuals' privacy and ensure the responsible use of health data. One of the primary concerns in precision health tracking is to ensure that individuals' health data is protected from unauthorized access. Wearable devices collect highly sensitive information about an individual's physiological state, which could be misused if it falls into the wrong hands (Filkins et al. 2016). For instance, data breaches could lead to identity theft, insurance fraud, or the unauthorized use of health data for marketing purposes. To address these concerns, digital health platforms must implement robust data privacy measures. One way to protect privacy is to anonymize health data before it is shared with third parties. Anonymization removes personally identifiable information from data, ensuring that individuals cannot be traced back to their health records (Seh et al. 2020). Anonymized data can still be used for research purposes or population health analytics without compromising individual privacy. Another way is obtaining informed consent from individuals before collecting and using their health data. Users should have the ability to control who has access to their data and for what purposes it is used. Digital health platforms must provide clear, transparent privacy policies that explain how data will be collected, stored, and shared. Data security is another critical concern in precision health tracking (Zuo et al. 2021). Digital health platforms must implement strong security protocols to protect health data from cyberattacks, hacking, and data breaches. Key security measures include end-to-end encryption, which is a process of encrypting data both during transmission and while at rest ensures that sensitive health information cannot be intercepted or accessed by unauthorized parties (Andrew et al. 2023). End-to-end encryption provides an extra layer of protection, ensuring that only the intended recipient can decrypt and view the data. Further, limiting access to health data through role-based access controls and multifactor authentication helps to prevent unauthorized access. Only individuals with the appropriate permissions—such as healthcare providers or approved researchers—should be able to view or modify health data. Furthermore, the digital health platforms should conduct regular security audits to identify vulnerabilities in their systems and ensure compliance with data protection regulations.

1.6 Conclusions and Outlook

In this chapter we discuss precision health monitoring and wellness tracking, highlighting its gradual transformation through bioelectronics and wearable biosensors. Precision health emphasizes proactive, individualized care using continuous, real-time data from wearable and implantable devices. These devices utilize electrochemical, optical, and mechanical biosensors to measure physiological parameters, including glucose levels, heart rate, stress biomarkers, and hydration status. This continuous monitoring facilitates early disease detection, personalized interventions, and preventive healthcare, fundamentally shifting healthcare away from reactive, episodic treatment toward predictive health management. Advancements in nanotechnology, AI, and flexible electronics promise enhanced capabilities and multimodal sensing, enabling a comprehensive understanding of health dynamics. Integrating bioelectronics with telemedicine, mobile health apps, cloud computing, and AI-powered analytics significantly improves patient engagement and clinical decision-making. However, challenges such as ensuring accuracy, protecting data privacy, achieving regulatory compliance, and system interoperability remain significant barriers. Successfully addressing these issues is essential for harnessing the full potential of precision healthcare technologies. Ultimately, precision health monitoring offers a powerful paradigm shift, fostering a patient-centered approach focused on disease prevention, early intervention, and personalized wellness management.

The future of precision health and wellness tracking is being redefined at the nexus of technological innovation, transformative healthcare models, and an increasing societal shift toward personalized and preventive care. Advancements in wearable technologies, AI, nanotechnology, and biosensing platforms are catalyzing a significant transition in healthcare delivery—from episodic, symptom-driven interventions to a proactive, predictive, and continuous monitoring framework. This paradigm shift heralds a new era in which individualized health insights are not only accessible but also actionable, empowering individuals and clinicians alike in the pursuit of optimized well-being (Huang et al. 2025). Figure 1.2 illustrates the comparison of the future of such technologies in context to their impact on precision health. As the field of bioelectronics continues to evolve, the development of next-generation wearable biosensors promises to revolutionize health surveillance by enabling early disease detection, precise monitoring, and real-time intervention. Future bioelectronic platforms will be underpinned by innovations in advanced materials, miniaturization, and multifunctional sensor integration. These systems are anticipated to seamlessly merge with daily life, transforming the body into a data-rich ecosystem capable of self-regulation and dynamic response to physiological deviations. One of the most transformative applications of emerging bioelectronic technologies lies in their potential to shift the clinical focus from reactive treatment to anticipatory diagnostics. Present-day healthcare models often rely on overt symptoms to prompt medical action. However, future wearable biosensors will facilitate subclinical detection of pathophysiological changes, allowing for timely interventions before disease progression occurs. For instance, chronic

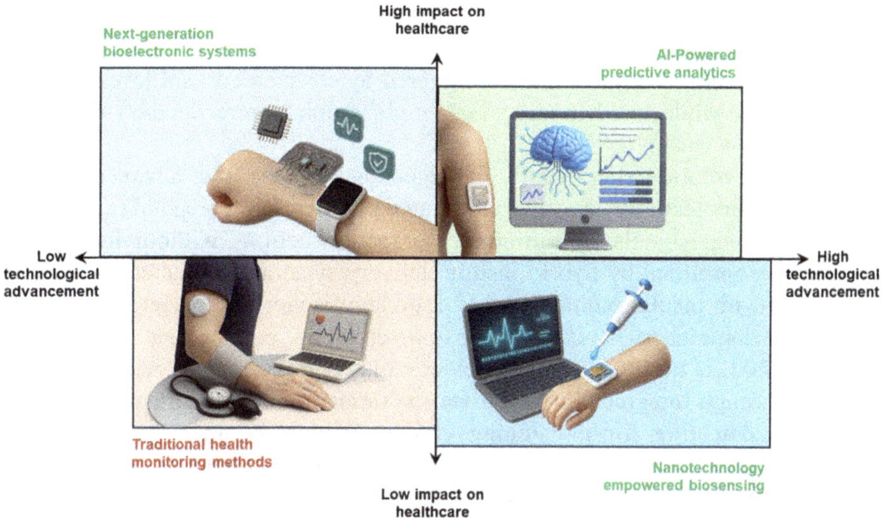

Fig. 1.2 Comparison metrics for the next-generation bioelectronic-based health monitoring technologies empowered by advanced analytics

diseases such as diabetes, cardiovascular disorders, and various cancers could be managed through continuous monitoring of specific biomarkers (Obianyo et al. 2024). While current CGMs provide essential insights for diabetes management, future iterations are expected to be non-invasive, more accurate, and capable of providing predictive alerts for glycemic excursions.

Similarly, wearable ECG devices of the future may offer enhanced sensitivity in detecting subclinical arrhythmias, facilitating earlier diagnosis of cardiac conditions such as atrial fibrillation or ischemic events. In oncology, the integration of biosensors capable of detecting circulating tumor cells or tumor-specific antigens could lead to early cancer detection, substantially improving prognoses through earlier and less aggressive treatment modalities (Abdelrazik et al. 2025). Such innovations represent a radical departure from conventional diagnostics, reconfiguring oncology and chronic disease management around prevention and early intervention. Moreover, future health monitoring platforms will increasingly rely on multimodal biosensors—integrated devices capable of measuring multiple physiological and biochemical signals simultaneously. Contemporary wearable devices are often limited to single or few metrics, such as heart rate, glucose levels, or physical activity (Hemdan et al. 2024). However, next-generation wearables will encompass a broader spectrum of biomarkers, including indicators of metabolic function, inflammatory states, hormonal fluctuations, and oxidative stress. For example, a single device might track heart rate, respiration, oxygen saturation, cortisol, glucose, and lactate levels, offering a comprehensive physiological profile in real time. The integration of multimodal data will enhance the granularity of health insights, supporting personalized wellness strategies. This will be particularly impactful in tailoring health recommendations based on individual physiological responses to environmental and behavioral stimuli (Mahato et al. 2024a). A user prone to stress-related disorders, for instance, could benefit from personalized alerts and interventions triggered by elevated cortisol levels or ANS dysregulation, while another individual might require more detailed metabolic monitoring.

Further revolutionizing health management will be the advancement of closed-loop bioelectronic systems, automated platforms that respond in real-time to physiological signals by initiating therapeutic actions without user input. Currently exemplified by hybrid insulin delivery systems for diabetes (integrating CGMs with insulin pumps), closed-loop approaches are expected to expand into other domains such as hypertension, epilepsy, and cardiac arrhythmias (Templer 2022). Future wearable devices may autonomously regulate blood pressure through integrated drug delivery systems or activate emergency protocols upon detecting life-threatening cardiac anomalies, thereby minimizing delays in critical care (Templer 2022). The development of such systems hinges on the integration of AI and ML algorithms, which are essential for real-time data processing and pattern recognition across vast health datasets. AI will not only enhance anomaly detection but also support predictive analytics, enabling clinicians to anticipate disease trajectories and personalize intervention strategies (Teymourian et al. 2022). For instance, AI models analyzing data from wearable ECGs, heart rate variability, sleep, and activity patterns could forecast cardiovascular events or detect early metabolic dysfunctions. Additionally, continuous learning algorithms can refine recommendations over time, adapting to an individual's changing physiology and lifestyle, and supporting highly personalized, evidence-based care plans. Nanotechnology represents another pivotal frontier in precision health. Nano-sensors devices capable of detecting molecular-level biomarkers at extremely low concentrations offer immense potential for non-invasive, continuous health monitoring (Shukla et al. 2025). Integrated into skin patches, textiles, or even contact lenses, nano-sensors can measure analytes in sweat, saliva, or interstitial fluid, circumventing the need for blood drawing and enabling long-term monitoring with minimal discomfort. These technologies are particularly promising in early disease detection; for example, nano-sensors capable of identifying beta-amyloid or tau proteins could signal preclinical stages of neurodegenerative diseases such as Alzheimer's (Chen et al. 2024) and Parkinson's diseases (Mahato et al. 2024b).

Concurrently, flexible and stretchable electronics are reshaping the form factor and usability of health-monitoring devices. Unlike rigid traditional components, flexible electronics conform to the body's topology, offering comfort and reducing motion artifacts. These materials support the development of epidermal electronics, devices worn like temporary tattoos, as well as implantable biosensors that remain embedded under the skin to track internal physiological variables (Teymourian et al. 2021). Such innovations are particularly suited for continuous monitoring during post-operative recovery or in the management of complex chronic diseases. Chronic conditions, which account for a significant proportion of global morbidity and healthcare expenditure, stand to benefit enormously from the maturation of continuous, non-invasive monitoring systems. Wearable devices capable of continuously tracking blood pressure through optical sensors or piezoelectric elements will provide real-time trends, alerting patients and clinicians to hypertensive events that may otherwise go undetected (Hu et al. 2023). Respiratory monitoring is also set to advance through wearable sensors that track oxygen saturation, respiratory rate, and airflow patterns. These devices will be invaluable for patients with asthma, COPD, and sleep apnea. For example, smart patches or integrated clothing with embedded sensors can detect early signs of respiratory distress, enable preemptive interventions, and reduce hospitalization (Vitazkova et al. 2024).

Despite these exciting prospects, the successful integration of wearable and AI-driven technologies into mainstream healthcare will depend on overcoming several systemic and regulatory challenges. Regulatory bodies such as the FDA and EMA must adapt to the evolving landscape, ensuring that wearable devices meet rigorous standards for accuracy, safety, and clinical utility (Mennella et al. 2024). Variability in sensor performance due to user-specific factors (e.g., skin tone, hydration, movement) necessitates robust validation across diverse populations and environments. Furthermore, AI algorithms used in health monitoring must be interpretable, bias-free, and clinically validated before they can be deployed for medical decision-making. Another major consideration is interoperability (Nashruddin et al. 2024). For wearable technologies to have maximal impact, the data they generate must be seamlessly integrated into EHRs and healthcare platforms. This requires standardized protocols and infrastructure capable of aggregating and synthesizing large volumes of real-time health data. Additionally, clinicians must be supported by intelligent systems that can distill actionable insights from continuous data streams, mitigating the risk of information overload (Ye et al. 2024). Lastly, the increasing granularity and volume of health data generated by wearable devices raises critical ethical and privacy concerns. Ensuring secure storage, transmission, and use of personal health data is paramount. Regulatory frameworks must emphasize individual ownership of data and enforce stringent cybersecurity standards, including encryption and multifactor authentication. Moreover, as AI becomes more embedded in clinical workflows, developers must ensure that models are trained on diverse datasets to avoid perpetuating health inequities through algorithmic bias. In summary, the convergence of bioelectronics, nanotechnology, AI, and flexible electronics is ushering in a transformative era for precision health and wellness tracking. These innovations are enabling continuous, personalized, and predictive health monitoring systems that promise to enhance disease prevention, optimize chronic disease management, and empower individuals in their health journeys. While technological and regulatory hurdles remain, the trajectory is clear that the precision health is transitioning from concept to reality, heralding a future in which healthcare is more responsive, individualized, and preventive than ever before.

References

Abay TY, Kyriacou PA (2018) Photoplethysmography for blood volumes and oxygenation changes during intermittent vascular occlusions. J Clin Monit Comput 32:447–455. https://doi.org/10.1007/s10877-017-0030-2

Abdelrazik A, Eldesouky M, Antoun I et al (2025) Wearable devices for arrhythmia detection: advancements and clinical implications. Sensors 25. https://doi.org/10.3390/S25092848

Andersen LW, Mackenhauer J, Roberts JC et al (2013) Etiology and therapeutic approach to elevated lactate levels. Mayo Clin Proc 88:1127–1140. https://doi.org/10.1016/j.mayocp.2013.06.012

Andrew J, Eunice RJ, Karthikeyan J (2023) An anonymization-based privacy-preserving data collection protocol for digital health data. Front Public Heal 11:1125011. https://doi.org/10.3389/fpubh.2023.1125011

Ates HC, Nguyen PQ, Gonzalez-Macia L et al (2022) End-to-end design of wearable sensors. Nat Rev Mater 7:887–907. https://doi.org/10.1038/s41578-022-00460-x

Baker LB (2019) Physiology of sweat gland function: The roles of sweating and sweat composition in human health. Temperature 6:211–259. https://doi.org/10.1080/23328940.2019.1632145

Baker LB, Baker LB, Baker LB et al (2020) Skin-interfaced microfluidic system with personalized sweating rate and sweat chloride analytics for sports science applications. Sci Adv 6:eabe3929. https://doi.org/10.1126/sciadv.abe3929

Banday MZ, Sameer AS, Nissar S (2020) Pathophysiology of diabetes: an overview. Avicenna J Med 10:174–188. https://doi.org/10.4103/AJM.AJM_53_20/ID/JR_70

Bhatnagar I, Mahato K, Ealla KKR et al (2018) Chitosan stabilized gold nanoparticle mediated self-assembled gliP nanobiosensor for diagnosis of Invasive Aspergillosis. Int J Biol Macromol 110:449–456. https://doi.org/10.1016/j.ijbiomac.2017.12.084

Block AJ, Boysen PG, Wynne JW, Hunt LA (1979) Sleep apnea, hypopnea and oxygen desaturation in normal subjects. N Engl J Med 300:513–517. https://doi.org/10.1056/nejm197903083001001

Canali S, Schiaffonati V, Aliverti A (2022) Challenges and recommendations for wearable devices in digital health: data quality, interoperability, health equity, fairness. PLoS Digit Health 1:e0000104. https://doi.org/10.1371/journal.pdig.0000104

Chen C, Ding S, Wang J (2023) Digital health for aging populations. Nat Med 29:1623–1630. https://doi.org/10.1038/s41591-023-02391-8

Chen CH, Liang HH, Wang CC et al (2024) Unlocking early detection of Alzheimer's disease: the emerging role of nanomaterial-based optical sensors. J Food Drug Anal 32:296–324. https://doi.org/10.38212/2224-6614.3520

Clement David-Olawade A, Olawade DB, Ojo IO et al (2024) Nursing in the digital age: harnessing telemedicine for enhanced patient care. Informatics Health 1:100–110. https://doi.org/10.1016/j.infoh.2024.07.003

Deniz-Garcia A, Fabelo H, Rodriguez-Almeida AJ et al (2023) Quality, usability, and effectiveness of mHealth apps and the role of artificial intelligence: current scenario and challenges. J Med Internet Res 25:e44030. https://doi.org/10.2196/44030

Dobkin BH (2013) Wearable motion sensors to continuously measure real-world physical activities. Curr Opin Neurol 26:602–608. https://doi.org/10.1097/WCO.0000000000000026

Filkins BL, Kim JY, Roberts B et al (2016) Privacy and security in the era of digital health: what should translational researchers know and do about it? Am J Transl Res 8:1560–1580

Gao F, Liu C, Zhang L et al (2023) Wearable and flexible electrochemical sensors for sweat analysis: a review. Microsyst Nanoeng 9:1–21. https://doi.org/10.1038/s41378-022-00443-6

Gu M, Guo CF, Song Y (2025) Multimodal bioelectronics: a pathway to digital health management. Matter 8:102048. https://doi.org/10.1016/j.matt.2025.102048

Haleem A, Javaid M, Singh RP, Suman R (2021) Telemedicine for healthcare: capabilities, features, barriers, and applications. Sensors Int 2:100117. https://doi.org/10.1016/j.sintl.2021.100117

Hemdan M, Ali MA, Doghish AS et al (2024) Innovations in biosensor technologies for healthcare diagnostics and therapeutic drug monitoring: applications, recent progress, and future research challenges. Sensors 24:5143. https://doi.org/10.3390/S24165143

Hickey BA, Chalmers T, Newton P et al (2021) Smart devices and wearable technologies to detect and monitor mental health conditions and stress: a systematic review. Sensors 21. https://doi.org/10.3390/s21103461

Hu H, Huang H, Li M et al (2023) A wearable cardiac ultrasound imager. Nature 613:667–675. https://doi.org/10.1038/s41586-022-05498-z

Huang G, Chen X, Liao C (2025) AI-Driven Wearable Bioelectronics in Digital Healthcare. Biosensors 15:410. https://doi.org/10.3390/BIOS15070410

Jafleh EA, Alnaqbi FA, Almaeeni HA et al (2024) The role of wearable devices in chronic disease monitoring and patient care: a comprehensive review. Cureus 16. https://doi.org/10.7759/cureus.68921

Jaime FJ, Muñoz A, Rodríguez-Gómez F, Jerez-Calero A (2023) Strengthening privacy and data security in biomedical microelectromechanical systems by IoT communication security and protection in smart healthcare. Sensors (Basel) 23:8944. https://doi.org/10.3390/s23218944

Kang HS, Exworthy M (2022) Wearing the future—wearables to empower users to take greater responsibility for their health and care: scoping review. JMIR Mhealth Uhealth 10:e35684. https://doi.org/10.2196/35684

Kim J, Jeerapan I, Sempionatto JR et al (2018) Wearable bioelectronics: enzyme-based body-worn electronic devices. Acc Chem Res 51:2820–2828. https://doi.org/10.1021/ACS.ACCOUNTS.8B00451/ASSET/IMAGES/MEDIUM/AR-2018-00451B_0004.GIF

Kim J, Campbell AS, de Ávila BEF, Wang J (2019) Wearable biosensors for healthcare monitoring. Nat Biotechnol 37:389–406. https://doi.org/10.1038/s41587-019-0045-y

Kumar A, Purohit B, Mahato K et al (2020) Bio-Nano-Interface engineering strategies of AuNPs passivation for next-generation biomedical applications. In: Biointerface engineering: prospects in medical diagnostics and drug delivery. Springer Singapore, Singapore, pp 215–231

Laochai T, Moonla C, Moon J-M et al (2024) Touch-based potentiometric sensors for simultaneous detection of urea and ammonium from fingertip sweat. Sensors Actuators B Chem 413:135898. https://doi.org/10.1016/j.snb.2024.135898

Leal-Junior A, Avellar L, Frizera A, Marques C (2020) Smart textiles for multimodal wearable sensing using highly stretchable multiplexed optical fiber system. Sci Rep 10:1–12. https://doi.org/10.1038/s41598-020-70880-8

Leppänen T, Kainulainen S, Korkalainen H et al (2022) Pulse oximetry: the working principle, signal formation, and applications. Adv Exp Med Biol 1384:205–218. https://doi.org/10.1007/978-3-031-06413-5_12

Li Q, Zhang Y, Fan H et al (2021) In vitro and in vivo detection of lactate with nanohybrid-functionalized Pt microelectrode facilitating assessment of tumor development. Biosens Bioelectron 191. https://doi.org/10.1016/J.BIOS.2021.113474

Li X, Zhan C, Huang Q et al (2022) Smart diaper based on integrated multiplex carbon nanotube-coated electrode array sensors for in situ urine monitoring. ACS Appl Nano Mater 5:4767–4778. https://doi.org/10.1021/ACSANM.1C04220/ASSET/IMAGES/LARGE/AN1C04220_0007.JPEG

Li K, Cardoso C, Moctezuma-Ramirez A et al (2023) Heart rate variability measurement through a smart wearable device: another breakthrough for personal health monitoring? Int J Environ Res Public Health 20:7146. https://doi.org/10.3390/ijerph20247146

Luks AM, Swenson ER (2020) Pulse oximetry for monitoring patients with COVID-19 at home potential pitfalls and practical guidance. Ann Am Thorac Soc 17:1040–1046. https://doi.org/10.1513/AnnalsATS.202005-418FR

Mahato K (2019) Design and development of proof-of-concept biosensors based on engineered surfaces and nanomaterials, Indian Institute of Technology Guwahati, India. http://hdl.handle.net/10603/419929

Mahato K (2024a) Implantable biosensors for personalized healthcare. In: Biosensors for personalized healthcare. Springer Nature Singapore, Singapore, pp 375–392

Mahato K (2024b) Introduction to biosensors for personalized health. In: Biosensors for personalized healthcare. Springer Nature Singapore, Singapore, pp 1–25

Mahato K, Chandra P (2019) Paper-based miniaturized immunosensor for naked eye ALP detection based on digital image colorimetry integrated with smartphone. Biosens Bioelectron 128:9–16. https://doi.org/10.1016/j.bios.2018.12.006
Mahato K, Chandra P (2022) Miniaturized biosensing devices. Springer Nature Singapore, Singapore
Mahato K, Chandra P (2024) Biosensors for personalized healthcare. Springer Nature Singapore, Singapore
Mahato K, Wang J (2021) Electrochemical sensors: from the bench to the skin. Sensors Actuators B Chem 344:130178. https://doi.org/10.1016/j.snb.2021.130178
Mahato K, Srivastava A, Chandra P (2017) Paper based diagnostics for personalized health care: emerging technologies and commercial aspects. Biosens Bioelectron 96:246–259. https://doi.org/10.1016/j.bios.2017.05.001
Mahato K, Maurya PK, Chandra P (2018a) Fundamentals and commercial aspects of nanobiosensors in point-of-care clinical diagnostics. 3 Biotech 8:1–14. https://doi.org/10.1007/s13205-018-1148-8
Mahato K, Kumar A, Maurya PK, Chandra P (2018b) Shifting paradigm of cancer diagnoses in clinically relevant samples based on miniaturized electrochemical nanobiosensors and microfluidic devices. Biosens Bioelectron 100:411–428. https://doi.org/10.1016/j.bios.2017.09.003
Mahato K, Kumar S, Srivastava A et al (2018c) Electrochemical immunosensors: fundamentals and applications in clinical diagnostics. In: Handbook of immunoassay technologies: approaches, performances, and applications. Elsevier, pp 359–414
Mahato K, Nagpal S, Shah MA et al (2019) Gold nanoparticle surface engineering strategies and their applications in biomedicine and diagnostics. 3 Biotech 9:57. https://doi.org/10.1007/s13205-019-1577-z
Mahato K, Purohit B, Kumar A, Chandra P (2020a) Clinically comparable impedimetric immunosensor for serum alkaline phosphatase detection based on electrochemically engineered Au-nano-Dendroids and graphene oxide nanocomposite. Biosens Bioelectron 148:111815. https://doi.org/10.1016/j.bios.2019.111815
Mahato K, Purohit B, Kumar A, Chandra P (2020b) Paper-based biosensors for clinical and biomedical applications: emerging engineering concepts and challenges. Compr Anal Chem. https://doi.org/10.1016/bs.coac.2020.02.001
Mahato K, Kumar A, Purohit B et al (2022) Onsite quality controls for food safety based on miniaturized biosensing. Springer Nature Singapore
Mahato K, Baranwal A, Kumari R et al (2023) Differential electrochemical behaviour of phytofabricated and chemically synthesized silver nanoparticles towards hydrogen peroxide sensing**. Electroanalysis 35:e202300094. https://doi.org/10.1002/elan.202300094
Mahato K, Saha T, Ding S et al (2024a) Hybrid multimodal wearable sensors for comprehensive health monitoring. Nat Electron 7:735–750. https://doi.org/10.1038/s41928-024-01247-4
Mahato K, Moon J, Moonla C et al (2024b) Biosensor strip for rapid on-site assessment of levodopa pharmacokinetics along with motor performance in Parkinson's disease. Angew Chem Int Ed 63:e202403583. https://doi.org/10.1002/anie.202403583
Mennella C, Maniscalco U, De Pietro G, Esposito M (2024) Ethical and regulatory challenges of AI technologies in healthcare: a narrative review. Heliyon 10:e26297. https://doi.org/10.1016/J.HELIYON.2024.E26297/ASSET/1DD7435A-9F70-4F4C-B1BE-A28E0C6DB767/MAIN.ASSETS/GR005.JPG
Menon S, Goldstein SL, Mottes T et al (2016) Urinary biomarker incorporation into the renal angina index early in intensive care unit admission optimizes acute kidney injury prediction in critically ill children: a prospective cohort study. Nephrol Dial Transplant 31:586–594. https://doi.org/10.1093/ndt/gfv457
Mudgal SK, Sharma SK, Chaturvedi J, Sharma A (2020) Brain computer interface advancement in neurosciences: applications and issues. Interdiscip Neurosurg Adv Tech Case Manag 20:100694. https://doi.org/10.1016/j.inat.2020.100694
Namvari M, Lipoth J, Knight S et al (2022) Photoplethysmography enabled wearable devices and stress detection: a scoping review. J Pers Med 12. https://doi.org/10.3390/jpm12111792

Nashruddin SNABM, Salleh FHM, Yunus RM, Zaman HB (2024) Artificial intelligence–powered electrochemical sensor: recent advances, challenges, and prospects. Heliyon 10:e37964. https://doi.org/10.1016/J.HELIYON.2024.E37964

Obianyo CM, Ezeamii VC, Idoko B, et al (2024) The future of wearable health technology: from monitoring to preventive healthcare. https://wjbphs.com/sites/default/files/WJBPHS-2024-0709.pdf 20:036–055. https://doi.org/10.30574/WJBPHS.2024.20.1.0709

Pohanka M (2018) Overview of piezoelectric biosensors, immunosensors and DNA sensors and their applications. Materials 11:448. https://doi.org/10.3390/MA11030448

Porumb M, Stranges S, Pescapè A, Pecchia L (2020) Precision medicine and artificial intelligence: a pilot study on deep learning for hypoglycemic events detection based on ECG. Sci Rep 10:1–16. https://doi.org/10.1038/s41598-019-56927-5

Prasad A, Mahato K, Chandra P et al (2016) Bioinspired composite materials: applications in diagnostics and therapeutics. J Mol Eng Mater 04:1640004. https://doi.org/10.1142/s2251237316400049

Priyadarshni N, Singh P, Mahato K (2023) Magnetic nanoparticle-based sensing strategies for clinical analysis and environmental safety assessment. In: Purohit B, Chandra P (eds) Surface engineering and functional nanomaterials for point-of-care analytical devices. Springer Nature Singapore, Singapore, pp 67–102

Purohit B, Kumar A, Mahato K et al (2019) Cancer cytosensing approaches in miniaturized settings based on advanced nanomaterials and biosensors. In: Nanotechnology in modern animal biotechnology: concepts and applications. Elsevier, pp 133–147. https://doi.org/10.1016/B978-0-12-818823-1.00009-0

Purohit B, Kumar A, Mahato K, Chandra P (2020) Smartphone-assisted personalized diagnostic devices and wearable sensors. Curr Opin Biomed Eng 13:42–50. https://doi.org/10.1016/j.cobme.2019.08.015

Purohit B, Kumar A, Mahato K, Chandra P (2022a) Continuous glucose monitoring for diabetes management based on miniaturized biosensors BT – miniaturized biosensing devices: fabrication and applications. In: Mahato K (ed) Chandra P. Springer Nature Singapore, Singapore, pp 149–175

Purohit B, Kumar A, Mahato K et al (2022b) Continuous glucose monitoring for diabetes management based on miniaturized biosensors. In: Miniaturized biosensing devices, pp 149–175. https://doi.org/10.1007/978-981-16-9897-2_7

Rehg JM, Murphy SA, Kumar S (2017) Mobile health: sensors, analytic methods, and applications. Springer Cham. https://doi.org/10.1007/978-3-319-51394-2

Sachdeva S, Bhatia S, Al Harrasi A et al (2024) Unraveling the role of cloud computing in health care system and biomedical sciences. Heliyon 10:e29044. https://doi.org/10.1016/j.heliyon.2024.e29044

Saha T, Del Caño R, Mahato K et al (2023) Wearable electrochemical glucose sensors in diabetes management: a comprehensive review. Chem Rev 123:7854–7889. https://doi.org/10.1021/acs.chemrev.3c00078

Sahu M, Gupta R, Ambasta RK, Kumar P (2022) Artificial intelligence and machine learning in precision medicine: a paradigm shift in big data analysis. In: Teplow DBBT-P in MB and TS (ed) Precision Medicine. Academic Press, pp 57–100

Samson C, Koh A (2020) Stress monitoring and recent advancements in wearable biosensors. Front Bioeng Biotechnol 8:1–8. https://doi.org/10.3389/fbioe.2020.01037

Scholte NTB, van Ravensberg AE, Shakoor A et al (2024) A scoping review on advancements in noninvasive wearable technology for heart failure management. npj Digit Med 7:279. https://doi.org/10.1038/s41746-024-01268-5

Seh AH, Zarour M, Alenezi M et al (2020) Healthcare data breaches: insights and implications. Healthcare 8:133. https://doi.org/10.3390/healthcare8020133

Shahid S, Iqbal M, Saeed H et al (2025) Diagnostic accuracy of apple watch electrocardiogram for atrial fibrillation: a systematic review and meta-analysis. JACC Adv 4:101538. https://doi.org/10.1016/j.jacadv.2024.101538

Shajari S, Kuruvinashetti K, Komeili A, Sundararaj U (2023) The emergence of AI-based wearable sensors for digital health technology: a review. Sensors 23:1–36. https://doi.org/10.3390/s23239498

Shei RJ, Holder IG, Oumsang AS et al (2022) Wearable activity trackers–advanced technology or advanced marketing? Eur J Appl Physiol 122:1975–1990. https://doi.org/10.1007/s00421-022-04951-1

Shim J, Fleisch E, Barata F (2023) Wearable-based accelerometer activity profile as digital biomarker of inflammation, biological age, and mortality using hierarchical clustering analysis in NHANES 2011–2014. Sci Rep 13:1–12. https://doi.org/10.1038/s41598-023-36062-y

Shukla M, Mahato K, Pandya A (2025) Introduction to emerging biosensing technologies. Prog Mol Biol Transl Sci 215:1–33. https://doi.org/10.1016/BS.PMBTS.2025.05.013

Sic A, Cvetkovic K, Manchanda E, Knezevic NN (2024) Neurobiological implications of chronic stress and metabolic dysregulation in inflammatory bowel diseases. Diseases 12:220. https://doi.org/10.3390/DISEASES12090220

Stoumpos AI, Kitsios F, Talias MA (2023) Digital transformation in healthcare: technology acceptance and its applications. Int J Environ Res Public Health 20:3407. https://doi.org/10.3390/ijerph20043407

Tang W, Yin L, Sempionatto JR et al (2021) Touch-based stressless cortisol sensing. Adv Mater 33:2008465. https://doi.org/10.1002/ADMA.202008465

Tehrani F, Teymourian H, Wuerstle B et al (2022) An integrated wearable microneedle array for the continuous monitoring of multiple biomarkers in interstitial fluid. Nat Biomed Eng 6:1214–1224. https://doi.org/10.1038/s41551-022-00887-1

Templer S (2022) Closed-loop insulin delivery systems: past, present, and future directions. Front Endocrinol (Lausanne) 13:919942. https://doi.org/10.3389/FENDO.2022.919942

Teymourian H, Tehrani F, Mahato K, Wang J (2021) Lab under the skin: microneedle based wearable devices. Adv Healthc Mater 10:2002255. https://doi.org/10.1002/ADHM.202002255

Teymourian H, Tehrani F, Longardner K et al (2022) Closing the loop for patients with Parkinson disease: where are we? Nat Rev Neurol 18:497–507. https://doi.org/10.1038/s41582-022-00674-1

Vashishth TK, Sharma V, Sharma KK et al (2025) Securing the cloud: strategies for protecting sensitive patient data in cloud-based healthcare recommender systems. In: Healthcare recommender systems, pp 185–220. https://doi.org/10.1007/978-3-031-80056-6_9

Vitazkova D, Foltan E, Kosnacova H et al (2024) Advances in respiratory monitoring: a comprehensive review of wearable and remote technologies. Biosensors 14:1–50. https://doi.org/10.3390/bios14020090

Vo DK, Trinh KTL (2024) Advances in wearable biosensors for healthcare: current trends, applications, and future perspectives. Biosensors 14:560. https://doi.org/10.3390/bios14110560

Walter JR, Xu S, Rogers JA (2024) From lab to life: how wearable devices can improve health equity. Nat Commun 15:1–4. https://doi.org/10.1038/s41467-023-44634-9

Wang J (2023) Analytical electrochemistry, 4th edn. John Willy and Sons, Inc

Wang C, He T, Zhou H et al (2023) Artificial intelligence enhanced sensors – enabling technologies to next-generation healthcare and biomedical platform. Bioelectron Med 9:1–34. https://doi.org/10.1186/s42234-023-00118-1

Wasilewski T, Kamysz W, Gębicki J (2024) AI-assisted detection of biomarkers by sensors and biosensors for early diagnosis and monitoring. Biosensors 14. https://doi.org/10.3390/bios14070356

Windmiller JR, Zhou N, Chuang MC et al (2011) Microneedle array-based carbon paste amperometric sensors and biosensors. Analyst 136:1846–1851. https://doi.org/10.1039/C1AN00012H

Wu J, Liu H, Chen W et al (2023) Device integration of electrochemical biosensors. Nat Rev Bioeng 1:346–360. https://doi.org/10.1038/s44222-023-00032-w

Yadav SK, Verma D, Yadav U et al (2023) Point-of-care devices for viral detection: COVID-19 pandemic and beyond. Micromachines 14. https://doi.org/10.3390/mi14091744

Ye J, Woods D, Jordan N, Starren J (2024) The role of artificial intelligence for the application of integrating electronic health records and patient-generated data in clinical decision support. AMIA Summits Transl Sci Proc 2024:459

Yin L, Cao M, Kim KN et al (2022) A stretchable epidermal sweat sensing platform with an integrated printed battery and electrochromic display. Nat Electron 5:694–705. https://doi.org/10.1038/s41928-022-00843-6

Zhang M, Wang X, Huang Z, Rao W (2020) Liquid metal based flexible and implantable biosensors. Biosensors 10. https://doi.org/10.3390/BIOS10110170

Zhang N, Zhu Y, Cao B (2025a) Effects of environmental temperature mimicking circadian rhythms on thermal comfort and sleep quality. Build Environ 281:113196. https://doi.org/10.1016/j.buildenv.2025.113196

Zhang Y, Chen H, Song Y (2025b) Wearable healthcare monitoring and therapeutic bioelectronics. Wearable Electron 2:18–22. https://doi.org/10.1016/j.wees.2024.12.004

Zhao C, Park J, Root SE, Bao Z (2024) Skin-inspired soft bioelectronic materials, devices and systems. Nat Rev Bioeng 2:671–690. https://doi.org/10.1038/s44222-024-00194-1

Zhou L, Guess M, Kim KR, Yeo WH (2024) Skin-interfacing wearable biosensors for smart health monitoring of infants and neonates. Commun Mater 5:1–13. https://doi.org/10.1038/s43246-024-00511-6

Zuo Z, Watson M, Budgen D et al (2021) Data anonymization for pervasive health care: systematic literature mapping study. JMIR Med Informatics 9:e29871. https://doi.org/10.2196/29871

Introduction to Nano-Bioelectronics for Precision Health Monitoring

2

Alok Tripathi and José Bonilla Cruz

Abstract

The demand for prompt disease detection methods in the early stages has increased in recent years. This leads to expanding research associated with the multi-sectorial subject of nano-bioelectronics, which integrates electronics, biology, and nanomaterials. It is fast gaining prominence and presents opportunities to address current problems linked with healthcare monitoring. Specifically, the electronic transducers can express biological characteristics in them and can lead to notable enhancements in sensitivity and biocompatibility, creating avenues for advancements in essential healthcare monitoring. This chapter highlights the latest developments in nano-bioelectronics made possible by semiconductor nanomaterials. These devices with flexible systems such as power supply units, data storage, displays, biosensing, and therapeutic stimulation are distinguished by their high sensitivity and quick detection speed. In addition, their natural compactness, mobility, and ease of use make them well-suited for the creation of flexible healthcare monitoring devices. Hence, to promote a deeper understanding and further advancement of this discipline, this chapter clarifies the idea and outlines potential directions for future research. This chapter also highlights new developments in the field, such as implantable sensors and stimulators. This chapter will significantly improve the area by offering a thorough overview of the developments and possible uses of nano-bioelectronics in healthcare. Finally,

A. Tripathi
Department of Chemical and Biochemical Engineering, School of Engineering, Indrashil University, Mahesana, Gujarat, India

J. B. Cruz (✉)
Advanced Functional Materials and Nanotechnology Group, Centro de Investigación en Materiales Avanzados S. C. (CIMAV-Subsede Monterrey), Alianza norte 202, Parque PIIT, Apodaca-Nuevo León, CP, Mexico
e-mail: jose.bonilla@cimav.edu.mx

K. Mahato, P. Chandra (eds.), *Nano-bioelectronics for Precision Health Monitoring*, https://doi.org/10.1007/978-981-95-2117-3_2

the discussion on current challenges associated with nano-bioelectronics concludes the chapter.

Keywords

Bioelectronics · Nanomaterials · Semiconductor · Biosensing · Healthcare monitoring

2.1 Introduction

The multidisciplinary subject of nano-bioelectronics, which integrates electronics, biology, and nanomaterials, is quickly growing and presents opportunities to address current bioelectronics difficulties (Yoo et al. 2020). Specifically, reducing the size of electronic transducers to the nanoscale and enhancing their biological characteristics can result in notable enhancements in sensitivity and biocompatibility, which in turn can lead to novel prospects in basic biology and medicine (García-Guzmán et al. 2021; Ferreira et al. 2019; Jaiswal et al. 2017). Herein, the combination of biological systems with electronics converts biological system signals into electrical impulses at the bio-electronic interface. Bioelectronics has led to the invention of essential biomedical devices, including deep brain stimulators, cardiac pacemakers, and blood glucose monitors (Benabid 2003). It should be acknowledged that, despite these devices' effectiveness, the electrical transducers have a significant size incompatibility with the biological mechanisms to which they are interfaced (Shim et al. 2021; Cha et al. 2019). Therefore, significantly reducing the size of electronic transducers and giving them more biological-looking characteristics may dramatically improve the susceptibility and biocompatibility of next-generation bioelectronics, which may enhance and/or create new opportunities in the fields of fundamental biology and healthcare (Bazaka and Jacob 2012).

In this sense, a range of nanomaterials have emerged over the past few decades, with significant advancements in their chemical synthesis, manufacturing, and characterization. These incorporate zero-dimensional (0D) nanoparticles, one-dimensional (1D) nanotubes and nanowires, and two-dimensional (2D) nanosheets (Dasgupta et al. 2014; Lieber 2003). These endeavors have aimed to clarify how unique electronic, optical, and magnetic features, such as quantum confinement framework in multiple dimensions, are caused by the size, structure, and composition of these nanostructures (Patolsky and Lieber 2005). Such nanoparticles have potentially exceptional biological prospects due to their improved and even rare physical characteristics. Precisely, the rapidly developing discipline of nano-bioelectronics blends biology, electronics, nanomaterials, and nanoscience (Patolsky et al. 2006). By doing so, it presents opportunities to solve current bioelectronics obstacles as well as investigate novel domains. An affinity-based biosensor, for instance, works with a biological analyte in solution to preferentially engage with it. It then produces an electrical signal that corresponds precisely to the analyte concentration (Patolsky et al. 2007). Such types of sensors are protein and DNA sensors. The primary component of a nano-bioelectronics apparatus is the nanostructure, which is employed to precisely capture or induce a biochemical process of significance (Chen et al. 2011). Nanostructures have intrinsic biological potential due to their tiny dimensions and substantial surface-to-volume ratios. Their elevated ratio of surface area to volume provides enhanced responsiveness to surface phenomena (Tian et al. 2012). Analyte molecules may be reduced in quantity and biosensors can be made smaller since just a limited number of them are required to provide a measurable electrical signal (Liu et al. 2013). Furthermore, the small scale of nanostructures may be likened to cellular components like proteins and nucleic acids, providing novel avenues for causing perturbations in living systems from the subatomic to tissue levels (Liu et al. 2015). Because biological building blocks and nanostructures have comparable sizes, it can be easier to integrate nanoelectronics with cells and tissues. This opens up new possibilities for biomedical prostheses and synthetic tissues (Xie et al. 2015).

This chapter also provides an overview of current developments in nanomaterial- and composite-based bioelectronics technology. Special attention has been given to the engineering and material elements of bioelectronics devices that function well electrically and are mechanically compatible with biological tissues.

2.2 Nanostructured Materials

One significant aspect influencing the flexible stiffness of devices is material size. As a result, an enormous amount of research has been placed into creating electronic nanostructured materials. There are many different methods for creating different kinds of nanomaterials, which may be divided into top-down and bottom-up approaches. The synthesis and manufacturing processes of nanomaterials determine their properties, including sizes and different shapes. Table 2.1 provides a summary of relevant instances of the production and synthesis processes for nanomaterials. Depending on their morphology, the resultant nanostructured materials may be divided into four groups: zero dimension (0D), one dimension (1D), two dimensions (2D), and three dimensions (3D) (Zhang et al. 2018; Zhou et al. 2020; Wang et al. 2020a; Garnett et al. 2019). Compared to their bulk versions, these nanoparticles have unique characteristics. The scale of nanomaterials has a significant impact on band gaps because electrons are confined in materials with nanometer-sized dimensions (Ning et al. 2017; Wang et al. 2011, 2021a). Additionally, the materials' nanoscale domain significantly raises their surface-area-to-volume ratio, which causes a sharp rise in their chemical responsiveness. Several fabrication and incorporation strategies for nanomaterials have been developed because of their inherent flexibility and special material characteristics, which may be used in bioelectronics devices when they are correctly embedded in the device platform (Royston et al. 2008; Gambhire et al. 2008; Moon et al. 2011).

Table 2.1 Summary of relevant instances of the production and synthesis processes for nanomaterials

Type of nanomaterial	Synthesis process	Size	Ref.
ZnO	Atomic layer deposition	5–6 Å	Johns et al. (2013)
Fe_2O_3 nanorods	Thermal deposition	12 nm	Wang and Yang (2009)
SnO_2	Sol-gel	20–50 nm	Ansari et al. (1997)
Si	Laser ablation	13 nm	Popovic et al. (2014)
Fe	Chemical reduction	6–110 nm	Huang and Ehrman (2007)
Cds/TiO_2 core-shell	Ultrasonication	10–90 nm	Alizadeh et al. (2019)
Ni	Microemulsion	4.2 nm	Chen and Wu (2000)
PAni	Electrochemical deposition	0.033 μm	Qin and Guo (2012)
ZnO nanowires	Spray pyrolysis	200 nm thick	Htay et al. (2009)
Co_3O_4 nanowires	Biosynthesis	2–3 nm diameter, and 10 cm length	Nam et al. (2006)
GaAs nanomembrane	Photolithography	200 nm	Yoon et al. (2010)
SnO_2	Ultrasonication	8–30 nm	Majumdar and Devi (2010)
Al	Ball milling	30 nm	Jiang et al. (2019)
Fe_3O_4/TaO_x core-shell	Sol-gel	21 nm	Lee et al. (2012a)
Au/SiO_2	Electrospinning/thermal decomposition	130–170 nm	Shi et al. (2009)
Si nanoribbon	Photolithography	80–300 nm	Ko et al. (2006)
Cu nanowires	Chemical reduction	66 nm diameter and 50 μm thick	Won et al. (2014)
SiO_2	Microemulsion	14 nm	Bagwe et al. (2004)
PEDOT	Electrochemical deposition	1.5 μm	Goda and Miyahara (2019)
SWCNT	Laser CVD	10–50 nm	Bondi et al. (2006)
Al_2O_3	Atomic layer deposition	300 nm	Ritala et al. (1999)

2.2.1 0D Nanomaterials

This category includes nanoparticles, such as fullerenes, metal nanoparticles, and quantum dots. The quantum confinement effect, which is present in these materials, is the result of electrons being completely enclosed within nanoparticles. This leads to the introduction of distinct material characteristics that were not seen in their bulk condition (Cho et al. 2022). Fullerene, for instance, is a carbon-based, spherical, hollow material, and out of various forms, C60 and C70, with their 60 and 70 carbon atoms, are the most stable and have been used in optics and electronics. They are employed in electrochemical systems like fuel cells due to their notable electrocatalytic properties. However, C60's cytotoxicity, when it aggregates in the plasma membrane or other cell components produces reactive oxygen species which limits its use in biomedical applications (Taylor and Walton 1993; Martín 2006). Usually, precursors of metals in colloidal solutions are used to create metallic nanoparticles. The high surface-to-volume ratio of the metal nanoparticles allows them to exhibit distinctive physiochemical characteristics and surface plasmon effects. Metal nanoparticles have a large capacity for tiny molecules to adsorb onto their surface. One of its examples is Au nanoparticles bind to proteins, including peptides and antibodies, which makes them useful for biosensing. Their applicability for different cancer treatments is made possible by their photothermal characteristics resulting from localized surface plasmon resonance (Salata 2004; Kreuter 2007; Ahamed et al. 2010; Zhao et al. 2013).

Semiconductor nanoparticles with special optical characteristics are called quantum dots (QDs). Consequently, light-emitting diodes (LEDs) and bioimaging have benefited from the use of QDs. By altering their size, QDs' emission wavelengths are readily regulated. When QD gets bigger, longer wavelength emission occurs and vice versa. One additional benefit of QDs is their remarkable color purity. Owing to the toxicity of cadmium, QDs with thick, biocompatible shells or QDs utilizing different elements are being researched for use in devices. The quantum efficiency is increased, and surface imperfections are decreased by the core/shell configuration (Lim et al. 2015; Jamieson et al. 2007).

2.2.2 1D Nanomaterials

Numerous applications in electronics, optics, and biology have been presented for 1D nanomaterials, such as nanotubes (Joselevich et al. 2008), nanowires (Duan and Lieber 2000; Lee et al. 2012b; Rackauskas et al. 2009), and nanorods (Liu et al. 2009; Heitsch et al. 2009; Lohse and Murphy 2013). Carbon materials in the form of carbon nanotubes (CNTs) are very strong, flexible, chemically and thermally stable, and have excellent electrical conductivity. Consequently, CNTs have found application in stretchy and flexible electronics, such as soft bioelectronics. Graphene sheets, or carbon sheets, are rolled on the side of a cylindrical shape to generate two different forms of carbon nanotubes (CNTs): single-walled and multiwalled (Wang et al. 2005; Nadeem et al. 2017; Kim et al. 2006). Depending on the rolling direction, SWCNTs can be metallic or semiconducting, but MWCNTs are normally conducting (Cheung et al. 2010).

Semiconductor or metal nanowires are another kind of 1D nanomaterial. Silicon nanowires (SiNWs) are exemplary semiconductor nanowires that have been utilized as nanoscale components in a variety of bioelectronic systems (Peng et al. 2002). To capture electrophysiological signals, for instance, SiNW transistors were connected with cells (capturing action potentials from neuronal cells). SiNW-based ultrathin and flexible mesh circuits have also been applied to the recording of in vivo electrophysiology (Zhang et al. 2001). Metal nanowires, on the other hand, show a lot of promise for elastic and flexible electrodes (Zheng et al. 2004). A densely permeated network made of metal nanowires, such as AgNWs, has been employed to create conductive channels even under extreme stresses. AgNW-based stretchy and flexible electrodes have also been utilized in soft electrical and optoelectronic devices (Pereira et al. 2010; Deignan and Goldthorpe 2017; De et al. 2009; Luu et al. 2011).

Similar to nanowires, but with somewhat lower aspect ratios, are nanorods such as Au and ZnO nanorods. The usual method for creating Au nanorods is a colloidal solution. Their application in photothermal therapy is documented. When exposed to the right wavelength of light, Au nanorods' surface plasmon resonance causes electrons to polarize, which results in localized hyperthermia that is used in photothermal therapy. Since they have a smaller aspect ratio and hence more intense percolation between them, nanorods are used in bioelectronics less frequently than nanowires (Sau and Murphy 2004). ZnO nanorods can be generated in the interim by colloidal synthesis or CVD techniques. The biodegradability of ZnO nanorods is one of its special qualities. The breakdown products of this process are biocompatible since zinc is a necessary mineral for human health. As a result, biodegradable implantable bioelectronics can make advantage of them (Wu et al. 2006).

2.2.3 2D Nanomaterials

2D nanomaterials are defined as those that have a sheet-like form. Due to their sheet-like structure, 2D nanomaterials are frequently utilized in the development of electronic components and devices (Sang et al. 2019; Kang et al. 2016). Graphene, a type of 2D nanomaterial, may be produced by chemical vapor deposition (CVD) processes or exfoliating from graphite via Scotch tape (Hernandez et al. 2008). The remarkable physical, electrical, and chemical characteristics of graphene, a layer of sp^2-hybridized carbons, include high electrical and thermal conductivities, superb transparency, and a high Young's modulus (Lotya et al. 2009; Parvez et al. 2014; Chang et al. 2019). Numerous research studies have been conducted to produce graphene on an extensive basis and at a low cost. One such study used CVD to synthesize graphene across a vast region. Graphene may be produced on a copper foil (catalytic substrate) in the CVD chamber at the appropriate temperature and with gaseous precursors. Graphene's inherent softness and excellent transparency make it a suitable material for sensitive optoelectronic devices (An et al. 2011; Muñoz and Gómez-Aleixandre 2013).

Molybdenum disulfide (MoS_2) 2D films also have received attention due to their semiconductivity (Perkgoz and Bay 2016). MoS_2 layers have an average thickness of 0.65 nm, and an assembly of a few layers is usually utilized as the working layer (Nath et al. 2001; Lee et al. 2012c; Wang et al. 2013). With a straight band gap and an optical transition band gap at 1.9 eV, this ultrathin and inherently soft 2D nanomaterial exhibits a high quantum yield and a high photo absorption coefficient (about 5×10^7 m^{-1}) (Lembke et al. 2015). Additionally, it can be bent to a radius of 0.75 mm deprived of losing its original electrical characteristics and stretched up to 11% without experiencing mechanical fractures. Due to the high van der Waals binding force, graphene and other 2D materials, such as MoS_2, may form strong bonds when layered. Thus, MoS_2 and graphene are frequently coupled to make channels and electrodes, respectively, and to constitute important 2D components for optoelectronics and soft electronics, respectively.

Bioelectronics with great performance has been implemented using the single-crystalline semiconductor nanomembrane. It depicts a very thin sheet, approximately a few hundred nanometer-thick, made of a single crystal substance. It also exhibits a high degree of flexibility not found in the initial bulk condition. The target substrate is coated with an ultrathin layer that has been manufactured from bulk-state single-crystal wafers. This procedure creates the nanomembrane. To build the Si nanomembrane, a silicon-on-insulator (SOI) wafer with an ultrathin Si layer above an oxide layer can be utilized. After the underlying oxide layer is removed, the top Si layer may be placed onto the substrates. Superior mechanical deformability appropriate for soft electronic devices is provided by the ultrathin dimension of the high-quality monocrystalline structure, which also gives outstanding electronic performance (Rogers et al. 2011).

Gallium nitride (GaN) is another kind of monocrystalline semiconductor nanomembrane. It is possible to create a standalone, large-area GaN nanomembrane that is thinner than 100 nm. Higher voltages can be sustained by GaN, which allows it to channel current more quickly. Furthermore, GaN significantly reduces energy loss, making GaN-based electronics energy-efficient. GaN may be used to create extensible blue light-producing devices since it has a larger band gap than GaAs (ElAfandy et al. 2014).

2.2.4 3D Nanomaterials

Usually, 1D or 2D nanomaterials may be assembled to create 3D nanomaterials (Cho et al. 2022). When compared to electron transport channels, 3D nanomaterials' huge surface area could significantly boost the volume of active reaction sites, which is the main justification for their use. 3D nanomaterials are useful for creating sensors and electrochemical devices because of their high reactivity. Mixing 1D CNT and 2D graphene can create a novel 3D carbon structure. Because of their huge aspect ratio, CNTs may entangle themselves, and the strong van der Waals force across layers causes graphene to clump together. The hierarchical structure is maintained, and the entanglement problems are resolved by the 3D carbon framework with graphene layers joined by CNT pillars (Duan et al. 2017). Additionally, this arrangement offers several CNT channels for quick electron transfer between graphene layers. It is possible to regulate the 3D carbon structure's density, size, and form. Moreover, zinc oxide nanoflowers can made up of several ZnO nanosheets put together in the form of a flower exhibiting great photocatalytic activity and high photoluminescence due to its enormous surface area within its 3D shape (Ya et al. 2017).

Excitons and surface plasmons interact inside 3D nanostructures to give rise to unique electrical, magnetic, and optical phenomena in 3D nanomaterials. The enokitake-shaped positioned gold nanowires exhibit exceptionally high elastic properties of around 90% without the need for extra structural engineering, such as buckling or serpentine-shaped designs, when they are chemically attached to an elastomer (Wang et al. 2018a). There are no topological restrictions on the growth of positioned AuNWs on polymer substrates like polydimethylsiloxane and polyethylene terephthalate. Since the upper nanoparticle layer appears stiffer than the underneath nanowire layer, the film of positioned AuNWs produces fractures under applied strain that have a characteristic form (V-shaped cracks). After the applied tension is released, these V-shaped fissures enable the stationary AuNWs to revert to their initial conductivity. The combination of the standing AuNWs can be worn as portable face expression recognition sensors or used as resilient conductors for extensible supercapacitors (Wang et al. 2015a). Another type of nanowire structure is a branching network called dendritic nanowires. Synthesized ZnO dendritic nanowires were used on the dye-sensitized solar cell's photoanode. The effectiveness of the solar cell was boosted by the dendritic nanowires' greater specific surface area when compared to the original 1D nanowires. This led to improvements in the dye load, transportation of electrons, and light-acquiring capability (Wang et al. 2018a; Zhou et al. 2019).

2.2.5 Techniques for Nanomaterial Processing and Integration

A crucial prerequisite for the development of high-performance soft bio-integrated electronic components is the simultaneous integration of nanostructured materials with extensible and resilient electronic components. When nanomaterials are integrated properly, their performance can be optimized and new device functions can be introduced that were not possible with earlier bulk-fabricated devices (Han et al. 2015; Hong et al. 2011). To accomplish the consistent, systematic, and extensive incorporation of nanomaterials into the devices, numerous techniques to facilitate processing and incorporation have been implemented, including inkjet printing, spin coating, blading, dip-pen lithography, transfer printing, Langmuir-Blodgett assembly, and dip coating as depicted in Fig. 2.1.

Dip and Spin Coating The dip coating process is one of the facile and efficient ways to assemble nanomaterials. A consistent thin layer of nanomaterials can be created by submerging a rootstock in a nanomaterial aqueous solution and then uplifting it upward. One benefit of this technique is that it is simple to get around the substrate topology's intricacy. Furthermore, without the need for expensive equipment, an extensive region on a substrate may be covered consistently and fast (Cho et al. 2022). Similar to dip coating in spin coating, the nanomaterial fluid mixture is applied to a rotating substrate. It is possible to modify the film thickness by adjusting significant variables such as rotation rate, hydrophilicity, and viscosity (Cho et al. 2022).

Blading Method The blading approach is an additional technique for creating thin films with a consistent thickness. This involves dropping the nanomaterial solution onto the substrate and using a blade to remove any surplus nanomaterials. Thin

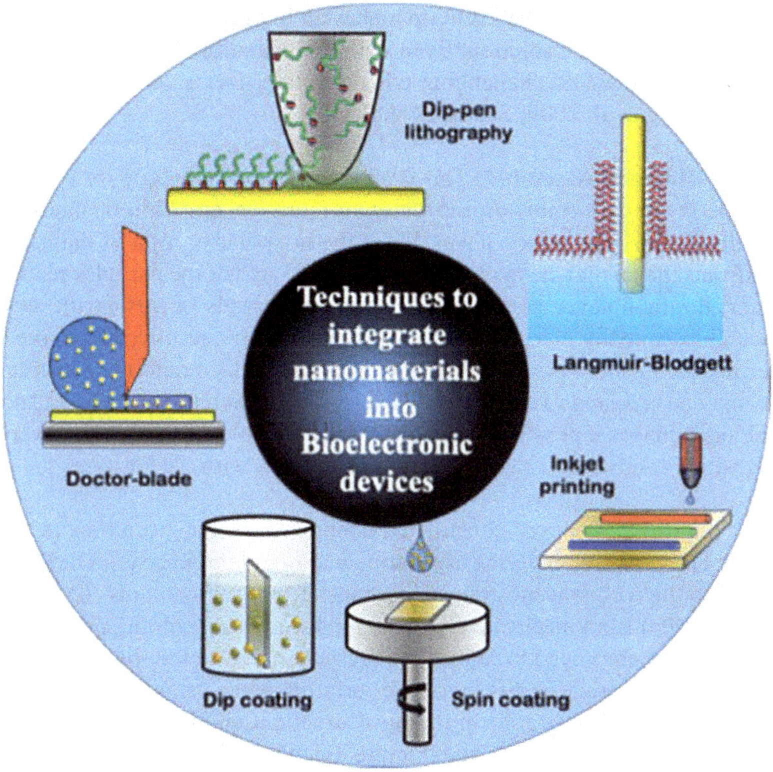

Fig. 2.1 Various techniques to integrate nanomaterials with bioelectronics

films with thicknesses ranging from tens to hundreds of micrometers can be coated using this technique. The solution's viscosity and the space across the blade and the substrate allow for adjustment of the thickness (Cho et al. 2022).

Inkjet Printing The methods indicated above work well for coating a broad surface area, however, to get more structured or organized designs in the coated film, further patterning steps need to be performed. To address this issue, techniques for reducing the number of extra patterning steps are addressed; one such method is inkjet printing. It is essential to use ink composed of nanoparticles at the ideal ratio and viscosity (Sirringhaus et al. 2000). Compared to dip/spin coating or blading processes, the inkjet printing method is less redundant than nanomaterial ink, causing it to make up for its cost-effectiveness. However, printing machinery is still required. Roughly 10–100 μm is the printing range for inkjet printing (Liu et al. 2014a).

Dip-Pen Lithography To increase the printing resolution even more, dip-pen lithography is frequently utilized. A nanomaterial-ink-coated atomic force microscopy tip is utilized for the dip-pen lithography (Piner et al. 1999). The motif is drawn on the surface when the inked tip makes an impact on it. Tens of nanometers is the upper limit of the resolution. Even with the incredibly high resolution, printing is extremely slow and challenging when printing over a substantial area (Huo et al. 2008; Wang et al. 2009; Salaita et al. 2007).

Langmuir–Blodgett Assembly This is a paradigmatic technique for the nanomaterial single-layered assembly. At the water-air interface, amphiphilic nanoparticles are initially formed into a monolayer. Since the nonpolar portion of the nanoparticles confronts the air and the hydrophilic polar head area of the particles rests within the water, the monolayer assembly of the nanomaterials occurs at the water-air interface. Subsequently, the substrate is submerged in and removed from the water, transferring the nanomaterials from the hydrophilic surface onto the substrate. The process may be repeated to create many levels as well (Son et al. 2014). In addition, this technique makes it possible to integrate and assemble nanomaterials over a vast surface, which is advantageous when creating devices with a wide area.

Transfer Printing A film of constructed nanomaterials has been transferred with great yield and precision in terms of location and orientation between the precursor substrate and the recipient substrate using transfer printing methods. To prevent the nanomaterial film from mechanically breaking during the stamping process, elastomeric stamps are employed to reduce the pressure given to the film. It is even feasible to get the necessary motifs in transfer printing when using a systematic stamp (Choi et al. 2016a). For transfer printing of excellent quality, there must be exact control over both surface energy and fracture dynamics. To successfully retrieve the nanomaterial film, the adhesion at the interface between the stamp and the film needs to be greater than the attachment between the nanomaterial film and the donor substrate. However, it ought to be less strong than the interfacial adhesion between the nanomaterial and the recipient during the printing process. The structured transfer prints may be produced with a far greater resolution using an altered kind of transfer printing called intaglio transfer printing (Kim et al. 2010a). In this instance, a flat stamp retrieves a layer of nanomaterials, which contacts the silicon wafer with designed pits and presses on it. The silicon wafer retains the nanomaterials imprinted on its surface, while those that are in the trench continue to stick to the stamp. Subsequently, the residual nanoparticles are moved to the intended substrate (Yoo et al. 2021a). The intaglio transfer printing process shows a substantially greater resolution and yield when compared to the patterned transfer printing employing a structured stamp.

2.3 Expandable Conductive Nanocomposites

Even though devices are more flexible and stretchable due to their extremely thin designs and unique deformable structures, the high Young's modulus of the electronic materials still results in an internal structural inconsistency at the interface between the device and biological tissue, which triggers inflammatory responses nearby (Gulino et al. 2019). The research states that one of the primary causes of inflammation is a variation in a mechanical parameter like Young's modulus. Moreover, ultrathin bio-integrated electronics are subject to severe or repeated stress and are susceptible to mechanical failures because of the local mechanical mismatch and the continually changing motion of the human body. To address these problems, inherently stretchy bio-integrated devices have been developed using instinctively stretchable nanocomposite materials, which have Young's modulus similar to that of tissue (Zhang et al. 2021; Lee et al. 2020). The nanocomposite is made up of functional materials, such as conducting filler material combined with naturally soft polymeric medium. Due to their ability to be incorporated into the resilient matrix, establish an infiltration network, and keep the network intact in the face of mechanical deformations, nanostructured materials have received special attention as filler materials. Many biocompatible and mechanically sensitive polymers, including hydrogels and elastomers, have been employed for the stretchy media. This section reviews the several kinds of elastic media that have cross-linked both physically and chemically (Park et al. 2014; Cho et al. 2021; Wang et al. 2018b; Lim et al. 2018).

2.3.1 Electroconductive Fillers for Nanostructures

Although the majority of elastic matrices are insulative, an infiltration network that acts as a conduit for charged carriers can be formed by the well-distributed and highly conductive filler elements in the elastic matrix, such as metal nanomaterials, carbon nanomaterials, and conducting polymers (Hu et al. 2008). The fillers' dimensions and surface qualities were altered, and the quantity and caliber of filler-to-filler interactions were enhanced. The elastic matrix is filled with a highly diffused network of filler materials that are concurrently highly conductive and stretchable. The electrical conductivity is determined by the quantity and quality of connections between filler components in the elastomer matrix as the electrical power travels through the permeation network. Stated differently, the conductivity of the nanocomposite is influenced not only by the inherent conductivity of the filler material but also by the configuration and order of dispersion of the filler materials within the polymeric matrix (Lee et al. 2012d; Rai and Moore 2017). Consequently, the mathematical notion known as the percolation threshold has been used to assess the integrity of the percolation network (Valentín et al. 2010). This idea suggests that the network's filler material must have a threshold content for long-range connection to emerge. Significant electrical conductivity in the nanocomposite with a high percolation threshold requires a lot of fillers (Wolff and Wang 1992; Lim et al. 2017). Changing the surface characteristics and filler dimensions are two methods for altering the filler materials. Elevating conductivity can be achieved by employing filler substances with an excellent aspect ratio since this minimizes conduction from connections between filler moieties while simultaneously increasing electrical conduction via a filler moiety. The total conductance rises because the interaction of the resistance between the filler moieties is significantly larger than the resistance of a filler moiety alone. Since the resistance is inversely correlated with thickness, conductance may be further increased by adding more high-aspect-ratio fillers (Han et al. 2014; Li et al. 2007; Sandler et al. 2003). Additionally, surface chemistry is crucial for the dispersion of fillers inside the polymeric matrix. By chemical and physical interactions such as hydrophilicity, van der Waals force, and electrostatic force, nanoscale filler materials tend to aggregate. The nanocomposite's electrical conductivity is reduced as a result of this aggregation because of the high resistance between the aggregates (Taherian 2014). When creating their percolation network, filler materials should be dispersed uniformly. Improved electrical conductivity can be achieved by progressively lowering the percolation threshold by localized microscale phase separation. A variety of surface modification approaches comprising hydrophilicity control, solvent exchange, and ligand exchange have been employed to precisely regulate the distribution of filler materials in the elastomeric medium (Liu et al. 2018; Zare et al. 2019). The total resistance of the nanocomposite is mostly determined by the interaction of resistance across the filler components throughout the percolation network. As a result, increasing the quantity and caliber of interactions is crucial. The easiest method to enhance the proportion of successful contacts among fillers is to add more fillers to the composite. On the other hand, overusing fillers results in increased manufacturing costs, cross-linked chemical failure, and mechanical stiffness in the process. Alternately, the effective contact between fillers can also be increased by changing the filler shape, such as changing the aspect ratio, curvature, and branching. A further common strategy is to use conducting polymers or the welding process to improve the quality of connections. This section provides a more detailed discussion of a variety of conductive filler materials, which comprises metal-based nanomaterials, carbon-based nanomaterials, conducting polymers, and hybrid-type fillers, all of which are based on the ideas of the percolation network (Homede et al. 2018; Hotze et al. 2010; Zhang 2014).

2.3.2 Conducting Polymers and Their Nanocomposites

Since they mix well with soft polymeric matrices made of hydrogels and elastomers, conducting polymers are the most typical filler substances. Because of their stability, low electrical impedance, and electrical conductivity, polymers such as polypyrrole (DS et al. 2012; Da Silva et al. 2013; Kurian et al. 2020), polystyrenesulfonate (PSS) (Si et al. 2018; Fan and Ouyang 2019; El Zein et al. 2017), polyaniline (PANI) (Molberg et al. 2010; Sreeja et al. 2006; Saritha Chandran and Narayanankutty 2008), and poly-3,4-ethylenedioxythiophene (PEDOT) (Si et al. 2018; Fan and Ouyang 2019; El Zein et al. 2017) are acknowledged as viable options for the conductive filler materials. Conducting polymers do not meet the criteria for being considered nanomaterials, but they can be typically mixed with nanoparticles to create a hybrid filler material. They are tiny enough as individual chains and form fibril architectures inside insulating elastomers. The use of hybrid-type filler materials in nanocomposites (i.e., nanoparticles mixed with conducting polymers) will be covered further. Simultaneous cross-linking of the conducting polymers can result in conductive hydrogels. Using 0.1 mol/L sulfuric acid at 90 °C, Yao. et al. treated the PEDOT/PSS dispersion to produce hydrogel having a conductivity of 46 S/m. After applying strong sulfuric acid to the hydrogel, the PSS component was largely eliminated, and the conductivity of PEDOT/PSS hydrogel concentration rose to 880 S/m. Nevertheless, the use of strong acids in the treatment procedure might not be appropriate for use in biological areas (Yao et al. 2017). Pure PEDOT/PSS hydrogels were created by Lu et al. by combining aqueous PEDOT and PSS with an organic solvent called dimethyl sulfoxide at high temperatures with reswelling in water. In phosphate-buffered saline and deionized water, the pure hydrogels had conductivities of 20 S/cm and 40 S/cm, respectively (Lu et al. 2019). The PEDOT/PSS hydrogel could be stretched up to 35% and had a modulus of about 2 MPa. Furthermore, a 400 μm resolution may be achieved by patterned printing of the hydrogel solution (Lu et al. 2019).

Polypyrrole (PPy) and PANI are marked as gelate conducting polymers, and the dopants can be inserted inside. PANI's amino groups can be protonated by a phytic acid dopant to interact with polyaniline. 3D hydrogel structures with pore diameters ranging from an angstrom to a nanometer and micron are created by the interaction of phytotoxic acid with several polyaniline chains. The conductivity of the hydrogel was 0.11 S/cm (Wang et al. 2015b). Moreover, PPy was cross-linked using a tetrafunctional dopant molecule called copper phthalocyanine-3,4′,4″,4‴-tetrasulfonic acid tetrasodium salt. With a conductivity of 0.07 S/cm, pure PPy tends to take on a granular shape. Copper phthalocyanine-3,4′,4″,4‴-tetrasulfonic acid tetrasodium salt dopant aligned PPy chains by hydrogen bonding and electrostatic interaction to create 1D PPy nanofibers. The conductivity of these fibers rose to 7.8 S/cm as a result of better PPy chain arrangement (Wang et al. 2015b). Hydrogel can be cross-linked onto conductive polymers in an attempt to include the conductive polymers in the hydrogel. Lee et al. substituted ethylene glycol for water while cross-linking polyacrylamide hydrogel in the presence of densely packed PEDOT/PSS sheets (Lee et al. 2016a). The organogel gained a stretchy quality from the polyacrylamide network, and conducting pathways were made possible via PEDOT/PSS. Increasing the content of PEDOT/PSS enhanced chain percolation and decreased organogel resistivity from 5529 Ω cm (5.49 wt%) to 133 Ω cm (9.06 wt%) (Lee et al. 2016a). Another example is when Kim et al. used UV light to cross-link alginate with Ca^{2+} and polyacrylamide. The consequent hydrogel was annealed at 150 °C and submerged in ethylene glycol to replace water. By altering the PEDOT/PSS's conformation through solvent replacement and subsequent thermal annealing, the conductivity of the hydrogel improved (Kim et al. 2019). In-place polymerization of conductive polymers within prefabricated hydrogel is an additional technique for adding conductive polymers to hydrogels. Tondera and co-workers synthesized a Laponite/polyacrylamide hydrogel, which was subsequently subjected to an oxidizing agent, ammonium persulfate, before being submerged in a mineral oil that included monomers of 3,4-ethylenedioxythiophene (Tondera et al. 2019). PEDOT underwent oxidative polymerization, starting at the surface. Doping of PEDOT chains, similar to PSS, is made possible by the Laponite crystals' negative charge. The hydrogel had a stretchability of 800%, and its conductivity was 26 S/m (Tondera et al. 2019). PEDOT is electropolymerized within an agarose hydrogel by Sekine et al. To develop a roughly 5-micrometer-thick pattern, 1.0 V was supplied to the micropatterned platinum electrode and the agarose hydrogel for 60 minutes. The sheet resistivity of the electropolymerized was approximately 11.0 kΩ/sq (Sekine et al. 2010). Due to the hydrophilic PSS component in PEDOT/PSS, the hydrophobic elastomer typically exhibits poor dispersion (Dauzon et al. 2020; Luo et al. 2020; Kayser and Lipomi 2019). Seyedin and colleagues created a conductive fiber by spinning a mixture of polyurethane, poly-3,4-ethylenedioxythiophene/PSS, and dimethyl sulfoxide. The resultant composite outperformed polyurethane in terms of stretchability and conductivity, whereas dimethyl sulfoxide enabled effective dispersion of poly-3,4-ethylenedioxythiophene/PSS, in polyurethane (Seyedin et al. 2014). The polyurethane's PEDOT/PSS ratio might be adjusted to modify the composite's physical characteristics. When a 5% strain was applied to pure polyurethane fibers, the stress was around 0.5 MPa (Seyedin et al. 2014). However, when a polyurethane nanocomposite fiber containing 25% PEDOT/PSS was added, the stress rose to approximately 9.1 MPa. Polyethylene glycol segments and polyols are examples of plasticizers that can be added to poly-3,4-ethylenedioxythiophene/PSS to increase its stretchability in the interim (Savagatrup et al. 2015; Meier et al. 2012; Li et al. 2014). These plasticizers allow the poly-3,4-ethylenedioxythiophene/PSS chains to move freely by penetrating the interchain gap between the polymers. To enhance electrical qualities, some plasticizers, such as xylitol or ionic liquid, can also act as a secondary dopant (Teo et al. 2017; Okuzaki et al. 2014). Kim et al. created a composite by mixing water-miscible ionic liquid with different anions into the mixture of water-borne polyurethane and poly-3,4-ethylenedioxythiophene/PSS. This composite demonstrated exceptional mechanical and electrical qualities. It has substantially enhanced properties over those of pure poly-3,4-ethylenedioxythiophene, including a conductivity of 140 S/cm, a modulus of around 7 MPa, and a stretchability of 500% (Kim et al. 2020a). Bioelectronics have made use of polydimethylsiloxane as a stretchy building component. Polydimethylsiloxane has been utilized in many experiments to create a composite with poly-3,4-ethylenedioxythiophene/polystyrenesulfonate. One such composite was created by Teng et al., which had a stretchability of more than 50%, as compared to the original poly-3,4-ethylenedioxythiophene/polystyrenesulfonate aerogel. The composite withstood 43% of strain stress and 60% of compressive stress because of the improved mechanical durability provided by the polydimethylsiloxane matrix. A 10% external strain could be tolerated by the composite up to around 5000 times if its sheet resistance was kept at 2.5 Ω/cm^2 (Teng et al. 2013).

2.3.3 Combinations of Conducting Polymers and Nanomaterials

The stretchable conductive nanocomposite may benefit from the synergistic effects of two different types of filler materials by combining them, which can enhance the material performance compared to the nanocomposite that uses just one type of filler material (DS et al. 2012; Da Silva et al. 2013; Sreeja et al. 2006; Saritha

Chandran and Narayanankutty 2008). These dual material characteristics can improve the overall material functionality of the nanocomposite such as magnetic properties, mechanical deformability, and electrical conductivity. Two distinct kinds of filler materials are selected to incorporate different features, such as various dimensions and mechanical properties (Park et al. 2016; Hong et al. 2017). For example, combining two distinct kinds of filler materials (e.g., conducting polymer-containing nanomaterials, dimension- and softness-varying nanomaterials) might improve the connection between conductive filler materials within an infiltration architecture. In these situations, the amount of nanoparticles required to achieve the same conductivity is reduced, improving the mechanical qualities such as high stretchability. Utilizing the hybrid-type filler technique, additional physical features, such as thermal or magnetic properties, may be introduced or improved (Min et al. 2019; Xiang et al. 2020; Kwon et al. 2018; Haghgoo et al. 2019). We will talk about the nanocomposite that uses a hybrid of two or more filler kinds in this part.

Hybrids of Nanomaterials and Conducting Polymers Employing a core/shell arrangement, wherein conducting polymers (shell) wrap conductive filler materials (core), is one method of combining two distinct kinds of filler materials. One way that the conducting polymer helps reduce the contact resistance between 1D filler materials is through its contribution (Si et al. 2018; Mao et al. 2021; Chen et al. 2005; Xie et al. 2018; Yuksel et al. 2018). Conductive fillers were encased by Čhakmak et al. using an electrochemical deposition tactic with a conductive composite consisting of polypyrrole and polydimethylsiloxane as a conducting polymer and an elastomer (Çakmak et al. 2004). This allowed for an improvement in the engagement region across conductive fillers inside the nanocomposite bundle and strengthened the mechanical stability of the infiltration network. Moreover, polypyrrole aids in lowering contact resistance. The polypyrrole and polydimethylsiloxane composite, with a thickness of about 20 μm, was deposited onto the conductive filler, which measured 5.5 μm in diameter, inside the framework of the nanocomposite bundles (Çakmak et al. 2004). Within the core/shell framework, they are entrapped with one another. In comparison to the situation employing only a single-type filler, the composite may have improved mechanical and electrical characteristics (~4.0 S/cm) due to the strong and persistent interaction across conductive filler materials (Çakmak et al. 2004). A nanocomposite including many filler components, such as magnetic, conductive, and conductive nanofibers covered in a conductive polymer, was studied by Moucůka et al. to increase the material's performance. Polypyrrole, conductive fillers, carbon black, and aluminum powder were employed as conductive filler materials, while sintered MnZn spinel-like ferrite particles (53.75, 26.10, and 20.15, mol% of Fe_2O_3, MnO, and ZnO) were applied as magnetic filler materials (Moučka et al. 2007). Polypyrrole functions as a binding agent for various conductive and magnetic filler materials. When this material is subjected to an alternating magnetic field, it loses energy as a result of the creation of eddy currents, and this phenomenon is known as magnetic loss. As the nanocomposite was supplemented with both conductive and magnetic fillers, the magnetic loss of the material was increased. The conductive framework of filler materials experiences eddy currents due to the alternating magnetic field, which amplifies the magnetic loss. An external magnetic field can be utilized to regulate localized wireless thermal stimulation through the usage of this phenomenon (Moučka et al. 2007). A nanocomposite of polypyrrole and SWCNTs was described by Liang et al. An intact oxidative polymerization technique was used to encapsulate polypyrrole around SWCNTs. There is a core/shell structure to this nanocomposite. This hybrid-sort filler approach may improve the nanocomposite's overall performance. The nanocomposite containing an optimized amount of SWCNTs (60 wt%) was able to achieve a power factor of 19.7 ± 0.8 μW/mK^2, Seebeck coefficient of 22.2 ± 0.1 μV/K, and electrical conductivity of 399 ± 14 S/cm. Conversely, the undecorated polypyrrole film's equivalent values were 0.041 ± 0.003 μW/mK^2, 9.4 ± 0.3 μV/K, and 4.6 ± 0.1 S/cm, respectively. While the extension rate at the rupture point was only around 2.6%, the nanocomposite demonstrated a significant degree of flexibility with a tensile modulus of about 418 MPa (Liang et al. 2016).

Hybrids of Nanomaterials with Various Softness Compared to bulk materials, nanoscale filler materials are softer, but they are still stiffer than the elastic matrix. The permeable channel may become detached because the filler material's softness is unable to succeed the elastic matrix's distortion in response to external stresses. At such disjointed areas of the channel, the use of a secondary filler material with a higher mechanical softness can offer substitute conductive paths. As a result, under external stresses, the conductance of a nanocomposite made of two fillers of varying softness might vary less than that of a nanocomposite made of one filler type (Choi et al. 2016a). In polydimethylsiloxane, Ho et al. created a nanocomposite of gold (diameter: about 2 nm) and silver nanowires (diameter: approximately 100 nm). Gold nanowires are substantially softer than silver nanowires due to their significantly smaller thickness. Idiosyncratically, gold nanowires were coated utilizing the Langmuir-Blodgett assembly method after the drop-casting of silver nanowires on a polydimethylsiloxane substrate (Ho et al. 2017). Ultimately, a second polydimethylsiloxane layer was applied on top. Silver nanowires, which are moderately thick had their percolation network strengthened by ultrathin gold nanowires (Ho et al. 2017). A nanocomposite consisting of 1.2 nm diameter CNTs and 150 nm diameter silver nanowires combined in silicone rubber was created by Lee et al. An improved optical transparency, electrical conductivity, and mechanical compliance were demonstrated by the nanocomposite. Initially, a CNT solution was passed through a membrane filter with a pore size of 0.2 μm, subsequently, a second filtering using the same filter was performed on a silver nanowire solution (Lee et al. 2014). As a result, the membrane filter created a layer of CNTs and silver nanowires, which were then deposited over a silicone rubber surface. Subsequently, a second silicone layer was placed between the transferred films. Silver nanowires have a significantly thicker permeation channel, and CNTs strengthen it, especially when there are external stresses. A sheet resistance of 24–27 Ω/sq. was demonstrated by the nanocomposite, it also demonstrated cyclic endurance (>10,000 cycles of stretching), a transparency of 80–93%, and a stretchability of about 460% (Lee et al. 2014).

Hybrids of Nanomaterials with Varying Dimensions It is possible to combine nanomaterials of various sizes to create hybrid filler materials. Combining multiple dimensions can improve the nanoparticles' connection and lower the percolation threshold. The advantages and disadvantages of each component having a distinct dimension can be balanced by combining fillers with various dimensions (Choi et al. 2016a).

Kim et al. created a nanocomposite with CNT (1D) and graphene flakes (2D) combined in polydimethylsiloxane. They then evaluated the material's performance in comparison to nanocomposites with alternative filler materials (graphite or carbon black with CNTs). Graphene combined with CNTs performed the best of all the configurations. The partly severed CNT network was restored when a small quantity of graphene was injected. Meanwhile, the graphene proportion rose, and the conductivity improvement effect vanished because more graphene disrupted the bulk of the CNTs' electrical walkways. At a ratio of 9:1, the nanocomposite containing graphene had the lowest resistance, measuring less than 100 Ω cm. After extending by 30%, the conductivity dropped by around 14% (Kim et al. 2016). A nanocomposite made of silver nanoparticles (0D) and silver fluoride (2D) distributed in fluorine rubber was created by Matsuhisa et al. This nanocomposite outperformed other nanocomposites, including a single-type nanomaterial, in terms of mechanical and electrical performance (Matsuhisa et al. 2017). When silver fluoride and fluorine rubber were heated in the close vicinity of a fluorine surfactant, silver nanoparticles were produced within the nanocomposite. In the presence of external stresses, the conductivity of this nanocomposite was 4919 S/cm and 719 S/cm, respectively, whereas the conductivity of the silver fluoride and fluorine rubber nanocomposite was 3727 S/cm in the absence of external strains and 297 S/cm in the presence of 300% strain. Silver nanoparticles might improve the connection between silver fluoride even in minute amounts, without compromising the nanocomposite's mechanical robustness or stretchability. Despite being renowned as highly conductive materials, silver nanowires or silver/gold core/shell nanowires can have their impedance additionally decreased by incorporating the proper low-impedance filler materials. Silver/gold core/shell nanowires, platinum black, and poly-styrene-butadiene-styrene, were combined to create a nanocomposite by Sunwoo et al. Platinum black and silver/gold nanowires were combined in poly-styrene-butadiene-styren to create the nanocomposite (Sunwoo et al. 2020). Toluene was employed as a solvent to improve the distribution of nanomaterials in the course of the blending process. The nanocomposite exhibited good electrochemical performance, high conductivity, and stretchability due to platinum black's huge surface area and platinum's low impedance. To be precise, the nanocomposite exhibited a high conductivity of 29,500 S/cm and an excellent stretchability of 720%, all while reducing the impedance by around 20 times. Platinum black added to the nanocomposite resulted in an approximately 3.3-fold increase in charge storage capacity and an 8.4-fold increase in charge injection limit (Sunwoo et al. 2020).

2.3.4 Elastic Grid for Nanocomposite Materials

Elastomers, often known as elastic polymers, are viscoelastic polymers with a low intermolecular force. The lengthy polymer chains that makeup elastomers are joined by comparatively weak bonds (Alemán et al. 2007). This corresponds to flexibility because, under external force, these weak links can be broken and repaired. Subject to the force that exists between molecules and the interactions between polymer chains, elastomers can have a wide range of characteristics (Plante and Dubowsky 2007). For instance, the elastomer is less elastic and mechanically robust if the polymer chains are not properly intertwined or if the force that exists is substantial and dense. On the other hand, the elastomer is soft or flexible if the force between the molecules is insufficient and infirm or if the polymer chains are highly intertwined (O'Halloran et al. 2008). Hydrogels consist of porous, hydrophilic three-dimensional polymer lattices that are sustained by many processes, such as ionic interactions, physical entanglement, and covalent cross-linking (Wang et al. 2019a; Cha et al. 2020a; Yu et al. 2020; Fang et al. 2013). Hydrogels have gained interest as promising candidates for materials in biomedical applications, including cell therapy, biosensing, drug delivery, tissue engineering, neural interfacing, regenerative medicine, and soft robotics, due to their profound attraction for water, biocompatibility, and physical characteristics that resemble biological soft tissues (Kim et al. 2010b, 2015; Eckhouse et al. 2014; Gajanayake et al. 2014; Zhang et al. 2015, 2020a; Huang et al. 2016; Li et al. 2020; Tybrandt et al. 2018; Yuk et al. 2019; Canales et al. 2015; Chiang et al. 2020; Park et al. 2019; Cho et al. 2019; Soh et al. 2017). Hydrogels may have their physical, chemical, and biological characteristics customized based on a wide range of building blocks, including both synthetic and natural materials, and different synthetic techniques. Moreover, special qualities of materials, such as conductivity, self-healing, hardness, and resistance to severe environments, may be tailored for their intended uses (Zhang et al. 2017a; Mano et al. 2007; Liu et al. 2020; Dong et al. 2020). These hydrogels and elastomers can be categorized in multiple manners. Herein, we categorize them according to the types of cross-linking: chemical, physical, and hybrid cross-linking. The intermolecular force influencing the elastic matrix's physical characteristics determines the type of cross-linking. The elastic matrix's physical characteristics are taken into consideration while choosing functional fillers. The nanocomposite's stretchability and conductivity are significantly impacted by the diffusion system that the functional fillers create.

Matrix Elasticity Through Physical Cross-linking Noncovalent connections (Chen et al. 2019; Monemian et al. 2014), including hydrophobic (Miwa et al. 2018; Weadock et al. 1995) and ionic interactions as well as electrostatic interactions (Wang et al. 2020b), are the main source of physical cross-linking. Certain kinds of elastic matrices may be created by host-guest synergies, wherein the molecules of the host and guest establish noncovalent links (Sinawang et al. 2020; Mantooth et al. 2019; Ooi et al. 2020; Zhang et al. 2020b; Hong and Song 2019). Additionally, certain matrices have a characteristic known as self-restorability, whereby bonding connections can be reversibly repaired in response to external mechanical pressures (Wietor and Sijbesma 2008; Khatib et al. 2020; Shin et al. 2021). While there are instances of chemical self-restoration processes through dynamic covalent cross-linkage, physical cross-linking is the primary basis for most self-healing systems. Chemical processes that alter the polymer confine structure are less important in physical cross-linking than the inherent qualities of the component polymers (Zhang and Khademhosseini 2017). The elastic matrix's characteristics' tunability may be limited by their reliance on inherent qualities. However, to initiate or stop the creation of chemical bonds, physical cross-linking lacks the need for chemical changes to the polymer chains. It is essential to remember that the physically cross-linked elastic matrix is stabilized more effectively by the integration of several noncovalent bonds than by a single kind of contact (Nele et al. 2020). Physically connected elastomers frequently feature a block-copolymer architecture with hard and soft domains. To achieve firmness and softness, respectively, deformability is associated with soft domains, whereas hard domains are in charge of physical contact (Kawarazaki et al. 2020; Wedler et al. 1995). For instance, the structure of the polystyrene-butadiene-styrene elastomer alternates between the locations of styrene and butadiene blocks along the chain. A soft domain is made up of butadiene chains, whereas an aggregation of styrene blocks forms a hard domain. By changing the dimension and kind of each block, the physically cross-linked elastomer's physical characteristics may be adjusted. If the hard domain is comparatively prolonged, the elastomer gets tougher, and vice versa. Physically cross-linked elastomers are more deformable than chemically cross-linked ones in response to external forces, but they also have lower resilience after deformation because they have a smaller intermolecular force. Furthermore, the physically cross-linked elastomer's porous structure can be penetrated by organic solvents or biological fluids, which can cause the elastomer to become denatured, especially at high temperatures. Its low endurance, however, suggests that, with the right activation, it can return to its former structure. Its ease of processing and recycling is another benefit. Physically cross-linked elastomers have a large variety of filler materials because the filler material may readily infiltrate the soft domains while the hard domains retain cross-linking. Electrostatic interactions between species with opposing charges create ionic cross-linking. Ionic bonds often develop between ionizable polymers and opposing ions, or among ionizable polymers and extra-blended oppositely charged polyelectrolytes (Moura et al. 2011; Sun et al. 2013). These systems are susceptible to changes in temperature, pH, and the proportion of polymers to opposing ions (Rizzo and Kehr 2021). The most typical instance of an ionic cross-linked polymer is the metal-alginate hydrogel. Trivalent cations (Al^{3+}, Fe^{3+}), as well as divalent cations (Ba^{2+}, Ca^{2+}, Mg^{2+}), can crosslink algae (Nele et al. 2020). Generally, trivalent cations produce narrow networks with metal-alginate and enhance the strength and recoverability of ionic connections. One significant kind of bonding in physical cross-linking is hydrogen bonding (Kouwer et al. 2013; Oh et al. 2004; Kimura et al. 2004; Wang et al. 2019b; Zhu et al. 2017). When one electronegative atom is bonded to another electronegative atom, hydrogen bonds are created across them. Hydrophilic groups like carboxyl acid ($-COOH$), amine ($-NH_2$), and hydroxyl ($-OH$) are commonly present in polymers that experience hydrogen bond cross-linking (Rizzo and Kehr 2021). Furthermore, they can be introduced using a syringe due to the shear-thinning nature of their weak hydrogen bonds. This is a non-Newtonian feature where fluid viscosity reduces under shear strain, such as during infusion, but returns to its mechanical properties complying with an infusion. Despite having good biocompatibility due to the lack of chemical cross-linkers, they are unstable when hydrated because water molecules interact with the hydrogen bonds holding the polymers together. Hydrophobic interactions cause physical cross-linking in amphiphilic polymers, which include both hydrophilic and hydrophobic domains. The most common type of hydrophobic component is an alkyl chain consisting of 12–22 carbon atoms (Pekař 2015; Xing et al. 2002; Tuncaboylu et al. 2011; Mihajlovic et al. 2017; Deng et al. 2018). Certain amphiphilic polymers dissolve at low temperatures but congregate as they warm up (Rizzo and Kehr 2021; Yang et al. 2020). The lower critical solution temperature is the term used to describe the transition temperature. By reducing the hydrophobic domains' surface area in contact with water, the hydrophobic domains collectively reduce the system's overall energy. Finally, host-guest chemistry may also be used to induce physical cross-linking, wherein the molecules of the host and the guest form insertion complexes. Macroscopic molecules including crown ether, cucurbituril, cyclodextrin, and calixarene are among the most prevalent kinds of host molecules. Certain guests, including proteins and tiny molecules, can be encased in the space of host molecules (Rizzo and Kehr 2021).

Matrix Elasticity Through Chemical Cross-linking Intense chemical connections, such as C–C, C–S, C–O, and Si–O bonds, bind each polymer chain together during chemical cross-linking. As an illustration of a chemically crosslinked elastomer, silicone rubber has chains with repeating Si–O linkages created by catalytic processes (Zhao et al. 2016). The level of cross-linkage may be adjusted to modify the physical characteristics of the chemically cross-linked elastomer, such as toughness and elasticity. Greater cross-linkage produces tougher and less elastic material. Chemically cross-linked elastomers have been utilized in medical implants because of their strong intermolecular force, which also contributes to their excellent mechanical resilience and chemical stability. It is challenging to process and recycle, though.

Additionally, as filler can obstruct elastomer chain cross-linkage, care should be taken while choosing the kind and quantity of filler material. Encasing curing chemicals into the elastomer matrix allows the chemically crosslinked elastomer to have a self-healing characteristic in the interim. Curing substances spread and reattach the broken polymer chains in elastomers that have mechanical cracks (Yu et al. 2015; Schacht 2004; Salleh et al. 2018). When hydrogels are chemically cross-linked, covalent connections are created using a variety of reaction processes, including photopolymerization, enzyme-mediated cross-linking, and click chemistry, which includes Schiff base reaction or thiol-Michael-type addition. One of the main chemical cross-linking techniques, click chemistry, was first described in the year 2001 by Sharpless and colleagues. Click chemistry produces crosslinks that are selective, fast, substantial reaction yield, stereospecific, and orthogonal. Click chemistry often operates in moderate settings and produces by-products with potential uses in biomedicine (Bao et al. 2019; Konuray et al. 2020; Yigit et al. 2011; Kolb et al. 2001). Hydrogels are most commonly created using Cu(I)-catalyzed azide-alkyne click cycloaddition. Still, traces of leftover Cu salts have the potential to be hazardous. Metal-free click reactions have been proposed to tackle this problem (Nair et al. 2014).

Owing to its biocompatibility, reversibility, pH sensitivity, and simplicity, Schiff base reaction has been utilized extensively in biomedical applications (Ozay et al. 2021; Jalalvandi et al. 2017; Xu et al. 2019). Hydrogels can repair themselves following mechanical injury because of the reversal of dynamic Schiff base connections. Oxime, hydrazine, acylhydrazone, or imine bonds are formed when carbonyl groups in aldehydes or ketones combine with primary aminooxys, hydrazides, acylhydrazides, or amines in the Schiff base reaction, which is a condensation reaction. Compared to imines, hydrazones and oximes are more chemically stable against pH shifts, while acylhydrazones are more hydrolytically stable. By adding benzene rings to the carbon-nitrogen double bond, Schiff base couplings can become more stable.

Additionally, photopolymerization—a process that results from radiation—can cause cross-linkage (Karaca et al. 2010; Sun et al. 2020; Mironi-Harpaz et al. 2012; Choi et al. 2019). Photopolymerizations may be classified into two types: step-growth reactions and chain-growth reactions (Zhu et al. 2020). The photoinitiator, which is responsible for producing radicals to start radical polymerization and absorbing photons from irradiation light, is crucial to both processes. Type I and type II photoinitiators are the two groups into which they fall. Radiation causes radicals to be produced by type I photoinitiators through intramolecular bond breakage and type II photoinitiators through hydrogen abstraction.

The -ene groups methacrylate, maleimide, acrylate, alkene, and norbornene are frequently utilized in step-growth reactions. When radical species are added to carbon-carbon double bonds, the chain length increases in the chain-growth reaction. The polymer chain then grows further when a new radical is produced at its terminus. The incredibly typical instance of chain-growth reactions is the polymerization of monomers based on methacrylate. Hydrogel creation is inexpensive, simple, and quick with this process. However, oxygen may quickly stifle the reaction and cause it to become non-uniform, leaving unreacted functional groups amidst structures (Miller et al. 2010; Shih and Lin 2012; Brandl et al. 2010; Jiang et al. 2020; Yokozawa and Ohta 2016). However, compared to the chain-growth reaction, the step-growth reaction is more biocompatible and controlled.

Matrix Elasticity Through Both Kinds of Cross-linking The physically or chemically cross-linked elastomers have advantages and disadvantages. As a result, elastomers with characteristics of both kinds were developed (Petrović et al. 1991; Liu et al. 2014b; Kontou et al. 1990). One type of structured polyurethane found in thermoplastic polyurethanes has both rigid and soft segments. Low molecular weight diol and aromatic di-isocyanate make up the hard portion, whereas polyester or polyether macroglycol makes up the soft section. Regarding physical cross-linking, hydrogen bonding amid neighboring hard domains is crucial for thermoplastic polyurethanes's flexibility (Kontou et al. 1990). Moreover, when there is extra di-isocyanate contentment, additional cross-linking can be added via the allophanate connections. The elasticity of thermoplastic polyurethanes is impacted by both chemical and physical cross-linking, as reported by Kontou et al. The impact of the ratio of hard domains to soft domains as well as the effects of chemical and physical cross-linking in thermoplastic polyurethanes were investigated. By adjusting the chemical chain width and the hard-to-soft ratio, it is possible to modify the mechanical characteristics of polyurethane-based elastomers (Kontou et al. 1990).

2.4 Fabrication Approaches for Stretchable Electronics

Technological developments in nanostructured materials for device consolidation have made bio-integrated electronics extremely flexible. Nonetheless, rather than allowing for multifaceted bending, the flexibility provided by employing nanostructured elements (i.e., incredibly thin devices) enables unilateral folding. Stretchability—a more advanced degree of deformability than flexibility—is necessary for the congruent incorporation of bioelectronics with the curvaceous, translucent, and dynamically moving organs. For instance, a multi-directionally bendable system might be conformally embedded on the surface of a spherical organ (heart, brain, etc.) (Lee et al. 2019). This section will address common techniques for adding stretchability based on non-traditional design, such as fractal design, buckling structures, serpentine design, and pop-up structures (Lee et al. 2021a).

2.4.1 Fractal Design

Greater extensible properties may be added by utilizing the fractal design. A highly structured framework of curving filamentary structures, or branch-like loops and lines. The greater the degree of fractal configuration the more stretchable the device becomes. The fractal orders are disassembled one after the other. A lower fractal order, such as first order, persists undeformed throughout stretching, but an elevated fractal order, such as second order, unravels first (Zhang et al. 2014; Yun et al. 2020; Fan et al. 2014). Although the loop design is often made in a serpentine shape, alternative shapes, such as a horseshoe or sinusoidal shape, can also be implemented. The application of this fractal architecture to soft bioelectronics resulted in an extremely reduced flexural stiffness for the device. As a result, the setup can conformally touch the surfaces of intricately shaped 3D biological organs or tissues. A wearable device that integrates a soft electrode array with a fractal pattern can be tangled on the surface of a human skin mimic (Fan et al. 2014).

2.4.2 Buckled Structures

While there is a limit to how far an ultrathin gadget can stretch, a device with the right architecture can. A wavy framework, which may be created via buckling, is the intended approach for the device. Applying an exterior tensile load along the path of the buckling allows the device's buckled framework to be expanded and straightened (Xu et al. 2014; Jiang et al. 2018; Koo et al. 2018a). Before creating this kind of buckling framework, a slim device must be attached to an elastomeric substrate that has been prestrained. The use of thermal expansion or mechanical substrate elongation is two methods for inducing restraint. An extremely thin device forms a chemical link with the prestrained substrate, and the prestrain gives way to create the buckled structure (Jiang et al. 2007).

By covalently connecting to a prestrained elastomeric platform and relieving the prestrain, a single crystalline silicon nanoribbon with a width of 100 nm was used to create the buckled structure. The buckling form (i.e., global or localized buckling), amplitude, and wavelength can be adjusted per the mechanical characteristics and dimensions of the electronic component and the elastomeric substrate (Wang et al. 2008). The amplitude and wavelength of the buckled nanostructures may be predicted using a variety of mechanical hypotheses. Typically, when the prestrain rises, the wavelength lowers, and the buckling amplitude raises (Jiang et al. 2007). Mechanical fractures are caused by high strain above the prestrain, even if an incredibly thin device could turn flexible with a buckled structure.

2.4.3 Serpentine Design

The prestretch value sets an upper stretchability constraint for both the pop-up and buckled structures. The buckling or pop-up structure would turn flat if the exterior strain applied equals the prestrain value. The device will fracture mechanically if an exterior strain is applied that is greater than the prestrain configuration (Fu et al. 2015; Zhang et al. 2013). Stretchability and the geometry of a serpentine structure are determined by parameters like arm length, arc angle, line width, and arc radius. There are two possible configurations for the serpentine structure: partly delaminated off the substrate or entirely attached to it. Compared to the bonded casing, the suspended serpentine architecture is more stretchable, yet it is also more susceptible to breaking down from mechanical stress or impact from the outside. Implantable devices that are incorporated into curved, continually moving organs, like the heart, can be built using this design (Park et al. 2016).

2.4.4 Pop-Up Structures

The device's buckling construction may cause strain within its active area, which may have an impact on how well the device performs in terms of strain-induced changes in electron circulation. A controlled delamination process has been developed to reduce these mechanical adverse effects on the functioning of the electronic device (Xue et al. 2020). With this technique, the metal integration, whose performance is mostly insensitive to mechanical strain, delaminates via the substrate resulting in an arch-shaped pop-up structure, while the active region of the device—such as the semiconductor channel region—remains on the substrate in a linear state. The device is initially moved to the prestrained elastomer substrate to create this structure. The active device region, also known as the island region, remains attached to the substrate upon the release of the prestrain, but the metal link region, also known as the bridge region, delaminates out of the substrate to produce a pop-up structure (Ma et al. 2018; Kwon et al. 2019). It is possible to reduce the prompted strain over the island zone since the pop-up structure releases the majority of the prestrain (Kim et al. 2008). The pop-up design enables the gadget to bend, twist, and stretch among other deformities. Using silicon that was around 300 nm thick and the island-bridge architecture, a series of extremely thin complementary metal-oxide-semiconductor inverters were created on a substrate that was preconstrained. The complementary metal-oxide-semiconductor inverter array's pop-up configuration was attained when the prestrain ($\sim$20%) was removed, allowing for its bending distortion with no mechanical breakdowns or distinctions in performance (Kim et al. 2008). Complex pop-up frameworks might be created by adjusting the areas of bonding and nonbonding zones and using different designs. Nevertheless, a substantial adjacent strain over the prestrain threshold might result in mechanical fractures in the device, much like the buckled structure. Consequently, to get better stretchability, alternative design techniques are required.

2.5 Device Components for Bio-Integrated Electronics

It is possible to construct the enclosed-loop soft bio-integrated electronic system (depicted in Fig. 2.2) by combining various soft device components, such as those for data storage and display, bio-signal attainment, power supply, and feedback stimulation. The primary ingredients required to create such device components are the nanostructured materials and their corresponding nanocomposites, which were covered in earlier sections. These device components' soft mechanical characteristics enable long-lasting, outstanding bio-interfacing as well as conforming contact within the device and the target tissue. In the ensuing segments, we will scrutinize the several categories of apparatus constituents for flexible bioelectronic systems, encompassing biosensors, stimulators, displays, memory, and energy devices. This part will provide a brief overview of the requirements and functioning ideas for each device component. The section that follows will provide application illustrations for the device elements.

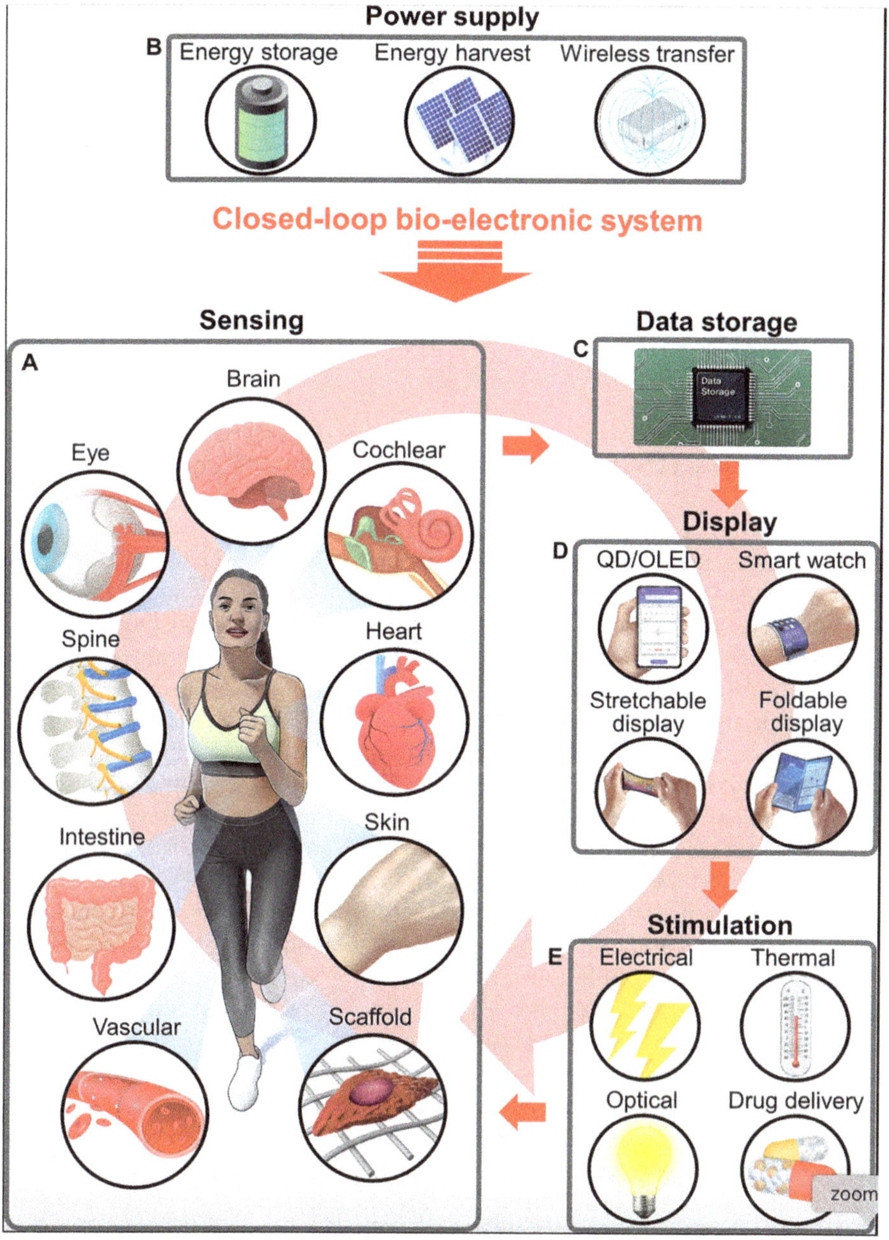

Fig. 2.2 Overview of the closed-loop soft bioelectronic system. (Copyright © 2024 Published by American Chemical Society under Creative Commons (CC). CC-BY 4.0 (Cho et al. 2022))

2.6 Applications of Biosensors

The initial stage in creating a closed-loop system conforms to biosensors. Details of the biosensor and closed loop have been explained in Chaps. 2 and 13, respectively. They identify bio-signals, which are vital indicators of disease disorders and health state. Reliable and dependable biosignal detection is made possible by the mechanical characteristics of biosensors constructed from nanomaterials and nanocomposites that may be adjusted to correspond with those of intended organs or tissues (Tripathi and Bonilla-Cruz 2023). The three categories of bio-signals that biosensors may identify are physical, biochemical, and electrophysiological. Their functioning fundamentals and operational requirements are explained briefly in the subsequent subsections.

2.6.1 Physical (Thermal, Optical, and Mechanical)

Since abrupt variations in body temperature typically signify sickness or autoimmune diseases, thermal sensors become crucial to bioelectronics. The common instruments for measuring temperature are infrared and resistive temperature detectors. In bioelectronics uses, resistive temperature detectors are more commonly utilized than infrared sensors because they can be easily produced into a miniature form. Their form is extremely thin and flexible, and they can be easily incorporated into portable and implantable biosensors (Tripathi and Bonilla-Cruz 2023).

Optical sensors are devices that use light to generate electrical impulses. One of its uses in healthcare is retinal prosthesis and photoplethysmography. Because they can continuously measure blood pressure without intrusive procedures, flexible and portable sensors made from photoplethysmography have been the subject of extensive research (Fortin et al. 2021; Kim et al. 2017; Jinno et al. 2021). By detecting the quantity of illumination received or rebounded by the blood artery, the photoplethysmography sensor visually tracks variations within the volume of blood beneath the skin (Tamura et al. 2014; Elgendi et al. 2019). Since red or near-infrared light is known to penetrate the epidermis deep down, it is frequently utilized as the illumination for transmissive photoplethysmography sensors. Light is less invasive than red or near-IR light, yet it exhibits a greater signal-to-noise ratio and the largest modulation of the reflected signals, even in the presence of transpiration disturbances (Tamura et al. 2014; Elgendi et al. 2019). As the level of oxygen saturation may be determined by the optical reception of hemoglobin, pulse oximetry may additionally be carried out utilizing photoplethysmography signals by exposing various wavelengths of light. Co-integrated electrodes provide appropriate feedback electrical stimulations to the optic nerves in the retinal prosthesis, while sensitive photodetectors are embedded on a hemispherical retina for visual sensing. The precise non-invasive detection of important signals from biological cells is made possible by optical/electrical sensors, but to create a constrained contact, they must be placed on unadorned skin (Yoon et al. 2020; Kim et al. 2020b; Choi et al. 2020).

In addition, current research has looked at the use of wearable technology and mechano-acoustic impulses for profound tissue monitoring. Acoustic sensors can monitor the crucial signals produced by the wave motion of bodily tissues and surrounding fluids. For example, sounds in the 1–1000 Hz range are produced by normal bodily functions such as blood flow through the heart valves, gastrointestinal tract motions, and vocal fold vibrations (Liu et al. 2024). Electrical changes brought on by mechanical deformations are commonly measured using mechanical (the amount of strain or pressure) sensors, which are two-terminal devices. Numerous electrical sensing techniques, including assessments involving changes in capacitance, piezo-electrical change, piezo-resistive change, discontinuation in percolation networks, rupture change, and triboelectric production, have all been employed (Wang et al. 2018c, 2021b). The movement of bodily components, such as joints, arms, and legs, may be detected by strain sensors, whereas the application of forces, such as on organs, cells, and tissues, can be detected by pressure sensors (Gupta et al. 2020). A rigid connection between the sensor and the intended tissue is required to obtain high sensitivity with low noise. Electrical changes brought on by mechanical deformations are commonly measured using mechanical (pressure or strain) sensors, which are usually two-terminal devices (Choi et al. 2015; Park et al. 2015a). A multiplicity of electrical sensing techniques has been employed, including measurements based on changes in the crack (Gong et al. 2019; Zhang et al. 2011; Lee et al. 2016b; Han et al. 2018; Kang et al. 2014), triboelectric production (Dhakar et al. 2016; Nguyen and Yang 2013; Garcia et al. 2018; Fan et al. 2012), disconnection in the percolation networks (Akar et al. 2001; McIntosh and Casada 2008), piezoelectrical change (Rajala et al. 2016; Ihn and Chang 2004; Ng and Liao 2005; Tzou and Tseng 1990), piezoresistive change (Fiorillo et al. 2018; Zhu et al. 2013; Huang et al. 2008), and capacitance change (Akar et al. 2001; McIntosh and Casada 2008). Body components including joints, arms, and legs may be detected by strain sensors, while forces acting on organs, tissues, and cells, can also be detected by pressure sensors (Gong et al. 2014; Shin et al. 2019).

Biochemical Sensor Evaluating the physiological health state associated with homeostasis requires knowledge of biochemical parameters such as sodium concentrations (Huang et al. 2011; Ang et al. 2008; Saari and Seitz 1982), glucose (Huang et al. 2011; Ang et al. 2008; Saari and Seitz 1982), lactate (Huang et al. 2011; Ang et al. 2008; Saari and Seitz 1982), pH (Huang et al. 2011; Ang et al. 2008; Saari and Seitz 1982), as well as other illness indicators. Traditionally, this kind of data has been recorded from blood, however, new technologies have been proposed that can detect it from other bodily fluids as well, such as saliva (Huang et al. 2011; Ang et al. 2008; Saari and Seitz 1982), tears (Huang et al. 2011; Ang et al. 2008; Saari and Seitz 1982), perspiration (Huang et al. 2011; Ang et al. 2008; Saari and Seitz 1982), and interstitial fluid (Huang et al. 2011; Ang et al. 2008; Saari and Seitz 1982). Several different detecting techniques are employed for these kinds of evaluations. Electrochemical alterations on the electrodes, resulting in electrical signal alterations, are detected by several processes, such as reactions caused by enzymes that transform substrates into products or antibody-antigen binding events.

Additionally, by measuring the potential changes between the sources and working electrodes, both the pH and ion concentrations of the biofluids may be determined. The analysis can be performed using a target-specific membrane that intercepts the appropriate kind of ions specifically (Hong et al. 2018; Lee et al. 2018). In recent years, wearable electrochemical detectors and electrodes incorporated with microfluidic channels for capturing targeted bodily fluids have gained attention due to their non-invasive nature. The specifications needed for further development include those for stability over time, ongoing tracking capability, sensor accuracy, and pleasant wearability.

Sensors with Electrophysiology Electrophysiology sensors can monitor the variations in electrical charge that arise from ionic currents generated by the tissues in different organs, including the cardiovascular system, brain, and muscles. Electrooculograms (Steinberg et al. 1985; Lee et al. 2017a; Usakli et al. 2010; Thilagaraj et al. 2021), electrocardiograms (Fye 1994; Barold 2003), electroencephalograms (Stevens et al. 2008; Freeman 2015; Ge and Liu 2016), and electromyograms (Hof 1984; Gerdle et al. 1991; Kamen and Caldwell 1996) are examples of bio-signals that provide essential data on the electrical responses of the heart (Cho et al. 2020), brain (Cho et al. 2019), muscles, and eyes, respectively (Sunwoo et al. 2020). Therefore, reliable biosignal surveillance is necessary for exact illness detection. The electrodes must achieve congruent contact with the desired tissues or skin to detect bio-signals accurately. Although direct signal measurement from the target tissue yields significantly better sensitivity, the majority of electrophysiology sensors are wearable varieties that collect signals from the intended tissue noninvasively via the skin (Sunwoo et al. 2019; Yang et al. 2015; Gong et al. 2015). Nevertheless, the limited responsiveness of the on-skin assessment is a drawback. Creating superior bio-interfacing across the epidermis and the devices is crucial to make up for the loss of sensibility. Stated differently, it is recommended that extremely fine or inherently soft electronic devices be integrated conformally on-skin to optimize the signal-to-noise ratio through the reduction of contact impedance.

Stimulators Based on the outcomes of biosensing, the stimulators usually offer therapeutic activities in the form of electrical (Sunwoo et al. 2019; Yang et al. 2015; Gong et al. 2015), chemical (Sunwoo et al. 2019; Yang et al. 2015; Gong et al. 2015), thermal (Sun et al. 2019; Zhang et al. 2017b), and optical feedback (Kim et al. 2013; Choi et al. 2013; Kim et al. 2020c). These stimulators can regulate the intended tissues' or organs' spatiotemporal functioning. For instance, personalized control over targeted medicine administration is made possible by implantable soft bioelectronics-based drug delivery, which reduces consequences to other organs and increases the concentration of drugs in the target location (Son et al. 2015a; Cha et al. 2020b; Joo et al. 2024; Lee et al. 2015, 2021b; Choi et al. 2016b). It is possible to increase thermal stimulations to regulate and quicken targeted medication administration. Electrical stimulations have been applied to regulate muscle and neuron actions (including cardiac function) to treat abnormalities in activity, reduce discomfort, or encourage the regrowth of tissue. Soft versions of these stimulators, in particular, are advantageous for persistent stimulations because they reduce impedance and increase charge injection efficiency (Lim et al. 2024). While electrical stimulations are typically employed to activate neurons or muscles, optical stimulations, or optogenetics, have recently attracted attention since light can only be focused inside specific target areas, while electrical signals can cover a sizable portion of the cells (Zheng et al. 2019; Park et al. 2015b). Wireless optical stimulation of certain neurons can be achieved by the use of implanted light-emitting diodes combined with transmitters.

Information Displays Real-time visualization of the collected biological signals for the user can be provided by an embedded data display. Because of mechanical disparities with the soft and stretched human anatomy, traditional static displays are inappropriate for these kinds of usage. As a result, new technologies with flexible and elastic shape characteristics must be designed (Koo et al. 2018a; Yang et al. 2021; Choi et al. 2018a). Numerous studies on the creation of deformation-resistant displays, such as flexible organic LEDs, inorganic micro-LEDs, and polymer LEDs, have been documented. Additionally, inorganic nanocrystals (QDs) have formed the basis of a stretchable display (Choi et al. 2018b). Impressive color integrity, great illumination, low power consumption, mechanical adaptability, and water/air durability are among the characteristics of this LED. All of these qualities are suitable for displaying biological signals, particularly heart rate or health indicators, on electronic devices using wearable sensors (Choi et al. 2018b). Furthermore, fabric-type LEDs have been created recently using displays made of yarns. These LEDs have the benefit of being able to be formed into different types of clothing, but their level of resolution is still less than that of stretchable OLEDs or QLEDs (Choi et al. 2018b).

Devices for Power Supplies On account of their use in biointegrated electronics, the power generation devices—energy harvesting, storing, and transferring devices—must be light, compact, extremely fine, and flexible. The advantages of using nanoparticles in devices that store energy are their huge specific surface area, superior energy storage qualities, and deformability (Tang et al. 2015; Niu et al. 2013). Soft energy storage systems that are highly stretchable, have a large power reserve, and are stable over an extended period are yet undeveloped, despite extensive research endeavors. The primary power source for traditional biomedical electronics is a battery, which inevitably has a restricted working life (Jang et al. 2012; Han et al. 2005; Lee et al. 2013). Batteries that are flexible or stretchy usually have poor capacity and consistency. Integrating harvesting devices with bioelectronics has been explored to address these problems (Jung et al. 2015). Numerous technologies, including triboelectric and piezoelectric nanogenerators, as well as solar and thermoelectric devices, have been created to harvest electricity from ambient energy. Triboelectric and piezoelectric nanogenerators transform mechanical energy from physical movements into electrical energy, which may be utilized for powering bioelectronics or recharging their batteries (Savagatrup et al. 2015; Meier et al. 2012; Kim et al. 2020a). Bioelectronics is powered by photoelectric and thermoelectric devices, which capture solar energy with the body's energy from heat. Usually on wearable systems, such harvesting technologies are combined with additional devices. Wireless energy transfer technologies have been employed in implantable devices to replenish the batteries within the implanted systems. Inductive or capacitive coupling is accomplished by electromagnetic fields in near-field wireless power transmission systems. Power has been sent independently with a considerable extent of penetration within the body using radio-frequency impulses and associated circuit elements. In addition, the implanted system may be controlled by the bio-integrated electronics by the use of wireless transmitted electromagnetic field energy (Yoo et al. 2021b). Nevertheless, further advancements are still needed in the effectiveness and dependability of wireless power transmission approaches before they can be used in genuine clinical settings.

Instruments for Storing Data An enormous quantity of data is gathered from the ongoing monitoring of physiological signals; they must be further processed, evaluated, and understood to determine their clinical significance. Storage mechanisms for data are required in the field of bioelectronics for this cause (Lee et al. 2017b). This kind of capability is particularly useful for implanted instruments, whose data extraction is sometimes problematic because of the difficulties in getting to the devices within the body. This has been accomplished through the employment of non-volatile storage technologies such as resistive random access memory and flash memory (Lee et al. 2017b). There have been attempts to manufacture these memory devices stretchy and flexible, with a particular emphasis on the creation of new materials (Son et al. 2015b; Kim et al. 2024). Using nanomaterials, significant advancements have been achieved in the development of stretchable data storage elements, such as wearable flash memory and wearable resistive random access memory. Oxide nanoparticles included in the metal-oxide-metal framework allowed a wearable resistive random access memory to operate at a very low voltage, while self-assembled metal nanoparticles used as the floating gate optimized the performance of the wearable flash memory.

2.7 Implantable Sensors and Stimulators

When choosing materials for implanted devices, attention must be taken to ensure that they are adhesion-resistant, biocompatible, and durable in vivo. In particular, a robust adhesion across soft embedded parts and the intended tissues is essential to both reducing inflammatory tissue harm from mechanical friction stress and ensuring the implantation devices work reliably (Nerger et al. 2024). Diverse bioadhesives, including chemically altered hydrogels and nanocomposites, have been created to offer robust bioelectronic adhesion to wet dynamic tissue surfaces. Noble metals have been utilized because of their remarkable electrical qualities and ability to resist oxidation, which include gold and platinum. These materials have undergone processing to give them nanoscale thicknesses, which enables them to conformally interface delicate, curved organs including the heart, brain, and peripheral nerves. Implantable devices have utilized soft nanocomposites containing different conducting nanoparticles in the recent past (Deng et al. 2021). These implanted devices will be examined in this section.

2.7.1 Cardiac Stimulations and Monitoring

To endure the rapid movements of the heart, implanted devices used in the diagnosis and treatment of cardiac disorders need to have extremely soft characteristics. In this context, Park et al. described the elasto-conductive epicardial mesh, a soft implanted technology for electromechanical cardioplasty (Park et al. 2016). The development of the device was done by the use of a composite of ligand exchange of silver nanowires and poly-styrene-butadiene-styrene elastomer. Following the nanocomposite's curing, the blade-coated liquid-phase ligand exchange of silver nanowires and poly-styrene-butadiene-styrene composite was removed from the polydimethylsiloxane mold. The mold had a mesh pattern. For efficient current supply to the whole myocardial, the device might conceal the whole ventricle. The purpose of the epicardial mesh is to provide mechanical assistance for the heart contraction in addition to encircling the heart without obstructing its expansion. For this purpose, the rat epicardium's mechanical features were nearly identical to those of the epicardial mesh, which was built as a serpentine framework. The electromechanical cardioplasty involved implanting the epicardial mesh onto the rat's heart. Ventricular pacing with a cycle duration of 280 ms was administered to 36 rats that had myocardial infarction. There was no discernible variation in the QRS duration in control rats. Nonetheless, in myocardial infarction rats, the baseline QRS length of 38.9 ± 3.6 ms could be effectively decreased by pacing to the baseline scale of the unaltered rats (26.1 ± 3.0 ms), indicating the successful electrical cardioplasty (Park et al. 2016). Biocompatibility is a crucial component for the long-time insertion of epicardial mesh devices.

The ligand exchange of silver nanowires was swapped out for gold-coated silver nanowires to improve the ability to survive off the tissue mesh. The prolonged cytotoxicity problem with silver nanowires was resolved by the seamless covering with a gold sheath, which stopped silver ions from leaching. To monitor and stimulate a pig heart, Choi and colleagues created the widespread epicardial mesh utilizing a nanocomposite of poly-styrene-butadiene-styrene with gold-silver nanowires (Choi et al. 2018c). Electrocardiography signals might be used to create a voltage mapping, and it would be possible to identify a significant voltage shift caused by acute ischemia in the anterior portion of the mid-apex left ventricular area.

The nanocomposite of poly-styrene-butadiene-styrene with silver-gold nanowires was combined with extra platinum black to create an epicardial mesh by Sunwoo et al. (2020). There was a considerable drop in the impedance of nanocomposite in addition to platinum black. The minimal impedance of nanocomposite leads to excellent ECG recordings and effective electrical stimulations. Furthermore, the delicate quality of the epicardial mesh (whose elongation at break was 720%) allowed it to be uniformly looped around the rat heart. Electrical stimulations might alter a rat's myocardium's functioning via the epicardial mesh. Additionally, the epicardial mesh was used to provide an electric shock to the heart that was experiencing ventricular fibrillation, allowing for effective defibrillation (Sunwoo et al. 2020). Analysis of the impact of electrical stimulations on the heart with myocardial infarction also showed that these stimulations considerably enhanced the contraction of the ventricular chamber during the systolic phase.

2.7.2 Peripheral Nerves

Limb control loss may result from trauma-related damage to the peripheral nerves. Implantable devices offer a treatment or a way around injured peripheral nerves. Since the nerves surrounded by muscles usually include dynamic motions, substance tenderness and device durability are the main factors to be taken into account while constructing such devices that interface together with the peripheral nerves (Lee et al. 2012b; Wenger et al. 2016). In light of these circumstances, Seo et al. proposed the development of a neuroprosthetic device that is soft, robust, and fatigue-resistant and that permits bilateral signaling on peripheral nerves (Seo et al. 2021). Silver fluoride enclosed with gold nanoshells dispensed in a strong, flexible, and self-repairing polymer matrix made of polydimethylsiloxane-4,4′-methylenebis(phenyl urea)-isophorone bisurea units. Due to the self-healing matrix's dynamic hydrogen bond re-crosslinking and low glass transition temperature, the nanocomposite may self-heal material degradations brought on by repeated deformations on peripheral nerves. The device's ability to self-heal allowed it to modulate the target nerve's electrical signal steadily. Intended for demonstration, the device was positioned on a rat's sciatic nerve. The electrical stimulation that the gadget administered to the sciatic nerve was accompanied by an observation of tetanic and twitch contractions of muscles. Rats using treadmills were given neural impulses from their sciatic nerves, which were then utilized to examine the motions of their joints (Seo et al. 2021).

According to Koo et al., an implanted electrical stimulator may improve peripheral nerve injury functioning and hasten its restoration to function. The bioresorbable electrical peripheral nerve stimulation instrument and associated circuitry elements make up the system (Koo et al. 2018b). The silicon nanomembrane about 320 nm thick radio frequency diode and the magnesium (approximately 300 nm thick) stimulation electrodes are the specific components of the implanted wireless stimulator. Additionally, a tubular construction made of hot-pressed poly-lactic-co-glycolic acid with a biocompatible metallic strip implanted in it was used as a cuff to implant the device close to the peripheral nerve. When the device was decomposed on a rat's sciatic nerve during 8 weeks over implantation, no indications of compressive axonopathy or nerve injury were found. After being inserted in the damaged nerve area, the wireless stimulator electronically prompted the nerve. Using surgical scissors, a rat's sciatic nerve was separated for testing purposes, and the cuff-type wireless stimulator was attached to it. Electrical stimulation improved muscular activation and sped up healing, according to EMG recordings made from the tibialis anterior muscles at different points after surgery (Koo et al. 2018b).

2.7.3 Brain Monitoring

The fields of neuro-engineering and neuro-science have made significant progress with the introduction of delicate forms of implantable devices for brain-computer interfaces. These tools make it possible to analyze potentials for action from distinct neurons and to continuously observe macroscopic brain activity (Liu et al. 2015). Inflammatory reactions and the creation of scars are brought on by the mechanical and physiological incompatibilities between neurological tissues with micromotions and traditional stiff implants. Different sensitive implantable medical devices have been created as a result. Liu et al., for instance, have created a pleasant injectable device that can be administered with a syringe into the brain cavity (Liu et al. 2015). It may be possible to avoid extensive surgery for the device installation by employing the syringe injection approach. The extremely fine gold electrodes with thickness < 100 nm of the mesh-shaped structure are covered in multiple coats of SU-8 epoxy. Initially, this mesh-shaped structure-type device with dispersed sensing electrodes is coiled up within the syringe. It may stretch out both during and after the injection. Electrode-conformal incorporation with the brain was made possible by a relatively simple injection of the mesh-type device within the hippocampal area of a living rodent. While the mesh electronic system can be merely invasively integrated with brain tissue utilizing syringe injection, reproducibility in targeting a particular area of the brain and spatial mapping of high-quality brain signals are challenging due to the difficult job of precisely and repeatedly controlling the position and shape of the network of electrodes during the procedure of injection (Liu et al. 2015).

A further instance is the development of an implanted gadget with bioresorbable silicon circuits. In this instance, long-term physical strain and foreign body reactions can be prevented since the device degrades in vivo (Yu et al. 2016). Electrodes for sensing were made from an extremely doped silicon monocrystalline nanomembrane. The silicon nanomembrane's characteristics were comparable to those of the traditional gold electrode. Because of its thinness, about 300 nm, silicon nanomembrane is easily biodegraded in vivo. The duration of the device's effectiveness throughout the body is determined by controlling the rate of silicon nanomembrane deterioration in biofluids. This rate is influenced by both the degree of thickness and loading concentration of the silicon nanomembrane (Yu et al. 2016).

An essential component of prolonged period brain surveillance is the implanted device's biomechanical compliance with brain tissue. Tybrandt et al. created a brain filing electrode matrix (neuro-grid) that is highly dense and stretchy. The electrodes are made of a soft nanocomposite consisting of polydimethylsiloxane and gold-coated TiO_2 nanowires (Tybrandt et al. 2018). When $HAuCl_4$ was added to the scattered TiO_2 nanowires together with a moderate reducing agent (such as hydroxylamine), the result was gold-capped TiO_2 nanowires. An additional polydimethylsiloxane layer was used to enclose the integrated and patterned gold-coated TiO_2 nanowires on a secured polydimethylsiloxane substrate. In its initial form, the nanocomposite had a resistance of $0.63 \pm 0.03\ \Omega/sq.$ (~16,000 S/cm), which grew to around 3 Ω/sq. at 50% strain. The nanocomposite was subjected to 1000 cycles of stretching at 100% strain. The neurogrid was able to conformally establish touch with curved surfaces because of its extensible properties. The electrode's impedance was reduced by further Pt coating. After implanting the neurogrid on a somatosensory cortex, 28 electrodes on a freely roaming rat were used to capture induced potentials. Additionally, longer than 3 months of continuous brain monitoring might be shown (Tybrandt et al. 2018).

In addition, a conductive hydrogel nanocomposite for epidural and intracortical neural recording was created using a supramolecular β-peptide. It is made up of CNTs, a self-assemblage β-peptide, and hexosaminidase (Nam et al. 2020). Helicoidally frameworks were formed by the hexosaminidase self-assembling into nanofibers by pairing with hydrogen, which firmly wrapped surrounding bundled CNTs. A homogenous distribution of hexosaminidase/CNT complexes throughout the hydrogel is achieved by the combining of hexosaminidase nanofibers and CNTs, which inhibits the CNTs from aggregating. As the concentration of hexosaminidase/CNTs rose, the nanocomposite's conductivity grew linearly. The addition of hexosaminidase/CNTs to a bare hydrogel and nanocomposite (0.02% hexosaminidase/CNTs) amid provoked epileptic seizures enhanced the neural signals, according to epidural ECoG measurements. Specifically, the strength of the highest-frequency seizure impulses (β as well as γ bands) increased threefold. For over 6 weeks, the brain surface stayed consistent after being coated with polydimethylsiloxane and the nanocomposite overlay. Additionally, the somatosensory cortex, which is located in the epileptic mice was administered with the 0.02% hexosaminidase/CNTs hydrogel to evaluate the local field potential. The resultant signal intensity was therefore about 2.4 times greater than those obtained from a simple electrode (Nam et al. 2020). For almost a month, the hydrogel that was injected remained persistent.

2.7.4 Brain Stimulation

Responses to brain monitoring can be employed by implantable devices to administer therapeutic stimulations with feedback. Just like brain monitoring devices, brain stimulation devices must also be harmonious and compatible with soft brain tissue both chemically and mechanically. As a result, soft mechanical components and chemically robust components have been employed. These items can stimulate the brain chemically, optically, and electrically (Park et al. 2018).

For instance, epilepsy has been treated with electrical stimulation techniques. Whenever the whole brain has a seizure, the seizure focal can receive tailored electrical stimulations from a neuro-stimulator placed in the skull (Park et al. 2018). Nevertheless, the thickness or rigidity of traditional electrode arrays reduces the effectiveness of electrically stimulating treatment. Since graphene has superior electrical, mechanical, and biological qualities, a very delicate and malleable microelectrode array composed of it was created. Ten monitoring pathways and 20 electrotherapy lines were made with an electrode array made of CVD-originated graphene. Epileptic impulses from the autistic mouse model could be recorded by the array, which was transplanted in the motor cortex region. Transfer of specific epileptiform brain activity to adjacent brain areas was inhibited by high frequencies of sinusoidal electrical stimulation impulses provided by the graphene electrode (Park et al. 2018). Extended in vivo electrical brain stimulations were also made possible by the supple, extremely thin graphene electrode.

Notwithstanding the advantages of electrical stimulation, connecting tissues and biofluids allowed the signals to propagate outside of the target location, making it impossible to precisely regulate local brain activity. In the meanwhile, targeted visual stimulation of the particular neuron is made possible by the optic stimulation of the genetically altered neuron. Sensitive bioelectronic gadgets have this kind of optical stimulation function integrated into them. Yu et al. created a multilayer device made up of a portable circuit, retractable microelectrode array, and extensible optical fibers (Yu et al. 2021). Utilizing the transfer printing procedure, the flexible microelectrode array—which is composed of loaded layers of pulse induction/copper/pulse induction—was conformally merged onto a curved optical fiber. Gold nanoparticles were also used to modify the outermost layer of the electrode to obtain low interface impedance and great chemical durability. The device was inserted in the basolateral amygdala of a genetically altered rat, causing it to react to visual cues. It is possible to track the action prospective elicited by visual stimulations. Both visual stimulation and recording of impulses were made possible at the same time by the combination of adaptive electrodes and an optic fiber (Yu et al. 2021).

Treating a variety of brain disorders with pharmaceuticals involves delivering the right kind of chemical stimulation. Nevertheless, in certain situations, such as brain tumors, oral or intravenous medication administration is ineffective because the blood-brain barrier impairs the effectiveness of the drug's distribution. Therefore, a targeted intracranial medication delivery strategy is required. To administer drugs intracranially to specific brain areas, Lee et al. created a flexible, adhesive, biodegradable, and wirelessly managed device. Featuring a drug-loaded polymer patch incorporated inside a wireless temperature sensor, this bioresorbable equipment is made of two parts (Yu et al. 2021). All components are bioresorbable in vivo, including magnesium for the electronics, polylactic acid for encapsulation, and oxidized starch for the drug reservoir. The primary ingredient of the patch is oxidized starch, into which doxorubicin, a chemotherapy medication, is inserted. Long-lasting persistent drug administration is made possible by the imine conjugation of doxorubicin with oxidized starch. Additionally, the hydrophilic nature of the oxidized starch layer permits robust and precise adherence to the surface of brain tissue. It can be challenging to remove the patch of the brain tissue after it has been affixed. A cerebral area has been surgically fitted with the biodegradable, wirelessly governed patch. The drug discharge quantity and drug absorption depth were enhanced by the wireless heat stimulation triggered via an RF magnetic field. The glioblastoma model in mice was used to verify the therapeutic's effectiveness. Drug delivery that is both continuous and remotely regulated has the potential to slow down tumor development and improve rates of survival (Yu et al. 2021).

2.8 Future Prospects and Challenges

Bioelectronics is developing quickly as a result of recent major advancements in nanostructured substances, which may be created from atoms and molecules or processed from existing electronic materials. The mechanical, biological, and electronic performances of the bioelectronic devices have all increased overall. Significant modifications have been made to the mechanical aspects in particular among those enhancements. The gadgets have extraordinary sensitivity due to the incorporation of nanoparticles, which are comparable to the softness of tissues or organs. In addition, naturally elastic and flexible materials like hydrogels and elastomers have been carefully combined with tailored fillers at the nanoscale to formulate soft nanocomposites. These have produced incredibly exclusive and linear device-tissue interactions that facilitate efficient input stimulation functions and high-quality biological data recording. Furthermore, the soft bioelectronic devices' deformability has been further enhanced by the use of novel design techniques, such as extremely thin and stretchy technology layouts, making wearable and implantable device applications functional for extended periods. Even with these advancements, there are still a lot of real-world problems. Although structurally engineered ultrathin devices show outstanding functioning and deformation ability, nanoscale active layers still lack appropriate durability over time, particularly in dynamic and wet in vivo conditions. Appropriate soft encapsulating technologies and techniques for coating exempt from the biofouling problem are required to address this problem. To achieve the proven efficiency of rigid electronic components, nanocomposites' electrical or electrochemical performance requires additional advancements. It is also necessary to conduct further studies on the creation of nanocomposites and equipment using nanomaterials that are biocompatible, such as gold, platinum, and non-cadmium QDs, to overcome the cytotoxicity and oxidation issues of many nanomaterials, for example, silver nanoparticles, silver nanowires, and cadmium-based QDs, which are obstacles to overcome for in vivo applications. Furthermore, even if the mechanical incompatibilities between electronics and substantial, rigid equipment have significantly decreased, the creation of insulative connective tissue surrounding the implanted electronics remains difficult. More research is required on the substances and therapeutic approaches to avoid and reduce immunological reactions, particularly for a long time implantation.

2.9 Conclusions

The commercialization of nanomaterial-based devices requires the development of efficient and cost-effective mass manufacturing systems. Additionally, since the bulk of earlier studies relied on small animal specimens, more clinical research based on big animal models including trials on human subjects is necessary. To create a closed-loop structure with all the functional elements needed for bioelectronics, such as the power supply, monitoring, storage of data, informational display, and input stimulation function, nanomaterials-based soft bioelectronics must meet certain requirements. To accomplish this aim, further research will have to be done in the future to combine high-performance flexible or stretchy sensors, stimulators, and display components in a monolithic manner, together with small and inherently soft power sources along with information transmission components. Furthermore, the integration of artificial intelligence with soft bioelectronics based on nanomaterials is crucial for the development of intelligent closed-loop systems. The area of bioelectronics has advanced significantly because of recent developments in nanostructures and soft devices, despite many difficulties. Close cooperation between medical professionals, electronics engineers, and materials researchers would advance the clinical integration of nanomaterial-based soft bioelectronics and ultimately lead to the resolution of unresolved clinical issues.

References

Ahamed M, AlSalhi MS, Siddiqui MKJ (2010) Silver nanoparticle applications and human health. Clin Chim Acta 411(23–24):1841–1848. https://doi.org/10.1016/J.CCA.2010.08.016

Akar O, Akin T, Najafi K (2001) A wireless batch sealed absolute capacitive pressure sensor. Sens Actuators A Phys 95(1):29–38. https://doi.org/10.1016/S0924-4247(01)00753-1

Alemán JV, Chadwick AV, He J, Hess M, Horie K, Jones RG, Kratochvíl P, Meisel I, Mita I, Moad G, Penczek S, Stepto RFT (2007) Definitions of terms relating to the structure and processing of sols, gels, networks, and inorganic-organic hybrid materials (IUPAC recommendations 2007). Pure Appl Chem 79(10):1801–1829. https://doi.org/10.1351/pac200779101801

Alizadeh S, Fallah N, Nikazar M (2019) An ultrasonic method for the synthesis, control and optimization of CdS/TiO2 core–shell nanocomposites. RSC Adv 9(8):4314–4324. https://doi.org/10.1039/C8RA10155H

An H, Lee WJ, Jung J (2011) Graphene synthesis on Fe foil using thermal CVD. Curr Appl Phys 11(4):S81–S85. https://doi.org/10.1016/J.CAP.2011.03.077

Ang PK, Chen W, Wee ATS, Loh KP (2008) Solution-gated epitaxial graphene as PH sensor. J Am Chem Soc 130(44):14392–14393. https://doi.org/10.1021/ja805090z

Ansari SG, Boroojerdian P, Sainkar SR, Karekar RN, Aiyer RC, Kulkarni SK (1997) Grain size effects on H2 gas sensitivity of thick film resistor using SnO2 nanoparticles. Thin Solid Films 295(1):271–276. https://doi.org/10.1016/S0040-6090(96)09152-3

Bagwe RP, Yang C, Hilliard LR, Tan W (2004) Optimization of dye-doped silica nanoparticles prepared using a reverse microemulsion method. Langmuir 20(19):8336–8342. https://doi.org/10.1021/la049137j

Bao Z, Xian C, Yuan Q, Liu G, Wu J (2019) Natural polymer-based hydrogels with enhanced mechanical performances: preparation, structure, and property. Adv Healthc Mater 8(17):1900670. https://doi.org/10.1002/adhm.201900670

Barold SS (2003) Willem Einthoven and the birth of clinical electrocardiography a hundred years ago. Card Electrophysiol Rev 7(1):99–104. https://doi.org/10.1023/A:1023667812925

Bazaka K, Jacob M (2012) Implantable devices: issues and challenges. Electronics (Basel) 2(1):1–34. https://doi.org/10.3390/electronics2010001

Benabid AL (2003) Deep brain stimulation for Parkinson's disease. Curr Opin Neurobiol 13(6):696–706. https://doi.org/10.1016/J.CONB.2003.11.001

Bondi SN, Lackey WJ, Johnson RW, Wang X, Wang ZL (2006) Laser assisted chemical vapor deposition synthesis of carbon nanotubes and their characterization. Carbon N Y 44(8):1393–1403. https://doi.org/10.1016/j.carbon.2005.11.023

Brandl F, Kastner F, Gschwind RM, Blunk T, Teßmar J, Göpferich A (2010) Hydrogel-based drug delivery systems: comparison of drug diffusivity and release kinetics. J Control Release 142(2):221–228. https://doi.org/10.1016/j.jconrel.2009.10.030

Çakmak G, Küçükyavuz Z, Küçükyavuz S, Çakmak H (2004) Mechanical, electrical and thermal properties of carbon fiber reinforced poly(dimethylsiloxane)/polypyrrole composites. Compos Part A Appl Sci Manuf 35(4):417–421. https://doi.org/10.1016/j.compositesa.2003.11.006

Canales A, Jia X, Froriep UP, Koppes RA, Tringides CM, Selvidge J, Lu C, Hou C, Wei L, Fink Y, Anikeeva P (2015) Multifunctional fibers for simultaneous optical, electrical and chemical interrogation of neural circuits in vivo. Nat Biotechnol 33(3):277–284. https://doi.org/10.1038/nbt.3093

Cha GD, Kang D, Lee J, Kim D-H (2019) Bioresorbable electronic implants: history, materials, fabrication, devices, and clinical applications. Adv Healthc Mater 8(11):1801660. https://doi.org/10.1002/adhm.201801660

Cha GD, Lee WH, Lim C, Choi MK, Kim D-H (2020a) Materials engineering, processing, and device application of hydrogel nanocomposites. Nanoscale 12(19):10456–10473. https://doi.org/10.1039/D0NR01456G

Cha GD, Kang T, Baik S, Kim D, Choi SH, Hyeon T, Kim D-H (2020b) Advances in drug delivery technology for the treatment of glioblastoma multiforme. J Control Release 328:350–367. https://doi.org/10.1016/j.jconrel.2020.09.002

Chang KE, Kim C, Yoo TJ, Kwon MG, Heo S, Kim S-Y, Hyun Y, Yoo JI, Ko HC, Lee BH (2019) High-responsivity near-infrared photodetector using gate-modulated graphene/germanium Schottky junction. Adv Electron Mater 5(6):1800957. https://doi.org/10.1002/aelm.201800957

Chen D-H, Wu S-H (2000) Synthesis of nickel nanoparticles in water-in-oil microemulsions. Chem Mater 12(5):1354–1360. https://doi.org/10.1021/cm991167y

Chen A, Kamata K, Nakagawa M, Iyoda T, Haiqiao, Li X (2005) Formation process of silver–polypyrrole coaxial nanocables synthesized by redox reaction between AgNO3 and pyrrole in the presence of Poly(Vinylpyrrolidone). J Phys Chem B 109(39):18283–18288. https://doi.org/10.1021/jp053247x

Chen KI, Li BR, Chen YT (2011) Silicon nanowire field-effect transistor-based biosensors for biomedical diagnosis and cellular recording investigation. Nano Today 6(2):131–154. https://doi.org/10.1016/J.NANTOD.2011.02.001

Chen J, Li F, Luo Y, Shi Y, Ma X, Zhang M, Boukhvalov DW, Luo Z (2019) A self-healing elastomer based on an intrinsic non-covalent cross-linking mechanism. J Mater Chem A Mater 7(25):15207–15214. https://doi.org/10.1039/C9TA03775F

Cheung W, Pontoriero F, Taratula O, Chen AM, He H (2010) DNA and carbon nanotubes as medicine. Adv Drug Deliv Rev 62(6):633–649. https://doi.org/10.1016/J.ADDR.2010.03.007

Chiang C-H, Won SM, Orsborn AL, Yu KJ, Trumpis M, Bent B, Wang C, Xue Y, Min S, Woods V, Yu C, Kim BH, Kim SB, Huq R, Li J, Seo KJ, Vitale F, Richardson A, Fang H, Huang Y, Shepard K, Pesaran B, Rogers JA, Viventi J (2020) Development of a neural interface for high-definition, long-term recording in rodents and nonhuman primates. Sci Transl Med 12(538):eaay4682. https://doi.org/10.1126/scitranslmed.aay4682

Cho KW, Kim SJ, Kim J, Song SY, Lee WH, Wang L, Soh M, Lu N, Hyeon T, Kim B-S, Kim D-H (2019) Large scale and integrated platform for digital mass culture of anchorage dependent cells. Nat Commun 10(1):4824. https://doi.org/10.1038/s41467-019-12777-3

Cho KW, Lee WH, Kim B-S, Kim D-H (2020) Sensors in heart-on-a-chip: a review on recent progress. Talanta 219:121269. https://doi.org/10.1016/j.talanta.2020.121269

Cho Y, Park S, Lee J, Yu KJ (2021) Emerging materials and technologies with applications in flexible neural implants: a comprehensive review of current issues with neural devices. Adv Mater 33(47):2005786. https://doi.org/10.1002/adma.202005786

Cho KW, Sunwoo SH, Hong YJ, Koo JH, Kim JH, Baik S, Hyeon T, Kim DH (2022) Soft bioelectronics based on nanomaterials. Chem Rev 122(5):5068–5143. https://doi.org/10.1021/acs.chemrev.1c00531

Choi M, Choi JW, Kim S, Nizamoglu S, Hahn SK, Yun SH (2013) Light-guiding hydrogels for cell-based sensing and optogenetic synthesis in vivo. Nat Photonics 7(12):987–994. https://doi.org/10.1038/nphoton.2013.278

Choi MK, Park I, Kim DC, Joh E, Park OK, Kim J, Kim M, Choi C, Yang J, Cho KW, Hwang J-H, Nam J-M, Hyeon T, Kim JH, Kim D-H (2015) Thermally controlled, patterned graphene trans-

fer printing for transparent and wearable electronic/optoelectronic system. Adv Funct Mater 25(46):7109–7118. https://doi.org/10.1002/adfm.201502956
Choi C, Choi MK, Hyeon T, Kim D-H (2016a) Nanomaterial-based soft electronics for healthcare applications. ChemNanoMat 2(11):1006–1017. https://doi.org/10.1002/cnma.201600191
Choi MK, Park OK, Choi C, Qiao S, Ghaffari R, Kim J, Lee DJ, Kim M, Hyun W, Kim SJ, Hwang HJ, Kwon S-H, Hyeon T, Lu N, Kim D-H (2016b) Cephalopod-inspired miniaturized suction cups for smart medical skin. Adv Healthc Mater 5(1):80–87. https://doi.org/10.1002/adhm.201500285
Choi MK, Yang J, Hyeon T, Kim D-H (2018a) Flexible quantum dot light-emitting diodes for next-generation displays. npj Flexible Electronics 2(1):10. https://doi.org/10.1038/s41528-018-0023-3
Choi MK, Yang J, Kim DC, Dai Z, Kim J, Seung H, Kale VS, Sung SJ, Park CR, Lu N, Hyeon T, Kim D-H (2018b) Extremely vivid, highly transparent, and ultrathin quantum dot light-emitting diodes. Adv Mater 30(1):1703279. https://doi.org/10.1002/adma.201703279
Choi S, Han SI, Jung D, Hwang HJ, Lim C, Bae S, Park OK, Tschabrunn CM, Lee M, Bae SY, Yu JW, Ryu JH, Lee S-W, Park K, Kang PM, Lee WB, Nezafat R, Hyeon T, Kim D-H (2018c) Highly conductive, stretchable and biocompatible Ag–Au Core–Sheath nanowire composite for wearable and implantable bioelectronics. Nat Nanotechnol 13(11):1048–1056. https://doi.org/10.1038/s41565-018-0226-8
Choi JR, Yong KW, Choi JY, Cowie AC (2019) Recent advances in photo-crosslinkable hydrogels for biomedical applications. BioTechniques 66(1):40–53. https://doi.org/10.2144/btn-2018-0083
Choi C, Leem J, Kim M, Taqieddin A, Cho C, Cho KW, Lee GJ, Seung H, Bae HJ, Song YM, Hyeon T, Aluru NR, Nam S, Kim D-H (2020) Curved neuromorphic image sensor array using a MoS2-organic heterostructure inspired by the human visual recognition system. Nat Commun 11(1):5934. https://doi.org/10.1038/s41467-020-19806-6
Da Silva AB, Marini J, Gelves G, Sundararaj U, Gregório R, Bretas RES (2013) Synergic effect in electrical conductivity using a combination of two fillers in PVDF hybrids composites. Eur Polym J 49(10):3318–3327. https://doi.org/10.1016/J.EURPOLYMJ.2013.06.039
Dasgupta NP, Sun J, Liu C, Brittman S, Andrews SC, Lim J, Gao H, Yan R, Yang P (2014) 25th anniversary article: semiconductor nanowires – synthesis, characterization, and applications. Adv Mater 26(14):2137–2184. https://doi.org/10.1002/adma.201305929
Dauzon E, Lin Y, Faber H, Yengel E, Sallenave X, Plesse C, Goubard F, Amassian A, Anthopoulos TD (2020) Stretchable and transparent conductive PEDOT:PSS-based electrodes for organic photovoltaics and strain sensors applications. Adv Funct Mater 30(28):2001251. https://doi.org/10.1002/adfm.202001251
De S, Higgins TM, Lyons PE, Doherty EM, Nirmalraj PN, Blau WJ, Boland JJ, Coleman JN (2009) Silver nanowire networks as flexible, transparent, conducting films: extremely high DC to optical conductivity ratios. ACS Nano 3(7):1767–1774. https://doi.org/10.1021/nn900348c
Deignan G, Goldthorpe IA (2017) The dependence of silver nanowire stability on network composition and processing parameters. RSC Adv 7(57):35590–35597. https://doi.org/10.1039/C7RA06524H
Deng Y, Hussain I, Kang M, Li K, Yao F, Liu S, Fu G (2018) Self-recoverable and mechanical-reinforced hydrogel based on hydrophobic interaction with self-healable and conductive properties. Chem Eng J 353:900–910. https://doi.org/10.1016/j.cej.2018.07.187
Deng J, Yuk H, Wu J, Varela CE, Chen X, Roche ET, Guo CF, Zhao X (2021) Electrical bioadhesive interface for bioelectronics. Nat Mater 20(2):229–236. https://doi.org/10.1038/s41563-020-00814-2
Dhakar L, Gudla S, Shan X, Wang Z, Tay FEH, Heng C-H, Lee C (2016) Large scale triboelectric nanogenerator and self-powered pressure sensor array using low cost roll-to-roll UV embossing. Sci Rep 6(1):22253. https://doi.org/10.1038/srep22253
Dong M, Shi B, Liu D, Liu J-H, Zhao D, Yu Z-H, Shen X-Q, Gan J-M, Shi B, Qiu Y, Wang C-C, Zhu Z-Z, Shen Q-D (2020) Conductive hydrogel for a photothermal-responsive stretchable arti-

ficial nerve and coalescing with a damaged peripheral nerve. ACS Nano 14(12):16565–16575. https://doi.org/10.1021/acsnano.0c05197

DS PD, Thekkedath J, Kutty SKN (2012) Synthesis and characterization of conducting composites of polypyrrole/polypyrrole-coated short nylon fiber and natural rubber. Polym Plast Technol Eng 51(8):823–831. https://doi.org/10.1080/03602559.2012.671414

Duan X, Lieber CM (2000) General synthesis of compound semiconductor nanowires. Adv Mater 12(4):298–302. https://doi.org/10.1002/(SICI)1521-4095(200002)12:4<298::AID-ADMA298>3.0.CO;2-Y

Duan K, Li L, Hu Y, Wang X (2017) Pillared graphene as an ultra-high sensitivity mass sensor. Sci Rep 7(1):14012. https://doi.org/10.1038/s41598-017-14182-6

Eckhouse SR, Purcell BP, McGarvey JR, Lobb D, Logdon CB, Doviak H, O'Neill JW, Shuman JA, Novack CP, Zellars KN, Pettaway S, Black RA, Khakoo A, Lee T, Mukherjee R, Gorman JH, Gorman RC, Burdick JA, Spinale FG (2014) Local hydrogel release of recombinant TIMP-3 attenuates adverse left ventricular remodeling after experimental myocardial infarction. Sci Transl Med 6(223):223ra21. https://doi.org/10.1126/scitranslmed.3007244

El Zein A, Huppé C, Cochrane C (2017) Development of a flexible strain sensor based on PEDOT:PSS for thin film structures. Sensors. https://doi.org/10.3390/s17061337

ElAfandy RT, Majid MA, Ng TK, Zhao L, Cha D, Ooi BS (2014) Exfoliation of threading dislocation-free, single-crystalline, ultrathin gallium nitride nanomembranes. Adv Funct Mater 24(16):2305–2311. https://doi.org/10.1002/adfm.201303001

Elgendi M, Fletcher R, Liang Y, Howard N, Lovell NH, Abbott D, Lim K, Ward R (2019) The use of photoplethysmography for assessing hypertension. NPJ Digit Med 2(1):60. https://doi.org/10.1038/s41746-019-0136-7

Fan Z, Ouyang J (2019) Thermoelectric properties of PEDOT:PSS. Adv Electron Mater 5(11):1800769. https://doi.org/10.1002/aelm.201800769

Fan F-R, Lin L, Zhu G, Wu W, Zhang R, Wang ZL (2012) Transparent triboelectric nanogenerators and self-powered pressure sensors based on micropatterned plastic films. Nano Lett 12(6):3109–3114. https://doi.org/10.1021/nl300988z

Fan JA, Yeo W-H, Su Y, Hattori Y, Lee W, Jung S-Y, Zhang Y, Liu Z, Cheng H, Falgout L, Bajema M, Coleman T, Gregoire D, Larsen RJ, Huang Y, Rogers JA (2014) Fractal design concepts for stretchable electronics. Nat Commun 5(1):3266. https://doi.org/10.1038/ncomms4266

Fang J, Mehlich A, Koga N, Huang J, Koga R, Gao X, Hu C, Jin C, Rief M, Kast J, Baker D, Li H (2013) Forced protein unfolding leads to highly elastic and tough protein hydrogels. Nat Commun 4(1):2974. https://doi.org/10.1038/ncomms3974

Ferreira PC, Ataíde VN, Silva Chagas CL, Angnes L, Tomazelli Coltro WK, Longo Cesar Paixão TR, Reis de Araujo W (2019) Wearable electrochemical sensors for forensic and clinical applications. TrAC Trends Anal Chem 119:115622. https://doi.org/10.1016/J.TRAC.2019.115622

Fiorillo AS, Critello CD, Pullano SA (2018) Theory, technology and applications of piezoresistive sensors: a review. Sens Actuators A Phys 281:156–175. https://doi.org/10.1016/j.sna.2018.07.006

Fortin J, Rogge DE, Fellner C, Flotzinger D, Grond J, Lerche K, Saugel B (2021) A novel art of continuous noninvasive blood pressure measurement. Nat Commun 12(1):1387. https://doi.org/10.1038/s41467-021-21271-8

Freeman WJ (2015) Mechanism and significance of global coherence in scalp EEG. Curr Opin Neurobiol 31:199–205. https://doi.org/10.1016/j.conb.2014.11.008

Fu H, Xu S, Xu R, Jiang J, Zhang Y, Rogers JA, Huang Y (2015) Lateral buckling and mechanical Stretchability of fractal interconnects partially bonded onto an elastomeric substrate. Appl Phys Lett 106(9):91902. https://doi.org/10.1063/1.4913848

Fye WB (1994) A history of the origin, evolution, and impact of electrocardiography. Am J Cardiol 73(13):937–949. https://doi.org/10.1016/0002-9149(94)90135-X

Gajanayake T, Olariu R, Leclère FM, Dhayani A, Yang Z, Bongoni AK, Banz Y, Constantinescu MA, Karp JM, Vemula PK, Rieben R, Vögelin E (2014) A single localized dose of enzyme-responsive hydrogel improves long-term survival of a vascularized composite allograft. Sci Transl Med 6(249):249ra110. https://doi.org/10.1126/scitranslmed.3008778

Gambhire AB, Lande MK, Kalokhe SB, Shirsat MD, Patil KR, Gholap RS, Arbad BR (2008) Synthesis and characterization of high surface area CeO2-doped SnO2 nanomaterial. Mater Chem Phys 112(3):719–722. https://doi.org/10.1016/J.MATCHEMPHYS.2008.06.022

Garcia C, Trendafilova I, Guzman de Villoria R, Sanchez del Rio J (2018) Self-powered pressure sensor based on the triboelectric effect and its analysis using dynamic mechanical analysis. Nano Energy 50:401–409. https://doi.org/10.1016/j.nanoen.2018.05.046

García-Guzmán JJ, Pérez-Ràfols C, Cuartero M, Crespo GA (2021) Microneedle based electrochemical (bio)sensing: towards decentralized and continuous health status monitoring. TrAC Trends Anal Chem 135:116148. https://doi.org/10.1016/J.TRAC.2020.116148

Garnett E, Mai L, Yang P (2019) Introduction: 1D nanomaterials/nanowires. Chem Rev 119(15):8955–8957. https://doi.org/10.1021/acs.chemrev.9b00423

Ge L, Liu X (2016) Electrical resonance with voltage-gated ion channels: perspectives from biophysical mechanisms and neural electrophysiology. Acta Pharmacol Sin 37(1):67–74. https://doi.org/10.1038/aps.2015.140

Gerdle B, Henriksson-Larsén K, Lorentzon R, Wretling M-L (1991) Dependence of the mean power frequency of the electromyogram on muscle force and fibre type. Acta Physiol Scand 142(4):457–465. https://doi.org/10.1111/j.1748-1716.1991.tb09180.x

Goda T, Miyahara Y (2019) Electrodeposition of Zwitterionic PEDOT films for conducting and antifouling surfaces. Langmuir 35(5):1126–1133. https://doi.org/10.1021/acs.langmuir.8b01492

Gong S, Schwalb W, Wang Y, Chen Y, Tang Y, Si J, Shirinzadeh B, Cheng W (2014) A wearable and highly sensitive pressure sensor with ultrathin gold nanowires. Nat Commun 5(1):3132. https://doi.org/10.1038/ncomms4132

Gong S, Lai DTH, Wang Y, Yap LW, Si KJ, Shi Q, Jason NN, Sridhar T, Uddin H, Cheng W (2015) Tattoolike polyaniline microparticle-doped gold nanowire patches as highly durable wearable sensors. ACS Appl Mater Interfaces 7(35):19700–19708. https://doi.org/10.1021/acsami.5b05001

Gong S, Yap LW, Zhu B, Zhai Q, Liu Y, Lyu Q, Wang K, Yang M, Ling Y, Lai DTH, Marzbanrad F, Cheng W (2019) Local crack-programmed gold nanowire electronic skin tattoos for in-plane multisensor integration. Adv Mater 31(41):1903789. https://doi.org/10.1002/adma.201903789

Gulino M, Kim D, Pané S, Santos SD, Pêgo AP (2019) Tissue response to neural implants: the use of model systems toward new design solutions of implantable microelectrodes. Front Neurosci 13(JUL):1–24. https://doi.org/10.3389/fnins.2019.00689

Gupta P, Moghimi MJ, Jeong Y, Gupta D, Inan OT, Ayazi F (2020) Precision wearable accelerometer contact microphones for longitudinal monitoring of mechano-acoustic cardiopulmonary signals. NPJ Digit Med 3(1):19. https://doi.org/10.1038/s41746-020-0225-7

Haghgoo M, Ansari R, Hassanzadeh-Aghdam MK (2019) Prediction of electrical conductivity of carbon fiber-carbon nanotube-reinforced polymer hybrid composites. Compos Part B Eng 167:728–735. https://doi.org/10.1016/j.compositesb.2019.03.046

Han S, Jang B, Kim T, Oh SM, Hyeon T (2005) Simple synthesis of hollow tin dioxide microspheres and their application to lithium-ion battery anodes. Adv Funct Mater 15(11):1845–1850. https://doi.org/10.1002/adfm.200500243

Han S, Hong S, Ham J, Yeo J, Lee J, Kang B, Lee P, Kwon J, Lee SS, Yang M-Y, Ko SH (2014) Fast plasmonic laser nanowelding for a Cu-nanowire percolation network for flexible transparent conductors and stretchable electronics. Adv Mater 26(33):5808–5814. https://doi.org/10.1002/adma.201400474

Han S, Hong S, Yeo J, Kim D, Kang B, Yang M-Y, Ko SH (2015) Nanorecycling: monolithic integration of copper and copper oxide nanowire network electrode through selective reversible photothermochemical reduction. Adv Mater 27(41):6397–6403. https://doi.org/10.1002/adma.201503244

Han Z, Liu L, Zhang J, Han Q, Wang K, Song H, Wang Z, Jiao Z, Niu S, Ren L (2018) High-performance flexible strain sensor with bio-inspired crack arrays. Nanoscale 10(32):15178–15186. https://doi.org/10.1039/C8NR02514B

Heitsch AT, Hessel CM, Akhavan VA, Korgel BA (2009) Colloidal silicon nanorod synthesis. Nano Lett 9(8):3042–3047. https://doi.org/10.1021/nl901520t

Hernandez Y, Nicolosi V, Lotya M, Blighe FM, Sun Z, De S, McGovern IT, Holland B, Byrne M, Gun'Ko YK, Boland JJ, Niraj P, Duesberg G, Krishnamurthy S, Goodhue R, Hutchison J, Scardaci V, Ferrari AC, Coleman JN (2008) High-yield production of graphene by liquid-phase exfoliation of graphite. Nat Nanotechnol 3(9):563–568. https://doi.org/10.1038/nnano.2008.215

Ho MD, Ling Y, Yap LW, Wang Y, Dong D, Zhao Y, Cheng W (2017) Percolating network of ultrathin gold nanowires and silver nanowires toward "invisible" wearable sensors for detecting emotional expression and apexcardiogram. Adv Funct Mater 27(25):1700845. https://doi.org/10.1002/adfm.201700845

Hof AL (1984) EMG and muscle force: an introduction. Hum Mov Sci 3(1):119–153. https://doi.org/10.1016/0167-9457(84)90008-3

Homede E, Zigelman A, Abezgauz L, Manor O (2018) Signatures of van Der Waals and electrostatic forces in the deposition of nanoparticle assemblies. J Phys Chem Lett 9(18):5226–5232. https://doi.org/10.1021/acs.jpclett.8b02052

Hong KH, Song S-C (2019) 3D hydrogel stem cell niche controlled by host-guest interaction affects stem cell fate and survival rate. Biomaterials 218:119338. https://doi.org/10.1016/j.biomaterials.2019.119338

Hong SW, Du F, Lan W, Kim S, Kim H-S, Rogers JA (2011) Monolithic integration of arrays of single-walled carbon nanotubes and sheets of graphene. Adv Mater 23(33):3821–3826. https://doi.org/10.1002/adma.201101955

Hong S, Lee J, Do K, Lee M, Kim JH, Lee S, Kim D-H (2017) Stretchable electrode based on laterally combed carbon nanotubes for wearable energy harvesting and storage devices. Adv Funct Mater 27(48):1704353. https://doi.org/10.1002/adfm.201704353

Hong YJ, Lee H, Kim J, Lee M, Choi HJ, Hyeon T, Kim D-H (2018) Multifunctional wearable system that integrates sweat-based sensing and vital-sign monitoring to estimate pre-/post-exercise glucose levels. Adv Funct Mater 28(47):1805754. https://doi.org/10.1002/adfm.201805754

Hotze EM, Phenrat T, Lowry GV (2010) Nanoparticle aggregation: challenges to understanding transport and reactivity in the environment. J Environ Qual 39(6):1909–1924. https://doi.org/10.2134/jeq2009.0462

Htay MT, Tani Y, Hashimoto Y, Ito K (2009) Synthesis of optical quality ZnO nanowires utilizing ultrasonic spray pyrolysis. J Mater Sci Mater Electron 20(1):341–345. https://doi.org/10.1007/s10854-008-9613-5

Hu N, Masuda Z, Yan C, Yamamoto G, Fukunaga H, Hashida T (2008) The electrical properties of polymer nanocomposites with carbon nanotube fillers. Nanotechnology 19(21):215701. https://doi.org/10.1088/0957-4484/19/21/215701

Huang K-C, Ehrman SH (2007) Synthesis of iron nanoparticles via chemical reduction with palladium ion seeds. Langmuir 23(3):1419–1426. https://doi.org/10.1021/la0618364

Huang C-T, Shen C-L, Tang C-F, Chang S-H (2008) A wearable yarn-based piezo-resistive sensor. Sens Actuators A Phys 141(2):396–403. https://doi.org/10.1016/j.sna.2007.10.069

Huang W-D, Cao H, Deb S, Chiao M, Chiao JC (2011) A flexible PH sensor based on the iridium oxide sensing film. Sens Actuators A Phys 169(1):1–11. https://doi.org/10.1016/j.sna.2011.05.016

Huang J-F, Zhong J, Chen G-P, Lin Z-T, Deng Y, Liu Y-L, Cao P-Y, Wang B, Wei Y, Wu T, Yuan J, Jiang G-B (2016) A hydrogel-based hybrid theranostic contact lens for fungal keratitis. ACS Nano 10(7):6464–6473. https://doi.org/10.1021/acsnano.6b00601

Huo F, Zheng Z, Zheng G, Giam LR, Zhang H, Mirkin CA (2008) Polymer Pen Lithography. Science (1979) 321(5896):1658–1660. https://doi.org/10.1126/science.1162193

Ihn J-B, Chang F-K (2004) Detection and monitoring of hidden fatigue crack growth using a built-in piezoelectric sensor/actuator network: I. Diagnostics. Smart Mater Struct 13(3):609. https://doi.org/10.1088/0964-1726/13/3/020

Jaiswal N, Tiwari I, Foster CW, Banks CE (2017) Highly sensitive amperometric sensing of nitrite utilizing bulk-modified MnO2 decorated graphene oxide nanocomposite screen-printed electrodes. Electrochim Acta 227:255–266. https://doi.org/10.1016/J.ELECTACTA.2017.01.007

Jalalvandi E, Hanton LR, Moratti SC (2017) Schiff-base based hydrogels as degradable platforms for hydrophobic drug delivery. Eur Polym J 90:13–24. https://doi.org/10.1016/j.eurpolymj.2017.03.003

Jamieson T, Bakhshi R, Petrova D, Pocock R, Imani M, Seifalian AM (2007) Biological applications of quantum dots. Biomaterials 28(31):4717–4732. https://doi.org/10.1016/J.BIOMATERIALS.2007.07.014

Jang B, Park M, Chae OB, Park S, Kim Y, Oh SM, Piao Y, Hyeon T (2012) Direct synthesis of self-assembled ferrite/carbon hybrid nanosheets for high performance lithium-ion battery anodes. J Am Chem Soc 134(36):15010–15015. https://doi.org/10.1021/ja305539r

Jiang H, Khang D-Y, Song J, Sun Y, Huang Y, Rogers JA (2007) Finite deformation mechanics in buckled thin films on compliant supports. Proc Natl Acad Sci 104(40):15607–15612. https://doi.org/10.1073/pnas.0702927104

Jiang Z, Fukuda K, Xu X, Park S, Inoue D, Jin H, Saito M, Osaka I, Takimiya K, Someya T (2018) Reverse-offset printed ultrathin ag mesh for robust conformal transparent electrodes for high-performance organic photovoltaics. Adv Mater 30(26):1707526. https://doi.org/10.1002/adma.201707526

Jiang A, Wang F, Xia D, Li M, Qiang L, Zhu Z, Wang P, Fan R, Lin K, Yang Y (2019) Aluminum nanoparticles manufactured using a ball-milling method with ammonium chloride as a grinding aid: achieving energy release at low temperature. New J Chem 43(4):1851–1856. https://doi.org/10.1039/C8NJ05356A

Jiang Z, Shaha R, McBride R, Jiang K, Tang M, Xu B, Goroncy AK, Frick C, Oakey J (2020) Crosslinker length dictates step-growth hydrogel network formation dynamics and allows rapid on-chip photoencapsulation. Biofabrication 12(3):35006. https://doi.org/10.1088/1758-5090/ab7ef4

Jinno H, Yokota T, Koizumi M, Yukita W, Saito M, Osaka I, Fukuda K, Someya T (2021) Self-powered ultraflexible photonic skin for continuous bio-signal detection via air-operation-stable polymer light-emitting diodes. Nat Commun 12(1):2234. https://doi.org/10.1038/s41467-021-22558-6

Johns JE, Alaboson JMP, Patwardhan S, Ryder CR, Schatz GC, Hersam MC (2013) Metal oxide nanoparticle growth on graphene via chemical activation with atomic oxygen. J Am Chem Soc 135(48):18121–18125. https://doi.org/10.1021/ja408248z

Joo H, Lee Y, Kim J, Yoo J-S, Yoo S, Kim S, Arya AK, Kim S, Choi SH, Lu N, Lee HS, Kim S, Lee S-T, Kim D-H (2024) Soft implantable drug delivery device integrated wirelessly with wearable devices to treat fatal seizures. Sci Adv 7(1):eabd4639. https://doi.org/10.1126/sciadv.abd4639

Joselevich E, Dai H, Hata K, Windle A (2008) Carbon nanotube synthesis and organization. Physics (College Park Md) 111:101–164. https://doi.org/10.1007/978-3-540-72865-8_4

Jung S, Hong S, Kim J, Lee S, Hyeon T, Lee M, Kim D-H (2015) Wearable fall detector using integrated sensors and energy devices. Sci Rep 5(1):17081. https://doi.org/10.1038/srep17081

Kamen G, Caldwell GE (1996) Physiology and interpretation of the electromyogram. J Clin Neurophysiol 13(5)

Kang D, Pikhitsa PV, Choi YW, Lee C, Shin SS, Piao L, Park B, Suh K-Y, Kim T, Choi M (2014) Ultrasensitive mechanical crack-based sensor inspired by the spider sensory system. Nature 516(7530):222–226. https://doi.org/10.1038/nature14002

Kang P, Wang MC, Nam S (2016) Bioelectronics with two-dimensional materials. Microelectron Eng 161:18–35. https://doi.org/10.1016/J.MEE.2016.04.003

Karaca N, Temel G, Karaca Balta D, Aydin M, Arsu N (2010) Preparation of hydrogels by photopolymerization of acrylates in the presence of type I and one-component type II photoinitiators. J Photochem Photobiol A Chem 209(1):1–6. https://doi.org/10.1016/j.jphotochem.2009.09.017

Kawarazaki I, Hayashi M, Takasu A (2020) Extraction of intrinsic cross-linking effects of a hard domains on segmental motion of B soft block for ABA triblock copolymer-based elastomers by utilizing photo cross-linking. Polymer (Guildf) 192:122343. https://doi.org/10.1016/j.polymer.2020.122343

Kayser LV, Lipomi DJ (2019) Stretchable conductive polymers and composites based on PEDOT and PEDOT:PSS. Adv Mater 31(10):1806133. https://doi.org/10.1002/adma.201806133

Khatib M, Zohar O, Saliba W, Srebnik S, Haick H (2020) Highly efficient and water-insensitive self-healing elastomer for wet and underwater electronics. Adv Funct Mater 30(22):1910196. https://doi.org/10.1002/adfm.201910196

Kim JH, Nam KW, Ma SB, Kim KB (2006) Fabrication and electrochemical properties of carbon nanotube film electrodes. Carbon N Y 44(10):1963–1968. https://doi.org/10.1016/J.CARBON.2006.02.002

Kim D-H, Song J, Choi WM, Kim H-S, Kim R-H, Liu Z, Huang YY, Hwang K-C, Zhang Y, Rogers JA (2008) Materials and noncoplanar mesh designs for integrated circuits with linear elastic responses to extreme mechanical deformations. Proc Natl Acad Sci 105(48):18675–18680. https://doi.org/10.1073/pnas.0807476105

Kim S, Wu J, Carlson A, Jin SH, Kovalsky A, Glass P, Liu Z, Ahmed N, Elgan SL, Chen W, Ferreira PM, Sitti M, Huang Y, Rogers JA (2010a) Microstructured elastomeric surfaces with reversible adhesion and examples of their use in deterministic assembly by transfer printing. Proc Natl Acad Sci 107(40):17095–17100. https://doi.org/10.1073/pnas.1005828107

Kim D-H, Viventi J, Amsden JJ, Xiao J, Vigeland L, Kim Y-S, Blanco JA, Panilaitis B, Frechette ES, Contreras D, Kaplan DL, Omenetto FG, Huang Y, Hwang K-C, Zakin MR, Litt B, Rogers JA (2010b) Dissolvable films of silk fibroin for ultrathin conformal bio-integrated electronics. Nat Mater 9(6):511–517. https://doi.org/10.1038/nmat2745

Kim T, McCall JG, Jung YH, Huang X, Siuda ER, Li Y, Song J, Song YM, Pao HA, Kim R-H, Lu C, Lee SD, Song I-S, Shin G, Al-Hasani R, Kim S, Tan MP, Huang Y, Omenetto FG, Rogers JA, Bruchas MR (2013) Injectable, cellular-scale optoelectronics with applications for wireless optogenetics. Science (1979) 340(6129):211–216. https://doi.org/10.1126/science.1232437

Kim SJ, Cho HR, Cho KW, Qiao S, Rhim JS, Soh M, Kim T, Choi MK, Choi C, Park I, Hwang NS, Hyeon T, Choi SH, Lu N, Kim D-H (2015) Multifunctional cell-culture platform for aligned cell sheet monitoring, transfer printing, and therapy. ACS Nano 9(3):2677–2688. https://doi.org/10.1021/nn5064634

Kim T, Park J, Sohn J, Cho D, Jeon S (2016) Bioinspired, highly stretchable, and conductive dry adhesives based on 1D–2D hybrid carbon nanocomposites for all-in-one ECG electrodes. ACS Nano 10(4):4770–4778. https://doi.org/10.1021/acsnano.6b01355

Kim T-H, Lee C-S, Kim S, Hur J, Lee S, Shin KW, Yoon Y-Z, Choi MK, Yang J, Kim D-H, Hyeon T, Park S, Hwang S (2017) Fully stretchable optoelectronic sensors based on colloidal quantum dots for sensing photoplethysmographic signals. ACS Nano 11(6):5992–6003. https://doi.org/10.1021/acsnano.7b01894

Kim, D.; Ahn, S.-K.; Yoon, J. Highly stretchable strain sensors comprising double network hydrogels fabricated by microfluidic devices. Adv Mater Technol 2019, 4 (7), 1800739. https://doi.org/10.1002/admt.201800739

Kim N, Lienemann S, Petsagkourakis I, Alemu Mengistie D, Kee S, Ederth T, Gueskine V, Leclère P, Lazzaroni R, Crispin X, Tybrandt K (2020a) Elastic conducting polymer composites in thermoelectric modules. Nat Commun 11(1):1424. https://doi.org/10.1038/s41467-020-15135-w

Kim M, Lee GJ, Choi C, Kim MS, Lee M, Liu S, Cho KW, Kim HM, Cho H, Choi MK, Lu N, Song YM, Kim D-H (2020b) An aquatic-vision-inspired camera based on a monocentric lens and a silicon nanorod photodiode array. Nat Electron 3(9):546–553. https://doi.org/10.1038/s41928-020-0429-5

Kim D, Yokota T, Suzuki T, Lee S, Woo T, Yukita W, Koizumi M, Tachibana Y, Yawo H, Onodera H, Sekino M, Someya T (2020c) Ultraflexible organic light-emitting diodes for optogenetic nerve stimulation. Proc Natl Acad Sci 117(35):21138–21146. https://doi.org/10.1073/pnas.2007395117

Kim J, Son D, Lee M, Song C, Song J-K, Koo JH, Lee DJ, Shim HJ, Kim JH, Lee M, Hyeon T, Kim D-H (2024) A wearable multiplexed silicon nonvolatile memory array using nanocrystal charge confinement. Sci Adv 2(1):e1501101. https://doi.org/10.1126/sciadv.1501101

Kimura M, Fukumoto K, Watanabe J, Ishihara K (2004) Hydrogen-bonding-driven spontaneous gelation of water-soluble phospholipid polymers in aqueous medium. J Biomater Sci Polym Ed 15(5):631–644. https://doi.org/10.1163/156856204323046898

Ko HC, Baca AJ, Rogers JA (2006) Bulk quantities of single-crystal silicon micro-/nanoribbons generated from bulk wafers. Nano Lett 6(10):2318–2324. https://doi.org/10.1021/nl061846p

Kolb HC, Finn MG, Sharpless KB (2001) Click chemistry: diverse chemical function from a few good reactions. Angew Chem Int Ed 40(11):2004–2021. https://doi.org/10.1002/1521-3773(20010601)40:11<2004::AID-ANIE2004>3.0.CO;2-5

Kontou E, Spathis G, Niaounakis M, Kefalas V (1990) Physical and chemical cross-linking effects in polyurethane elastomers. Colloid Polym Sci 268(7):636–644. https://doi.org/10.1007/BF01410405

Konuray O, Fernández-Francos X, De la Flor S, Ramis X, Serra À (2020) The use of click-type reactions in the preparation of thermosets. Polymers. https://doi.org/10.3390/polym12051084

Koo JH, Kim DC, Shim HJ, Kim T-H, Kim D-H (2018a) Flexible and stretchable smart display: materials, fabrication, device design, and system integration. Adv Funct Mater 28(35):1801834. https://doi.org/10.1002/adfm.201801834

Koo J, MacEwan MR, Kang S-K, Won SM, Stephen M, Gamble P, Xie Z, Yan Y, Chen Y-Y, Shin J, Birenbaum N, Chung S, Kim SB, Khalifeh J, Harburg DV, Bean K, Paskett M, Kim J, Zohny ZS, Lee SM, Zhang R, Luo K, Ji B, Banks A, Lee HM, Huang Y, Ray WZ, Rogers JA (2018b) Wireless bioresorbable electronic system enables sustained nonpharmacological neuroregenerative therapy. Nat Med 24(12):1830–1836. https://doi.org/10.1038/s41591-018-0196-2

Kouwer PHJ, Koepf M, Le Sage VAA, Jaspers M, van Buul AM, Eksteen-Akeroyd ZH, Woltinge T, Schwartz E, Kitto HJ, Hoogenboom R, Picken SJ, Nolte RJM, Mendes E, Rowan AE (2013) Responsive biomimetic networks from polyisocyanopeptide hydrogels. Nature 493(7434):651–655. https://doi.org/10.1038/nature11839

Kreuter J (2007) Nanoparticles—a historical perspective. Int J Pharm 331(1):1–10. https://doi.org/10.1016/J.IJPHARM.2006.10.021

Kurian AS, Souri H, Mohan VB, Bhattacharyya D (2020) Highly stretchable strain sensors based on polypyrrole-silicone rubber composites for human motion detection. Sens Actuators A Phys 312:112131. https://doi.org/10.1016/J.SNA.2020.112131

Kwon J, Suh YD, Lee J, Lee P, Han S, Hong S, Yeo J, Lee H, Ko SH (2018) Recent progress in silver nanowire based flexible/wearable optoelectronics. J Mater Chem C Mater 6(28):7445–7461. https://doi.org/10.1039/C8TC01024B

Kwon KY, Shin YJ, Shin JH, Jeong C, Jung YH, Park B, Kim T (2019) Stretchable, patch-type calorie-expenditure measurement device based on pop-up shaped nanoscale crack-based sensor. Adv Healthc Mater 8(19):1801593. https://doi.org/10.1002/adhm.201801593

Lee N, Cho HR, Oh MH, Lee SH, Kim K, Kim BH, Shin K, Ahn T-Y, Choi JW, Kim Y-W, Choi SH, Hyeon T (2012a) Multifunctional Fe3O4/TaOx core/shell nanoparticles for simultaneous magnetic resonance imaging and X-Ray computed tomography. J Am Chem Soc 134(25):10309–10312. https://doi.org/10.1021/ja3016582

Lee J, Lee P, Lee H, Lee D, Lee SS, Ko SH (2012b) Very long Ag nanowire synthesis and its application in a highly transparent, conductive and flexible metal electrode touch panel. Nanoscale 4(20):6408–6414. https://doi.org/10.1039/C2NR31254A

Lee Y-H, Zhang X-Q, Zhang W, Chang M-T, Lin C-T, Chang K-D, Yu Y-C, Wang JT-W, Chang C-S, Li L-J, Lin T-W (2012c) Synthesis of large-area MoS2 atomic layers with chemical vapor deposition. Adv Mater 24(17):2320–2325. https://doi.org/10.1002/adma.201104798

Lee P, Lee J, Lee H, Yeo J, Hong S, Nam KH, Lee D, Lee SS, Ko SH (2012d) Highly stretchable and highly conductive metal electrode by very long metal nanowire percolation network. Adv Mater 24(25):3326–3332. https://doi.org/10.1002/adma.201200359

Lee SH, Yu S-H, Lee JE, Jin A, Lee DJ, Lee N, Jo H, Shin K, Ahn T-Y, Kim Y-W, Choe H, Sung Y-E, Hyeon T (2013) Self-assembled Fe3O4 nanoparticle clusters as high-performance anodes for lithium ion batteries via geometric confinement. Nano Lett 13(9):4249–4256. https://doi.org/10.1021/nl401952h

Lee P, Ham J, Lee J, Hong S, Han S, Suh YD, Lee SE, Yeo J, Lee SS, Lee D, Ko SH (2014) Highly stretchable or transparent conductor fabrication by a hierarchical multiscale hybrid nanocomposite. Adv Funct Mater 24(36):5671–5678. https://doi.org/10.1002/adfm.201400972

Lee H, Lee Y, Song C, Cho HR, Ghaffari R, Choi TK, Kim KH, Lee YB, Ling D, Lee H, Yu SJ, Choi SH, Hyeon T, Kim D-H (2015) An endoscope with integrated transparent bioelectronics and theranostic nanoparticles for colon cancer treatment. Nat Commun 6(1):10059. https://doi.org/10.1038/ncomms10059

Lee Y-Y, Kang H-Y, Gwon SH, Choi GM, Lim S-M, Sun J-Y, Joo Y-C (2016a) A strain-insensitive stretchable electronic conductor: PEDOT:PSS/acrylamide organogels. Adv Mater 28(8):1636–1643. https://doi.org/10.1002/adma.201504606

Lee T, Choi YW, Lee G, Pikhitsa PV, Kang D, Kim SM, Choi M (2016b) Transparent ITO mechanical crack-based pressure and strain sensor. J Mater Chem C Mater 4(42):9947–9953. https://doi.org/10.1039/C6TC03329F

Lee K-R, Chang W-D, Kim S, Im C-H (2017a) Real-Time "Eye-Writing" Recognition Using Electrooculogram. IEEE Trans Neural Syst Rehabil Eng 25(1):37–48. https://doi.org/10.1109/TNSRE.2016.2542524

Lee J, Yoo B, Lee H, Cha GD, Lee H-S, Cho Y, Kim SY, Seo H, Lee W, Son D, Kang M, Kim HM, Park YI, Hyeon T, Kim D-H (2017b) Ultra-wideband multi-dye-sensitized upconverting nanoparticles for information security application. Adv Mater 29(1):1603169. https://doi.org/10.1002/adma.201603169

Lee H, Hong YJ, Baik S, Hyeon T, Kim D-H (2018) Enzyme-based glucose sensor: from invasive to wearable device. Adv Healthc Mater 7(8):1701150. https://doi.org/10.1002/adhm.201701150

Lee M, Shim HJ, Choi C, Kim D-H (2019) Soft high-resolution neural interfacing probes: materials and design approaches. Nano Lett 19(5):2741–2749. https://doi.org/10.1021/acs.nanolett.8b04895

Lee D, Park K, Seo J (2020) Recent advances in anti-inflammatory strategies for implantable biosensors and medical implants. Biochip J 14(1):48–62. https://doi.org/10.1007/s13206-020-4105-7

Lee W, Yun H, Song J-K, Sunwoo S-H, Kim D-H (2021a) Nanoscale materials and deformable device designs for bioinspired and biointegrated electronics. Acc Mater Res 2(4):266–281. https://doi.org/10.1021/accountsmr.1c00020

Lee Y, Kang T, Cho HR, Lee GJ, Park OK, Kim S, Lee B, Kim HM, Cha GD, Shin Y, Lee W, Kim M, Kim H, Song YM, Choi SH, Hyeon T, Kim D-H (2021b) Localized delivery of theranostic nanoparticles and high-energy photons using microneedles-on-bioelectronics. Adv Mater 33(24):2100425. https://doi.org/10.1002/adma.202100425

Lembke D, Bertolazzi S, Kis A (2015) Single-layer MoS2 electronics. Acc Chem Res 48(1):100–110. https://doi.org/10.1021/ar500274q

Li J, Ma PC, Chow WS, To CK, Tang BZ, Kim J-K (2007) Correlations between percolation threshold, dispersion state, and aspect ratio of carbon nanotubes. Adv Funct Mater 17(16):3207–3215. https://doi.org/10.1002/adfm.200700065

Li Y, Tanigawa R, Okuzaki H (2014) Soft and flexible PEDOT/PSS films for applications to soft actuators. Smart Mater Struct 23(7):74010. https://doi.org/10.1088/0964-1726/23/7/074010

Li L, Zeng Z, Chen Z, Gao R, Pan L, Deng J, Ye X, Zhang J, Zhang S, Mei C, Yu J, Feng Y, Wang Q, Yu A-Y, Yang M, Huang J (2020) Microenvironment-triggered degradable hydrogel for imaging diagnosis and combined treatment of intraocular choroidal melanoma. ACS Nano 14(11):15403–15416. https://doi.org/10.1021/acsnano.0c06000

Liang L, Gao C, Chen G, Guo C-Y (2016) Large-area, stretchable, super flexible and mechanically stable thermoelectric films of polymer/carbon nanotube composites. J Mater Chem C Mater 4(3):526–532. https://doi.org/10.1039/C5TC03768A

Lieber CM (2003) Nanoscale science and technology: building a big future from small things. MRS Bull 28(7):486–491. https://doi.org/10.1557/mrs2003.144

Lim SY, Shen W, Gao Z (2015) Carbon quantum dots and their applications. Chem Soc Rev 44(1):362–381. https://doi.org/10.1039/C4CS00269E

Lim G-H, Ahn K, Bok S, Nam J, Lim B (2017) Curving silver nanowires using liquid droplets for highly stretchable and durable percolation networks. Nanoscale 9(26):8938–8944. https://doi.org/10.1039/C7NR02615C

Lim C, Shin Y, Jung J, Kim JH, Lee S, Kim D-H (2018) Stretchable conductive nanocomposite based on alginate hydrogel and silver nanowires for wearable electronics. APL Mater 7(3):31502. https://doi.org/10.1063/1.5063657

Lim C, Hong YJ, Jung J, Shin Y, Sunwoo S-H, Baik S, Park OK, Choi SH, Hyeon T, Kim JH, Lee S, Kim D-H (2024) Tissue-like skin-device interface for wearable bioelectronics by using ultrasoft, mass-permeable, and low-impedance hydrogels. Sci Adv 7(19):eabd3716. https://doi.org/10.1126/sciadv.abd3716

Liu Q, Yan Z, Henderson NL, Bauer JC, Goodman DW, Batteas JD, Schaak RE (2009) Synthesis of CuPt nanorod catalysts with tunable lengths. J Am Chem Soc 131(16):5720–5721. https://doi.org/10.1021/ja810151r

Liu J, Xie C, Dai X, Jin L, Zhou W, Lieber CM (2013) Multifunctional three-dimensional macroporous nanoelectronic networks for smart materials. Proc Natl Acad Sci USA 110(17):6694–6699. https://doi.org/10.1073/pnas.1305209110

Liu M, Wang J, He M, Wang L, Li F, Jiang L, Song Y (2014a) Inkjet printing controllable footprint lines by regulating the dynamic wettability of coalescing ink droplets. ACS Appl Mater Interfaces 6(16):13344–13348. https://doi.org/10.1021/am5042548

Liu Y, Chen J, Li G, Liu X, Liao X, Yang Q (2014b) Effect of physical and chemical crosslinking structure on fatigue behavior of styrene butadiene elastomer. J Appl Polym Sci 131(20) https://doi.org/10.1002/app.40917

Liu J, Fu T-M, Cheng Z, Hong G, Zhou T, Jin L, Duvvuri M, Jiang Z, Kruskal P, Xie C, Suo Z, Fang Y, Lieber CM (2015) Syringe-injectable electronics. Nat Nanotechnol 10(7):629–636. https://doi.org/10.1038/nnano.2015.115

Liu Z, Peng W, Zare Y, Hui D, Rhee KY (2018) Predicting the electrical conductivity in polymer carbon nanotube nanocomposites based on the volume fractions and resistances of the nanoparticle, interphase, and tunneling regions in conductive networks. RSC Adv 8(34):19001–19010. https://doi.org/10.1039/C8RA00811F

Liu X, Zhang Q, Gao G (2020) Solvent-resistant and nonswellable hydrogel conductor toward mechanical perception in diverse liquid media. ACS Nano 14(10):13709–13717. https://doi.org/10.1021/acsnano.0c05932

Liu Y, Norton JJS, Qazi R, Zou Z, Ammann KR, Liu H, Yan L, Tran PL, Jang K-I, Lee JW, Zhang D, Kilian KA, Jung SH, Bretl T, Xiao J, Slepian MJ, Huang Y, Jeong J-W, Rogers JA (2024) Epidermal mechano-acoustic sensing electronics for cardiovascular diagnostics and human-machine interfaces. Sci Adv 2(11):e1601185. https://doi.org/10.1126/sciadv.1601185

Lohse SE, Murphy CJ (2013) The quest for shape control: a history of gold nanorod synthesis. Chem Mater 25(8):1250–1261. https://doi.org/10.1021/cm303708p

Lotya M, Hernandez Y, King PJ, Smith RJ, Nicolosi V, Karlsson LS, Blighe FM, De S, Wang Z, McGovern IT, Duesberg GS, Coleman JN (2009) Liquid phase production of graphene by exfoliation of graphite in surfactant/water solutions. J Am Chem Soc 131(10):3611–3620. https://doi.org/10.1021/ja807449u

Lu B, Yuk H, Lin S, Jian N, Qu K, Xu J, Zhao X (2019) Pure PEDOT:PSS hydrogels. Nat Commun 10(1):1043. https://doi.org/10.1038/s41467-019-09003-5

Luo R, Li H, Du B, Zhou S, Zhu Y (2020) A simple strategy for high stretchable, flexible and conductive polymer films based on PEDOT:PSS-PDMS blends. Org Electron 76:105451. https://doi.org/10.1016/j.orgel.2019.105451

Luu QN, Doorn JM, Berry MT, Jiang C, Lin C, May PS (2011) Preparation and optical properties of silver nanowires and silver-nanowire thin films. J Colloid Interface Sci 356(1):151–158. https://doi.org/10.1016/J.JCIS.2010.12.077

Ma R, Wu C, Wang ZL, Tsukruk VV (2018) Pop-up conducting large-area biographene kirigami. ACS Nano 12(10):9714–9720. https://doi.org/10.1021/acsnano.8b04507

Majumdar S, Devi PS (2010) Synthesis of SnO2 nanoparticles using ultrasonication. AIP Conf Proc 1276(1):1–7. https://doi.org/10.1063/1.3504298

Mano N, Yoo JE, Tarver J, Loo Y-L, Heller A (2007) An electron-conducting cross-linked polyaniline-based redox hydrogel, formed in one step at PH 7.2, wires glucose oxidase. J Am Chem Soc 129(22):7006–7007. https://doi.org/10.1021/ja071946c

Mantooth SM, Munoz-Robles BG, Webber MJ (2019) Dynamic hydrogels from host–guest supramolecular interactions. Macromol Biosci 19(1):1800281. https://doi.org/10.1002/mabi.201800281

Mao Y, Xie J, Liu H, Hu W (2021) Hierarchical core-shell Ag@Ni(OH)2@PPy nanowire electrode for ultrahigh energy density asymmetric supercapacitor. Chem Eng J 405:126984. https://doi.org/10.1016/j.cej.2020.126984

Martín N (2006) New challenges in fullerene chemistry. Chem Commun 20:2093–2104. https://doi.org/10.1039/B601582B

Matsuhisa N, Inoue D, Zalar P, Jin H, Matsuba Y, Itoh A, Yokota T, Hashizume D, Someya T (2017) Printable elastic conductors by in situ formation of silver nanoparticles from silver flakes. Nat Mater 16(8):834–840. https://doi.org/10.1038/nmat4904

McIntosh RB, Casada ME (2008) Fringing field capacitance sensor for measuring the moisture content of agricultural commodities. IEEE Sensors J 8(3):240–247. https://doi.org/10.1109/JSEN.2007.913140

Meier R, Birkenstock C, Palumbiny CM, Müller-Buschbaum P (2012) Efficiency-improved organic solar cells based on plasticizer assisted soft embossed PEDOT:PSS layers. Phys Chem Chem Phys 14(43):15088–15098. https://doi.org/10.1039/C2CP42918G

Mihajlovic M, Staropoli M, Appavou M-S, Wyss HM, Pyckhout-Hintzen W, Sijbesma RP (2017) Tough supramolecular hydrogel based on strong hydrophobic interactions in a multiblock segmented copolymer. Macromolecules 50(8):3333–3346. https://doi.org/10.1021/acs.macromol.7b00319

Miller JS, Shen CJ, Legant WR, Baranski JD, Blakely BL, Chen CS (2010) Bioactive hydrogels made from step-growth derived PEG–peptide macromers. Biomaterials 31(13):3736–3743. https://doi.org/10.1016/j.biomaterials.2010.01.058

Min S-H, Lee G-Y, Ahn S-H (2019) Direct printing of highly sensitive, stretchable, and durable strain sensor based on silver nanoparticles/multi-walled carbon nanotubes composites. Compos Part B Eng 161:395–401. https://doi.org/10.1016/j.compositesb.2018.12.107

Mironi-Harpaz I, Wang DY, Venkatraman S, Seliktar D (2012) Photopolymerization of cell-encapsulating hydrogels: crosslinking efficiency versus cytotoxicity. Acta Biomater 8(5):1838–1848. https://doi.org/10.1016/j.actbio.2011.12.034

Miwa Y, Kurachi J, Kohbara Y, Kutsumizu S (2018) Dynamic ionic crosslinks enable high strength and ultrastretchability in a single elastomer. Commun Chem 1(1):5. https://doi.org/10.1038/s42004-017-0004-9

Molberg M, Crespy D, Rupper P, Nüesch F, Månson J-AE, Löwe C, Opris DM (2010) High breakdown field dielectric elastomer actuators using encapsulated polyaniline as high dielectric constant filler. Adv Funct Mater 20(19):3280–3291. https://doi.org/10.1002/adfm.201000486

Monemian S, Jang K-S, Ghassemi H, Korley LTJ (2014) Probing the interplay of ultraviolet cross-linking and noncovalent Interactions in supramolecular elastomers. Macromolecules 47(16):5633–5642. https://doi.org/10.1021/ma501183a

Moon GD, Ko S, Min Y, Zeng J, Xia Y, Jeong U (2011) Chemical transformations of nanostructured materials. Nano Today 6(2):186–203. https://doi.org/10.1016/J.NANTOD.2011.02.006

Moučka R, Lopatin AV, Kazantseva NE, Vilčáková J, Sáha P (2007) Enhancement of magnetic losses in hybrid polymer composites with MnZn-ferrite and conductive fillers. J Mater Sci 42(22):9480–9490. https://doi.org/10.1007/s10853-007-2081-0

Moura MJ, Faneca H, Lima MP, Gil MH, Figueiredo MM (2011) In situ forming chitosan hydrogels prepared via ionic/covalent co-cross-linking. Biomacromolecules 12(9):3275–3284. https://doi.org/10.1021/bm200731x

Muñoz R, Gómez-Aleixandre C (2013) Review of CVD synthesis of graphene. Chem Vap Depos 19(10-11–12):297–322. https://doi.org/10.1002/cvde.201300051

Nadeem S, Khan AU, Hussain ST (2017) Model based study of SWCNT and MWCNT thermal conductivities effect on the heat transfer due to the oscillating wall conditions. Int J Hydrog Energy 42(48):28945–28957. https://doi.org/10.1016/J.IJHYDENE.2017.09.114

Nair DP, Podgórski M, Chatani S, Gong T, Xi W, Fenoli CR, Bowman CN (2014) The Thiol-Michael addition click reaction: a powerful and widely used tool in materials chemistry. Chem Mater 26(1):724–744. https://doi.org/10.1021/cm402180t

Nam KT, Kim D-W, Yoo PJ, Chiang C-Y, Meethong N, Hammond PT, Chiang Y-M, Belcher AM (2006) Virus-enabled synthesis and assembly of nanowires for lithium ion battery electrodes. Science (1979) 312(5775):885–888. https://doi.org/10.1126/science.1122716

Nam J, Lim H-K, Kim NH, Park JK, Kang ES, Kim Y-T, Heo C, Lee O-S, Kim S-G, Yun WS, Suh M, Kim YH (2020) Supramolecular peptide hydrogel-based soft neural interface augments brain signals through a three-dimensional electrical network. ACS Nano 14(1):664–675. https://doi.org/10.1021/acsnano.9b07396

Nath M, Govindaraj A, Rao CNR (2001) Simple synthesis of MoS2 and WS2 nanotubes. Adv Mater 13(4):283–286. https://doi.org/10.1002/1521-4095(200102)13:4<283::AID-ADMA283>3.0.CO;2-H

Nele V, Wojciechowski JP, Armstrong JPK, Stevens MM (2020) Tailoring gelation mechanisms for advanced hydrogel applications. Adv Funct Mater 30(42):2002759. https://doi.org/10.1002/adfm.202002759

Nerger BA, Brun P-T, Nelson CM (2024) Marangoni flows drive the alignment of fibrillar cell-laden hydrogels. Sci Adv 6(24):eaaz7748. https://doi.org/10.1126/sciadv.aaz7748

Ng TH, Liao WH (2005) Sensitivity analysis and energy harvesting for a self-powered piezoelectric sensor. J Intell Mater Syst Struct 16(10):785–797. https://doi.org/10.1177/1045389X05053151

Nguyen V, Yang R (2013) Effect of humidity and pressure on the triboelectric nanogenerator. Nano Energy 2(5):604–608. https://doi.org/10.1016/j.nanoen.2013.07.012

Ning C-Z, Dou L, Yang P (2017) Bandgap engineering in semiconductor alloy nanomaterials with widely tunable compositions. Nat Rev Mater 2(12):17070. https://doi.org/10.1038/natrevmats.2017.70

Niu Z, Zhang L, Liu L, Zhu B, Dong H, Chen X (2013) All-solid-state flexible ultrathin micro-supercapacitors based on graphene. Adv Mater 25(29):4035–4042. https://doi.org/10.1002/adma.201301332

O'Halloran A, O'Malley F, McHugh P (2008) A review on dielectric elastomer actuators, technology, applications, and challenges. J Appl Phys 104(7):71101. https://doi.org/10.1063/1.2981642

Oh KS, Han SK, Choi YW, Lee JH, Lee JY, Yuk SH (2004) Hydrogen-bonded polymer gel and its application as a temperature-sensitive drug delivery system. Biomaterials 25(12):2393–2398. https://doi.org/10.1016/j.biomaterials.2003.09.008

Okuzaki H, Takagi S, Hishiki F, Tanigawa R (2014) Ionic liquid/polyurethane/PEDOT:PSS composites for electro-active polymer actuators. Sens Actuators B Chem 194:59–63. https://doi.org/10.1016/j.snb.2013.12.059

Ooi HW, Kocken JMM, Morgan FLC, Malheiro A, Zoetebier B, Karperien M, Wieringa PA, Dijkstra PJ, Moroni L, Baker MB (2020) Multivalency enables dynamic supramolecular host–guest hydrogel formation. Biomacromolecules 21(6):2208–2217. https://doi.org/10.1021/acs.biomac.0c00148

Ozay H, Ilgin P, Ozay O (2021) Novel hydrogels based on crosslinked chitosan with formylphosphazene using Schiff-base reaction. Int J Polym Mater Polym Biomater 70(4):246–255. https://doi.org/10.1080/00914037.2019.1706514

Park G, Chung H-J, Kim K, Lim SA, Kim J, Kim Y-S, Liu Y, Yeo W-H, Kim R-H, Kim SS, Kim J-S, Jung YH, Kim T, Yee C, Rogers JA, Lee K-M (2014) Immunologic and tissue biocompatibility of flexible/stretchable electronics and optoelectronics. Adv Healthc Mater 3(4):515–525. https://doi.org/10.1002/adhm.201300220

Park M, Do K, Kim J, Son D, Koo JH, Park J, Song J-K, Kim JH, Lee M, Hyeon T, Kim D-H (2015a) Oxide nanomembrane hybrids with enhanced mechano- and thermo-sensitivity for semitransparent epidermal electronics. Adv Healthc Mater 4(7):992–997. https://doi.org/10.1002/adhm.201500097

Park SI, Brenner DS, Shin G, Morgan CD, Copits BA, Chung HU, Pullen MY, Noh KN, Davidson S, Oh SJ, Yoon J, Jang K-I, Samineni VK, Norman M, Grajales-Reyes JG, Vogt SK, Sundaram SS, Wilson KM, Ha JS, Xu R, Pan T, Kim T, Huang Y, Montana MC, Golden JP, Bruchas MR,

Gereau RW, Rogers JA (2015b) Soft, stretchable, fully implantable miniaturized optoelectronic systems for wireless optogenetics. Nat Biotechnol 33(12):1280–1286. https://doi.org/10.1038/nbt.3415

Park J, Choi S, Janardhan AH, Lee S-Y, Raut S, Soares J, Shin K, Yang S, Lee C, Kang K-W, Cho HR, Kim SJ, Seo P, Hyun W, Jung S, Lee H-J, Lee N, Choi SH, Sacks M, Lu N, Josephson ME, Hyeon T, Kim D-H, Hwang HJ (2016) Electromechanical cardioplasty using a wrapped elasto-conductive epicardial mesh. Sci Transl Med 8(344):344ra86. https://doi.org/10.1126/scitranslmed.aad8568

Park S-W, Kim J, Kang M, Lee W, Park BS, Kim H, Choi S-Y, Yang S, Ahn J-H, Yang S (2018) Epidural electrotherapy for epilepsy. Small 14(30):1801732. https://doi.org/10.1002/smll.201801732

Park S, Park H-H, Sun K, Gwon Y, Seong M, Kim S, Park T-E, Hyun H, Choung Y-H, Kim J, Jeong HE (2019) Hydrogel nanospike patch as a flexible anti-pathogenic scaffold for regulating stem cell behavior. ACS Nano 13(10):11181–11193. https://doi.org/10.1021/acsnano.9b04109

Parvez K, Wu Z-S, Li R, Liu X, Graf R, Feng X, Müllen K (2014) Exfoliation of graphite into graphene in aqueous solutions of inorganic salts. J Am Chem Soc 136(16):6083–6091. https://doi.org/10.1021/ja5017156

Patolsky F, Lieber CM (2005) Nanowire nanosensors. Mater Today 8(4):20–28. https://doi.org/10.1016/S1369-7021(05)00791-1

Patolsky F, Zheng G, Lieber CM (2006) Nanowire-based biosensors. Anal Chem 78(13):4260–4269. https://doi.org/10.1021/ac069419j

Patolsky F, Timko BP, Zheng G, Lieber CM (2007) Nanowire-based Nanoelectronic devices in the life sciences. MRS Bull 32(2):142–149. https://doi.org/10.1557/mrs2007.47

Pekař M (2015) Hydrogels with micellar hydrophobic (nano)domains. Front Mater 1:35

Peng K-Q, Yan Y-J, Gao S-P, Zhu J (2002) Synthesis of large-area silicon nanowire arrays via self-assembling nanoelectrochemistry. Adv Mater 14(16):1164–1167. https://doi.org/10.1002/1521-4095(20020816)14:16<1164::AID-ADMA1164>3.0.CO;2-E

Pereira LFC, Rocha CG, Latgé A, Ferreira MS (2010) A computationally efficient method for calculating the maximum conductance of disordered networks: application to one-dimensional conductors. J Appl Phys 108(10):103720. https://doi.org/10.1063/1.3514007

Perkgoz NK, Bay M (2016) Investigation of Single-Wall MoS2 monolayer flakes grown by chemical vapor deposition. Nanomicro Lett 8(1):70–79. https://doi.org/10.1007/s40820-015-0064-2

Petrović ZS, Ilavský M, Dušek K, Vidaković M, Javni I, Banjanin B (1991) The effect of cross-linking on properties of polyurethane elastomers. J Appl Polym Sci 42(2):391–398. https://doi.org/10.1002/app.1991.070420211

Piner RD, Zhu J, Xu F, Hong S, Mirkin CA (1999) "Dip-Pen" nanolithography. Science (1979) 283(5402):661–663. https://doi.org/10.1126/science.283.5402.661

Plante J-S, Dubowsky S (2007) On the performance mechanisms of dielectric elastomer actuators. Sens Actuators A Phys 137(1):96–109. https://doi.org/10.1016/j.sna.2007.01.017

Popovic DM, Chai JS, Zekic AA, Trtica M, Stasic J, Sarvan MZ (2014) The influence of applying the additional continuous laser on the synthesis of silicon-based nanoparticles by picosecond laser ablation in liquid. Laser Phys Lett 11(11):116101. https://doi.org/10.1088/1612-2011/11/11/116101

Qin Q, Guo Y (2012) Preparation and characterization of nano-polyaniline film on ITO conductive glass by electrochemical polymerization. J Nanomater 2012(1):519674. https://doi.org/10.1155/2012/519674

Rackauskas S, Nasibulin AG, Jiang H, Tian Y, Kleshch VI, Sainio J, Obraztsova ED, Bokova SN, Obraztsov AN, Kauppinen EI (2009) A novel method for metal oxide nanowire synthesis. Nanotechnology 20(16):165603. https://doi.org/10.1088/0957-4484/20/16/165603

Rai A, Moore AL (2017) Enhanced thermal conduction and influence of interfacial resistance within flexible high aspect ratio copper nanowire/polymer composites. Compos Sci Technol 144:70–78. https://doi.org/10.1016/J.COMPSCITECH.2017.03.020

Rajala S, Siponkoski T, Sarlin E, Mettänen M, Vuoriluoto M, Pammo A, Juuti J, Rojas OJ, Franssila S, Tuukkanen S (2016) Cellulose nanofibril film as a piezoelectric sensor material. ACS Appl Mater Interfaces 8(24):15607–15614. https://doi.org/10.1021/acsami.6b03597

Ritala M, Leskelä M, Dekker J-P, Mutsaers C, Soininen PJ, Skarp J (1999) Perfectly conformal TiN and Al2O3 films deposited by atomic layer deposition. Chem Vap Depos 5(1):7–9. https://doi.org/10.1002/(SICI)1521-3862(199901)5:1<7::AID-CVDE7>3.0.CO;2-J

Rizzo F, Kehr NS (2021) Recent advances in injectable hydrogels for controlled and local drug delivery. Adv Healthc Mater 10(1):2001341. https://doi.org/10.1002/adhm.202001341

Rogers JA, Lagally MG, Nuzzo RG (2011) Synthesis, assembly and applications of semiconductor nanomembranes. Nature 477(7362):45–53. https://doi.org/10.1038/nature10381

Royston E, Ghosh A, Kofinas P, Harris MT, Culver JN (2008) Self-assembly of virus-structured high surface area nanomaterials and their application as battery electrodes. Langmuir 24(3):906–912. https://doi.org/10.1021/la7016424

Saari LA, Seitz WR (1982) PH sensor based on immobilized Fluoresceinamine. Anal Chem 54(4):821–823. https://doi.org/10.1021/ac00241a052

Salaita K, Wang Y, Mirkin CA (2007) Applications of dip-pen nanolithography. Nat Nanotechnol 2(3):145–155. https://doi.org/10.1038/nnano.2007.39

Salata OV (2004) Applications of nanoparticles in biology and medicine. J Nanobiotechnol 2(1):3. https://doi.org/10.1186/1477-3155-2-3

Salleh KM, Zakaria S, Sajab MS, Gan S, Chia CH, Jaafar SNS, Amran UA (2018) Chemically crosslinked hydrogel and its driving force towards superabsorbent behaviour. Int J Biol Macromol 118:1422–1430. https://doi.org/10.1016/j.ijbiomac.2018.06.159

Sandler JKW, Kirk JE, Kinloch IA, Shaffer MSP, Windle AH (2003) Ultra-low electrical percolation threshold in carbon-nanotube-epoxy composites. Polymer (Guildf) 44(19):5893–5899. https://doi.org/10.1016/S0032-3861(03)00539-1

Sang M, Shin J, Kim K, Yu KJ (2019) Electronic and thermal properties of graphene and recent advances in graphene based electronics applications. Nanomaterials. https://doi.org/10.3390/nano9030374

Saritha Chandran A, Narayanankutty SK (2008) An elastomeric conducting composite based on polyaniline coated nylon fiber and chloroprene rubber. Eur Polym J 44(7):2418–2429. https://doi.org/10.1016/J.EURPOLYMJ.2008.05.012

Sau TK, Murphy CJ (2004) Seeded high yield synthesis of short Au nanorods in aqueous solution. Langmuir 20(15):6414–6420. https://doi.org/10.1021/la049463z

Savagatrup S, Chan E, Renteria-Garcia SM, Printz AD, Zaretski AV, O'Connor TF, Rodriquez D, Valle E, Lipomi DJ (2015) Plasticization of PEDOT:PSS by common additives for mechanically robust organic solar cells and wearable sensors. Adv Funct Mater 25(3):427–436. https://doi.org/10.1002/adfm.201401758

Schacht EH (2004) Polymer chemistry and hydrogel systems. J Phys Conf Ser 3(1):22. https://doi.org/10.1088/1742-6596/3/1/004

Sekine S, Ido Y, Miyake T, Nagamine K, Nishizawa M (2010) Conducting polymer electrodes printed on hydrogel. J Am Chem Soc 132(38):13174–13175. https://doi.org/10.1021/ja1062357

Seo H, Han SI, Song K-I, Seong D, Lee K, Kim SH, Park T, Koo JH, Shin M, Baac HW, Park OK, Oh SJ, Han H-S, Jeon H, Kim Y-C, Kim D-H, Hyeon T, Son D (2021) Durable and fatigue-resistant soft peripheral neuroprosthetics for in vivo bidirectional signaling. Adv Mater 33(20):2007346. https://doi.org/10.1002/adma.202007346

Seyedin MZ, Razal JM, Innis PC, Wallace GG (2014) Strain-responsive polyurethane/PEDOT:PSS elastomeric composite fibers with high electrical conductivity. Adv Funct Mater 24(20):2957–2966. https://doi.org/10.1002/adfm.201303905

Shi W, Lu W, Jiang L (2009) The fabrication of photosensitive self-assembly au nanoparticles embedded in silica nanofibers by electrospinning. J Colloid Interface Sci 340(2):291–297. https://doi.org/10.1016/j.jcis.2009.09.011

Shih H, Lin C-C (2012) Cross-linking and degradation of step-growth hydrogels formed by thiol–ene photoclick chemistry. Biomacromolecules 13(7):2003–2012. https://doi.org/10.1021/bm300752j

Shim HJ, Sunwoo S-H, Kim Y, Koo JH, Kim D-H (2021) Functionalized elastomers for intrinsically soft and biointegrated electronics. Adv Healthc Mater 10(17):2002105. https://doi.org/10.1002/adhm.202002105

Shin J, Yan Y, Bai W, Xue Y, Gamble P, Tian L, Kandela I, Haney CR, Spees W, Lee Y, Choi M, Ko J, Ryu H, Chang J-K, Pezhouh M, Kang S-K, Won SM, Yu KJ, Zhao J, Lee YK, MacEwan MR, Song S-K, Huang Y, Ray WZ, Rogers JA (2019) Bioresorbable pressure sensors protected with thermally grown silicon dioxide for the monitoring of chronic diseases and healing processes. Nat Biomed Eng 3(1):37–46. https://doi.org/10.1038/s41551-018-0300-4

Shin W, Kim JS, Choi HJ, Kim H, Park S, Lee HJ, Choi MK, Chung K (2021) 3D antidrying antifreezing artificial skin device with self-healing and touch sensing capability. Macromol Rapid Commun 42(9):2100011. https://doi.org/10.1002/marc.202100011

Si P, Trinidad J, Chen L, Lee B, Chen A, Persic J, Lyn R, Leonenko Z, Zhao B (2018) PEDOT:PSS nano-gels for highly electrically conductive silver/epoxy composite adhesives. J Mater Sci Mater Electron 29(3):1837–1846. https://doi.org/10.1007/s10854-017-8093-x

Sinawang G, Osaki M, Takashima Y, Yamaguchi H, Harada A (2020) Biofunctional hydrogels based on host–guest interactions. Polym J 52(8):839–859. https://doi.org/10.1038/s41428-020-0352-7

Sirringhaus H, Kawase T, Friend RH, Shimoda T, Inbasekaran M, Wu W, Woo EP (2000) High-resolution inkjet printing of all-polymer transistor circuits. Science (1979) 290(5499):2123–2126. https://doi.org/10.1126/science.290.5499.2123

Soh M, Kang D-W, Jeong H-G, Kim D, Kim DY, Yang W, Song C, Baik S, Choi I-Y, Ki S-K, Kwon HJ, Kim T, Kim CK, Lee S-H, Hyeon T (2017) Ceria–Zirconia nanoparticles as an enhanced multi-antioxidant for sepsis treatment. Angew Chem Int Ed 56(38):11399–11403. https://doi.org/10.1002/anie.201704904

Son D, Lee J, Qiao S, Ghaffari R, Kim J, Lee JE, Song C, Kim SJ, Lee DJ, Jun SW, Yang S, Park M, Shin J, Do K, Lee M, Kang K, Hwang CS, Lu N, Hyeon T, Kim D-H (2014) Multifunctional wearable devices for diagnosis and therapy of movement disorders. Nat Nanotechnol 9(5):397–404. https://doi.org/10.1038/nnano.2014.38

Son D, Lee J, Lee DJ, Ghaffari R, Yun S, Kim SJ, Lee JE, Cho HR, Yoon S, Yang S, Lee S, Qiao S, Ling D, Shin S, Song J-K, Kim J, Kim T, Lee H, Kim J, Soh M, Lee N, Hwang CS, Nam S, Lu N, Hyeon T, Choi SH, Kim D-H (2015a) Bioresorbable electronic stent integrated with therapeutic nanoparticles for endovascular diseases. ACS Nano 9(6):5937–5946. https://doi.org/10.1021/acsnano.5b00651

Son D, Koo JH, Song J-K, Kim J, Lee M, Shim HJ, Park M, Lee M, Kim JH, Kim D-H (2015b) Stretchable carbon nanotube charge-trap floating-gate memory and logic devices for wearable electronics. ACS Nano 9(5):5585–5593. https://doi.org/10.1021/acsnano.5b01848

Sreeja R, Sharma PD, Najidha S, Jayan SR, Predeep P (2006) Semi-interpenetrating conducting elastomer composite thin films from polyaniline. Polym Polym Compos 14(3):261–269. https://doi.org/10.1177/096739110601400305

Steinberg RH, Linsenmeier RA, Griff ER (1985) Chapter 2 retinal pigment epithelial cell contributions to the electroretinogram and electrooculogram. Prog Retin Res 4:33–66. https://doi.org/10.1016/0278-4327(85)90004-5

Stevens C, Fanning J, Coch D, Sanders L, Neville H (2008) Neural mechanisms of selective auditory attention are enhanced by computerized training: electrophysiological evidence from language-impaired and typically developing children. Brain Res 1205:55–69. https://doi.org/10.1016/j.brainres.2007.10.108

Sun TL, Kurokawa T, Kuroda S, Ihsan A, Bin, Akasaki T, Sato K, Haque MA, Nakajima T, Gong JP (2013) Physical hydrogels composed of polyampholytes demonstrate high toughness and viscoelasticity. Nat Mater 12(10):932–937. https://doi.org/10.1038/nmat3713

Sun W-J, Xu L, Jia L-C, Zhou C-G, Xiang Y, Yin R-H, Yan D-X, Tang J-H, Li Z-M (2019) Highly conductive and stretchable carbon nanotube/thermoplastic polyurethane composite for wearable heater. Compos Sci Technol 181:107695. https://doi.org/10.1016/j.compscitech.2019.107695

Sun A, He X, Li L, Li T, Liu Q, Zhou X, Ji X, Li W, Qian Z (2020) An injectable photopolymerized hydrogel with antimicrobial and biocompatible properties for infected skin regeneration. NPG Asia Mater 12(1):25. https://doi.org/10.1038/s41427-020-0206-y

Sunwoo SH, Lee JS, Bae S, Shin YJ, Kim CS, Joo SY, Choi HS, Suh M, Kim SW, Choi YJ, Kim T (2019) Chronic and acute stress monitoring by electrophysiological signals from adrenal gland. Proc Natl Acad Sci 116(4):1146–1151. https://doi.org/10.1073/pnas.1806392115

Sunwoo S-H, Han SI, Kang H, Cho YS, Jung D, Lim C, Lim C, Cha M, Lee S-P, Hyeon T, Kim D-H (2020) Stretchable low-impedance nanocomposite comprised of Ag–Au Core–Shell nanowires and Pt Black for epicardial recording and stimulation. Adv Mater Technol 5(3):1900768. https://doi.org/10.1002/admt.201900768

Taherian R (2014) Development of an equation to model electrical conductivity of polymer-based carbon nanocomposites. ECS J Solid State Sci Technol 3(6):M26. https://doi.org/10.1149/2.023406jss

Tamura T, Maeda Y, Sekine M, Yoshida M (2014) Wearable photoplethysmographic sensors—past and present. Electronics:282–302. https://doi.org/10.3390/electronics3020282

Tang Y, Zhang Y, Li W, Ma B, Chen X (2015) Rational material design for ultrafast rechargeable lithium-ion batteries. Chem Soc Rev 44(17):5926–5940. https://doi.org/10.1039/C4CS00442F

Taylor R, Walton DRM (1993) The chemistry of fullerenes. Nature 363(6431):685–693. https://doi.org/10.1038/363685a0

Teng C, Lu X, Zhu Y, Wan M, Jiang L (2013) Polymer in situ embedding for highly flexible, stretchable and water stable PEDOT:PSS composite conductors. RSC Adv 3(20):7219–7223. https://doi.org/10.1039/C3RA41124A

Teo MY, Kim N, Kee S, Kim BS, Kim G, Hong S, Jung S, Lee K (2017) Highly stretchable and highly conductive PEDOT:PSS/ionic liquid composite transparent electrodes for solution-processed stretchable electronics. ACS Appl Mater Interfaces 9(1):819–826. https://doi.org/10.1021/acsami.6b11988

Thilagaraj M, Arunkumar N, Ramkumar S, Hariharasitaraman S (2021) Electrooculogram signal identification for elderly disabled using Elman network. Microprocess Microsyst 82:103811. https://doi.org/10.1016/j.micpro.2020.103811

Tian B, Liu J, Dvir T, Jin L, Tsui JH, Qing Q, Suo Z, Langer R, Kohane DS, Lieber CM (2012) Macroporous nanowire nanoelectronic scaffolds for synthetic tissues. Nat Mater 11(11):986–994. https://doi.org/10.1038/nmat3404

Tondera C, Akbar TF, Thomas AK, Lin W, Werner C, Busskamp V, Zhang Y, Minev IR (2019) Highly conductive, stretchable, and cell-adhesive hydrogel by nanoclay doping. Small 15(27):1901406. https://doi.org/10.1002/smll.201901406

Tripathi A, Bonilla-Cruz J (2023) Review on healthcare biosensing nanomaterials. ACS Appl Nano Mater 6(7):5042–5074. https://doi.org/10.1021/acsanm.3c00941

Tuncaboylu DC, Sari M, Oppermann W, Okay O (2011) Tough and self-healing hydrogels formed via hydrophobic Interactions. Macromolecules 44(12):4997–5005. https://doi.org/10.1021/ma200579v

Tybrandt K, Khodagholy D, Dielacher B, Stauffer F, Renz AF, Buzsáki G, Vörös J (2018) High-density stretchable electrode grids for chronic neural recording. Adv Mater 30(15):1706520. https://doi.org/10.1002/adma.201706520

Tzou HS, Tseng CI (1990) Distributed piezoelectric sensor/actuator design for dynamic measurement/control of distributed parameter systems: a piezoelectric finite element approach. J Sound Vib 138(1):17–34. https://doi.org/10.1016/0022-460X(90)90701-Z

Usakli AB, Gurkan S, Aloise F, Vecchiato G, Babiloni F (2010) On the use of electrooculogram for efficient human computer interfaces. Comput Intell Neurosci 2010(1):135629. https://doi.org/10.1155/2010/135629

Valentín JL, Mora-Barrantes I, Carretero-González J, López-Manchado MA, Sotta P, Long DR, Saalwächter K (2010) Novel experimental approach to evaluate filler−elastomer interactions. Macromolecules 43(1):334–346. https://doi.org/10.1021/ma901999j

Wang Y, Yang H (2009) Synthesis of iron oxide nanorods and nanocubes in an imidazolium ionic liquid. Chem Eng J 147(1):71–78. https://doi.org/10.1016/j.cej.2008.11.043

Wang MS, Wang JY, Chen Q, Peng L-M (2005) Fabrication and electrical and mechanical properties of carbon nanotube interconnections. Adv Funct Mater 15(11):1825–1831. https://doi.org/10.1002/adfm.200500215

Wang S, Song J, Kim D-H, Huang Y, Rogers JA (2008) Local versus global buckling of thin films on elastomeric substrates. Appl Phys Lett 93(2):23126. https://doi.org/10.1063/1.2956402

Wang WM, LeMieux MC, Selvarasah S, Dokmeci MR, Bao Z (2009) Dip-pen nanolithography of electrical contacts to single-walled carbon nanotubes. ACS Nano 3(11):3543–3551. https://doi.org/10.1021/nn900984w

Wang J-J, Hu J-S, Guo Y-G, Wan L-J (2011) Eco-friendly visible-wavelength photodetectors based on bandgap engineerable nanomaterials. J Mater Chem 21(44):17582–17589. https://doi.org/10.1039/C1JM12173A

Wang X, Feng H, Wu Y, Jiao L (2013) Controlled synthesis of highly crystalline MoS2 flakes by chemical vapor deposition. J Am Chem Soc 135(14):5304–5307. https://doi.org/10.1021/ja4013485

Wang MC, Chun S, Han RS, Ashraf A, Kang P, Nam S (2015a) Heterogeneous, three-dimensional texturing of graphene. Nano Lett 15(3):1829–1835. https://doi.org/10.1021/nl504612y

Wang Y, Shi Y, Pan L, Ding Y, Zhao Y, Li Y, Shi Y, Yu G (2015b) Dopant-enabled supramolecular approach for controlled synthesis of nanostructured conductive polymer hydrogels. Nano Lett 15(11):7736–7741. https://doi.org/10.1021/acs.nanolett.5b03891

Wang Y, Gong S, Wang SJ, Yang X, Ling Y, Yap LW, Dong D, Simon GP, Cheng W (2018a) Standing enokitake-like nanowire films for highly stretchable elastronics. ACS Nano 12(10):9742–9749. https://doi.org/10.1021/acsnano.8b05019

Wang S, Xu J, Wang W, Wang G-JN, Rastak R, Molina-Lopez F, Chung JW, Niu S, Feig VR, Lopez J, Lei T, Kwon S-K, Kim Y, Foudeh AM, Ehrlich A, Gasperini A, Yun Y, Murmann B, Tok JB-H, Bao Z (2018b) Skin electronics from scalable fabrication of an intrinsically stretchable transistor array. Nature 555(7694):83–88. https://doi.org/10.1038/nature25494

Wang C, Li X, Hu H, Zhang L, Huang Z, Lin M, Zhang Z, Yin Z, Huang B, Gong H, Bhaskaran S, Gu Y, Makihata M, Guo Y, Lei Y, Chen Y, Wang C, Li Y, Zhang T, Chen Z, Pisano AP, Zhang L, Zhou Q, Xu S (2018c) Monitoring of the central blood pressure waveform via a conformal ultrasonic device. Nat Biomed Eng 2(9):687–695. https://doi.org/10.1038/s41551-018-0287-x

Wang L, Zhang X, Xia Y, Zhao X, Xue Z, Sui K, Dong X, Wang D (2019a) Cooking-inspired versatile design of an ultrastrong and tough polysaccharide hydrogel through programmed supramolecular interactions. Adv Mater 31(41):1902381. https://doi.org/10.1002/adma.201902381

Wang YJ, Zhang XN, Song Y, Zhao Y, Chen L, Su F, Li L, Wu ZL, Zheng Q (2019b) Ultrastiff and tough supramolecular hydrogels with a dense and robust hydrogen bond network. Chem Mater 31(4):1430–1440. https://doi.org/10.1021/acs.chemmater.8b05262

Wang Z, Hu T, Liang R, Wei M (2020a) Application of zero-dimensional nanomaterials in biosensing. Front Chem 8(April):1–19. https://doi.org/10.3389/fchem.2020.00320

Wang X, Liang D, Cheng B (2020b) Preparation and research of intrinsic self-healing elastomers based on hydrogen and ionic bond. Compos Sci Technol 193:108127. https://doi.org/10.1016/j.compscitech.2020.108127

Wang Y, Wang L, Zhang X, Liang X, Feng Y, Feng W (2021a) Two-dimensional nanomaterials with engineered bandgap: synthesis, properties, applications. Nano Today 37:101059. https://doi.org/10.1016/J.NANTOD.2020.101059

Wang C, Qi B, Lin M, Zhang Z, Makihata M, Liu B, Zhou S, Huang Y, Hu H, Gu Y, Chen Y, Lei Y, Lee T, Chien S, Jang K-I, Kistler EB, Xu S (2021b) Continuous monitoring of deep-tissue haemodynamics with stretchable ultrasonic phased arrays. Nat Biomed Eng 5(7):749–758. https://doi.org/10.1038/s41551-021-00763-4

Weadock KS, Miller EJ, Bellincampi LD, Zawadsky JP, Dunn MG (1995) Physical crosslinking of collagen fibers: comparison of ultraviolet irradiation and dehydrothermal treatment. J Biomed Mater Res 29(11):1373–1379. https://doi.org/10.1002/jbm.820291108

Wedler W, Tang W, Winter HH, MacKnight WJ, Farris RJ (1995) A segmented polyurethane elastomer with liquid crystalline hard segments. 2. Rheological study. Macromolecules 28(2):512–516. https://doi.org/10.1021/ma00106a014

Wenger N, Moraud EM, Gandar J, Musienko P, Capogrosso M, Baud L, Le Goff CG, Barraud Q, Pavlova N, Dominici N, Minev IR, Asboth L, Hirsch A, Duis S, Kreider J, Mortera A, Haverbeck O, Kraus S, Schmitz F, DiGiovanna J, van den Brand R, Bloch J, Detemple P, Lacour SP, Bézard E, Micera S, Courtine G (2016) Spatiotemporal neuromodulation therapies engaging muscle synergies improve motor control after spinal cord injury. Nat Med 22(2):138–145. https://doi.org/10.1038/nm.4025

Wietor J-L, Sijbesma RP (2008) A self-healing elastomer. Angew Chem Int Ed 47(43):8161–8163. https://doi.org/10.1002/anie.200803072

Wolff S, Wang M-J (1992) Filler—Elastomer Interactions. Part IV. The effect of the surface energies of fillers on elastomer reinforcement. Rubber Chem Technol 65(2):329–342. https://doi.org/10.5254/1.3538615

Won Y, Kim A, Lee D, Yang W, Woo K, Jeong S, Moon J (2014) Annealing-free fabrication of highly oxidation-resistive copper nanowire composite conductors for photovoltaics. NPG Asia Mater 6(6):e105–e105. https://doi.org/10.1038/am.2014.36

Wu L, Wu Y, Pan X, Kong F (2006) Synthesis of ZnO nanorod and the annealing effect on its photoluminescence property. Opt Mater (Amst) 28(4):418–422. https://doi.org/10.1016/j.optmat.2005.03.007

Xiang D, Zhang X, Harkin-Jones E, Zhu W, Zhou Z, Shen Y, Li Y, Zhao C, Wang P (2020) Synergistic effects of hybrid conductive nanofillers on the performance of 3D printed highly elastic strain sensors. Compos Part A Appl Sci Manuf 129:105730. https://doi.org/10.1016/j.compositesa.2019.105730

Xie C, Liu J, Fu T-M, Dai X, Zhou W, Lieber CM (2015) Three-dimensional macroporous nanoelectronic networks as minimally invasive brain probes. Nat Mater 14(12):1286–1292. https://doi.org/10.1038/nmat4427

Xie A, Zhang K, Sun M, Xia Y, Wu F (2018) Facile growth of coaxial Ag@polypyrrole nanowires for highly tunable electromagnetic waves absorption. Mater Des 154:192–202. https://doi.org/10.1016/j.matdes.2018.05.039

Xing B, Yu C-W, Chow K-H, Ho P-L, Fu D, Xu B (2002) Hydrophobic interaction and hydrogen bonding cooperatively confer a vancomycin hydrogel: a potential candidate for biomaterials. J Am Chem Soc 124(50):14846–14847. https://doi.org/10.1021/ja028539f

Xu P, Kang J, Choi J-B, Suhr J, Yu J, Li F, Byun J-H, Kim B-S, Chou T-W (2014) Laminated ultrathin chemical vapor deposition graphene films based stretchable and transparent high-rate supercapacitor. ACS Nano 8(9):9437–9445. https://doi.org/10.1021/nn503570j

Xu J, Liu Y, Hsu S (2019) Hydrogels based on Schiff base linkages for biomedical applications. Molecules. https://doi.org/10.3390/molecules24163005

Xue Z, Song H, Rogers JA, Zhang Y, Huang Y (2020) Mechanically-guided structural designs in stretchable inorganic electronics. Adv Mater 32(15):1902254. https://doi.org/10.1002/adma.201902254

Ya Y, Jiang C, Li T, Liao J, Fan Y, Wei Y, Yan F, Xie L (2017) A zinc oxide nanoflower-based electrochemical sensor for trace detection of sunset yellow. Sensors. https://doi.org/10.3390/s17030545

Yang S, Chen Y-C, Nicolini L, Pasupathy P, Sacks J, Su B, Yang R, Sanchez D, Chang Y-F, Wang P, Schnyer D, Neikirk D, Lu N (2015) "Cut-and-paste" manufacture of multiparametric epidermal sensor systems. Adv Mater 27(41):6423–6430. https://doi.org/10.1002/adma.201502386

Yang J, Bai R, Chen B, Suo Z (2020) Hydrogel adhesion: a supramolecular synergy of chemistry, topology, and mechanics. Adv Funct Mater 30(2):1901693. https://doi.org/10.1002/adfm.201901693

Yang J, Choi MK, Yang UJ, Kim SY, Kim YS, Kim JH, Kim D-H, Hyeon T (2021) Toward full-color electroluminescent quantum dot displays. Nano Lett 21(1):26–33. https://doi.org/10.1021/acs.nanolett.0c03939

Yao B, Wang H, Zhou Q, Wu M, Zhang M, Li C, Shi G (2017) Ultrahigh-conductivity polymer hydrogels with arbitrary structures. Adv Mater 29(28):1700974. https://doi.org/10.1002/adma.201700974

Yigit S, Sanyal R, Sanyal A (2011) Fabrication and functionalization of hydrogels through "click" chemistry. Chem Asian J 6(10):2648–2659. https://doi.org/10.1002/asia.201100440

Yokozawa T, Ohta Y (2016) Transformation of step-growth polymerization into living chain-growth polymerization. Chem Rev 116(4):1950–1968. https://doi.org/10.1021/acs.chemrev.5b00393

Yoo SS, Kim SY, Kim KS, Hong S, Oh MJ, Nam MG, Kim WJ, Park J, Chung CH, Choe WS, Yoo PJ (2020) Controlling inter-sheet-distance in reduced graphene oxide electrodes for highly sensitive electrochemical impedimetric sensing of myoglobin. Sens Actuators B Chem 305:127477. https://doi.org/10.1016/J.SNB.2019.127477

Yoo JI, Kim SH, Ko HC (2021a) Stick-and-play system based on interfacial adhesion control enhanced by micro/nanostructures. Nano Res 14(9):3143–3158. https://doi.org/10.1007/s12274-021-3533-6

Yoo S, Lee J, Joo H, Sunwoo S-H, Kim S, Kim D-H (2021b) Wireless power transfer and telemetry for implantable bioelectronics. Adv Healthc Mater 10(17):2100614. https://doi.org/10.1002/adhm.202100614

Yoon J, Jo S, Chun IS, Jung I, Kim H-S, Meitl M, Menard E, Li X, Coleman JJ, Paik U, Rogers JA (2010) GaAs photovoltaics and optoelectronics using releasable multilayer epitaxial assemblies. Nature 465(7296):329–333. https://doi.org/10.1038/nature09054

Yoon, J.; Bae, G.-Y.; Yoo, S.; Yoo, J.I; You, N.-H.; Hong, W.-K.; Ko, H. C. Deep-ultraviolet sensing characteristics of transparent and flexible IGZO thin film transistors. J Alloys Compd 2020, 817, 152788. https://doi.org/10.1016/j.jallcom.2019.152788

Yu Y, Lau LCM, Lo AC, Chau Y (2015) Injectable chemically crosslinked hydrogel for the controlled release of bevacizumab in vitreous: a 6-month in vivo study. Transl Vis Sci Technol 4(2):5. https://doi.org/10.1167/tvst.4.2.5

Yu KJ, Kuzum D, Hwang S-W, Kim BH, Juul H, Kim NH, Won SM, Chiang K, Trumpis M, Richardson AG, Cheng H, Fang H, Thompson M, Bink H, Talos D, Seo KJ, Lee HN, Kang S-K, Kim J-H, Lee JY, Huang Y, Jensen FE, Dichter MA, Lucas TH, Viventi J, Litt B, Rogers JA (2016) Bioresorbable silicon electronics for transient spatiotemporal mapping of electrical activity from the cerebral cortex. Nat Mater 15(7):782–791. https://doi.org/10.1038/nmat4624

Yu HC, Zheng SY, Fang L, Ying Z, Du M, Wang J, Ren K-F, Wu ZL, Zheng Q (2020) Reversibly transforming a highly swollen polyelectrolyte hydrogel to an extremely tough one and its application as a tubular grasper. Adv Mater 32(49):2005171. https://doi.org/10.1002/adma.202005171

Yu J, Ling W, Li Y, Ma N, Wu Z, Liang R, Pan H, Liu W, Fu B, Wang K, Li C, Wang H, Peng H, Ning B, Yang J, Huang X (2021) A multichannel flexible optoelectronic fiber device for distributed implantable neurological stimulation and monitoring. Small 17(4):2005925. https://doi.org/10.1002/smll.202005925

Yuk H, Varela CE, Nabzdyk CS, Mao X, Padera RF, Roche ET, Zhao X (2019) Dry double-sided tape for adhesion of wet tissues and devices. Nature 575(7781):169–174. https://doi.org/10.1038/s41586-019-1710-5

Yuksel R, Alpugan E, Unalan HE (2018) Coaxial silver nanowire/polypyrrole nanocomposite supercapacitors. Org Electron 52:272–280. https://doi.org/10.1016/j.orgel.2017.10.012

Yun J, Lee H, Song C, Jeong YR, Park JW, Lee JH, Kim DS, Keum K, Kim MS, Jin SW, Lee YH, Kim JW, Zi G, Ha JS (2020) A fractal-designed stretchable and transparent microsupercapacitor as a skin-attachable energy storage device. Chem Eng J 387:124076. https://doi.org/10.1016/j.cej.2020.124076

Zare Y, Rhee KY, Park SJ (2019) A developed equation for electrical conductivity of polymer carbon nanotubes (CNT) nanocomposites based on Halpin-Tsai Model. Results Phys 14:102406. https://doi.org/10.1016/J.RINP.2019.102406

Zhang W (2014) Nanoparticle aggregation: principles and modeling BT - nanomaterial: impacts on cell biology and medicine. In: Capco DG, Chen Y (eds) . Springer Netherlands, Dordrecht, pp 19–43. https://doi.org/10.1007/978-94-017-8739-0_2

Zhang YS, Khademhosseini A (2017) Advances in engineering hydrogels. Science (1979) 356(6337):eaaf3627. https://doi.org/10.1126/science.aaf3627

Zhang X-Y, Zhang L-D, Meng G-W, Li G-H, Jin-Phillipp N-Y, Phillipp F (2001) Synthesis of ordered single crystal silicon nanowire arrays. Adv Mater 13(16):1238–1241. https://doi.org/10.1002/1521-4095(200108)13:16<1238::AID-ADMA1238>3.0.CO;2-3

Zhang J, Liu J, Zhuang R, Mäder E, Heinrich G, Gao S (2011) Single MWNT-glass fiber as strain sensor and switch. Adv Mater 23(30):3392–3397. https://doi.org/10.1002/adma.201101104

Zhang Y, Xu S, Fu H, Lee J, Su J, Hwang K-C, Rogers JA, Huang Y (2013) Buckling in serpentine microstructures and applications in Elastomer-supported ultra-stretchable electronics with high areal coverage. Soft Matter 9(33):8062–8070. https://doi.org/10.1039/C3SM51360B

Zhang Y, Fu H, Xu S, Fan JA, Hwang K-C, Jiang J, Rogers JA, Huang Y (2014) A hierarchical computational model for stretchable interconnects with fractal-inspired designs. J Mech Phys Solids 72:115–130. https://doi.org/10.1016/j.jmps.2014.07.011

Zhang S, Ermann J, Succi MD, Zhou A, Hamilton MJ, Cao B, Korzenik JR, Glickman JN, Vemula PK, Glimcher LH, Traverso G, Langer R, Karp JM (2015) An inflammation-targeting hydrogel for local drug delivery in inflammatory bowel disease. Sci Transl Med 7(300):300ra128. https://doi.org/10.1126/scitranslmed.aaa5657

Zhang K, Feng X, Ye C, Hempenius MA, Vancso GJ (2017a) Hydrogels with a memory: dual-responsive, organometallic poly(ionic liquid)s with hysteretic volume-phase transition. J Am Chem Soc 139(29):10029–10035. https://doi.org/10.1021/jacs.7b04920

Zhang M, Wang C, Liang X, Yin Z, Xia K, Wang H, Jian M, Zhang Y (2017b) Weft-knitted fabric for a highly stretchable and low-voltage wearable heater. Adv Electron Mater 3(9):1700193. https://doi.org/10.1002/aelm.201700193

Zhang H, Chhowalla M, Liu Z (2018) 2D nanomaterials: graphene and transition metal dichalcogenides. Chem Soc Rev 47(9):3015–3017. https://doi.org/10.1039/C8CS90048E

Zhang Y, Qin M, Xing C, Zhao C, Dou X, Feng C (2020a) Redox-driven in situ helix reversal of graphene-based hydrogels. ACS Nano 14(12):17151–17162. https://doi.org/10.1021/acsnano.0c06938

Zhang X, Liu Y, Wen J, Zhao Z, Chen H, Liu X, Liu S (2020b) Host–guest interaction-mediated fabrication of a hybrid microsphere-structured supramolecular hydrogel showing high mechanical strength. Soft Matter 16(14):3416–3424. https://doi.org/10.1039/D0SM00271B

Zhang D, Chen Q, Shi C, Chen M, Ma K, Wan J, Liu R (2021) Dealing with the foreign-body response to implanted biomaterials: strategies and applications of new materials. Adv Funct Mater 31(6):2007226. https://doi.org/10.1002/adfm.202007226

Zhao P, Li N, Astruc D (2013) State of the art in gold nanoparticle synthesis. Coord Chem Rev 257(3–4):638–665. https://doi.org/10.1016/J.CCR.2012.09.002

Zhao J, Xu R, Luo G, Wu J, Xia H (2016) A self-healing, re-moldable and biocompatible cross-linked polysiloxane elastomer. J Mater Chem B 4(5):982–989. https://doi.org/10.1039/C5TB02036K

Zheng G, Lu W, Jin S, Lieber CM (2004) Synthesis and fabrication of high-performance n-type silicon nanowire transistors. Adv Mater 16(21):1890–1893. https://doi.org/10.1002/adma.200400472

Zheng H, Zhang Z, Jiang S, Yan B, Shi X, Xie Y, Huang X, Yu Z, Liu H, Weng S, Nurmikko A, Zhang Y, Peng H, Xu W, Zhang J (2019) A shape-memory and spiral light-emitting device for precise multisite stimulation of nerve bundles. Nat Commun 10(1):2790. https://doi.org/10.1038/s41467-019-10418-3

Zhou Y, Wang J, Zhou H, Xiang F, Yang H, Cai X, Liao H, Gu L, Wang Y (2019) An electrochemical approach towards the controllable synthesis of highly ordered and hierarchical zinc oxide dendritic crystals composed of hexagonal nanosheets: some insights into the stacking-assembly of the hierarchical architecture. CrystEngComm 21(26):3919–3929. https://doi.org/10.1039/C9CE00342H

Zhou X, Liu B, Chen Y, Guo L, Wei G (2020) Carbon nanofiber-based three-dimensional nanomaterials for energy and environmental applications. Mater Adv 1(7):2163–2181. https://doi.org/10.1039/D0MA00492H

Zhu S-E, Krishna Ghatkesar M, Zhang C, Janssen GCAM (2013) Graphene based piezoresistive pressure sensor. Appl Phys Lett 102(16):161904. https://doi.org/10.1063/1.4802799

Zhu L, Qiu J, Sakai E (2017) A high modulus hydrogel obtained from hydrogen bond reconstruction and its application in vibration damper. RSC Adv 7(69):43755–43763. https://doi.org/10.1039/C7RA08272J

Zhu H, Yang H, Ma Y, Lu TJ, Xu F, Genin GM, Lin M (2020) Spatiotemporally controlled photoresponsive hydrogels: design and predictive modeling from processing through application. Adv Funct Mater 30(32):2000639. https://doi.org/10.1002/adfm.202000639

Fundamentals of Biomarkers and Biosensors for Precision Healthcare

3

Chen Chen and Linhan Li

Abstract

Biomarkers are critical indicators of biological processes, disease progression, and therapeutic responses, playing a key role in precision healthcare. They enable early disease detection, monitoring, and personalized treatment strategies. Biosensors, as advanced analytical tools, allow rapid and specific biomarker detection across clinical and point-of-care settings. Their advantages, including cost-effectiveness, user-friendliness, high sensitivity, and real-time monitoring, make them invaluable in modern diagnostics and health management. This chapter introduces the fundamental concepts of biomarkers, discussing their classifications and biological significance. Additionally, we explore biosensing technologies designed for biomarker detection, focusing on their applications in disease diagnosis, prognosis, and treatment monitoring. By bridging biomarker research with biosensor advancements, this chapter provides insights into their transformative impact on precision healthcare and encourages further exploration in this evolving field.

Keywords

Biomarkers · Biosensors · Biomarkers for wearables · Precision healthcare · Biomarker discovery

C. Chen (✉)
School of Life Science and Technology, Xidian University, Xi'an, China
e-mail: chenc@xidian.edu.cn

L. Li
Faculty of Biotechnology, Universitat Autònoma Barcelona, Barcelona, Spain

K. Mahato, P. Chandra (eds.), *Nano-bioelectronics for Precision Health Monitoring*, https://doi.org/10.1007/978-981-95-2117-3_3

3.1 Introduction

Biomarkers and biosensing technologies have become integral components of modern healthcare, revolutionizing the way diseases are diagnosed, monitored, and treated (Singh et al. 2023; Khan et al. 2023; Bag and Mandal n.d.). Biomarkers, which are biological molecules found in blood, other body fluids, or tissues, serve as indicators of normal or pathogenic processes and responses to therapeutic interventions (Aronson and Ferner 2017; Strimbu and Tavel 2010). Biosensors, on the other hand, are sophisticated analytical devices designed to detect specific biomarkers and convert biological interactions into quantifiable signals (Vigneshvar et al. 2016; Coulet 2019). These devices have revolutionized healthcare by enabling rapid, accurate, and non-invasive testing, which is crucial for timely medical intervention. The synergy between biomarkers and biosensors has paved the way for advancements in personalized medicine, allowing for treatments tailored to the unique genetic and molecular profiles of individual patients (Kim et al. 2023a; Haleem et al. 2021; Altintas 2017). The integration of biosensors into healthcare has significantly enhanced the precision and personalization of medical care (Ahmed et al. 2014; Chandra 2023). Over the past few decades, the advancement of biosensing technologies has allowed for the rapid, sensitive, and specific detection of a wide range of biomarkers. This has enabled early diagnosis of diseases, real-time monitoring of treatment efficacy, and the development of targeted therapies tailored to individual patient profiles. Biomarker-based biosensors are particularly evident in the fields of oncology, cardiology, infectious diseases, and metabolic disorders, where they have contributed to significant improvements in patient outcomes (Rasooly and Jacobson 2006; Fracchiolla et al. 2013; Tang et al. 2023; Sin et al. 2014; Chu et al. 2022). As healthcare continues to evolve toward a more personalized and precision-based approach, the role of biomarkers and biosensing is expected to expand. Innovations in biosensor design, materials science, and data analytics are driving the development of next-generation biosensors that are more portable, cost-effective, and capable of integrating with digital health platforms (Mukherjee et al. 2024; Chen et al. 2022a; Tu et al. 2020). These advancements are poised to further enhance the accessibility and effectiveness of biomarker-based diagnostics and monitoring, making them indispensable tools in the future of healthcare. In this context, understanding the fundamental principles of biomarkers and biosensing is crucial for healthcare professionals, researchers, and technology developers. This chapter aims to introduce the knowledge of biomarkers and biosensing in healthcare. In the first section, we mainly focus on the fundamental principle of biomarker, including the definition, classification, characteristics, and discovery of the biomarker. In the second part, we will introduce the basic knowledge of biosensors (mainly the electrochemical sensors) and their application in the healthcare sector, Finally, we offer our insights on the future of biomarker-based biosensing in healthcare.

3.1.1 Definition of Biomarker

Biomarker, an abbreviation term of "biological marker" was defined by the working group of the National Institutes of Health Director's Initiative on Biomarkers and Surrogate Endpoints (Biomarkers Definitions Working Group 2001). A biomarker is an indicator that enables objective measurement and assessment of the normal biological process, pathogenic activities, or pharmacologic reaction to therapy. Moreover, the Food and Drug Administration (FDA) organization emphasizes the biomarker's contribution to the disease process and the treatment development (Karsdal et al. 2009). Biomarkers present great value in disease detection including early diagnosis of disease, determination of disease stage, prediction of disease prognosis, and monitoring of health status. Comprehensively, biomarkers are closely related to normal biological processes and disease occurrence and progression, which is accessible to disease detection and treatment, and health safety evaluation (MacNamara et al. 2015). Thus, biomarker detection is of great significance in the healthcare sector.

3.1.2 Classification of Biomarker

Biomarkers can be classified into different types, e.g., molecular biomarkers (including nucleic acid, protein, enzyme, and chemical molecule), imaging biomarkers (positron emission tomography (PET), computed tomography (CT), and magnetic resonance imaging (MRI)), etc. (Wishart et al. 2021; Dhama et al. 2019; Firestein 2006). Here, we prefer the classification method provided by the FDA recommendation (Berk 2015). There are diagnostic biomarkers, prognostic biomarkers, predictive biomarkers, monitoring biomarkers, response biomarkers, and safety biomarkers (Table 3.1).

Diagnostic biomarkers are used for detecting or confirming the presence of a specific disease or medication condition in medical practice. This type of biomarkers can determine whether an individual should receive a clinical trial for a particular disease study or whether a patient has a particular medication condition. For example, hemoglobin A1c (HbA1c) for the confirmation of Type II diabetes mellitus, cardiac troponin for the diagnosis of cardiac muscle injury, sweat chloride molecule for the determination of cystic fibrosis (Farrell et al. 2008), 3-hydroxy-fatty acid for the identification of planctomycetes, galactomannan for the identification of invasive aspergillosis, glutamate for the determination of visceral obesity and altered metabolism, catestatin for psychological stress related to heart disease, cystatin-C for the estimation of the severity of the renal injury, etc. (Siu and U S Preventive Services Task Force 2015; Farrell et al. 2008).

Prognostic biomarkers can be used to predict the likelihood of a future clinical event including the disease recurrence or development of a new medical condition, disease progression, and death. In oncology, for example, tumor cells (number and size), lymph nodes (number), and the presence of metastasis are used to indicate prognosis. Until recently, a variety of biomolecules and chemical molecules have

Table 3.1 Summary of the biomarker type and its clinical application

Biomarker type	Examples
Diagnostic biomarkers	HbA1c for Type II diabetes mellitus Cardiac troponin for cardiac muscle injury Sweat chloride for cystic fibrosis 3-hydroxy-fatty acid for planctomycetes identification Galactomannan for invasive aspergillosis Glutamate for visceral obesity and altered metabolism Cystatin-C for renal injury severity
Prognostic biomarkers	BRCA1/2 mutations for breast cancer Chromosome 17p deletion and TP53 mutations for chronic lymphocytic leukemia PSA for prostate cancer progression C-reactive protein for recurrent coronary artery disease
Predictive biomarkers	Human leukocyte antigen allele genotype for HIV patients before abacavir treatment Cystic fibrosis transmembrane conductance regulator for cystic fibrosis treatment Thiopurine methyltransferase genotype for 6-mercaptopurine or azathioprine treatment response
Monitoring biomarkers	Hepatitis C virus RNA for chronic hepatitis C treatment response The monoclonal protein level in blood for blood cancer PSA and cancer antigen 125 for prostate and ovarian cancer, respectively HIV-RNA for antiretroviral therapy guidance B-type natriuretic peptide for pediatric pulmonary hypertension
Response biomarkers	Circulating B lymphocytes for B-lymphocyte stimulator inhibitor in lupus patients Fluoroestradiol F 18 for estrogen receptor-positive breast cancer response C-reactive protein, interleukin-6, fibrinogen, and F2-isoprostane for inflammation from tobacco use HbA1c reduction for diabetes treatment efficacy LDL cholesterol for cardiovascular drug approval HIV-RNA reduction for HIV drug approval
Safety biomarkers	Serum potassium for safety evaluation in diuretics, ACE inhibitors, angiotensin receptor blockers, or aldosterone antagonists Hyperkalemia for dose adjustment of aldosterone antagonist Hypokalemia for potassium supplementation needs Metabolizing enzyme defects to identify toxicity risk or treatment inefficacy

been discovered and identified as prognostic biomarkers, e.g., BRCA1/2 gene mutations for evaluating the condition of women's breast cancer (Basu et al. 2015), chromosome 17p deletion and TP53 mutations for evaluating the development of the chronic lymphocytic leukemia (Delgado et al. 2012; Shanafelt et al. 2006), increasing prostate-specific antigen for assessing the likelihood of prostate cancer progression (Roberts et al. 2001), C-reactive protein level for identifying patients with the likelihood of recurrent coronary artery disease (Arroyo-Espliguero et al. 2004). It is noteworthy to mention that prognostic biomarkers are crucial factors in drug development (Majkić-Singh 2011). Because they are eligibility criteria in clinical tests to identify individuals who are probably involved in specific medical conditions or disease progression. As a result, these biomarkers help to determine and work out a treatment protocol.

Predictive biomarkers enable individuals to confirm whether they are more likely to experience a positive or negative effect from exposure to a medical product or environmental agent. There is a conceptual overlap between predictive biomarkers and diagnostic biomarkers, which as indicators for the incidence or progression of a particular disease. The predictive biomarkers emphasize the possibility of both improved survival and adverse effects as the individuals respond to being exposed to a medical product in the clinical trial. Predictive biomarkers enable doctors to evaluate the specific therapy or the drug for the patients which helps to work out the desirable therapy for individuals or certain groups in most clinical cases. For example, the human leukocyte antigen allele genotype may be used to assess human immunodeficiency virus patients before abacavir treatment (AIDSinfo, A 2013). Cystic fibrosis transmembrane conductance regulator can be used to evaluate treatment for cystic fibrosis to identify patients who respond positive effects to the treatment (Davies et al. 2013). The thiopurine methyltransferase genotype has been used to evaluate the response of 6-mercaptopurine or azathioprine treatment to identify the patients at high risk for toxicity due to drug concentration (Relling et al. 2011). The utility of these types of biomarkers is not limited to clinical treatment. They inform patients' care decisions, i.e., determining who probably benefits from the specific therapy and verse vice to optimize the treatment protocol.

Monitoring biomarkers are used for evaluating disease conditions and treatment efficacy repeatedly over time. Specifically, they can assess the disease progression including the incidence of new symptoms, worsening of previous condition, or change in disease severity, and the individual responding to either favorable or negative treatment protocol. In the case of the clinical care setting, monitoring biomarkers are measured during more than one period of one individual's clinical course, i.e., the period during which a disease has been diagnosed before intervention, or the period during which intervention is being provided, or the period after delivery of the treatment. In addition, monitoring biomarkers can be used for drug development, especially in therapeutic or prevention trials of new medical products. Monitoring biomarkers used in the course of an intervention manages to determine the drug metabolism of a patient, as well as monitor therapeutic effects or side effects following treatment. There are a variety of monitoring biomarkers that have been applied in the healthcare sector, such as hepatitis C virus ribonucleic acid for evaluating treatment response with chronic hepatitis C (Mukherjee et al. 2015; Kotha et al. 2018), monoclonal protein level in blood as signs of some types of blood cancer (Kyle and Rajkumar 2003), prostate-specific antigen and cancer antigen 125 respectively used to assess disease condition with prostate cancer and ovarian cancer (Freedland and Moul 2007; Gundogdu et al. 2011), HIV-RNA for guiding treatment with antiretroviral therapy (Orsega 2007), B-type natriuretic peptide for determining pulmonary hypertension in pediatric patients (Cuna et al. 2013), etc. In summary, monitoring biomarkers serve as an "observer" in the surveillance of individuals in the presence of disease or medication conditions. In some cases, the monitored individuals do not show any clinical symptoms, while in other cases, they may have some medical condition that makes them easily get some new disease. Thus, it is necessary for healthy individuals or individuals without apparent symptoms to undergo routine physical examination based on monitoring biomarkers detection.

Response biomarkers are used to present the biological response as the patient has been exposed to a medical product or an environmental agent. They can be divided into subtypes, pharmacodynamic biomarkers, and surrogate endpoint biomarkers. A pharmacodynamic biomarker normally indicates the biological response of a therapeutic protocol or medical agent without a necessary conclusion of the efficacy or treatment result or necessarily relating the response to any established mechanism. Hence, pharmacodynamic biomarkers benefit the establishment of the proof-of-concept, which can be used for the guide dose-response study. For example, circulating B lymphocytes can be used to evaluate the efficacy of the B-lymphocytes stimulator inhibitor in patients with systemic lupus erythematosus (Stohl and Hilbert 2012). Fluoroestradiol F 18 detected by PET can be used to analyze the response of estrogen receptor-positive lesions to endocrine treatment for recurrent and metastatic breast cancer (Liao et al. 2016). When assessing inflammation related to tobacco consumption, several components may be used as pharmacodynamic biomarkers, including C-reactive protein (Eraly et al. 2014), interleukin-6 (Sunyer et al. 2009), fibrinogen (Van Dijk et al. 2013), and F2-isoprostane molecule (Basu 2010). Surrogate endpoints are normally used as an alternative for direct detection of how a patient feels, functions, or survives, which is often implemented in the final stage of clinical trials. A surrogate endpoint biomarker is expected to predict clinical benefit or harm according to pathophysiologic and therapeutic evidence, rather than directly providing the clinical result of interest (Aronson 2005; Buyse et al. 2010). For example, hemoglobin A1c decreasing has a direct relation with the reduction of microvascular complications in the case of diabetes mellitus, which can be used for approval of potential drugs (Sandler and McDonnell 2016). Low-density lipoprotein (LDL) cholesterol level is associated with cardiovascular events and has been used as the key parameter for approval of LDL-lowering drugs, e.g., statins and PCSK 9 inhibitors (Stone and Lloyd-Jones 2015). HIV-RNA reduction manifests the HIV under control and has been used for approval of HIV drug (Health, U. D. o., and Services, H. Panel on Antiretroviral Guidelines for Adults and Adolescents 2019).

A safety biomarker serves as an indicator for measuring the presence of an adverse effect before or after exposure to a medical intervention or an environmental agent. These biomarkers are required due to the possible undesirable and potentially harmful effects or even over-toxicity caused by the medication product in the clinical trial. In some cases, the toxicity can be identified via the measurement of change in a biomarker, allowing dosage adjustment and treatment termination before the situation worsens. For example, serum potassium can evaluate the safety while delivering diuretics, angiotensin-converting-enzyme inhibitors, angiotensin receptor blockers, or aldosterone antagonists (Shah et al. 2014; Roush and Sica 2016). In other cases, safety biomarkers can tell whether a treatment is required. For example, hyperkalemia associated with an aldosterone antagonist manifests the requirement of dose modification, while hypokalemia associated with a diuretic signals the need for potassium supplementation (Chen et al. 2011). Safety biomarkers should be monitored periodically to detect the potential toxicity of the medical product. Theoretically, safety biomarkers are expected to "report" the harmful effect before clinical symptoms and before irreversible damage. Moreover, safety biomarkers can assist in initiating a unique treatment protocol via removing the potential safety risks. Metabolizing enzyme defects can identify persons at risk for toxicity in the case of reduced dosage or identify patients who have no response to a vital treatment because they are not able to produce the active metabolite.

3.1.3 Discovery of Biomarker

This section discusses the methods for the discovery of molecular biomarkers. So far, various expression techniques have been widely applied for biomarker determination, for example, genomics, proteomics, metabolomics, pharmacogenetics, and integrative approaches (Majkić-Singh 2011; Zhou et al. 2012; Rundle et al. 2012; Gosho et al. 2012). Microarray and polymerase chain reactions are commonly used for gene expression analysis in genomics and pharmacogenomics. Genomics screens microRNA molecules, and pharmacogenetics mainly analyzes the single nucleotide polymorphism molecules, which results in the production of genetic biomarkers (Umelo et al. 2018; Tringe and Rubin 2005). Proteomic technologies enable to separation and identify a set of proteins, allowing to characterization of the specific protein (Tomasina et al. 2013; Hudler et al. 2014; Noronha et al. 2020). Unlike the fixed genome, the proteome is more complicated, which benefits the protein analysis to consider posttranslational modifications allowing change in the protein's function and activity. Metabolites are compounds with low molecular weight as biological systems responding to specific stimuli or medical products (Horgan and Kenny 2011; Weckwerth and Morgenthal 2005; García-Gutiérrez et al. 2020). Metabolites are in a wide range including lipids, amino acids, peptides, nucleic acids, and organic acids, which could serve as biomarkers to identify disease or signal the efficacy or toxicity of the particular drug. The metabolomics approach allows to detection of the metabolomics biomarkers by using spectroscopy and chromatography (Zhang et al. 2016a; Gika et al. 2014). Integration of genomic and proteomic techniques manages to obtain more comprehensive and superior information through the analysis of gene expression data and proteomic data together. For example, researchers identify and confirm a potential biomarker for the diagnosis of colon and ovarian cancers by the combination of genomics and proteomics (Tainsky 2009).

The procedure of a biomarker from discovery in a lab to application for practical use follows the five stages (Majkić-Singh 2011). Initially, preclinical investigation is implemented in the laboratory aiming to identify the target biomarker. The potential biomarker is evaluated by comparison of healthy and abnormal cases to establish the correlation with the biological phenomena. In the second stage, it is required to check whether the investigated biomarker is capable of identifying pathological conditions in comparison with the reference and the gold standard. In other words, researchers aim to assess the validity of the targeted biomarker at this stage. The third stage involves a retrospective investigation, mainly studying the screening and predictive value of the biomarker. In the fourth stage, the prospective clinical manifests the relation between biomarkers and the incidence of disease. In the last stage, a controlled study is performed to check whether the targeted biomarker benefits the modification of treatment in comparison with the previous one. Before the investigated biomarker enters into practical use, several issues should be emphasized. One is the distribution of the biomarker in the general and observed population, and the dependence on the demographic properties, including gender, age, nationality, and race. Moreover, we should consider whether the targeted biomarker has a relation with the known risk factors or pathophysiological mechanism of the corresponding disease. The biomarker's sensitivity and specificity should be disclosed, while its limitation should be taken into account. The prognostic value (increasing or decreasing) should be considered as well.

3.2 Biosensor: Basic Principle and Its Application in the Healthcare

3.2.1 A Basic Principle of Biosensor

3.2.1.1 Fundamentals of Biosensor

The biosensor can be defined as an analytical device incorporating a biological sensing element consisting of a biological recognition element, signal transducer, and signal processing element (Chen and Wang 2020). The biological recognition element refers to the capture of a specific biological process. The biological process refers to the recognition of biological elements and a variety of analytes, such as biomolecules (protein and nucleic acid) and materials (e.g., virus, bacteria, cell, tissue, and organ). The transducer converts a biological process and its change into a measurable signal, such as an electrical signal, which can be further processed and displayed via a data processor (Chen et al. 2020, 2022b, 2024). Detailed classification of the biosensors and their transducers is discussed in Chap. 1. The data processor used in the biosensor usually displays the readable result or data. This part is usually custom-designed and made to satisfy different working principles of biosensors. The general performance of a biosensor is usually determined by the biological recognition element, particularly the surface architectures, which connect the sensing element to the sample in the order of nanometer and the capability of a transducer. In general, the surface modification technique (or biofunctionalization strategy), the choice of the recognition receptor, and a transducer mechanism have a comprehensive impact on the ultimate sensing performance. To construct a highly qualified biosensor, several conditions should be satisfied. The receptor should be highly specific and affinity to the target analyte. The commonly used receptor is an antibody, aptamer, nucleic acid, cell, and hormone. The interaction between the receptor and analyte should not be influenced by environmental factors, such as temperature, pH value, humidity, etc. In this way, minimum noise will be introduced to the sensing system. Moreover, the response of the transducer should be accurate, reproducible, and linear over the concentration range of interest. It should be independent of the electrical (for an electronic transducer) signal in the environment. For the healthcare application, the recognition element should be biocompatible, non-toxic, or antigenic to the normal cell (Mohankumar et al. 2021). To develop as a real product, the biosensor should be economical, portable, and easier to operate.

The performance of a biosensor is estimated by the parameters, including sensitivity, limit of detection, specificity, affinity, accuracy, etc. The biological recognition element, which refers to the bioreceptor/analyte interaction mainly determines the specificity and affinity (Chen et al. 2017; Bozzola et al. 2017). Bioreceptors can be any molecule, such as antibody, aptamer, nucleic acid, or biomaterial, such as cells and tissue. Bioreceptors are always ready to specifically discover and recognize their partner analytes, which leads to further interaction with high affinity. Moreover, biological recognition interaction also influences biosensing sensitivity. An appropriate approach for biological recognition element incorporating with the transducer is a crucial step for creating a biosensor. Therefore, enormous efforts have been devoted to formulating novel strategies for bioreceptor immobilization. Good immobilization should ensure the efficient coverage of the sensor surface with receptors while maintaining their natural properties, but their stability should be guaranteed for storage and regeneration. In the healthcare sector, a well-defined surface with biomolecules is required to prevent nonspecific adsorption during the analysis of a complex real sample. It should be noted that the chemical activation of a sensor surface is always needed before immobilization. Organosilanes with different structures, lengths, functionalities, and physicochemical properties have become popular in the market, among which the most frequently used are those with a short alkyl chain, ending with amino, thiol, epoxy, or carboxylic groups. Concerning biofunctionalization strategies, the following immobilization methods have already been reported (Estevez et al. 2012): (1) physical adsorption, (2) covalent binding, (3) physical entrapment in a hydrogel, and (4) chemical crosslinking.

Physical adsorption is determined by the hydrophobic and electrostatic interactions between the receptors and the surface. The operating process is simple. However, receptors can get easily rinsed off under flowing conditions. The nature of the active receptors might be destroyed by an organic solvent or salt solution at higher concentrations and higher pH values when a regeneration cocktail is uploaded to the sensor. Due to the poor adsorption of receptors and the possibility of damage to natural biomolecules during the regeneration process, receptors can become undesirable despite their simplicity. Via covalent bonds, biomolecules can be immobilized onto the transducer surface. The exposure of the binding sites of biomolecules is required to subsequently recognize the target molecule with high specificity and affinity. Several chemical groups are used to form covalent bonds. Amino, carboxylic, or thiol groups are preferred to couple protein molecules. It is possible to incorporate a chemical group at the end of the DNA sequence benefiting from the versatility of nuclear acid synthesis. Antibodies/antigens should be attached in an oriented way to leave the binding sites free, which can be fulfilled by affinity proteins, e.g., A or G proteins (Prieto-Simon et al. 2008). Alternatively, bioreceptors can be physically entrapped in a hydrogel. The preferred hydrogel is sol-gel, a glassy silica compound generated by the polymerization of silicate monomers in the presence of a biological element (Yetisen et al. 2016; Kang et al. 2008). The features of a hydrogel, in terms of stability, wearability, and strength, make it an attractive material for use in the development of biosensors. By using bifunctional agents, biomolecules can be immobilized on the sensor surface, too (Górecka and Jastrzębska 2011; Miletić and Loos 2009; Mateo et al. 2004). Agents, such as glutaraldehyde, toluene diisocyanate, and diazobiphenyl, are tightly and directly bound with the bioreceptors. However, the complex process associated with the possible destruction of native molecules hinders its application and development. Overall, a desirable immobilization strategy needs to be formulated to exploit the full potential of an optical biosensor.

3.2.1.2 Fundamentals of Electrochemical Sensor

Among various transducers, an electrochemical sensor directly converts the biochemical reaction into a measurable electric signal. The electrochemical sensor makes use of electrodes as transducer elements in the presence of analytes (Grieshaber et al. 2008). A typical electrochemical sensor consists of three electrodes, i.e., working electrode, a counter electrode (or auxiliary electrode), and a reference electrode. The biorecognition element is constructed on the working electrode and the oxidation and reduction reaction of the species occurred there. The material of the working electrode is chosen based on the specific application and the type of electrochemical reaction. Gold (Au) and carbon-based materials (e.g., glassy carbon and graphite) are widely used materials due to high conductivity, chemical stability, and biocompatibility (Chaubey and Malhotra 2002; Ali et al. 2006). Sometimes, the surface of the working electrode is modified with materials like enzymes, nanoparticles, or polymers to enhance selectivity and sensitivity (Patolsky et al. 2006; Nair and Alam 2007; Wang 2006; Park et al. 2002; Wanekaya et al. 2006). For example, in glucose sensors, the working electrode is often coated with glucose oxidase, which catalyzes the oxidation of glucose to produce an electrical signal (Kang et al. 2009). The counter electrode serves to complete the electric circuit of the electrochemical system. It is involved in the redox reaction by providing a pathway for the current that flows due to the electrochemical reaction at the working electrode. The counter electrode does not need to participate directly in the analyte reaction but ensures that the current flows through the cell. The counter electrode is typically made of inert materials that can sustain the necessary current without undergoing significant reactions (Carter et al. 1997). For example, platinum (Pt) is commonly used due to its inertness and excellent conductivity (Jerkiewicz 2022). Compared with the working electrode, the counter electrode possesses a much larger surface area to carry the required current without limiting the reaction. The reference electrode provides a stable and known potential against which the potential of the working electrode can be measured (Sophocleous and Atkinson 2017). It is crucial to ensure that the applied potential is accurately controlled during the electrochemical reaction. The reference electrode is usually composed of a

Table 3.2 List of the main electrochemical methods and the respective monitored electrical properties and the units

Electrochemical methods	Monitored electrical properties	Units
Potentiometry	Potential difference	V
Conductometry	Resistance	Ω
Amperometry and voltammetry	Current as a function of applied potential	I
Coulometry	Current as a function of time	C = I·s
Capacitance	Potential load	$F = C \cdot V^{-1}$

metal/metal salt combination immersed in an electrolyte which maintains a constant potential. For example, silver/silver chloride (Ag/AgCl) is a versatile and widely used reference electrode in electrochemical biosensors, such as glucose sensors, where it provides a stable reference potential for accurate current or potential measurement. Its stability ensures reliable detection of analytes in complex biological matrices (Yoon et al. 2019; Ito et al. 1995). What's more, saturated calomel electrodes (SCE), made of mercury/mercury chloride, have become less commonly used today due to toxicity concerns (Kolthoff and Thomas 1965). The hydrogen electrode as a reference electrode is mainly considered the primary standard but less practical for routine use (Isse and Gennaro 2010).

The electrochemical sensor presents unique advantages, such as low-cost production, robustness, easy miniaturization, excellent detection limit, small volume, and ability to be used in turbid fluid samples, etc. Table 3.2 lists some of the main electrochemical methods and the corresponding measured electrical signal. In general, electrochemical responses measured by different strategies are based mainly on potential, resistance, and current (Grieshaber et al. 2008). Furthermore, electrochemical sensors can be classified according to the type of electrochemical reaction they take (Thévenot et al. 2001). The potentiometric sensor measures the potential difference between a working electrode and a reference electrode without current flowing. In the pH meter, the potential difference for the pH electrode is related to the hydrogen ion concentration in a solution. The amperometric sensor measures the current generated when a potential is applied between the working electrode and a reference electrode. The current is proportional to the concentration of the analyte undergoing oxidation or reduction. For a glucose sensor, glucose oxidase catalyzes the oxidation of glucose, and the resulting current is measured. The conductometric sensor measures the change in the conductivity of a solution due to the interaction between the analyte and the sensor. For example, the presence of ions increases the conductivity of the solution in a conductivity meter. The voltammetric sensor measures the current as a function of the applied potential. The current response provides information about the concentration of the analyte. Voltammetric sensor based on cyclic voltammetry is often used for studying electroactive species, trance metal detection, and organic compound analysis. Impedimetric sensor measures the impedance of a system, which is the opposition to alternating current. The impedance changes with the presence of the analyte. The coulometric sensor measures the total charge passed through the electrochemical cell during the complete electrolysis of the analyte. A coulometric sensor is usually used in titration methods.

3.2.2 Electrochemical Sensors for Healthcare Monitoring

Electrochemical sensors have become essential tools in healthcare monitoring, such as diagnosing, monitoring, and managing various medical conditions (Baranwal et al. 2022; Lu et al. 2021; Hernández-Rodríguez et al. 2020). This section describes various such applications in healthcare, along with examples.

3.2.2.1 Blood Glucose Monitoring

The most well-known application of electrochemical sensors in healthcare is in blood glucose monitoring, which is critical for managing diabetes. The sensors typically use amperometry, where the enzyme glucose oxidase is immobilized on the working electrode. Glucose in the blood reacts with glucose oxidase to produce gluconic acid and hydrogen peroxide. The hydrogen peroxide is then oxidized at the electrode, generating a current proportional to the glucose concentration. Accu-Chek and OneTouch are two of the most widely recognized brands of blood glucose meters, designed to help individuals with diabetes monitor their blood glucose levels (Vallera et al. 1991; Owiredu et al. 2009). The working principle is based on the amperometry method. The strips are designed with a capillary action that draws in the blood sample. The enzyme on the strip oxidizes the glucose, producing gluconic acid and hydrogen peroxide. The hydrogen peroxide is then involved in an electrochemical reaction at the electrode, generating a current that the meter measures. The accuracy of both brands meets stringent accuracy standards, typically within ±15% of laboratory results. Moreover, they are user-friendly, offering large displays, simple interfaces, access to smartphone apps, and minimal blood sample requirements. Therefore, Accu-Chek and OneTouch are trusted by millions of users worldwide for managing diabetes.

Continuous glucose monitor (CGM) is another advanced device used to continuously track glucose levels in real time, providing critical insights for managing diabetes (Owiredu et al. 2009). Unlike traditional blood glucose meters, CGM offers a continuous stream of glucose data, helping individuals make more informed decisions about their health (Gross and Mastrototaro 2000; Rodbard 2016). CGM contains a tiny electrode that measures the glucose levels in the interstitial fluid rather than directly in the blood. It is inserted just under the skin, typically on the abdomen or the upper arm. The sensor is usually worn for 7–14 days before it needs to be replaced. CGM offers numerous advantages over the traditional ones. By providing continuous data, CGM helps users identify trends and patterns, making it easier to manage blood glucose levels effectively. This can lead to better glycemic control and fewer complications. CGM can also help reduce the risk of hypoglycemia by providing real-time alerts before blood sugar levels drop too low, allowing for timely interventions. What's more, the convenience of not having to perform multiple daily finger sticks, combined with the peace of mind from real-time monitoring, can significantly improve the quality of life for people with diabetes. Valuable data from CGM can be used by patients and healthcare providers to make informed decisions about medication, diet, and lifestyle adjustments. However, this type of electrochemical sensor still suffers from some drawbacks (Boland et al. 2001). The cost is high, making it affordable for only a small portion of the population. Some users may experience skin irritation or allergy and some users suffer from adhesion problems (i.e., keeping the sensor securely attached to the skin can be challenging, especially in hot or humid conditions, during exercise, or for people with oily or sweaty skin). As mentioned before, CGM measures glucose levels in the interstitial fluid, not directly in the blood. This can result in a lag of about 5–15 minutes, especially during rapid glucose changes, potentially leading to discrepancies between CGM readings and actual blood glucose levels. High cost, discomfort, and lag time are among the main hurdles that users may face. Understanding these challenges is crucial for both users and healthcare providers to maximize the benefits of CGM while mitigating potential downsides.

3.2.2.2 Infectious Disease Detection

Electrochemical sensors are employed for the rapid detection of pathogens, such as viruses and bacteria, which is crucial for controlling infectious diseases (Simoska and Stevenson 2019). These sensors often use immunoassays or nucleic acid-based methods where antibodies, antigens, or DNA/RNA probes are immobilized on the electrode. Upon binding with the target pathogen, a measurable electrochemical signal (such as a change in current or potential) is generated (Monzó et al. 2015). For example, GenMark Diagnostics (USA) proposed the first electrochemical COVID-19 detection scheme in the early stages of the COVID-19 pandemic (Biswas et al. 2022). The detection technology enables voltammetric detection of viral RNA by the competitive DNA hybridization technique, The viral RNA is extracted from the sample and amplified by using transcription polymerase chain reaction (RT-PCR). In the detection scheme, the target RNA binds to the complementary DNA probes that have already immobilized on the electrode surface, which results in a change of the voltage or current. The test demonstrated a limit of detection of 1×10^5 copies/mL (0.17 fM) for SARS-CoV-2 RNA, comparable to the authorized RT-PCR methods approved for emergency use authorization (EUA). While this detection platform shows great promise, it does require a longer assay time of 2 hours. Recently, clustered regularly interspaced short palindromic repeats (CRISPR) technology has been extensively used for the detection of pathogens by programmed recognition and cutting of the pathogens' nucleic acid sequences with CRISPR-associated enzymes (such as Cas9, Cas12, and Cas13). CRISPR-assisted electrochemical sensors have been reported for quick and selective detection of COVID-19 (Chiu 2018; Kellner et al. 2019; Nouri et al. 2021). For example, Puig et al. introduced a CRISPR/Cas-based electrochemical platform capable of detecting SARS-CoV-2 RNA with 100% accuracy within 1 hour (de Puig et al. 2021). The gold electrodes with EDC/NHS-modified graphene oxide nanocomposite and target-specific single-strand DNA bound by an amine-terminated peptide nucleic acid linker, which allows it to detect viral RNA at the single-molecule level. The cyclic voltammetry method was taken for signal transduction. However, the proposed technology required 30 minutes of nucleic acid amplification at 65 °C. To overcome this issue, amplification-free CRISPR technology was demonstrated to incorporate into the electrochemical sensor (Hajian et al. 2019). For example,

Kashefi-Kheyrabadi et al. demonstrated an amplification-free multiplex electrochemical biosensor utilizing four-way junction (4-WJ) hybridization for detecting the SARS-CoV-2 RNA in clinical samples (Kashefi-Kheyrabadi et al. 2022). This biosensor achieved a remarkable LOD of two to three copies/μL (approximately 4.98 fM) within just 1 hour. A 4-WJ structure was assembled on double screen-printed gold electrodes (SPGEs) using a Universal DNA-Hairpin (UDH) probe hybridized with two adaptor strands and a SARS-CoV-2 RNA target. To enable electrochemical analysis, one of the adaptor strands was modified with a redox mediator. The SPGEs were further enhanced with electrodeposited 3D gold nanoneedles, significantly boosting the sensor's hybridization capacity and sensitivity. This single-step biosensor is capable of distinguishing single-nucleotide substitutions, making it highly specific and ideal for on-site infection diagnostics.

3.2.2.3 Cardiac Monitoring

Cardiovascular disease (CVD) remains one of the most widespread non-communicable diseases globally (Lakshmanakumar et al. 2019; Kaptoge et al. 2019). According to data from the World Health Organization, there was a 50% increase in global CVD-related mortality in 2020. Projections indicate that by 2030, over 23.3 million people will die annually from cardiovascular conditions (Şahin and İlgün 2022). Biomarkers including cardiac troponin I (cTnI), creatine kinase-muscle/brain (CK-MB), procalcitonin (PCT), and C-reactive protein (CRP) are acknowledged as the gold standard tests for diagnosing CVD (Ghormade et al. 2014; Luna et al. 1982; Stewart et al. 1984). For example, Boonkaew et al. reported an electrochemical biosensor for the simultaneous detection of multiple cardiac biomarkers (i.e., cTnI, PCT, and CRP) (Boonkaew et al. 2021). The sensor was constructed by layering wax-patterned paper with a transparency film, secured using laser-cut double-sided adhesive. This design effectively defines the sample inlet, three distinct detection zones for each biomarker, and the fluidic channels that connect the inlet to the detection zones. Antibodies were immobilized on a GO-modified carbon electrode. When immunocomplexes (antigen-antibody) were produced on the electrode surface, they created a barrier that reduced the electron transfer rate of the redox solution, leading to a decrease in the electrochemical signal. The result showed a good linearity (R2 > 0.99), low DL (0.38 ng/mL for CRP, 0.16 pg/mL for cTn1, 0.27 pg/mL for PCT), and good reproducibility (RSD < 5%). Aptamer (i.e., single-stranded nucleic acids), capable of selectively recognizing target nucleic acids, proteins, and peptides, has been adopted for constructing the recognition element in the biosensor (Zhang et al. 2016b; Lv et al. 2019; Zeng et al. 2019). For example, Jo et al. selected DNA aptamers against cTnI biomarker with a very low dissociation constants value (270 pM) and incorporated the aptamers with gold electrodes for building an electrochemical biosensor (Jo et al. 2015). The proposed sensor demonstrated exceptional analytical performance for cTnI, offering a wide linear detection range of 1–10,000 pM in a buffer and an impressive DL of 1.0 pM, which is well below the standard clinical cutoff values of 70–400 pg/mL. Besides, the short linear binding peptide was also extensively utilized as a molecular recognition element in the sensing event of cardiac biomarkers (Wu et al. 2010). For example, Qi et al. have developed a series of innovative electrochemical peptide-based sensors specifically designed for the detection of cTnI biomarkers. In these sensors, a carefully selected peptide served as the biological recognition element, exhibiting a strong affinity for binding with cTnI molecules (Wang et al. 2012, 2016; Qi et al. 2013; Shan et al. 2014; Dong et al. 2015; Liu et al. 2017). Upon binding, this interaction created a barrier that impedes the mass transfer of the redox couple $[Fe(CN)_6]^{4-/3-}$ to the electrode surface. As a result, the electrochemical impedance spectroscopy (EIS) response is directly influenced by the concentration of cTnI biomarkers. What's more, molecularly imprinted polymers (MIPs) were used for recognition element construction because they contain binding sites that are complementary in shape and functionality to their target analytes (Regan et al. 2019; Zeng et al. 2010). Zuo et al. established the first MIP-based electrochemical sensor for cTnI biomarkers by utilizing the voltammetry method within 5 minutes (Zuo et al. 2016). It showed good dynamic linearity at a concentration range from 0.05 to 5 nM and a low DL of 0.027 nM. Yola and Atar combined the advantages of boron nitride quantum dots (ODs) and MIPs technique to create a novel electrochemical biosensor (Yola and Atar 2019). ODs were modified on the glass carbon electrodes to enhance the sensor's performance by improving its conductivity and surface area, which in turn increased the sensitivity and detection capabilities of the electrode. The use of cyclic voltammetry (CV)-assisted electro-polymerization of pyrrole enabled the creation of highly specific molecular cavities within the polymer matrix. These cavities were designed to be complementary in shape and size to the cTnI molecules, allowing for selective recognition and binding. This method of molecular imprinting on a boron nitride quantum dot-modified electrode not only provided high selectivity for cTnI but also offered rapid response times and low detection limits, making it a powerful tool for the detection of cardiac biomarkers in clinical diagnostics.

3.3 Conclusions and Future Trends

The future of biomarker-based biosensing is poised to revolutionize healthcare, diagnostics, and personalized medicine. As technology advances, biosensors are expected to become more sensitive, specific, and accessible, enabling earlier and more accurate detection of diseases. Here are some key trends and developments that will shape the future of biomarker-based biosensing:

In the future, an increase in wearable biosensors capable of continuously monitoring biomarkers in real time is a typical trend (Empelen et al. 2021; Kim et al. 2023b). Biosensors will integrate seamlessly with smartphones and digital health platforms, providing patients and healthcare providers with instant access to vital health data. Data from wearable biosensors will be used to generate personalized health insights, allowing for more tailored treatment plans and proactive management of chronic conditions. Another future focus will be on creating rapid, cost-effective biosensors that can deliver results within minutes, making them ideal for point-of-care settings in both developed and resource-limited regions (Guest et al. 2016; Dai and Liu 2019). The development of portable, user-friendly biosensing devices will allow patients to perform diagnostic tests at home, reducing the need for frequent hospital visits and enabling timely intervention. AI and machine learning algorithms will be integrated into biosensing platforms to analyze complex datasets, identify patterns, and provide more accurate diagnoses based on biomarker levels. AI-driven biosensors will be able to predict disease progression or the likelihood of developing certain conditions based on historical biomarker data, enabling preventative healthcare measures (Guo et al. 2024). Biomarker-based biosensing is a cornerstone of personalized and precision medicine, offering the potential for more accurate diagnoses, targeted therapies, and individualized treatment plans (Ahmad et al. 2023; Liu et al. 2023). For personalized diagnostics, biosensor enables the identification of genetic and molecular biomarkers that are specific to an individual's risk profile, allowing for individual risk assessment. Moreover, biosensors can detect specific mutations or protein expressions that characterize different cancer subtypes. This allows for a more precise diagnosis, enabling oncologists to tailor treatment plans based on the exact molecular profile of a patient's tumor. Biomarker-based biosensing in environmental and lifestyle impact monitoring represents a cutting-edge approach to understanding how external factors influence human health (Kumari and Dubey 2023). By detecting specific biomarkers associated with environmental exposures and lifestyle choices, these biosensors provide valuable insights that can guide both individual behaviors and public health strategies. Biomarkers, for example, are related to environmental exposures, such as pollutants or allergens, providing insights into how these factors impact health or are associated with diet, exercise, stress, and sleep, offering personalized recommendations for maintaining optimal health and wellness.

By integrating advanced biosensing technologies into public health strategies, global health initiatives are better equipped to address pressing health challenges, particularly in resource-limited settings (Adsul et al. 2024). Biosensors are increasingly being integrated into global health monitoring networks, which track the spread of diseases across different regions. These networks use data from biosensors to map disease prevalence and trends, helping health organizations allocate resources and coordinate responses more effectively. As biosensors collect and transmit sensitive health data, ensuring the privacy and security of this information will be a top priority (Pateraki et al. 2020). The ethical implications of using biomarker data in healthcare will be closely examined, with policies developed to protect individuals' rights and autonomy. In conclusion, the future of biomarker-based biosensing in healthcare is promising, with a focus on integration with digital health and wearables, point-of-care and home diagnostics, AI and machine learning integration, personalized and precision medicine, environmental and lifestyle impact monitoring, integration into global health initiatives, and addressing ethical and privacy concerns. These advancements will not only improve the quality of healthcare but also make it more accessible and effective globally.

References

Adsul P et al (2024) Challenges and opportunities for paving the road to global health equity through implementation science. Annu Rev Immunol 45:27

Ahmad A, Imran M, Ahsan H (2023) Biomarkers as biomedical bioindicators: approaches and techniques for the detection, analysis, and validation of novel biomarkers of diseases. Pharmaceutics 15:1630

Ahmed MU, Saaem I, Wu PC, Brown AS (2014) Personalized diagnostics and biosensors: a review of the biology and technology needed for personalized medicine. Crit Rev Biotechnol 34:180–196

AIDSinfo, A (2013) Guidelines for the use of antiretroviral agents in HIV-1-infected adults and adolescents

Ali MB et al (2006) Formaldehyde assay by capacitance versus voltage and impedance measurements using bi-layer bio-recognition membrane. Biosens Bioelectron 22:575–581

Altintas Z (2017) Biosensors and nanotechnology: applications in health care diagnostics. John Wiley & Sons

Aronson JK (2005) Biomarkers and surrogate endpoints. Br J Clin Pharmacol 59:491

Aronson JK, Ferner RE (2017) Biomarkers—a general review. Curr Protoc Pharmacol 76:9–23

Arroyo-Espliguero R et al (2004) C-reactive protein elevation and disease activity in patients with coronary artery disease. Eur Heart J 25:401–408

Bag S, Mandal D. Revolutionizing healthcare: the role of biosensors in modern medicine, CRC Press

Baranwal J, Barse B, Gatto G, Broncova G, Kumar A (2022) Electrochemical sensors and their applications: a review. Chem 10:363

Basu S (2010) Bioactive eicosanoids: role of prostaglandin F 2α and F 2-isoprostanes in inflammation and oxidative stress related pathology. Mol Cells 30:383–391

Basu N et al (2015) Risk of contralateral breast cancer in BRCA1 and BRCA2 mutation carriers: a 30-year semi-prospective analysis. Fam Cancer 14:531–538

Berk M (2015) The classification of biomarkers. JAMA Psychiatry 72:1056–1057

Biomarkers Definitions Working Group (2001) Biomarkers and surrogate endpoints: preferred definitions and conceptual framework. Clin Pharmacol Ther 69:89–95

Biswas GC, Choudhury S, Rabbani MM, Das J (2022) A review on potential electrochemical point-of-care tests targeting pandemic infectious disease detection: COVID-19 as a reference. Chem 10:269

Boland E et al (2001) Limitations of conventional methods of self-monitoring of blood glucose: lessons learned from 3 days of continuous glucose sensing in pediatric patients with type 1 diabetes. Diabetes Care 24:1858–1862

Boonkaew S et al (2021) Electrochemical paper-based analytical device for multiplexed, point-of-care detection of cardiovascular disease biomarkers. Sensors Actuators B Chem 330:129336

Bozzola A, Perotto S, De Angelis F (2017) Hybrid plasmonic–photonic whispering gallery mode resonators for sensing: a critical review. Analyst 142:883–898

Buyse M, Sargent DJ, Grothey A, Matheson A, De Gramont A (2010) Biomarkers and surrogate end points—the challenge of statistical validation. Nat Rev Clin Oncol 7:309–317

Carter N et al (1997) Electrochemical sensor. Biosens Bioelectron 9:vi

Chandra P (2023) Personalized biosensors for point-of-care diagnostics: from bench to bedside applications. Nanotheranostics 7:210

Chaubey A, Malhotra B (2002) Mediated biosensors. Biosens Bioelectron 17:441–456

Chen C, Wang J (2020) Optical biosensors: an exhaustive and comprehensive review. Analyst 145:1605–1628

Chen X-H, Huang S, Kerr D (2011) Biomarkers in clinical medicine. IARC Sci Publ 163:303–322

Chen C, Hou X, Si J (2017) Protein analysis by Mach-Zehnder interferometers with a hybrid plasmonic waveguide with nano-slots. Opt Express 25:31294–31308

Chen C, Hou X, Wang J (2020) A novel hybrid plasmonic resonator with high quality factor and large free spectral range. IEEE Sensors J 21:1644–1654

Chen C, Wang K, Luo L (2022a) AuNPs and 2D functional nanomaterial-assisted SPR development for the cancer detection: a critical review. Cancer Nanotechnol 13:29
Chen C, Wang K, Hou X (2022b) Protein conformation characterization via a silicon resonator-based optical sensor based on the combination of wavelength interrogation and dual polarization detection. Opt Express 30:44472–44486
Chen C, Wagn K, Hou X (2024) Characterizing amyloid protein conformation with a multi-resonant metasurface based biosensor in the infrared window. Opt Express 32:5287–5300
Chiu C (2018) Cutting-edge infectious disease diagnostics with CRISPR. Cell Host Microbe 23:702–704
Chu SS, Nguyen HA, Zhang J, Tabassum S, Cao H (2022) Towards multiplexed and multimodal biosensor platforms in real-time monitoring of metabolic disorders. Sensors 22:5200
Coulet PR (2019) What is a biosensor? In: Biosensor principles and applications. CRC Press pp 1–6
Cuna A, Kandasamy J, Fineberg N, Sims B (2013) B-type natriuretic peptide is a biomarker for pulmonary hypertension in preterm infants with bronchopulmonary dysplasia. Res Rep Neonatol 33–36
Dai Y, Liu CC (2019) Recent advances on electrochemical biosensing strategies toward universal point-of-care systems. Angew Chem 131:12483–12496
Davies J et al (2013) Assessment of clinical response to ivacaftor with lung clearance index in cystic fibrosis patients with a G551D-CFTR mutation and preserved spirometry: a randomised controlled trial. Lancet Respir Med 1:630–638
de Puig H et al (2021) Simultaneous detection of SARS-CoV-2 RNA and host antibodies enabled by a multiplexed electrochemical sensor platform. medRxiv
Delgado J et al (2012) Chronic lymphocytic leukaemia with 17p deletion: a retrospective analysis of prognostic factors and therapy results. Br J Haematol 157:67–74
Dhama K et al (2019) Biomarkers in stress related diseases/disorders: diagnostic, prognostic, and therapeutic values. Front Mol Biosci 6:91
Dong M et al (2015) Electrogenerated chemiluminescence peptide-based biosensing method for cardiac troponin I using peptide-integrating Ru (bpy) 3 2+-functionalized gold nanoparticles as nanoprobe. Gold Bull 48:21–29
Empelen PV, Otten WL, Molema H, Keijsers J, Mooij R (2021) Chapter 2 - Digital health in the era of personalized healthcare: opportunities and challenges for bringing research and patient care to a new level. In: Digital health: mobile and wearable devices for participatory health applications. Elsevier pp 7–31, ISBN 978012820077
Eraly SA et al (2014) Assessment of plasma C-reactive protein as a biomarker of posttraumatic stress disorder risk. JAMA Psychiatry 71:423–431
Estevez MC, Alvarez M, Lechuga LM (2012) Integrated optical devices for lab-on-a-chip biosensing applications. Laser Photonics Rev 6:463–487
Farrell PM et al (2008) Guidelines for diagnosis of cystic fibrosis in newborns through older adults: Cystic Fibrosis Foundation consensus report. J Pediatr 153:S4–S14
Firestein GS (2006) A biomarker by any other name. Nat Clin Pract Rheumatol 2:635–635
Fracchiolla NS, Artuso S, Cortelezzi A (2013) Biosensors in clinical practice: focus on oncohematology. Sensors 13:6423–6447
Freedland SJ, Moul JW (2007) Prostate specific antigen recurrence after definitive therapy. J Urol 177:1985–1991
García-Gutiérrez MS et al (2020) Biomarkers in psychiatry: concept, definition, types and relevance to the clinical reality. Front Psych 11:432
Ghormade PS, Kumar NB, Tingne CV, Keoliya AN (2014) Distribution & diagnostic efficacy of cardiac markers CK-MB & LDH in pericardial fluid for postmortem diagnosis of ischemic heart disease. J Forensic Leg Med 28:42–46
Gika HG, Theodoridis GA, Plumb RS, Wilson ID (2014) Current practice of liquid chromatography–mass spectrometry in metabolomics and metabonomics. J Pharm Biomed Anal 87:12–25
Górecka E, Jastrzębska M (2011) Immobilization techniques and biopolymer carriers. Biotechnol Food Sci 75:65–86

Gosho M, Nagashima K, Sato Y (2012) Study designs and statistical analyses for biomarker research. Sensors 12:8966–8986
Grieshaber D, MacKenzie R, Vörös J, Reimhult E (2008) Electrochemical biosensors-sensor principles and architectures. Sensors 8:1400–1458
Gross TM, Mastrototaro JJ (2000) Efficacy and reliability of the continuous glucose monitoring system. Diabetes Technol Ther 2:19–26
Guest F, Guest P, Martins-de-Souza D (2016) The emergence of point-of-care blood-based biomarker testing for psychiatric disorders: enabling personalized medicine. Biomark Med 10: 431–443. Mol Psychiatry 15:1088–1100
Gundogdu F et al (2011) The role of serum CA-125 levels and CA-125 tissue expression positivity in the prediction of the recurrence of stage III and IV epithelial ovarian tumors (CA-125 levels and tissue CA-125 in ovarian tumors). Arch Gynecol Obstet 283:1397–1402
Guo L et al (2024) Electrochemical protein biosensors for disease marker detection: progress and opportunities. Microsyst Nanoeng 10:65
Hajian R et al (2019) Detection of unamplified target genes via CRISPR–Cas9 immobilized on a graphene field-effect transistor. Nat Biomed Eng 3:427–437
Haleem A, Javaid M, Singh RP, Suman R, Rab S (2021) Biosensors applications in medical field: a brief review. Sensors International 2:100100
Health, U. D. o. & Services, H. Panel on Antiretroviral Guidelines for Adults and Adolescents (2019) Guidelines for the use of antiretroviral agents in HIV-1-infected adults and adolescents, 2021
Hernández-Rodríguez JF, Rojas D, Escarpa A (2020) Electrochemical sensing directions for next-generation healthcare: trends, challenges, and frontiers. Anal Chem 93:167–183
Horgan RP, Kenny LC (2011) 'Omic'technologies: genomics, transcriptomics, proteomics and metabolomics. Obstet Gynaecol 13:189
Hudler P, Kocevar N, Komel R (2014) Proteomic approaches in biomarker discovery: new perspectives in cancer diagnostics. Sci World J 2014:260348
Isse AA, Gennaro A (2010) Absolute potential of the standard hydrogen electrode and the problem of interconversion of potentials in different solvents. J Phys Chem B 114:7894–7899
Ito S, Hachiya H, Baba K, Asano Y, Wada H (1995) Improvement of the silver/silver chloride reference electrode and its application to pH measurement. Talanta 42:1685–1690
Jerkiewicz G (2022) Applicability of platinum as a counter-electrode material in electrocatalysis research. ACS Catal 12:2661–2670
Jo H et al (2015) Electrochemical aptasensor of cardiac troponin I for the early diagnosis of acute myocardial infarction. Anal Chem 87:9869–9875
Kang JH et al (2008) Thermoresponsive hydrogel photonic crystals by three-dimensional holographic lithography. Adv Mater 20:3061–3065
Kang X et al (2009) Glucose oxidase–graphene–chitosan modified electrode for direct electrochemistry and glucose sensing. Biosens Bioelectron 25:901–905
Kaptoge S et al (2019) World Health Organization cardiovascular disease risk charts: revised models to estimate risk in 21 global regions. Lancet Glob Health 7:e1332–e1345
Karsdal M et al (2009) Biochemical markers and the FDA critical path: how biomarkers may contribute to the understanding of pathophysiology and provide unique and necessary tools for drug development. Biomarkers 14:181–202
Kashefi-Kheyrabadi L et al (2022) Rapid, multiplexed, and nucleic acid amplification-free detection of SARS-CoV-2 RNA using an electrochemical biosensor. Biosens Bioelectron 195:113649
Kellner MJ, Koob JG, Gootenberg JS, Abudayyeh OO, Zhang F (2019) SHERLOCK: nucleic acid detection with CRISPR nucleases. Nat Protoc 14:2986–3012
Khan H, Shah MR, Barek J, Malik MI (2023) Cancer biomarkers and their biosensors: a comprehensive review. TrAC Trends Anal Chem 158:116813
Kim ER, Joe C, Mitchell RJ, Gu MB (2023a) Biosensors for healthcare: current and future perspectives. Trends Biotechnol 41:374–395
Kim H et al (2023b) Recent advances in multiplexed wearable sensor platforms for real-time monitoring lifetime stress: a review. Biosensors 13:470

Kolthoff I, Thomas F (1965) Electrode potentials in acetonitrile. Estimation of the liquid junction potential between acetonitrile solutions and the aqueous saturated calomel electrode1. J Phys Chem 69:3049–3058

Kotha S, Neong S, Patel K (2018) Serum biomarkers for diagnosis and monitoring viral hepatitis and hepatocellular carcinoma. Expert Rev Mol Diagn 18:713–722

Kumari N, Dubey A (2023) Environment and human health: beyond disciplinary confinement. Mind Soc 12:47–53

Kyle RA, Rajkumar SV (2003) Monoclonal gammopathies of undetermined significance: a review. Immunol Rev 194:112–139

Lakshmanakumar M et al (2019) Functionalized graphene quantum dot interfaced electrochemical detection of cardiac Troponin I: an antibody free approach. Sci Rep 9:17348

Liao GJ, Clark AS, Schubert EK, Mankoff DA (2016) 18F-fluoroestradiol PET: current status and potential future clinical applications. J Nucl Med 57:1269–1275

Liu X et al (2017) Highly sensitive electrochemiluminescence assay for cardiac troponin I and adenosine triphosphate by using supersandwich amplification and bifunctional aptamer. ChemElectroChem 4:1708–1713

Liu Y et al (2023) Revolutionizing precision medicine: exploring wearable sensors for therapeutic drug monitoring and personalized therapy. Biosensors 13:726

Lu H, He B, Gao B (2021) Emerging electrochemical sensors for life healthcare. Eng Regen 2:175–181

Luna A, Villanueva E, Castellano M, Jimenez G (1982) The determination of CK, LDH and its isoenzymes in pericardial fluid and its application to the post-mortem diagnosis of myocardial infarction. Forensic Sci Int 19:85–91

Lv S, Zhang K, Zhu L, Tang D (2019) ZIF-8-assisted NaYF4: Yb, Tm@ ZnO converter with exonuclease III-powered DNA walker for near-infrared light responsive biosensor. Anal Chem 92:1470–1476

MacNamara J, Eapen DJ, Quyyumi A, Sperling L (2015) Novel biomarkers for cardiovascular risk assessment: current status and future directions. Futur Cardiol 11:597–613

Majkić-Singh N (2011) What is A biomarker? From its discovery to clinical application. J Med Biochem 30:186

Mateo C, Palomo JM, Van Langen LM, Van Rantwijk F, Sheldon RA (2004) A new, mild cross-linking methodology to prepare cross-linked enzyme aggregates. Biotechnol Bioeng 86:273–276

Miletić N, Loos K (2009) Over-stabilization of chemically modified and cross-linked Candida Antarctica lipase B using various epoxides and diepoxides. Aust J Chem 62:799–805

Mohankumar P, Ajayan J, Mohanraj T, Yasodharan R (2021) Recent developments in biosensors for healthcare and biomedical applications: a review. Measurement 167:108293

Monzó J, Insua I, Fernandez-Trillo F, Rodriguez P (2015) Fundamentals, achievements and challenges in the electrochemical sensing of pathogens. Analyst 140:7116–7128

Mukherjee R et al (2015) Diagnosis and management of hepatitis C virus infection. J Lab Autom 20:519–538

Mukherjee S et al (2024) 2D graphene-based advanced nanoarchitectonics for electrochemical biosensors: applications in cancer biomarker detection. Biosens Bioelectron 116050:116050

Nair PR, Alam MA (2007) Dimensionally frustrated diffusion towards fractal adsorbers. Phys Rev Lett 99:256101

Noronha V et al (2020) Gefitinib versus gefitinib plus pemetrexed and carboplatin chemotherapy in EGFR-mutated lung cancer. J Clin Oncol 38:124–136

Nouri R et al (2021) CRISPR-based detection of SARS-CoV-2: A review from sample to result. Biosens Bioelectron 178:113012

Orsega S (2007) Treatment of adult HIV infection: antiretroviral update and overview. J Nurse Pract 3:612–624

Owiredu W, Amegatcher G, Amidu N (2009) Precision and accuracy of three blood glucose meters: accu-chek advantage, one touch horizon, and sensocard. J Med Sci 9:185–193

Park S-J, Taton TA, Mirkin CA (2002) Array-based electrical detection of DNA with nanoparticle probes. Science 295:1503–1506

Pateraki M et al (2020) Biosensors and internet of things in smart healthcare applications: challenges and opportunities. In: Wearable and implantable medical devices, pp 25–53
Patolsky F, Zheng G, Lieber CM (2006) Nanowire-based biosensors. ACS Publications
Prieto-Simon B, Campas M, Marty J-L (2008) Biomolecule immobilization in biosensor development: tailored strategies based on affinity interactions. Protein Pept Lett 15:757–763
Qi H et al (2013) Ultrasensitive electrogenerated chemiluminescence peptide-based method for the determination of cardiac troponin I incorporating amplification of signal reagent-encapsulated liposomes. Anal Chem 85:3886–3894
Rasooly A, Jacobson J (2006) Development of biosensors for cancer clinical testing. Biosens Bioelectron 21:1851–1858
Regan B, Boyle F, O'Kennedy R, Collins D (2019) Evaluation of molecularly imprinted polymers for point-of-care testing for cardiovascular disease. Sensors 19:3485
Relling M et al (2011) Clinical pharmacogenetics implementation consortium guidelines for thiopurine methyltransferase genotype and thiopurine dosing. Clin Pharmacol Ther 89:387–391
Roberts SG, Blute ML, Bergstralh EJ, Slezak JM, Zincke H (2001) Mayo Clinic Proceedings. Elsevier, pp 576–581
Rodbard D (2016) Continuous glucose monitoring: a review of successes, challenges, and opportunities. Diabetes Technol Ther 18:S2-3-S2-13
Roush GC, Sica DA (2016) Diuretics for hypertension: a review and update. Am J Hypertens 29:1130–1137
Rundle A, Ahsan H, Vineis P (2012) Better cancer biomarker discovery through better study design. Eur J Clin Investig 42:1350–1359
Şahin B, İlgün G (2022) Risk factors of deaths related to cardiovascular diseases in World Health Organization (WHO) member countries. Health Soc Care Community 30:73–80
Sandler CN, McDonnell ME (2016) The role of hemoglobin A1c in the assessment of diabetes and cardiovascular risk. Cleve Clin J Med 83:S4–S10
Shah M et al (2014) Association between angiotensin converting enzyme inhibitor or angiotensin receptor blocker use prior to major elective surgery and the risk of acute dialysis. BMC Nephrol 15:1–9
Shan M et al (2014) Sensitive electrogenerated chemiluminescence peptide-based biosensor for the determination of troponin I with gold nanoparticles amplification. Gold Bull 47:57–64
Shanafelt TD et al (2006) Prospective evaluation of clonal evolution during long-term follow-up of patients with untreated early-stage chronic lymphocytic leukemia. J Clin Oncol 24:4634–4641
Simoska O, Stevenson KJ (2019) Electrochemical sensors for rapid diagnosis of pathogens in real time. Analyst 144:6461–6478
Sin ML, Mach KE, Wong PK, Liao JC (2014) Advances and challenges in biosensor-based diagnosis of infectious diseases. Expert Rev Mol Diagn 14:225–244
Singh B, Ma S, Hara TO, Singh S (2023) Nanomaterials-based biosensors for the detection of prostate cancer biomarkers: recent trends and future perspective. Adv Mater Technol 8:2201860
Siu AL, U S Preventive Services Task Force (2015) U. P. S. T. Screening for abnormal blood glucose and type 2 diabetes mellitus: US Preventive Services Task Force recommendation statement. Ann Intern Med 163:861–868
Sophocleous M, Atkinson JK (2017) A review of screen-printed silver/silver chloride (Ag/AgCl) reference electrodes potentially suitable for environmental potentiometric sensors. Sensors Actuators A Phys 267:106–120
Stewart RV, Zumwalt RE, Hirsch CS, Kaplan L (1984) Postmortem diagnosis of myocardial disease by enzyme analysis of pericardial fluid. Am J Clin Pathol 82:411–417
Stohl W, Hilbert DM (2012) The discovery and development of belimumab: the anti-BLyS–lupus connection. Nat Biotechnol 30:69–77
Stone NJ, Lloyd-Jones DM (2015) Mass Med Soc 372:1564–1565
Strimbu K, Tavel JA (2010) What are biomarkers? Curr Opin HIV AIDS 5:463–466
Sunyer J et al (2009) Interaction between smoking and the interleukin-6 gene affects systemic levels of inflammatory biomarkers. Nicotine Tob Res 11:1347–1353

Tainsky MA (2009) Genomic and proteomic biomarkers for cancer: a multitude of opportunities. Biochim Biophys Acta (BBA) 1796:176–193
Tang L, Yang J, Wang Y, Deng R (2023) Recent advances in cardiovascular disease biosensors and monitoring technologies. ACS Sensors 8:956–973
Thévenot DR, Toth K, Durst RA, Wilson GS (2001) Electrochemical biosensors: recommended definitions and classification. Biosens Bioelectron 16:121–131
Tomasina J, Lheureux S, Gauduchon P, Rault S, Malzert-Fréon A (2013) Nanocarriers for the targeted treatment of ovarian cancers. Biomaterials 34:1073–1101
Tringe SG, Rubin EM (2005) Metagenomics: DNA sequencing of environmental samples. Nat Rev Genet 6:805–814
Tu J, Torrente-Rodríguez RM, Wang M, Gao W (2020) The era of digital health: a review of portable and wearable affinity biosensors. Adv Funct Mater 30:1906713
Umelo IA, Costanza B, Castronovo V (2018) Innovative methods for biomarker discovery in the evaluation and development of cancer precision therapies. Cancer Metastasis Rev 37:125–145
Vallera DA, Bissell MG, Barron W (1991) Accuracy of portable blood glucose monitoring: effect of glucose level and prandial state. Am J Clin Pathol 95:247–252
Van Dijk WD et al (2013) The acute effect of cigarette smoking on the high-sensitivity CRP and fibrinogen biomarkers in chronic obstructive pulmonary disease patients. Biomark Med 7:211–219
Vigneshvar S, Sudhakumari C, Senthilkumaran B, Prakash H (2016) Recent advances in biosensor technology for potential applications–an overview. Front Bioeng Biotechnol 4:11
Wanekaya AK, Chen W, Myung NV, Mulchandani A (2006) Nanowire-based electrochemical biosensors. Electroanalysis 18:533–550
Wang J (2006) Electrochemical biosensors: towards point-of-care cancer diagnostics. Biosens Bioelectron 21:1887–1892
Wang C, Qi H, Qiu X, Gao Q, Zhang C (2012) Homogeneous electrogenerated chemiluminescence peptide-based method for determination of troponin I. Anal Methods 4:2469–2474
Wang B, Jing R, Qi H, Gao Q, Zhang C (2016) Label-free electrochemical impedance peptide-based biosensor for the detection of cardiac troponin I incorporating gold nanoparticles modified carbon electrode. J Electroanal Chem 781:212–217
Weckwerth W, Morgenthal K (2005) Metabolomics: from pattern recognition to biological interpretation. Drug Discov Today 10:1551–1558
Wishart DS et al (2021) MarkerDB: an online database of molecular biomarkers. Nucleic Acids Res 49:D1259–D1267
Wu J, Cropek DM, West AC, Banta S (2010) Development of a troponin I biosensor using a peptide obtained through phage display. Anal Chem 82:8235–8243
Yetisen AK, Butt H, Yun S-H (2016) Photonic crystal flakes. Acs Sensors 1:493–497
Yola ML, Atar N (2019) Development of cardiac troponin-I biosensor based on boron nitride quantum dots including molecularly imprinted polymer. Biosens Bioelectron 126:418–424
Yoon D, Baggett A, Phongikaroon S, King JA, Marsden K (2019) Fundamental data acquisition toward silver-silver chloride reference electrode. J Electrochem Soc 166:E159
Zeng Z, Hoshino Y, Rodriguez A, Yoo H, Shea KJ (2010) Synthetic polymer nanoparticles with antibody-like affinity for a hydrophilic peptide. ACS Nano 4:199–204
Zeng R et al (2019) Palindromic molecular beacon based Z-scheme BiOCl-Au-CdS photoelectrochemical biodetection. Anal Chem 91:2447–2454
Zhang A, Sun H, Yan G, Wang P, Wang X (2016a) Mass spectrometry-based metabolomics: applications to biomarker and metabolic pathway research. Biomed Chromatogr 30:7–12
Zhang Y, Sun Z, Tang L, Zhang H, Zhang G-J (2016b) Aptamer based fluorescent cocaine assay based on the use of graphene oxide and exonuclease III-assisted signal amplification. Microchim Acta 183:2791–2797
Zhou C et al (2012) Statistical considerations of optimal study design for human plasma proteomics and biomarker discovery. J Proteome Res 11:2103–2113
Zuo J et al (2016) A new molecularly imprinted polymer (MIP)-based electrochemical sensor for monitoring cardiac troponin I (cTnI) in the serum. Electroanalysis 28:2044–2049

Biochemical, Biophysical, and Digital Biomarkers in Precision Healthcare

4

Ravish Sharma, Gairika Ghosh, Amal Abbas,
and Kuldeep Mahato

Abstract

Biomarkers are essential tools in modern healthcare, offering measurable indicators of biological processes, disease states, and therapeutic responses. This chapter explores the pivotal role of biomarkers in precision healthcare, categorizing them into molecular, biophysical, and digital types, and emphasizing their importance in enabling personalized medicine. Molecular biomarkers, such as deoxyribonucleic acid (DNA), ribonucleic acid (RNA), and protein-based markers, are central to disease detection and treatment stratification, with applications in cancer, cardiovascular diseases, and neurodegenerative disorders. Physiological biomarkers, including vital signs like heart rate, blood pressure, and oxygen saturation, are increasingly monitored in real time through wearable devices and bioelectronic sensors, transforming chronic disease management. Inflammatory and immune biomarkers, such as cytokines and C-reactive protein (CRP), provide critical insights into the immune response and inflammation, aiding in the diagnosis and management of autoimmune and infectious diseases. The integration of digital biomarkers, derived from mobile health platforms, wearables, and AI-driven algorithms, further enables continuous monitoring of mental health, lifestyle factors, and sleep patterns, offering a comprehensive view of individual health. This chapter also highlights emerging trends in biomarker-based moni-

R. Sharma · G. Ghosh
Department of Chemistry and Biochemistry, University of California San Diego, La Jolla, CA, USA

A. Abbas
Aiiso Yufeng Li Family Department of Chemical and NanoEngineering, University of California San Diego, La Jolla, CA, USA

K. Mahato (✉)
Department of Biotechnology, Motilal Nehru National Institute of Technology Allahabad, Prayagraj, Uttar Pradesh, India

K. Mahato, P. Chandra (eds.), *Nano-bioelectronics for Precision Health Monitoring*, https://doi.org/10.1007/978-981-95-2117-3_4

toring, including advances in biosensor technologies, multi-omics integration, and the convergence of artificial intelligence and bioelectronics. It concludes by addressing the regulatory and ethical challenges inherent in biomarker-driven healthcare and envisions a future where predictive and preventive health monitoring redefines personalized care. This comprehensive overview underscores the transformative potential of biomarkers in improving health outcomes and advancing precision medicine.

Keywords

Biomarkers · Precision healthcare · Molecular biomarkers · Wearable biosensors · Digital health monitoring

4.1 Introduction

In the modern healthcare landscape, biomarkers have emerged as crucial tools in disease diagnosis, monitoring, and management. A biomarker is defined as a biological molecule found in blood, other body fluids, or tissues that signifies a normal or abnormal process, or a condition or disease. Biomarkers can be anything from genes, proteins, hormones, or enzymes, to metabolites, cellular structures, or whole physiological processes (Mahato et al. 2018a). For a detailed overview of the fundamentals of biomarkers, refer to Chap. 3. They provide measurable indicators of a biological state or condition, offering essential clues about the health or disease state of an individual. The importance of biomarkers in healthcare cannot be overstated. They play a pivotal role in precision medicine, where treatments and healthcare strategies are tailored to individual patients (Mahato et al. 2022). Biomarkers provide an objective way to diagnose diseases earlier and more accurately than traditional diagnostic methods, often before symptoms appear (Mahato et al. 2018b; Mahato and Chandra 2022, 2024). In disease management, biomarkers are used to assess how well a treatment is working or to monitor the progression of a disease (Saha et al. 2023). They can help predict disease risk, evaluate the likelihood of disease recurrence, or determine the need for lifestyle interventions. As such, biomarkers are not only critical in diagnosing and treating diseases but also in preventive healthcare (Mahato et al. 2024). Biomarkers can be classified into several types based on their nature and clinical utility (Zoccali et al. 2024).

- *Molecular biomarkers*: These are biological molecules, such as DNA, RNA, proteins, and metabolites, which provide information at a molecular level. Molecular biomarkers are widely used for diagnostic purposes and to predict the outcome or response to treatment. Genetic mutations (e.g., BRCA mutations in breast cancer) (John et al. 2024) or protein expression levels (Mahato et al. 2018a) (e.g., prostate-specific antigen, PSA, in prostate cancer) (Kachuri et al. 2023) are classic examples.
- *Physiological biomarkers*: These are indicators related to physiological states or processes. They can be as simple as heart rate (Avram et al. 2019), blood pres-

sure (Higashi 2023), body temperature (Windred et al. 2024), or more complex measurements such as glucose levels (Tehrani et al. 2022) or lung function. Physiological biomarkers are increasingly monitored using wearable technologies for real-time health monitoring (Mahato and Chandra 2024).

- *Digital biomarkers*: As healthcare embraces digital transformation, the concept of digital biomarkers has emerged. Digital biomarkers refer to health-related information collected through digital devices such as smartphones, wearables, and sensors. These devices capture data like physical activity, sleep patterns, or mood fluctuations (Sempionatto et al. 2022). Digital biomarkers hold immense promise in preventive healthcare, offering a way to continuously track and monitor individuals' health states.

Personalized medicine, often referred to as precision medicine, relies heavily on the use of biomarkers to customize healthcare for individual patients. Personalized medicine shifts away from the "one-size-fits-all" approach of traditional medicine and instead uses biomarkers to tailor treatment plans based on the genetic, molecular, and environmental characteristics of the patient (Mathur and Sutton 2017). This shift is driven by advancements in genomics, bioinformatics, and bioelectronics. Biomarkers allow for the stratification of patients into subgroups based on their risk of disease, their likely response to a particular treatment, or their likelihood of experiencing adverse reactions to a drug. For example, biomarkers like HER2 (human epidermal growth factor receptor 2) are used to determine whether a patient with breast cancer is likely to respond to HER2-targeted therapies, such as trastuzumab (Herceptin) leading to more effective treatments, fewer side effects, and eventually better patient outcomes (Xu et al. 2024). Additionally, biomarkers are central to pharmacogenomics, which studies how genetic variations affect an individual's response to drugs (Patrinos 2020). Pharmacogenomic biomarkers help in selecting the appropriate drugs and dosages for patients, minimizing the risk of adverse drug reactions and enhancing therapeutic efficacy (Chenchula et al. 2024). For instance, the genetic marker CYP2C9 influences how patients metabolize warfarin, a common blood thinner. By identifying this biomarker, healthcare providers can adjust warfarin doses to prevent bleeding complications while ensuring the drug's effectiveness (Fung et al. 2012). The process of identifying new biomarkers and validating their clinical utility is a complex and multistage effort (Chau et al. 2008). Biomarker discovery involves identifying new molecules or measurable biological indicators that are associated with specific diseases or health conditions. This process is often driven by advances in omics technologies, such as genomics, proteomics, metabolomics, and transcriptomics. These technologies allow scientists to explore the entire genome, proteome, or metabolome of patients to identify specific changes that may serve as biomarkers. For example, researchers might compare the gene expression profiles of patients with and without a certain disease to identify genes that are upregulated or downregulated in the diseased state (Yu et al. 2021; Lv et al. 2024). These genes could then be investigated further to determine whether they serve as reliable biomarkers for the disease. Once potential biomarkers have been discovered, they must be rigorously validated before they can be used in clinical settings. Biomarker validation involves confirming that the biomarker is consistently and specifically associated with the disease or condition of interest. Validation requires large-scale studies, often involving diverse patient populations, to ensure that the biomarker accurately reflects the disease state in different individuals and under different conditions.

4.2 Molecular Biomarkers in Disease Detection

Molecular biomarkers have become indispensable in modern medicine for the early detection, diagnosis, and prognosis of diseases. Unlike traditional diagnostic methods, which often rely on symptomatic presentation, molecular biomarkers offer the ability to detect diseases at their earliest stages by revealing changes at the genetic, proteomic, or metabolic levels (Doroszkiewicz et al. 2022). Molecular biomarkers are generally classified into three main categories: genomic, proteomic, and metabolomic biomarkers (Sempionatto et al. 2022; Smelik et al. 2024). Genomic biomarkers encompass DNA- and RNA-based indicators that reveal alterations at the genetic or transcriptional level, including mutations, gene amplifications, or epigenetic modifications. Proteomic biomarkers are protein-based markers that represent changes in protein expression or posttranslational modifications linked to pathological processes. Metabolomic biomarkers consist of small-molecule metabolites that reflect the biochemical status of an organism, offering valuable insights into metabolic disturbances associated with various diseases. These biomarkers enable clinicians and researchers to identify disease predispositions, predict responses to treatments, and monitor disease progression or therapeutic efficacy (Jain and Jain 2010). In precision healthcare, molecular biomarkers are particularly valuable as they offer a more personalized approach, guiding therapeutic decisions based on the individual's unique molecular profile (Graig et al. 2016).

4.2.1 DNA-Based Biomarkers

Among the most widely recognized molecular biomarkers are DNA-based, especially in the context of cancer. These include genetic mutations, alterations in DNA copy number, and epigenetic modifications such as DNA methylation (Herceg and Hainaut 2007). Mutations in oncogenes and tumor suppressor genes can drive tumorigenesis, and detecting these mutations can aid in diagnosing specific cancers or identifying individuals at high risk of developing cancer. For example, mutations in the BRCA1 and BRCA2 genes are associated with a significantly increased risk of developing breast and ovarian cancer (Mersch et al. 2015). Detecting these mutations allows for preventive measures, such as increased surveillance or prophylactic surgery, to be taken before the cancer develops. Another notable example is the detection of mutations in the KRAS gene, which plays a crucial role in cell signaling pathways regulating growth and survival (Huang et al. 2021). KRAS mutations are commonly found in cancers such as colorectal, pancreatic, and lung cancer (Lee et al. 2022). The identification of these mutations helps oncologists tailor treatment plans, as patients with specific KRAS mutations may not respond well to certain therapies, such as EGFR inhibitors (Liu et al. 2024). Beyond genetic mutations, epigenetic changes, reversible modifications to DNA that do not alter the genetic code itself, also serve as important molecular biomarkers for disease detection. DNA methylation, the addition of a methyl group to cytosine residues in the DNA sequence, can silence gene expression and is frequently altered in cancer (Liu et al. 2023). For example, hypermethylation of the MGMT (O6-methylguanine-DNA methyltransferase) promoter is a biomarker used in glioblastoma. MGMT is responsible for repairing DNA damage caused by chemotherapy, and its methylation can predict the patient's response to treatment. Patients with MGMT hypermethylation are more likely to respond to alkylating agents like temozolomide, as the silenced MGMT gene cannot repair the chemotherapy-induced DNA damage (Oldrini et al. 2020). Another exciting advancement in molecular biomarkers is the use of circulating tumor DNA (ctDNA) for noninvasive cancer detection and monitoring (Mishra et al. 2024). This ctDNA consists of small fragments of tumor DNA that are shed into the bloodstream by cancer cells. Liquid biopsy, a blood test that detects ctDNA, offers a noninvasive alternative to traditional tissue biopsy for identifying cancer mutations, tracking tumor progression, and monitoring the effectiveness of treatments. For instance, in lung cancer, ctDNA testing can identify mutations in the EGFR gene, allowing clinicians to determine if patients are likely to benefit from EGFR-targeted therapies such as erlotinib or gefitinib. Liquid biopsies are particularly useful for monitoring patients during treatment, as they can detect the emergence of resistance mutations or disease recurrence earlier than imaging techniques or traditional biomarkers.

4.2.2 RNA-Based Biomarkers

RNA-based biomarkers offer another promising avenue for early disease detection and monitoring (Mahato and Chandra 2024). Unlike DNA, which remains relatively stable throughout life, RNA is dynamic and reflects real-time changes in gene expression that may be linked to disease processes (Teles and Fonseca 2008). One of the most researched classes of RNA biomarkers is microRNAs (miRNAs), small noncoding RNA molecules that regulate gene expression at the posttranscriptional level. These miRNAs can either promote or inhibit the translation of messenger RNA (mRNA) into proteins, and their dysregulation has been linked to numerous diseases, including cancer, cardiovascular diseases, and neurodegenerative disorders (Ratti et al. 2020). In neurodegenerative diseases such as Alzheimer's disease (AD), Parkinson's disease (PD), and amyotrophic lateral sclerosis (ALS), specific miRNA expression profiles have been identified as biomarkers (Grasso et al. 2014). For example, miR-29a and miR-29b are downregulated in the brains of Alzheimer's disease patients and are involved in the regulation of amyloid precursor protein (APP) expression, which plays a central role in the development of AD (Pereira et al. 2016). Similarly, miR-133b has been implicated in Parkinson's disease as it is involved in the regulation of dopamine neuron function. Changes in miRNA levels can be detected in cerebrospinal fluid or blood, offering potential noninvasive biomarkers for neurodegenerative diseases (Shaheen et al. 2024). Another class of RNA biomarkers gaining attention in disease detection is long noncoding RNAs (lncRNAs), which are transcripts longer than 200 nucleotides that do not code for proteins but have regulatory roles in gene expression. Aberrant expression of lncRNAs has been linked to various diseases, including cancer, cardiovascular, and neurodegenerative diseases (Uppaluri et al. 2023). For instance, in Alzheimer's disease, the lncRNA BACE1-AS has been shown to stabilize the expression of BACE1, a key enzyme involved in the production of amyloid-beta plaques, which are hallmarks of Alzheimer's pathology. Detecting changes in lncRNA levels could offer new avenues for early diagnosis and understanding the progression of neurodegenerative diseases (Wu et al. 2024).

4.2.3 Protein-Based Biomarkers

Proteins are the workhorses of the cell, and their levels or modifications often reflect the underlying physiological or pathological processes in the body. Protein-based biomarkers are widely used in disease detection including cardiovascular, infectious diseases, etc. (Guo et al. 2024). For instance, in cardiovascular diseases, certain proteins are released into the bloodstream during heart damage or stress. The most widely recognized protein biomarker for cardiovascular disease is troponin, a regulatory protein involved in muscle contraction (Babuin and Jaffe 2005). During a myocardial infarction (heart attack), cardiac cells are damaged, and troponins (particularly troponin I and troponin T) are released into the bloodstream. Elevated levels of troponin are considered the gold standard for diagnosing acute myocardial infarction, as they indicate heart muscle damage (Babuin and Jaffe 2005). The sensitivity of troponin assays has improved significantly over the years, with high-sensitivity troponin (hs-troponin) assays capable of detecting even very low levels of troponin in the blood. This has allowed for earlier detection of heart attacks and better risk stratification of patients presenting with chest pain (Lazar et al. 2022). In addition to troponin, other protein biomarkers such as B-type natriuretic peptide (BNP) and its precursor, N-terminal proBNP (NT-proBNP), are commonly used in the diagnosis and management of heart failure. BNP is released by the ventricles of the heart in response to increased pressure, which occurs during heart failure. Elevated levels of BNP or NT-proBNP are indicative of heart failure and can be used to assess the severity of the condition, guide treatment decisions, and predict patient outcomes. The use of BNP and NT-proBNP in clinical practice extends beyond diagnosis; these biomarkers are also valuable for monitoring the effectiveness of treatment and detecting worsening heart failure (Cao et al. 2019). Serial measurements of BNP can help clinicians adjust therapies, such as diuretics and vasodilators, to optimize heart function and prevent hospitalization.

4.2.4 Metabolomic Biomarkers

Metabolomic biomarkers, which encompass small molecules involved in metabolic pathways, provide insights into the metabolic state of an individual and have significant potential in diagnosing and managing metabolic disorders (Qiu et al. 2023). However, glucose levels alone do not provide a complete picture of disease progression or risk, leading to the exploration of additional metabolomic biomarkers. One of the most important protein biomarkers in diabetes management is glycated hemoglobin (HbA1c), which reflects the average blood glucose levels over the past 2–3 months (Ortiz-Martínez et al. 2022). HbA1c is a product of the nonenzymatic glycation of hemoglobin and serves as a reliable indicator of long-term glycemic control. In clinical practice, HbA1c is used to diagnose diabetes and to monitor the effectiveness of treatment, as it correlates with the risk of diabetes-related complications such as retinopathy, nephropathy, and cardiovascular disease (Sherwani et al. 2016). Recent advances in metabolomics have enabled the identification of additional biomarkers for diabetes and metabolic disorders. For example, changes in levels of branched-chain amino acids (BCAAs), such as leucine, isoleucine, and valine, have been associated with insulin resistance and an increased risk of developing type 2 diabetes (Shahisavandi et al. 2023). Similarly, lipidomic profiling reveals alterations in lipid species, such as ceramides and phospholipids, in individuals with insulin resistance, providing potential biomarkers for early detection and stratification of metabolic disorders (Chen et al. 2023).

4.3 Biophysical Biomarkers

Biophysical biomarkers are indicators of normal biological processes, pathological states, or pharmacological responses to therapeutic interventions, measured through easily accessible physical parameters such as heart rate, blood pressure, respiratory rate, and glucose levels (Ayres 2020). Unlike molecular biomarkers, which often require invasive procedures or complex assays for detection, physiological biomarkers can be measured noninvasively, making them ideal for continuous, real-time health monitoring.

In precision healthcare, the continuous monitoring of biophysical biomarkers presents several distinct advantages. It enables early detection of disease by identifying abnormal fluctuations in physiological parameters in real time, often before clinical symptoms become evident. This proactive approach supports timely interventions. Furthermore, continuous assessment allows for personalized monitoring, generating individualized data that can inform optimized treatment plans, customized lifestyle modifications, and preventive strategies to mitigate health risks. Additionally, biophysical biomarkers offer dynamic feedback on treatment efficacy, facilitating prompt therapeutic adjustments to enhance clinical outcomes (Choi et al. 2022). The rise of wearable technologies, mobile health (mHealth) platforms, and biosensors has revolutionized how we monitor biophysical biomarkers, enabling real-time data collection and analysis (Huhn et al. 2022). These devices have the potential to transform healthcare by shifting from reactive care, where treatment is initiated after symptoms appear, to proactive care, where preventive measures can be taken based on real-time data. This section discusses various biophysical biomarkers for precision healthcare.

4.3.1 Biomarkers for Cardiovascular Health Monitoring

The cardiovascular system provides a wealth of biophysical biomarkers that can be monitored in real time to assess heart function, detect early signs of cardiovascular disease, and track the effectiveness of interventions. These include heart rate, heart rate variability, blood pressure, and oxygen saturation, all of which are routinely monitored in clinical settings and increasingly in home care through wearable devices (Mbewu and Mbanya 2006). One of the most commonly monitored biophysical biomarkers is heart rate. It reflects the number of times the heart beats per minute and is a key indicator of overall cardiovascular health. Elevated resting heart rate is associated with an increased risk of cardiovascular disease and mortality, while lower heart rate, particularly during exercise, indicates better cardiovascular fitness (Olshansky et al. 2023). Heart rate variability (HRV), the variation in time intervals between successive heartbeats, is another crucial biomarker, providing insights into the autonomic nervous system and the body's ability to adapt to stress. Higher HRV is associated with better cardiovascular health, greater resilience to stress, and improved recovery after exercise, while lower HRV may indicate chronic stress, fatigue, or heart disease (Shaffer and Ginsberg 2017). Wearable devices, such as smartwatches or chest straps, can now continuously monitor heart rate and HRV in real time, providing valuable insights for early detection of cardiac arrhythmias, such as atrial fibrillation, and for optimizing physical performance and recovery in athletes.

Blood pressure is a critical biophysical biomarker that reflects the force of blood against the walls of the arteries as the heart pumps. High blood pressure, or hypertension, is a major risk factor for cardiovascular diseases, including heart attack, stroke, and heart failure. Traditionally, blood pressure is measured intermittently using a sphygmomanometer in clinical settings, but this provides only a snapshot of the individual's blood pressure at a given moment (Fuchs and Whelton 2020). With the advent of ambulatory blood pressure monitoring and wearable blood pressure cuffs, it's now possible to monitor blood pressure continuously over 24 hours, capturing variations in blood pressure that occur throughout the day and night. This is particularly important for detecting white coat hypertension (elevated blood pressure only in a clinical setting) or masked hypertension (normal blood pressure in the clinic but elevated outside), both of which carry significant cardiovascular risk but may be missed with traditional blood pressure measurements. Continuous blood pressure monitoring allows for the assessment of treatment effectiveness in hypertension management and helps identify patients who are at risk of serious complications, such as hypertensive crises or cardiovascular events (Stergiou et al. 2014).

Oxygen saturation (SpO2) is a biophysical biomarker that measures the percentage of oxygen-carrying hemoglobin in the blood, providing an indication of respiratory and cardiovascular function (Shapiro et al. 2023). Normal oxygen saturation levels range from 95% to 100%, while lower levels may indicate respiratory or cardiac dysfunction, such as in cases of chronic obstructive pulmonary disease (COPD) or heart failure (Soguel Schenkel et al. 1996). Wearable pulse oximeters, integrated into devices such as smartwatches or fitness trackers, allow for continuous monitoring of SpO2, particularly in patients with respiratory diseases. During the COVID-19 pandemic, pulse oximeters were widely used to monitor patients at home for early signs of respiratory distress, enabling timely medical interventions. Similarly, respiratory rate, the number of breaths taken per minute, is a key indicator of respiratory function. An elevated respiratory rate can indicate respiratory distress, infection, or heart failure (Nicolò et al. 2020). Wearable devices that monitor respiratory rate in real time offer an opportunity for early detection of respiratory failure or exacerbations of chronic respiratory diseases, reducing the need for hospitalization and improving outcomes.

4.3.2 Biomarkers for Sleep Monitoring and Sleep Disorders

Sleep is a vital physiological process, and its disruption is associated with a wide range of health issues, including cardiovascular disease, obesity, diabetes, and mental health disorders. Sleep disorders, such as sleep apnea and insomnia, are common but often go undiagnosed, as they may not produce obvious symptoms during the day (Jackson et al. 2015; Buysse 2014). Sleep monitoring offers an opportunity to assess sleep quality and detect sleep disorders. Traditionally, sleep studies (polysomnography) are conducted in sleep labs, where multiple biophysical biomarkers, including brain waves (EEG), heart rate, respiratory rate, and oxygen saturation, are measured overnight (Kwon et al. 2021). However, wearable devices are increasingly being used to monitor sleep patterns at home, providing more convenient and accessible ways to assess sleep health. Wearable sleep trackers, such as those integrated into smartwatches or fitness bands, use a combination of actigraphy (movement detection), heart rate monitoring, and sometimes oxygen saturation to estimate sleep duration, sleep stages (light, deep, and REM sleep), and disruptions in sleep. These devices can help identify individuals with sleep disorders, such as sleep apnea, by detecting abnormal patterns in breathing or oxygen levels during sleep (De Zambotti et al. 2019). The continuous monitoring of sleep can provide valuable insights for both clinical and personal health purposes. For example, in individuals with obstructive sleep apnea (OSA), wearable devices can detect periods of oxygen desaturation and disrupted breathing, leading to earlier diagnosis and treatment with therapies such as continuous positive airway pressure (CPAP). Additionally, real-time feedback on sleep quality can encourage individuals to adopt healthier sleep habits, leading to improvements in overall health and well-being.

4.3.3 Biomarkers for Stress, Mental Health, and Autonomic Monitoring

Mental health and stress are increasingly recognized as key components of overall health, and biophysical biomarkers can provide valuable insight into an individual's stress levels and mental health status (De Zambotti et al. 2019; Djuric et al. 2008). The autonomic nervous system (ANS) plays a central role in regulating stress responses, and biophysical biomarkers such as heart rate variability (HRV), galvanic skin response (GSR), and cortisol levels are used to assess autonomic function and stress (Yugar et al. 2023). As mentioned earlier, heart rate variability (HRV) reflects the variation in time intervals between successive heartbeats, which is influenced by the balance between the sympathetic and parasympathetic branches of the autonomic nervous system. During periods of stress, sympathetic activity increases, reducing HRV, while parasympathetic activity, which promotes relaxation, increases HRV (Shaffer and Ginsberg 2017; Gullett et al. 2023). Continuous monitoring of HRV through wearable devices provides a real-time indicator of stress levels, allowing individuals to manage stress more effectively through techniques such as mindfulness, meditation, or biofeedback (Gullett et al. 2023). The GSR, also known as skin conductance, measures the electrical conductance of the skin, which increases with sweat gland activity (Nagai et al. 2019). GSR is closely linked to emotional arousal and is commonly used in stress monitoring. Wearable devices that monitor GSR can detect subtle changes in skin conductance that occur during periods of stress or anxiety, providing real-time feedback on emotional arousal. This data can be used to identify triggers for stress and develop personalized strategies for stress management. Cortisol, a hormone produced by the adrenal glands, is a key biomarker of stress. Cortisol levels follow a diurnal rhythm, peaking in the morning and declining throughout the day (Azmi et al. 2021). However, chronic stress can disrupt this rhythm, leading to elevated cortisol levels that are associated with a range of health issues, including cardiovascular disease, obesity, and depression. Traditional methods of measuring cortisol involve collecting blood, saliva, or urine samples at specific times of the day. However, wearable biosensors are being developed to continuously monitor cortisol levels in real time, providing a more dynamic picture of stress and its impact on health (Iqbal et al. 2023). For example, sweat-based cortisol sensors are being explored as noninvasive tools for monitoring stress in real time, offering potential applications in mental health management and workplace stress reduction (Singh et al. 2023).

4.4 Inflammatory and Immune Biomarkers

Inflammation is the body's natural response to injury, infection, or harmful stimuli, aimed at protecting and repairing tissues. While acute inflammation is a crucial defense mechanism, chronic inflammation is associated with various diseases, including autoimmune disorders, cardiovascular disease, cancer, and neurodegenerative diseases (Chen et al. 2018). Monitoring inflammation is essential for diagnosing and managing these conditions, and inflammatory biomarkers provide valuable insight into the immune system's activity and the extent of tissue damage. Immune biomarkers are molecules or cells that indicate the status of the immune system and its ability to mount responses against infections, diseases, or immunological challenges. These biomarkers can identify immune activation, dysfunction, or suppression, and they play a critical role in detecting autoimmune diseases, monitoring immune therapies, and predicting responses to treatments like immunotherapy (Kaushik et al. 2025). Inflammatory and immune biomarkers are critical for diagnosing diseases, predicting disease progression, and personalizing treatments, particularly in the context of precision healthcare. In this section, we explore key inflammatory and immune biomarkers, the technologies used to detect them, and their applications in various diseases. CRP is one of the most commonly used inflammatory biomarkers in clinical practice. Produced by the liver in response to inflammation, CRP levels rise rapidly following acute inflammation or infection (Sproston and Ashworth 2018). It is a key biomarker in identifying systemic inflammation and is widely used to assess the severity of inflammation in conditions such as bacterial infections, autoimmune diseases, and cardiovascular disease. CRP testing is particularly useful in diagnosing and monitoring rheumatoid arthritis and systemic lupus erythematosus (SLE). Elevated CRP levels in these patients often indicate active disease flares, allowing for early therapeutic interventions. In the context of cardiovascular disease, CRP is a predictor of cardiovascular events, with high-sensitivity CRP (hs-CRP) being used to assess the risk of heart attacks and strokes. Patients with elevated hs-CRP levels are more likely to develop cardiovascular events, making it a valuable biomarker for preventive care (Kamath et al. 2015). Cytokines are small proteins that play a central role in cell signaling and immune regulation. They can be pro-inflammatory or anti-inflammatory, depending on their role in promoting or resolving inflammation. Key pro-inflammatory cytokines include interleukin-6 (IL-6), interleukin-1 beta (IL-1β), tumor necrosis factor-alpha (TNF-α), and interferon-gamma (IFN-γ). These cytokines are secreted by immune cells such as macrophages and T cells in response to infections, tissue damage, or chronic inflammatory diseases (Bhol et al. 2024). Elevated levels of cytokines are associated with a wide range of inflammatory and autoimmune diseases. For example, IL-6 plays a critical role in the pathogenesis of chronic inflammatory conditions, including rheumatoid arthritis, multiple sclerosis, and Crohn's disease. Elevated levels of IL-6 are frequently associated with increased disease severity and progression (Alhendi and Naser 2023). Similarly, TNF-α is a key mediator of inflammation in autoimmune disorders such as rheumatoid arthritis and psoriasis (Jang et al. 2021). Therapeutic agents targeting TNF-α, including infliximab and adalimumab, are commonly used in clinical practice, and monitoring TNF-α levels can provide valuable insights into treatment response and therapeutic efficacy. Cytokine measurement is also used in cancer immunotherapy, where cytokine levels can indicate immune activation or suppression in response to therapies like immune checkpoint inhibitors (Majerowski and Gordon 2020). For instance, elevated IFN-γ levels are a sign of immune activation and a potential biomarker of response to immunotherapy. The erythrocyte sedimentation rate (ESR) is another widely used biomarker of inflammation. It measures the rate at which red blood cells settle at the bottom of a test tube over time. A faster-than-normal ESR indicates inflammation in the body. Though less specific than CRP, ESR is often used to monitor disease activity in inflammatory conditions such as rheumatoid arthritis, SLE, and vasculitis. ESR is typically used alongside other inflammatory biomarkers like CRP to provide a comprehensive view of inflammatory status. Individuals with high ESR and CRP levels may be undergoing an active inflammatory process, prompting the physician to investigate the underlying cause and adjust treatments accordingly.

4.5 Digital Biomarkers

In the rapidly advancing field of healthcare, digital biomarkers are emerging as powerful tools that are transforming how health data is captured, monitored, and analyzed. Unlike traditional biomarkers, measurable biological molecules that reflect physiological or pathological processes, digital biomarkers are defined as objective, quantifiable, physiological, and behavioral data collected and measured by digital devices, such as wearables, smartphones, and other connected technologies (Arya et al. 2023). Digital biomarkers allow for the continuous, real-time monitoring of health parameters, offering insights into various aspects of health, including physical activity, sleep, heart rate, and mental health (Smokovski et al. 2024). They are particularly valuable in precision healthcare because they provide personalized data over time, allowing clinicians to track the health of individuals and make data-driven decisions regarding treatment and lifestyle interventions (Johnson et al. 2021). Key to the utility of digital biomarkers is the integration of sensors, wearables, artificial intelligence (AI), and machine learning algorithms. These tools not only collect vast amounts of data but also analyze it to generate actionable insights. As digital health technologies continue to evolve, digital biomarkers are expected to play an even greater role in healthcare, enhancing disease prevention, early diagnosis, and personalized treatment (Shajari et al. 2023).

Digital biomarkers can be classified into several categories based on the source of data and the physiological or behavioral parameters they measure. These include passively collected data (e.g., movement, heart rate) and actively collected data (e.g., patient-reported outcomes). Each type of digital biomarker provides unique insights into health, enabling a comprehensive understanding of an individual's well-being. Wearable devices, such as smartwatches, fitness trackers, and biosensors, are some of the most common sources of digital biomarkers (Arya et al. 2023;

Jeong et al. 2022). These devices are equipped with various sensors that can continuously monitor physiological parameters such as heart rate, physical activity, sleep patterns, and body temperature. Smartphones are ubiquitous and offer a wide array of sensors that can be leveraged to collect digital biomarkers (Kukreti et al. 2025). Smartphones can passively monitor activity levels, location data, and even behavioral cues that are associated with health. Beyond wearables and smartphones, various connected health devices contribute to the growing pool of digital biomarkers. These include smart scales, connected blood pressure monitors, glucometers, and CGMs, which provide real-time data on key health metrics (Alonso et al. 2024).

Digital biomarkers are transforming healthcare by enabling continuous, real-time monitoring, early detection of disease, and personalized intervention. These biomarkers are particularly useful in the management of chronic diseases, mental health conditions, and preventive healthcare. Cardiovascular diseases (CVDs) are the leading cause of death globally, and digital biomarkers provide valuable insights into heart health. By monitoring biomarkers such as heart rate, heart rate variability, and blood pressure, clinicians can detect early signs of cardiovascular dysfunction and intervene before more serious events occur (Smokovski et al. 2024; Jabara et al. 2024). For example, wearable devices that track heart rate and rhythm can detect atrial fibrillation (AFib), a common arrhythmia associated with an increased risk of stroke (Iqbal et al. 2024). Continuous monitoring of HRV through wearables can also predict heart failure exacerbations or worsening cardiovascular health, allowing for timely treatment adjustments. In patients with hypertension, digital blood pressure monitors can track blood pressure trends over time, helping to optimize antihypertensive therapies. This real-time monitoring is particularly valuable for individuals with resistant hypertension, where traditional clinic-based measurements may not accurately reflect daily blood pressure variations (Tsao et al. 2023).

Despite the tremendous potential of digital biomarkers, several challenges must be addressed to fully integrate them into mainstream healthcare. The continuous collection of health data through wearables, smartphones, and connected devices raises significant concerns about data privacy and security. Digital biomarkers involve sensitive personal health information, and it is crucial to ensure that this data is protected from unauthorized access and breaches (Roos and Slavich 2023). Regulations like the General Data Protection Regulation (GDPR) in Europe and the Health Insurance Portability and Accountability Act (HIPAA) in the United States provide frameworks for safeguarding digital health data, but ongoing efforts are needed to address emerging security challenges. The quality of digital biomarkers depends on the accuracy and reliability of the sensors used to collect data (Conduah et al. 2025). Variability in device performance, sensor calibration, and data processing algorithms can impact on the consistency and validity of digital biomarkers. Standardization of measurement protocols and device certification is essential to ensure that digital biomarkers are accurate and comparable across different platforms and technologies. Moreover, the interpretation of digital biomarkers requires context (Hanna et al. 2025). For example, a low HRV reading might indicate stress in one individual but could be normal for someone with a well-conditioned heart. Personalized baselines and contextual analysis are necessary to avoid misinterpretation of data. To fully realize the potential of digital biomarkers in clinical care, it is important to integrate them with electronic health records (EHRs) (Abul-Husn and Kenny 2019). This integration allows healthcare providers to access real-time data alongside traditional medical records, enabling more informed clinical decisions. However, interoperability between wearable devices, mobile apps, and EHR systems remains a challenge, as many platforms use proprietary data formats that are not easily integrated into existing healthcare systems. Before digital biomarkers can be widely adopted in clinical practice, they must undergo rigorous validation to ensure their accuracy, reliability, and clinical utility (Khatiwada et al. 2024). Regulatory agencies like the US Food and Drug Administration (FDA) require that digital health technologies demonstrate safety and effectiveness through clinical trials and real-world studies. The process of validating digital biomarkers can be time-consuming and expensive, but it is critical for ensuring that they provide meaningful and actionable insights for healthcare providers and patients.

As digital health technologies continue to evolve, several trends are shaping the future of digital biomarkers in precision healthcare. The integration of artificial intelligence (AI) and machine learning (ML) with digital biomarkers is enabling more sophisticated analysis of health data. AI algorithms can process vast amounts of digital biomarker data in real time, identifying patterns and correlations that would be difficult for humans to detect (Hanna et al. 2025). For example, AI-driven analysis of wearable data can predict heart failure exacerbations based on subtle changes in heart rate variability, respiratory rate, and physical activity levels. The future of digital biomarkers lies in the integration of multiple data streams, including genomic, proteomic, and digital health data. Combining molecular biomarkers with digital biomarkers provides a more comprehensive understanding of an individual's health (Medhi et al. 2024). Integrating genetic risk factors for cardiovascular disease with continuous monitoring of heart rate, blood pressure, and activity levels could provide highly personalized risk assessments and preventive strategies (Cowie and Lam 2021). The COVID-19 pandemic has accelerated the adoption of remote patient monitoring (RPM) and telehealth, and digital biomarkers are at the core of these innovations. RPM platforms equipped with wearables and connected devices allow patients to be monitored from the comfort of their homes, reducing the need for in-person visits and improving access to care. As telehealth continues to grow, digital biomarkers will play an increasingly important role in virtual care delivery (Mann and Lawrence 2022). Digital biomarkers enable a shift from reactive healthcare, where treatment is initiated after symptoms arise, to predictive and preventive healthcare, where interventions are made based on early warning signs. By continuously monitoring physiological and behavioral data, digital biomarkers can detect early changes that precede disease, enabling preventive measures such as lifestyle modifications, medication adjustments, or early medical interventions.

4.6 Biomarkers for Future Health Monitoring

Biomarkers have long played a central role in diagnostics and disease management, but their application is expanding dramatically with advances in technology, data analytics, and personalized medicine. In the future, biomarkers will not only be used for diagnosing diseases but also for predicting disease risk, monitoring real-time health status, and guiding individualized treatment plans. As healthcare shifts from a reactive model to a more predictive and preventive approach, biomarkers will be key to driving this transformation. Several trends are emerging in biomarker-based health monitoring that will reshape how we approach disease detection, management, and prevention. These trends include the integration of multi-omics data, the growing importance of wearable biosensors and noninvasive testing, the role of artificial intelligence (AI) and machine learning (ML) in biomarker discovery, and the development of personalized, predictive healthcare models. In this section, we will explore these trends in detail and discuss how they are likely to impact the future of healthcare.

4.6.1 Multi-omics Approaches and Systems Biology

As our understanding of biology advances, it is becoming increasingly clear that no single type of biomarker provides a complete picture of an individual's health. Instead, a more holistic approach, one that integrates data from multiple biological layers, including genomics, proteomics, metabolomics, and transcriptomics, is needed. This approach is known as multi-omics, and it is transforming the field of biomarker discovery (Babu and Snyder 2023). Genomic biomarkers, such as DNA mutations or gene expression profiles, have already proven invaluable in precision medicine, particularly in oncology. Genetic sequencing technologies, including next-generation sequencing (NGS), have made it possible to identify genetic mutations that drive disease, predict patient responses to treatments, and tailor therapies to individuals' genetic profiles. For example, in cancer treatment, genomic biomarkers such as BRCA1/BRCA2 mutations in breast cancer or EGFR mutations in lung cancer guide the selection of targeted therapies (Ghoreyshi et al. 2025; Purohit et al. 2019). However, the future of genomics goes beyond identifying static genetic mutations. Emerging technologies, such as single cell sequencing and epigenetic profiling, will provide a more dynamic view of how genes are regulated and expressed in different cell types over time. These technologies will help researchers uncover new biomarkers that reveal how genetic and environmental factors interact to influence disease progression.

While genomics provides information about an individual's genetic predisposition to disease, it is the proteome, the collection of all proteins expressed by a cell, that provides a more immediate reflection of biological activity. Proteomic biomarkers, such as protein expression levels or posttranslational modifications, offer valuable insights into disease mechanisms, drug responses, and disease progression (Paraskar et al. 2025). For example, in Alzheimer's disease, researchers are identifying proteomic biomarkers, such as tau and beta-amyloid proteins, which can be used to detect the disease in its early stages (Pradeepkiran et al. 2024). Similarly, in cardiovascular disease, proteomic biomarkers such as troponin and B-type natriuretic peptide (BNP) are used to predict heart attacks and heart failure (Mani et al. 2022). Metabolomics, the study of small molecules involved in metabolism, provides another layer of information about an individual's health status. Metabolomic biomarkers reflect real-time changes in metabolism and are often used to assess metabolic disorders, such as diabetes and obesity. In the future, metabolomic profiling will help identify novel biomarkers for a wide range of diseases, including cancer, neurodegenerative diseases, and autoimmune conditions (Qiu et al. 2023).

The future of biomarker discovery will increasingly rely on systems biology, an interdisciplinary approach that integrates multi-omics data with computational models to understand the complex interactions between biological systems. By analyzing how different molecular layers (e.g., genes, proteins, and metabolites) interact with each other, researchers can gain a more comprehensive understanding of disease mechanisms and identify new biomarkers that would be difficult to detect using a single-omics approach. Multi-omics integration will also play a key role in developing personalized treatment plans. For example, combining genomic data with proteomic and metabolomic profiles will allow clinicians to better predict how patients will respond to specific therapies and adjust treatment plans in real time based on changes in biomarker levels.

4.6.2 Multiparameter Monitoring

Wearable biosensors are becoming increasingly sophisticated, allowing for continuous, real-time monitoring of biophysical and biochemical biomarkers. These devices are transforming health monitoring by providing dynamic data on key health metrics such as heart rate, glucose levels, and oxygen saturation (Mukherjee et al. 2025). In the future, wearable biosensors will play an even greater role in disease prevention and personalized care. Early wearable devices, such as fitness trackers and smartwatches, were primarily focused on tracking physical activity and heart rate (Mahato et al. 2024). However, the next generation of wearable biosensors is designed to monitor a significantly broader spectrum of biophysical and biochemical biomarkers. These advanced devices include CGMs for real-time glucose tracking in diabetes management, as well as sensors capable of detecting lactate and cortisol levels to assess physical stress and psychological strain. Additionally, modern wearables can monitor respiratory parameters such as respiratory rate and oxygen saturation, providing critical insights into conditions like COPD and asthma. ECG monitoring is also integrated into many devices, allowing for the detection of arrhythmias and other cardiovascular abnormalities with high precision. One of the most exciting developments in wearable biosensors is the use of sweat-based biosensors, which can noninvasively detect biomarkers in sweat, such as electrolytes, glucose, and metabolites. These biosensors are being developed for a range of applications, including monitoring dehydration, metabolic disorders, and stress levels.

For example, microfluidic biosensors that collect and analyze sweat in real time are being explored for use in athletes to monitor hydration and electrolyte balance. These sensors could also be adapted for patients with chronic diseases, such as heart failure, where changes in electrolyte levels could indicate disease exacerbations (Park et al. 2025). Noninvasive biomarker detection is a key focus in the future of healthcare. Traditional methods of biomarker detection, such as blood tests, are often invasive and require frequent clinic visits. In contrast, noninvasive methods, such as biosensors, breath analysis, and saliva testing, offer a more convenient and patient-friendly alternative. For example, saliva-based diagnostics are being developed to detect a wide range of biomarkers, including hormones, antibodies, and inflammatory markers. Saliva testing has the potential to replace more invasive blood tests for conditions such as diabetes, infectious diseases, and hormonal imbalances (Cui et al. 2022). Similarly, breath analysis is being explored as a noninvasive way to detect metabolic biomarkers, such as volatile organic compounds (VOCs), which are linked to diseases such as lung cancer, asthma, and liver disease.

The rise of remote health monitoring and telehealth is enabling patients to be monitored continuously outside of clinical settings, reducing the need for in-person visits and improving patient outcomes. Remote monitoring platforms that integrate data from wearable biosensors, digital biomarkers, and connected health devices allow clinicians to track patients in real time and intervene early if abnormal changes are detected. For example, in heart failure patients, wearable devices that monitor heart rate, respiratory rate, and oxygen saturation can detect early signs of disease exacerbation, allowing for timely interventions such as medication adjustments or hospitalization, thereby preventing severe complications. As telehealth becomes more widespread, the integration of biomarker data into EHRs will be essential for providing clinicians with a comprehensive view of a patient's health. This real-time data flow will enable healthcare providers to make more informed decisions and offer more personalized care, based on continuous biomarker monitoring.

4.6.3 Novel Biomarker Discovery Using AI/ML

The AI and ML are revolutionizing biomarker discoveries by enabling the analysis of large datasets, identifying patterns, and predicting disease outcomes. The sheer volume of data generated by multi-omics platforms, wearable biosensors, and digital biomarkers require sophisticated computational tools to extract meaningful insights (Alum 2025). AI algorithms are particularly well-suited for identifying predictive biomarkers that can be used to forecast disease onset or progression before clinical symptoms appear. These algorithms can analyze complex datasets, such as genetic sequences, proteomic profiles, and physiological data, to uncover patterns that might be invisible to human researchers (Farooq and Banka 2025). For example, in cancer research, machine learning algorithms are being used to identify gene expression patterns that predict patient responses to immunotherapies or chemotherapy. Similarly, in cardiology, AI is being used to develop predictive models that use digital biomarkers (e.g., heart rate variability) to forecast heart failure exacerbations. Machine learning algorithms can also be used to personalize treatment by identifying which patients are most likely to benefit from specific therapies. In the future, AI-driven analysis of biomarker data will allow clinicians to tailor treatment plans in real time based on an individual's unique biological profile and their response to therapy. For example, in oncology, AI is being used to analyze circulating tumor DNA (ctDNA) and other liquid biopsy biomarkers to monitor tumor evolution and adjust treatment strategies accordingly (Moon et al. 2025). By continuously analyzing biomarker data, AI algorithms can detect emerging drug resistance or disease recurrence earlier than traditional imaging techniques, allowing for more timely interventions. AI and machine learning are also playing an increasingly important role in clinical trials, particularly in the discovery and validation of new biomarkers. By analyzing data from thousands of patients, AI can identify novel biomarkers that predict disease risk, treatment response, or adverse events. This helps accelerate the development of precision therapies and improves the likelihood of success in clinical trials (Arya et al. 2023). In addition, AI-driven platforms are being developed to optimize clinical trial design by identifying patient subgroups that are most likely to respond to experimental therapies. This adaptive trial design approach reduces the time and cost of drug development, while increasing the chances of regulatory approval for new therapies.

4.7 Conclusions

This chapter discusses the integral role of biomarkers in advancing comprehensive health monitoring and facilitating precision medicine. The chapter classifies biomarkers into molecular, biophysical, inflammatory, immune, and digital categories, elucidating their diverse roles in diagnosing, monitoring, and managing various diseases and health conditions. Through the use of biomarkers, healthcare practice is increasingly shifting from generalized diagnostic approaches toward highly individualized, precision-based interventions. Molecular biomarkers, encompassing genomic, proteomic, metabolomic, and transcriptomic markers, are examined extensively for their contributions to early disease detection, prognosis, and treatment stratification. DNA-based biomarkers, such as mutations in the BRCA genes and KRAS gene alterations, demonstrate significant diagnostic utility, particularly in oncology. Similarly, RNA-based biomarkers, including microRNAs and long noncoding RNAs, provide valuable insights into complex disease processes, notably in neurodegenerative and cardiovascular disorders. Additionally, protein biomarkers such as cardiac troponins and BNPs are discussed as essential tools for cardiovascular diagnosis and prognosis. Metabolomic biomarkers, such as HbA1c and lipid profiles, are also detailed for their clinical relevance in managing diabetes and metabolic syndromes. Biophysical biomarkers represent another critical category, measured through essential parameters such as heart rate, blood pressure, respiratory rate, and oxygen saturation. The chapter emphasizes the growing importance of wearable sensor technology for continuous, real-time monitoring of these parameters. These wearable technologies have significantly enhanced the detection of cardiovascular anomalies, respiratory dysfunctions, and autonomic system disturbances. Physiological biomarkers related to sleep patterns and mental health status, including HRV and GSR, are also presented, demonstrating their capacity to extend biomarker monitoring into behavioral and psychological dimensions of health.

Moreover, inflammatory and immune biomarkers, particularly CRP and cytokines (IL-6, TNF-α), are discussed for their clinical utility in diagnosing and managing inflammatory and autoimmune diseases. These biomarkers are vital indicators of immune system activity and inflammation, thus playing a critical role in patient stratification and therapeutic guidance, especially in immune-targeted therapies. The emergence and integration of digital biomarkers is another focal point of this chapter. Digital biomarkers, derived from data captured by wearable devices, smartphones, and artificial intelligence-driven analytical platforms, offer continuous, noninvasive, and real-time monitoring of health parameters. This capability holds substantial promises for chronic disease management, mental health monitoring, and preventive care strategies. However, the chapter also addresses several challenges associated with digital biomarkers, including data privacy, security concerns, device accuracy, and the need for standardization and seamless integration with EHRs. Emerging trends in biomarker research, notably the integration of multi-omics data and advanced computational methods involving AI and ML, are explored as critical components of future healthcare innovation. Systems biology approaches, which integrate data from genomics, proteomics, and metabolomics, are highlighted for their potential to reveal complex biological interactions underlying disease states. Additionally, the chapter discusses advancements in wearable biosensors and noninvasive detection methods, emphasizing their future role in enhancing personalized and remote health monitoring. In summary, this chapter underscores the transformative potential of biomarkers in the evolution of healthcare, moving from a predominantly reactive model toward a predictive, preventive, and personalized approach. Continued research and technological advancement in biomarker discovery, validation, and integration into clinical practice promise significant improvements in health outcomes and the advancement of precision medicine.

References

Abul-Husn NS, Kenny EE (2019) Personalized medicine and the power of electronic health records. Cell 177:58. https://doi.org/10.1016/J.CELL.2019.02.039

Alhendi A, Naser SA (2023) The dual role of interleukin-6 in Crohn's disease pathophysiology. Front Immunol 14:1295230. https://doi.org/10.3389/fimmu.2023.1295230

Alonso AKM, Hirt J, Woelfle T et al (2024) Definitions of digital biomarkers: a systematic mapping of the biomedical literature. BMJ Heal Care Informatics 31:e100914. https://doi.org/10.1136/BMJHCI-2023-100914

Alum EU (2025) AI-driven biomarker discovery: enhancing precision in cancer diagnosis and prognosis. Discov Oncol 16:313. https://doi.org/10.1007/S12672-025-02064-7

Arya SS, Dias SB, Jelinek HF et al (2023) The convergence of traditional and digital biomarkers through AI-assisted biosensing: a new era in translational diagnostics? Biosens Bioelectron 235:115387. https://doi.org/10.1016/J.BIOS.2023.115387

Avram R, Tison GH, Aschbacher K et al (2019) Real-world heart rate norms in the Health eHeart study. npj Digit Med 2. https://doi.org/10.1038/s41746-019-0134-9

Ayres JS (2020) The biology of physiological health. Cell 181:250–269. https://doi.org/10.1016/j.cell.2020.03.036

Azmi NASM, Juliana N, Azmani S et al (2021) Cortisol on circadian rhythm and its effect on cardiovascular system. Int J Environ Res Public Health 18:1–15. https://doi.org/10.3390/ijerph18020676

Babu M, Snyder M (2023) Multi-omics profiling for health. Mol Cell Proteomics 22:100561. https://doi.org/10.1016/J.MCPRO.2023.100561

Babuin L, Jaffe AS (2005) Troponin: the biomarker of choice for the detection of cardiac injury. C Can Med Assoc J 173:1191–1202. https://doi.org/10.1503/cmaj/051291

Bhol NK, Bhanjadeo MM, Singh AK et al (2024) The interplay between cytokines, inflammation, and antioxidants: mechanistic insights and therapeutic potentials of various antioxidants and anti-cytokine compounds. Biomed Pharmacother 178:117177. https://doi.org/10.1016/j.biopha.2024.117177

Buysse DJ (2014) Sleep health: can we define it? Does it matter? Sleep 37:9–17. https://doi.org/10.5665/sleep.3298

Cao Z, Jia Y, Zhu B (2019) BNP and NT-proBNP as diagnostic biomarkers for cardiac dysfunction in both clinical and forensic medicine. Int J Mol Sci 20:1820. https://doi.org/10.3390/ijms20081820

Chau CH, Rixe O, McLeod H, Figg WD (2008) Validation of analytic methods for biomarkers used in drug development. Clin Cancer Res 14:5967–5976. https://doi.org/10.1158/1078-0432.CCR-07-4535

Chen L, Deng H, Cui H et al (2018) Inflammatory responses and inflammation-associated diseases in organs. Oncotarget 9:7204–7218. https://doi.org/10.18632/oncotarget.23208

Chen L, Mir SA, Bendt AK et al (2023) Plasma lipidomic profiling reveals metabolic adaptations to pregnancy and signatures of cardiometabolic risk: a preconception and longitudinal cohort study. BMC Med 21:1–24. https://doi.org/10.1186/s12916-023-02740-x

Chenchula S, Atal S, Uppugunduri CRS (2024) A review of real-world evidence on preemptive pharmacogenomic testing for preventing adverse drug reactions: a reality for future health care. Pharmacogenomics J 24:1–6. https://doi.org/10.1038/s41397-024-00326-1

Choi A, Chung K, Chung SP et al (2022) Advantage of vital sign monitoring using a wireless wearable device for predicting septic shock in febrile patients in the emergency department: a machine learning-based analysis. Sensors 22:7054. https://doi.org/10.3390/s22187054

Conduah AK, Ofoe S, Siaw-Marfo D (2025) Data privacy in healthcare: global challenges and solutions. Digit Health 11:20552076251343960. https://doi.org/10.1177/20552076251343959

Cowie MR, Lam CSP (2021) Remote monitoring and digital health tools in CVD management. Nat Rev Cardiol 18:457. https://doi.org/10.1038/S41569-021-00548-X

Cui Y, Yang M, Zhu J et al (2022) Developments in diagnostic applications of saliva in human organ diseases. Med Nov Technol Devices 13:100115. https://doi.org/10.1016/J.MEDNTD.2022.100115

De Zambotti M, Cellini N, Goldstone A et al (2019) Wearable sleep technology in clinical and research settings. Med Sci Sports Exerc 51:1538–1557. https://doi.org/10.1249/MSS.0000000000001947

Djuric Z, Bird CE, Furumoto-Dawson A et al (2008) Biomarkers of psychological stress in health disparities research. Open Biomark J 1:7–19. https://doi.org/10.2174/1875318300801010007

Doroszkiewicz J, Groblewska M, Mroczko B (2022) Molecular biomarkers and their implications for the early diagnosis of selected neurodegenerative diseases. Int J Mol Sci 23. https://doi.org/10.3390/ijms23094610

Farooq M, Banka AA (2025) Artificial intelligence in biomarker discovery and disease diagnosis-1. In: Artificial intelligence in human health and diseases, pp 93–109. https://doi.org/10.1007/978-981-96-8176-1_6

Fuchs FD, Whelton PK (2020) High blood pressure and cardiovascular disease. Hypertension 75:285–292. https://doi.org/10.1161/HYPERTENSIONAHA.119.14240

Fung E, Patsopoulos NA, Belknap SM et al (2012) Effect of genetic variants, especially CYP2C9 and VKORC1, on the pharmacology of warfarin. Semin Thromb Hemost 38:893–904. https://doi.org/10.1055/s-0032-1328891
Ghoreyshi N, Heidari R, Farhadi A et al (2025) Next-generation sequencing in cancer diagnosis and treatment: clinical applications and future directions. Discov Oncol 16:578. https://doi.org/10.1007/S12672-025-01816-9
Graig LA, Phillips JK, Moses HL (2016) Biomarker tests for molecularly targeted therapies: key to unlocking precision medicine. National Academies Press Washington (DC), https://doi.org/10.17226/21860
Grasso M, Piscopo P, Confaloni A, Denti MA (2014) Circulating miRNAs as biomarkers for neurodegenerative disorders. Molecules 19:6891–6910. https://doi.org/10.3390/molecules19056891
Gullett N, Zajkowska Z, Walsh A et al (2023) Heart rate variability (HRV) as a way to understand associations between the autonomic nervous system (ANS) and affective states: a critical review of the literature. Int J Psychophysiol 192:35–42. https://doi.org/10.1016/j.ijpsycho.2023.08.001
Guo L, Zhao Y, Huang Q et al (2024) Electrochemical protein biosensors for disease marker detection: progress and opportunities. Microsyst Nanoeng 10:1–20. https://doi.org/10.1038/s41378-024-00700-w
Hanna MG, Pantanowitz L, Dash R et al (2025) Future of artificial intelligence—machine learning trends in pathology and medicine. Mod Pathol 38:100705. https://doi.org/10.1016/J.MODPAT.2025.100705
Herceg Z, Hainaut P (2007) Genetic and epigenetic alterations as biomarkers for cancer detection, diagnosis and prognosis. Mol Oncol 1:26–41. https://doi.org/10.1016/j.molonc.2007.01.004
Higashi Y (2023) Era of biomarker-based disease risk management. Hypertens Res 46:893–895. https://doi.org/10.1038/s41440-023-01187-0
Huang L, Guo Z, Wang F, Fu L (2021) KRAS mutation: from undruggable to druggable in cancer. Signal Transduct Target Ther 6:1–20. https://doi.org/10.1038/s41392-021-00780-4
Huhn S, Axt M, Gunga HC et al (2022) The impact of wearable technologies in health research: scoping review. JMIR Mhealth Uhealth 10:e34384. https://doi.org/10.2196/34384
Iqbal T, Elahi A, Wijns W, Shahzad A (2023) Cortisol detection methods for stress monitoring in connected health. Health Sci Rev 6:100079. https://doi.org/10.1016/j.hsr.2023.100079
Iqbal SMA, Leavitt MA, Mahgoub I, Asghar W (2024) Advances in cardiovascular wearable devices. Biosensors 14:525. https://doi.org/10.3390/BIOS14110525
Jabara M, Kose O, Perlman G et al (2024) Artificial intelligence-based digital biomarkers for type 2 diabetes: a review. Can J Cardiol 40:1922–1933. https://doi.org/10.1016/J.CJCA.2024.07.028
Jackson CL, Redline S, Emmons KM (2015) Sleep as a potential fundamental contributor to disparities in cardiovascular health. Annu Rev Public Health 36:417–440. https://doi.org/10.1146/annurev-publhealth-031914-122838
Jain KK, Jain KK (2010) The handbook of biomarkers. Springer
Jang DI, Lee AH, Shin HY et al (2021) The role of tumor necrosis factor alpha (Tnf-α) in autoimmune disease and current tnf-α inhibitors in therapeutics. Int J Mol Sci 22:1–16. https://doi.org/10.3390/ijms22052719
Jeong H, Jeong YW, Park Y et al (2022) Applications of deep learning methods in digital biomarker research using noninvasive sensing data. Digit Health 8:20552076221136640. https://doi.org/10.1177/20552076221136642
John AO, Singh A, Yadav P et al (2024) The BRCA mutation spectrum among breast and ovarian cancers in India: highlighting the need to screen BRCA1 185delAG among South Indians. Eur J Hum Genet. https://doi.org/10.1038/s41431-024-01596-w
Johnson KB, Wei WQ, Weeraratne D et al (2021) Precision medicine, AI, and the future of personalized health care. Clin Transl Sci 14:86–93. https://doi.org/10.1111/cts.12884
Kachuri L, Hoffmann TJ, Jiang Y et al (2023) Genetically adjusted PSA levels for prostate cancer screening. Nat Med 29:1412–1423. https://doi.org/10.1038/s41591-023-02277-9

Kamath DY, Xavier D, Sigamani A, Pais P (2015) High sensitivity C-reactive protein (hsCRP) & cardiovascular disease: an Indian perspective. Indian J Med Res 142:261–268. https://doi.org/10.4103/0971-5916.166582
Kaushik D, Xu B, Kumar M (2025) Biomarkers in immunology: their impact on immune function and response. Adv Biomark Sci Technol 7:95–110. https://doi.org/10.1016/j.abst.2025.03.001
Khatiwada P, Yang B, Lin JC, Blobel B (2024) Patient-Generated Health Data (PGHD): understanding, requirements, challenges, and existing techniques for data security and privacy. J Pers Med 14:282. https://doi.org/10.3390/JPM14030282
Kukreti S, Lu MT, Yeh CY, Ko NY (2025) Physiological sensors equipped in wearable devices for management of long COVID persisting symptoms: scoping review. J Med Internet Res 27:e69506. https://doi.org/10.2196/69506
Kwon S, Kim H, Yeo WH (2021) Recent advances in wearable sensors and portable electronics for sleep monitoring. iScience 24:102461. https://doi.org/10.1016/j.isci.2021.102461
Lazar DR, Lazar FL, Homorodean C et al (2022) High-sensitivity troponin: a review on characteristics, assessment, and clinical implications. Dis Markers 2022:9713326. https://doi.org/10.1155/2022/9713326
Lee JK, Sivakumar S, Schrock AB et al (2022) Comprehensive pan-cancer genomic landscape of KRAS altered cancers and real-world outcomes in solid tumors. npj Precis Oncol 6:1–14. https://doi.org/10.1038/s41698-022-00334-z
Liu R, Zhao E, Yu H et al (2023) Methylation across the central dogma in health and diseases: new therapeutic strategies. Signal Transduct Target Ther 8. https://doi.org/10.1038/s41392-023-01528-y
Liu B, Zhou H, Tan L et al (2024) Exploring treatment options in cancer: tumor treatment strategies. Signal Transduct Target Ther 9. https://doi.org/10.1038/s41392-024-01856-7
Lv Y, Yuan Y, Zhong X, et al (2024) Exploration and practice of potential association prediction between diseases and drugs based on Swanson framework and bioinformatics. p. 1–15
Mahato K, Chandra P (2022) Miniaturized biosensing devices. Springer Nature Singapore, Singapore
Mahato K, Chandra P (2024) Biosensors for personalized healthcare. Springer Nature Singapore, Singapore
Mahato K, Kumar A, Maurya PK, Chandra P (2018a) Shifting paradigm of cancer diagnoses in clinically relevant samples based on miniaturized electrochemical nanobiosensors and microfluidic devices. Biosens Bioelectron 100:411–428. https://doi.org/10.1016/j.bios.2017.09.003
Mahato K, Maurya PK, Chandra P (2018b) Fundamentals and commercial aspects of nanobiosensors in point-of-care clinical diagnostics. 3 Biotech 8:1–14. https://doi.org/10.1007/s13205-018-1148-8
Mahato K, Kumar A, Purohit B et al (2022) Onsite quality controls for food safety based on miniaturized biosensing. Springer Nature Singapore
Mahato K, Saha T, Ding S et al (2024) Hybrid multimodal wearable sensors for comprehensive health monitoring. Nat Electron 7:735–750. https://doi.org/10.1038/s41928-024-01247-4
Majerowski J, Gordon KB (2020) Tumor necrosis factor inhibitors. In: Comprehensive dermatologic drug therapy, Fourth Ed, pp 287–301.e7. https://doi.org/10.1016/B978-0-323-61211-1.00026-7
Mani V, Durmus C, Khushaim W et al (2022) Multiplexed sensing techniques for cardiovascular disease biomarkers – a review. Biosens Bioelectron 216:114680. https://doi.org/10.1016/j.bios.2022.114680
Mann DM, Lawrence K (2022) Reimagining connected care in the era of digital medicine. JMIR Mhealth Uhealth 10:e34483. https://doi.org/10.2196/34483
Mathur S, Sutton J (2017) Personalized medicine could transform healthcare. Biomed Reports 7:3–5. https://doi.org/10.3892/br.2017.922
Mbewu A, Mbanya J (2006) Cardiovascular disease cardiovascular disease. Dis Mortal Sub-Saharan Africa 349:1–22. https://doi.org/10.1056/NEJMRA035098
Medhi D, Kamidi SR, Sree KPM et al (2024) Artificial intelligence and its role in diagnosing heart failure: a narrative review. Cureus 16:e59661. https://doi.org/10.7759/CUREUS.59661

Mersch J, Jackson MA, Park M et al (2015) Cancers associated with BRCA1 and BRCA2 mutations other than breast and ovarian. Cancer 121:269–275. https://doi.org/10.1002/cncr.29041
Mishra M, Ahmed R, Das DK et al (2024) Recent advancements in the application of circulating tumor DNA as biomarkers for early detection of cancers. ACS Biomater Sci Eng 10:4740–4756. https://doi.org/10.1021/acsbiomaterials.4c00606
Moon GY, Dalkiran B, Park HS et al (2025) Dual biomarker strategies for liquid biopsy: integrating circulating tumor cells and circulating tumor DNA for enhanced tumor monitoring. Biosensors 15:74. https://doi.org/10.3390/BIOS15020074
Mukherjee MD, Gupta P, Kumari V et al (2025) Wearable biosensors in modern healthcare: emerging trends and practical applications. Talanta Open 12:100486. https://doi.org/10.1016/J.TALO.2025.100486
Nagai Y, Jones CI, Sen A (2019) Galvanic skin response (GSR)/electrodermal/skin conductance biofeedback on epilepsy: a systematic review and meta-analysis. Front Neurol 10:435908. https://doi.org/10.3389/fneur.2019.00377
Nicolò A, Massaroni C, Schena E, Sacchetti M (2020) The importance of respiratory rate monitoring: from healthcare to sport and exercise. Sensors (Switzerland) 20:1–45. https://doi.org/10.3390/s20216396
Oldrini B, Vaquero-Siguero N, Mu Q et al (2020) MGMT genomic rearrangements contribute to chemotherapy resistance in gliomas. Nat Commun 11. https://doi.org/10.1038/s41467-020-17717-0
Olshansky B, Ricci F, Fedorowski A (2023) Importance of resting heart rate. Trends Cardiovasc Med 33:502–515. https://doi.org/10.1016/j.tcm.2022.05.006
Ortiz-Martínez M, González-González M, Martagón AJ et al (2022) Recent developments in biomarkers for diagnosis and screening of type 2 diabetes mellitus. Curr Diab Rep 22:95–115. https://doi.org/10.1007/s11892-022-01453-4
Paraskar G, Bhattacharya S, Kuttiappan A (2025) The role of proteomics and genomics in the development of colorectal cancer diagnostic tools and potential new treatments. ACS Pharmacol Transl Sci 8:1227–1250. https://doi.org/10.1021/ACSPTSCI.4C00686/ASSET/IMAGES/MEDIUM/PT4C00686_0006.GIF
Park S, Kim S, Lee S et al (2025) Advanced microfluidic-based wearable electrochemical sensors for continuous biochemical monitoring. Adv Electron Mater 11:2500010. https://doi.org/10.1002/AELM.202500010
Patrinos GP (2020) Sketching the prevalence of pharmacogenomic biomarkers among populations for clinical pharmacogenomics. Eur J Hum Genet 28:1–3. https://doi.org/10.1038/s41431-019-0499-x
Pereira PA, Tomás JF, Queiroz JA et al (2016) Recombinant pre-miR-29b for Alzheimer's disease therapeutics. Sci Rep 6:1–11. https://doi.org/10.1038/srep19946
Pradeepkiran JA, Baig J, Islam MA et al (2024) Amyloid-β and phosphorylated tau are the key biomarkers and predictors of Alzheimer's disease. Aging Dis 16:658. https://doi.org/10.14336/AD.2024.0286
Purohit B, Kumar A, Mahato K et al (2019) Cancer cytosensing approaches in miniaturized settings based on advanced nanomaterials and biosensors. In: Nanotechnology in modern animal biotechnology, pp 133–147. https://doi.org/10.1016/B978-0-12-818823-1.00009-0
Qiu S, Cai Y, Yao H et al (2023) Small molecule metabolites: discovery of biomarkers and therapeutic targets. Signal Transduct Target Ther 8:1–37. https://doi.org/10.1038/s41392-023-01399-3
Ratti M, Lampis A, Ghidini M et al (2020) MicroRNAs (miRNAs) and long non-coding RNAs (lncRNAs) as new tools for cancer therapy: first steps from bench to bedside. Target Oncol 15:261–278. https://doi.org/10.1007/s11523-020-00717-x
Roos LG, Slavich GM (2023) Wearable technologies for health research: opportunities, limitations, and practical and conceptual considerations. Brain Behav Immun 113:444–452. https://doi.org/10.1016/J.BBI.2023.08.008
Saha T, Del Caño R, Mahato K et al (2023) Wearable electrochemical glucose sensors in diabetes management: a comprehensive review. Chem Rev 123:7854–7889. https://doi.org/10.1021/acs.chemrev.3c00078

Sempionatto JR, Lasalde-Ramírez JA, Mahato K et al (2022) Wearable chemical sensors for biomarker discovery in the omics era. Nat Rev Chem 6:899–915. https://doi.org/10.1038/s41570-022-00439-w

Shaffer F, Ginsberg JP (2017) An overview of heart rate variability metrics and norms. Front Public Heal 5:258. https://doi.org/10.3389/fpubh.2017.00258

Shaheen N, Shaheen A, Osama M et al (2024) MicroRNAs regulation in Parkinson's disease, and their potential role as diagnostic and therapeutic targets. npj Park Dis 10:1–11. https://doi.org/10.1038/s41531-024-00791-2

Shahisavandi M, Wang K, Ghanbari M, Ahmadizar F (2023) Exploring Metabolomic patterns in type 2 diabetes mellitus and response to glucose-lowering medications—review. Genes (Basel) 14:1464. https://doi.org/10.3390/genes14071464

Shajari S, Kuruvinashetti K, Komeili A, Sundararaj U (2023) The emergence of AI-based wearable sensors for digital health technology: a review. Sensors 23:9498. https://doi.org/10.3390/s23239498

Shapiro I, Stein J, MacRae C, O'Reilly M (2023) Pulse oximetry values from 33,080 participants in the Apple Heart & Movement Study. npj Digit Med 6:1–13. https://doi.org/10.1038/s41746-023-00851-6

Sherwani SI, Khan HA, Ekhzaimy A et al (2016) Significance of HbA1c test in diagnosis and prognosis of diabetic patients. Biomark Insights 11:95–104. https://doi.org/10.4137/Bmi.s38440

Singh NK, Chung S, Chang AY et al (2023) A non-invasive wearable stress patch for real-time cortisol monitoring using a pseudoknot-assisted aptamer. Biosens Bioelectron 227:115097. https://doi.org/10.1016/j.bios.2023.115097

Smelik M, Zhao Y, Li X et al (2024) An interactive atlas of genomic, proteomic, and metabolomic biomarkers promotes the potential of proteins to predict complex diseases. Sci Rep 14:1–10. https://doi.org/10.1038/s41598-024-63399-9

Smokovski I, Steinle N, Behnke A et al (2024) Digital biomarkers: 3PM approach revolutionizing chronic disease management—EPMA 2024 position. EPMA J 152(15):149–162. https://doi.org/10.1007/S13167-024-00364-6

Soguel Schenkel N, Burdet L, De Muralt B, Fitting JW (1996) Oxygen saturation during daily activities in chronic obstructive pulmonary disease. Eur Respir J 9:2584–2589. https://doi.org/10.1183/09031936.96.09122584

Sproston NR, Ashworth JJ (2018) Role of C-reactive protein at sites of inflammation and infection. Front Immunol 9:754. https://doi.org/10.3389/fimmu.2018.00754

Stergiou GS, Asayama K, Thijs L et al (2014) Prognosis of white-coat and masked hypertension: international database of home blood pressure in relation to cardiovascular outcome. Hypertension 63:675–682. https://doi.org/10.1161/HYPERTENSIONAHA.113.02741

Tehrani F, Teymourian H, Wuerstle B et al (2022) An integrated wearable microneedle array for the continuous monitoring of multiple biomarkers in interstitial fluid. Nat Biomed Eng 6:1214–1224. https://doi.org/10.1038/S41551-022-00887-1

Teles FRR, Fonseca LP (2008) Trends in DNA biosensors. Talanta 77:606–623. https://doi.org/10.1016/j.talanta.2008.07.024

Tsao CW, Aday AW, Almarzooq ZI et al (2023) Heart disease and stroke statistics – 2023 update: a report from the American Heart Association. Circulation 147:E93–E621. https://doi.org/10.1161/CIR.0000000000001123

Uppaluri KR, Challa HJ, Gaur A et al (2023) Unlocking the potential of non-coding RNAs in cancer research and therapy. Transl Oncol 35:101730. https://doi.org/10.1016/j.tranon.2023.101730

Windred DP, Anderson C, Jeppe KJ et al (2024) Higher central circadian temperature amplitude is associated with greater metabolite rhythmicity in humans. Sci Rep 14:1–9. https://doi.org/10.1038/s41598-024-67297-y

Wu X, Xia P, Yang L et al (2024) The roles of long non-coding RNAs in Alzheimer's disease diagnosis, treatment, and their involvement in Alzheimer's disease immune responses. Non-coding RNA Res 9:659–666. https://doi.org/10.1016/j.ncrna.2024.03.008

Xu L, Xie Y, Gou Q et al (2024) HER2-targeted therapies for HER2-positive early-stage breast cancer: present and future. Front Pharmacol 15:1–16. https://doi.org/10.3389/fphar.2024.1446414

Yu H, Lu L, Chen M et al (2021) Genome-wide discovery of hidden genes mediating known drug-disease association using KDDANet. npj Genomic Med 6. https://doi.org/10.1038/s41525-021-00216-6

Yugar LBT, Yugar-Toledo JC, Dinamarco N et al (2023) The role of heart rate variability (HRV) in different hypertensive syndromes. Diagnostics 13:785. https://doi.org/10.3390/diagnostics13040785

Zoccali C, Tripepi G, Stel V et al (2024) Biomarkers in clinical epidemiology studies. Clin Kidney J 17:1–6. https://doi.org/10.1093/ckj/sfae130

Exploring Novel Healthcare Monitoring Modalities: Wearable, Ingestible, and Implantable Biosensors

5

Desiree Tamara Scheidt, Marcos Lazaro Moreli, Elsa Maria Materón, Larissa Meneghin Michilini, Emanuel Carrilho, and Laís Canniatti Brazaca

Abstract

Recent medical advances are closely tied to real-time healthcare monitoring, revolutionizing precise diagnosis, personalized treatments, and effective disease prevention. By continuously tracking the body's internal processes—such as biosignals, biochemical parameters, and biomarkers in biofluids—biosensors generate vital health information that enhances patient care. These sensors, available in wearable, implantable, and ingestible forms, measure specific chemical or biological reactions, producing real-time signals that reflect the concentration of targeted analytes. Beyond biochemical monitoring, physical sensors are crucial in tracking physiological metrics such as body temperature, heart rate, pressure, and movement. Integrating biochemical and physical monitoring offers a comprehensive health-tracking system, improving diagnostic accuracy and optimizing treatment efficacy. In recent years, significant investments from scientific and commercial sectors have driven the development of body-integrated technologies. However, key challenges remain in optimizing sensor design, selecting suitable materials, and ensuring biocompatibility and biodegradability. This chapter presents an in-depth exploration of the current state of sensor technology, high-

D. T. Scheidt · L. M. Michilini · E. Carrilho · L. C. Brazaca (✉)
Instituto de Química de São Carlos, Universidade de São Paulo, São Carlos, SP, Brazil

Instituto Nacional de Ciência e Tecnologia de Bioanalítica – Lauro Kubota (INCTBio-LK), Campinas, SP, Brazil
e-mail: lais.brazaca@usp.br

M. L. Moreli
Instituto de Química de São Carlos, Universidade de São Paulo, São Carlos, SP, Brazil

Instituto de Ciências da Saúde - Universadade Federal de Jataí, Jataí, GO, Brazil

E. M. Materón
Instituto de Física de São Carlos, Universidade de São Paulo, São Carlos, SP, Brazil

K. Mahato, P. Chandra (eds.), *Nano-bioelectronics for Precision Health Monitoring*, https://doi.org/10.1007/978-981-95-2117-3_5

lighting the technical challenges and their transformative impact on modern medicine.

Keywords

Wearable sensor · Implantable sensor · Ingestible sensor · Health monitoring · Biosensors · Personalized healthcare

5.1 Overview of Health Monitoring Modalities

5.1.1 Importance of Continuous Health Monitoring

Everyone's health is uniquely influenced by their genetic traits, lifestyle, and environmental contexts, which affect their well-being in specific ways. Continuous monitoring provides individualized information over time, enabling a personalized approach to healthcare. This customization improves the effectiveness of treatments, as interventions can be tailored to each patient's unique needs. With technological advancements, there is an increasing potential to enhance the precision and efficiency of these systems, allowing for faster and more effective interventions. Therefore, health monitoring is essential for maintaining and promoting population well-being. This practice involves the continuous collection, analysis, and interpretation of health data to detect, control, and prevent diseases and other health problems (Song et al. 2023; Shrivastava et al. 2020; Gambhir et al. 2021). Advanced technologies, such as wearable devices, ingestible biosensors, implantable devices, and digital platforms, enable real-time monitoring of various health indicators, promoting more individualized health management (Lu et al. 2023a; Mukherjee et al. 2022; Faust et al. 2020). Continuous monitoring allows for the early detection of anomalies or changes in an individual's health patterns, which is crucial for preventing chronic diseases like diabetes, hypertension, and cardiovascular diseases, which often present subtle symptoms in their early stages (Noura et al. 2022; Gupta et al. 2021; Eidi 2023; Yun et al. 2019). By identifying changes in these indicators, individuals can take preventive measures, such as lifestyle modifications or seeking medical treatment before the condition worsens (Shrivastava et al. 2020; Faust et al. 2020).

Continuous health monitoring is an essential component of the future of medicine, offering numerous opportunities to improve healthcare and patient quality of life (Gambhir et al. 2021; Mukherjee et al. 2022). Despite challenges related to privacy, accessibility, and data management, the advantages of these technologies are undeniable. Therefore, it is crucial to develop robust security protocols and comply with data protection regulations to ensure patient confidentiality (Handler et al. 2017; Awotunde et al. 2021). Additionally, accessibility poses a challenge, as continuous monitoring technologies can be costly and inaccessible to low-income individuals or those living in remote areas. To ensure everyone can benefit, policies must be implemented to guarantee equitable access and subsidize costs for those in need. With the ongoing advancement of technology and the implementation of policies promoting equal access, continuous monitoring has the potential to transform the healthcare system, making it more agile, personalized, and efficient (Gambhir et al. 2021; Malasinghe et al. 2019).

5.1.2 Evolution of Monitoring Technologies

Early health monitoring devices, although simple, were fundamental to technological advancements. In 1714, Daniel Gabriel Fahrenheit invented the mercury thermometer, a device that allowed for accurate measurement of body temperature, essential for assessing fevers and other febrile conditions. Though rudimentary compared to modern standards, this thermometer represented a significant breakthrough, enabling more precise diagnoses (Grodzinsky and Sund Levander 2020). Later, in 1816, René Laënnec developed the stethoscope, which revolutionized medicine by allowing for the precise auscultation of internal body sounds, such as lungs and heart, enhancing diagnosis capabilities. With advances in semiconductors and electronics, devices like the Holter monitor, developed by Norman J. Holter in 1961, allowed for the recording of electrocardiograms over extended periods, revolutionizing the evaluation of arrhythmias (Choudry et al. 2022; Donoso and Arriagada 2020; Xu et al. 2022). With advances in electronics, biocompatible materials, and nanomaterials, it became possible to create implantable devices that transformed medical diagnosis and prognosis through miniature biomedical sensors, elevating the quality and efficiency of healthcare. The first implantable medical device, the pacemaker, was introduced in 1958 to treat cardiac arrhythmias (Xu et al. 2024). Since then, implantable devices have been continuously refined and developed, such as the implantable Doppler for monitoring microvascular anastomoses, which uses an ultrasonic probe mounted directly on the blood vessel. However, this method relies on wired connections, limiting its use to hospital settings. To overcome this limitation, a biodegradable, wireless, and battery-free pressure sensor was developed, designed to monitor arterial blood flow during various postoperative periods. With its thin and flexible structure, the sensor can easily wrap around arteries of different sizes. Furthermore, as the device is entirely made from biodegradable materials, it is naturally absorbed by the body within a few months, eliminating the need for surgical removal and minimizing the risk of vascular trauma (Boutry et al. 2019). In another recent study, Gharbi et al. (2023) developed a sensor made of implantable and biodegradable bioactive glass, whose chemical composition, enriched with fluoride, proved sensitive to hydration. This sensitivity induces changes in the material's electrical properties, which can be reversed by dehydration. Due to these characteristics, fluoride-enriched bioactive glass has the potential to be used as a moisture sensor, capable of measuring cerebral edema and alerting to increased intracranial pressure. In parallel with these developments, technological advances have also facilitated the creation of wearable devices that provide continuous, real-time monitoring, seamlessly integrated into everyday accessories like watches, glasses, and clothing. Over time, these devices have been enhanced to include signal processors and integrated batteries, allowing for autonomous data processing, communication, and power supply (Park et al. 2016; Surantha et al. 2021). Among wearable devices, smartwatches stand out as one of the most popular options. The GlucoWatch® (Cygnus Inc., Redwood City, CA, USA) was the first non-invasive glucose monitoring device to receive approval from the Food and Drug Administration (FDA) (Potts et al. 2002). Another example is a device for the rapid and accurate detection of specific biomarkers for Parkinson's disease, using a biosensor attached to a smartwatch, enabling continuous monitoring and early detection of the disease (Ko et al. 2022).

Ingestible sensors are made of materials commonly found in dietary supplements, allowing them to be naturally eliminated by the body. After ingestion, these sensors are activated upon contact with stomach fluid, initiating data collection on health, which can be accessed through apps or websites. A unique feature of these sensors is that, as they pass through the digestive tract, they can provide relevant information about Gut health, disorders, and the effects of food, medical supplements, and environmental changes on the gastrointestinal tract (Kalantar-Zadeh et al. 2017a; Alipour et al. 2020). Additionally, they can be used to monitor cardiovascular diseases, infections, transplants, and more (Alipour et al. 2020). The use of artificial intelligence (AI) is at the forefront of innovation in health monitoring. Machine learning algorithms and large-scale data analysis are revolutionizing how health data is interpreted and applied. For example, AI can be used as a tool to assist in the analysis of medical images, such as X-rays and magnetic resonance imaging (MRI) (Mathesul et al. 2023; Zokaeinikoo et al. 2021; Suarez-Ibarrola et al. 2022). A study by Makino et al. (2019) demonstrated that a pre-predictive model could be used for diagnosing diabetic kidney disease (DKD) using AI. The study employed natural language processing techniques and data analysis with machine learning applied to large datasets. The database used consisted of electronic medical records of patients with diabetes. The AI analyzed information from the six months before the reference period and identified key factors to detect temporal patterns related to disease progression. As a result, the model achieved 71% accuracy in predicting the advancement of DKD. This AI-based model shows promise in detecting disease progression, potentially leading to more effective interventions and reducing the need for dialysis (Makino et al. 2019). The outlook is promising with the continued evolution of health monitoring technologies. Advances in AI, biometric sensors, and digital connectivity promise to further transform healthcare, offering new opportunities for disease prevention, diagnosis, and treatment. These innovations must be accompanied by appropriate regulations and careful consideration of ethical and social impacts.

5.2 Wearable Sensors

5.2.1 Definition and Overview

Wearable, non-invasive sensors represent a significant leap in healthcare technology, offering compact, miniaturized devices that can be seamlessly integrated into everyday accessories, clothing, or patches (Mukherjee et al. 2022; Ometov et al.

2021). These sensors enable continuous monitoring of vital body parameters, addressing the growing demand for less invasive diagnostics and real-time health tracking (Promphet et al. 2021). By minimizing the need for frequent blood tests and reducing the number of in-person doctor visits, wearable sensors empower patients and healthcare providers to monitor treatments and the progression of disease symptoms remotely. This real-time data collection provides a more efficient and less intrusive approach to ongoing care, revolutionizing health management (Heikenfeld et al. 2018). The development of wearable health monitors originated during the early years of space exploration (Launius 2023; Erdem et al. 2024). During the Mercury, Gemini, and Apollo missions, astronauts were frequently exposed to extreme physical conditions, necessitating real-time health monitoring. To address this need, astronauts were equipped with various monitoring devices. Electrodes were used to collect electrocardiograms (ECGs), capturing the classic heartbeat waveform, while a semiautomatic sphygmomanometer, coupled with a microphone to record pulse sounds, measured blood pressure. Additionally, astronauts wore impedance pneumograph equipment to monitor their breathing. All vital signs collected on those sensors were continuously monitored and transmitted back to Earth, ensuring that mission control could track the astronauts' health in real time. By the 1980s, the Polar "Sport Tester PE2000" emerged as one of the first wearable devices designed for heart rate monitoring, primarily targeting elite athletes. The device utilized a chest strap equipped with electrodes based on ECG technology, which detected the electrical activity of the heart. This data was then transmitted wirelessly to a wrist-mounted unit, where the heart rate could be displayed and analyzed in real-time, providing athletes with valuable physiological insights during exercise (Team P 2023). These early wearable devices paved the way for the development of more advanced technologies. With the rise of smartphones and the expanding market for wearable sensors, activity trackers capable of monitoring heart rate and step count were quickly integrated into consumer electronics. Modern devices now extend beyond basic activity tracking, enabling users to monitor critical health parameters such as sleep patterns, heart rate variability, stress levels, body temperature, and body movements, among others (Erdem et al. 2024). This widespread adoption highlights the growing consumer demand for personal health monitoring solutions. To gain a deeper understanding of the technologies employed in these devices, as well as the associated challenges and their applications in human health monitoring, wearable sensors are generally categorized into two main categories: physical wearable sensors and chemical wearable sensors (Choi et al. 2023). In the following sections, we will explore each category in detail.

5.2.2 Technologies for Wearable Sensor Manufacturing

Wearable sensors have the potential to significantly enhance the quality of life by continuously sensing, collecting, and transmitting physiological data from the human body (Seneviratne et al. 2017; Wang et al. 2019; Dunn et al. 2018). In recent years, these devices have not only been extensively studied in academic research but have also become widely available in the consumer market. Although many devices are primarily focused on tracking physical activity (McCallum et al. 2018), advancements in sensor technology are paving the way for real-time monitoring of chemical health parameters, marking the next frontier in wearable innovation. To understand the full scope of this technology, it is essential to explore how these sensors operate and the key technologies driving their functionality, starting with physical wearable sensors.

5.2.2.1 Physical Wearable Sensors

As the name implies, physical wearable sensors measure changes in the body's physical properties, such as pressure, heart rate, body movement, respiration, and temperature. These sensors are essential for continuous health monitoring and can be classified into three primary categories based on the technology they use: inertial (mechanical), optical, and electrical.

- *Inertial wearable sensors (mechanical sensors)*: Wearable mechanical sensors are specifically designed to measure body motion and orientation (Al-Daraghmeh and Stone 2023; Wang et al. 2017). These sensors utilize inertial measurement units (IMUs), which consist of accelerometers, gyroscopes, and sometimes magnetometers, to track changes in acceleration, rotational velocity, and magnetic field orientation. Accelerometers detect linear movement, gyroscopes capture rotational motion, and magnetometers provide orientation relative to Earth's magnetic field (Wang et al. 2019; Dunn et al. 2018; Al-Daraghmeh and Stone 2023). These IMUs are commonly integrated into devices such as fitness trackers and smartwatches, playing a key role in monitoring physical activity, posture, fall detection, and sleep biomechanics (Buckley et al. 2022). There are four main types of transducers used in wearable mechanical sensors: piezoresistive, capacitive, iontronic, and piezoelectric. Each transducer type has its unique mechanism for converting physical movement into electrical signals, further expanding the capabilities of wearable devices.
- *Optical wearable sensors*: Wearable optical sensors are classified into two main categories: optical fiber sensors and fiber-free optical sensors. Optical fiber sensors use embedded optical fibers in flexible, stretchable substrates to detect changes in environmental conditions such as pressure, strain, or angle shifts (Li et al. 2024a). Factors like temperature variations or respiratory rate changes can cause deformations (e.g., bending or stretching), which alter the light transmission properties in the fibers. These variations are processed by a photodetector and demodulator, converting them into measurable data (Li et al. 2024a; Pan et al. 2020; Shao et al. 2024). In contrast, fiber-free optical sensors measure physiological parameters by using light interactions with tissues. The choice of light wavelength (ranging from UV to infrared) depends on the tissue's absorption characteristics and desired penetration depth. As light enters the skin, it scatters and is absorbed, with the reflected light captured by an optical detector (e.g.,

photodiode, avalanche photodetector, or photomultiplier tube) (Heikenfeld et al. 2018). A well-known application of fiber-free optical sensors is photoplethysmography (PPG), which measures heart rate by detecting changes in light absorption in tissues (Dunn et al. 2018; Shilaih et al. 2017). This technology is commonly used in smartwatches and other commercial wearables for real-time health monitoring.

- *Electrical wearable sensor*: These sensors, designed with electrical interfaces, detect and measure biopotentials—electrical signals generated by the activity of living cells, particularly nerve and muscle cells. Electrical transducers, which respond to changes in electrical resistance or capacitance within tissues, are employed to capture these biopotentials. This enables continuous monitoring of key physiological functions, such as brain activity through electroencephalography (EEG), heart rhythms via electrocardiography (ECG), and muscle activity through electromyography (EMG). Such sensors provide crucial real-time insights into neurological, cardiovascular, and muscular health (Heikenfeld et al. 2018; Al-Daraghmeh and Stone 2023; Peng 2023). Physical wearable sensors can be coupled with different technologies to improve their accuracy. As an example, wearable sensors are commonly used for heart rate monitoring, with many devices relying on PPG technology. However, the accuracy of PPG often requires the integration of motion sensors, such as accelerometers or gyroscopes, to account for motion artifacts. These sensors provide a reference for body movements, allow motion-related noise to be subtracted from the PPG signal, thereby ensuring more accurate heart rate measurements (Wang et al. 2019).

5.2.2.2 Chemical Wearable Sensor

Most commercially available wearable sensors focus on monitoring physical parameters, such as those discussed in the previous section. However, for a more comprehensive assessment of a user's health, wearable chemical sensors offer the ability to track chemical biomarkers in real time, providing continuous insights into the body's biochemical processes (Promphet et al. 2021). Before diving into the details of these sensors, it is important to understand what a chemical sensor is. According to the International Union of Pure and Applied Chemistry (IUPAC), a chemical sensor can be described as a device that can convert chemical information—such as the concentration of a specific stance—into a measurable signal (Sempionatto and Gao 2022; Hulanicki et al. 1991). Therefore, wearable chemical sensors can monitor various biochemical and chemical signals produced by the body, such as electrolytes (e.g., $Na+$, $K+$, $Cl-$), metabolites (e.g., glucose, lactate), and hormones (e.g., cortisol, estradiol). In addition, they can detect external substances present in the body, such as drugs, heavy metals, or alcohol (Sempionatto and Gao 2022; Bandodkar et al. 2016). The detection and quantification of these analytes are made possible by their specific interactions with biomolecules integrated into the wearable sensor. These biomolecules, often enzymes, antibodies, or receptors, selectively bind to target analytes, triggering a measurable response that the sensor can capture and process in real-time (Heikenfeld et al. 2018). The most commonly employed transduction methods for converting chemical interactions into measurable signals are electrochemical (e.g., voltammetric, potentiometric, and impedimetric) and optical (e.g., colorimetric and fluorescence) techniques. Electrochemical methods detect changes in electrical properties resulting from chemical reactions, while optical methods rely on changes in light properties, such as color or fluorescence, to quantify the presence of specific analytes (Sempionatto and Gao 2022; Hulanicki et al. 1991).

5.2.3 Application of Wearable Sensors in Health Monitoring

A growing global population and an increasing number of aging individuals primarily drive the rising cost of healthcare (Ali et al. 2023). Wearable devices offer promising solutions for improving healthcare delivery across various settings, from hospitals and clinics to home care and remote regions, including rural and low-resource environments (Dunn et al. 2018). In this section, we will explore the latest advancements in wearable sensor technology, focusing on diseases that can significantly benefit from continuous monitoring. For clarity, the discussion will be categorized into physical and chemical wearable sensors.

5.2.3.1 Physical Wearable Sensor

Continuous health monitoring is vital for managing heart diseases like arrhythmias, hypertension, and heart failure, which often require real-time data to detect abnormalities before they become life-threatening. As reported by the World Health Organization (WHO), cardiovascular diseases (CVDs) continue to be the world's leading cause of death, accounting for an estimated 17.9 million deaths each year—representing 32% of all deaths worldwide, with 85% of these resulting from heart attacks and strokes. The rise in CVD-related deaths over recent decades is largely attributed to the aging population and lifestyle factors such as poor diet and lack of physical activity (Organization, W.H. 2023). It is important to note that up to 80% of heart diseases and 70% of strokes are preventable through timely interventions and preventive strategies (Casson et al. 2016). However, limited access to healthcare in certain regions contributes to higher mortality rates, emphasizing the need for continuous health monitoring to improve outcomes, especially in under-resourced areas. Although there are smartwatches on the market capable of monitoring heart rhythms, stress levels, and respiratory rates—parameters closely linked to heart health—the price range for these devices varies from around \$70 to \$500 (Shah 2023). This high cost makes them less accessible to everyone. In their effort to develop a more affordable sensor, Li and colleagues (2024b) introduced a wearable device based on a drop-shaped small-core fiber (DSSCF), known for its high sensitivity to pressure and deformation. The optical fiber was embedded in a flexible, biocompatible silica gel layer, forming a "fabric-DSSCF-silica gel sandwich structure," which makes the sensor soft and suitable for continuous wear. The sensor relies on optical fiber technology combined with the photoelastic effect. When external forces—such as heartbeats or breathing—apply pressure to the sensor, they cause mechanical deformations in the fiber, altering light transmission. The sensor

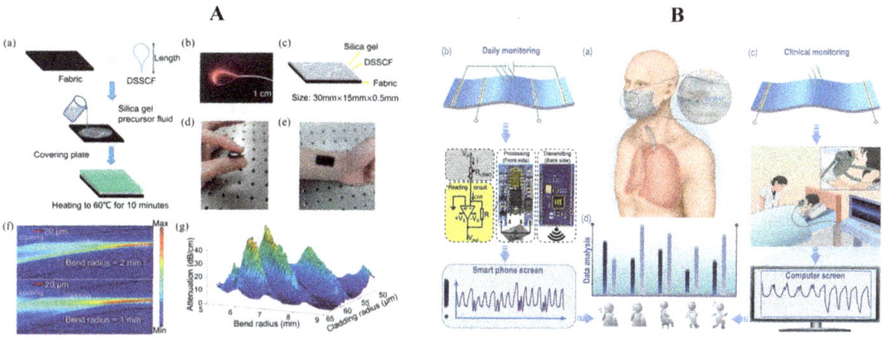

Fig. 5.1 (**a**) (a) illustration of the steps for constructing the DSSCF-based sensor. (b) Image of an optical signal (650 nm) going through the DSSCF fiber. (c) Sensor structure, configured as a fabric/DSSCF/silica gel sandwich. (d) Demonstration of the biocompatibility of the sensor when applied to the skin. (e) Mechanical analysis of the flexible sensor. (f) The distribution of electromagnetic field intensity was simulated for fibers bent. (g) The result shows a bend loss coefficient as a function of varying bend radius and cladding radius; (**b**) (a) Illustration of the wearable respiration sensor integrated into a face mask. (b) Representation of the wireless mode configuration for breath monitoring sensors. (c) Schematic of the same sensor in a wired mode configuration. (d) Illustration of the data acquisition and processing system used for analyzing the captured breath data. (Reprinted with permission from Li et al. (2024b), Ye et al. (2023))

was tested on the wrist, chosen for its accessibility and proximity to superficial blood vessels, which makes it an ideal location for detecting cardiovascular signals (Fig. 5.1a). It successfully captured detailed pulse waveforms, including the P-wave (arterial pulse during systole), T-wave (elastic recoil of vessels), and D-wave (blood flow hitting the aortic valve during diastole). With a fast response time of 5 ms and high sensitivity, the sensor proves highly effective for real-time monitoring of cardiovascular abnormalities. In a separate study, Shi and colleagues (2022) applied optical fiber technology to develop a wearable sensor. In this case, fiber bracket grating (FBG) technology was used to monitor respiration and heartbeat simultaneously. The sensor's design consists of an elastic arc-shaped surface structure, a suspended silicone membrane, an optical fiber inscribed with an FBG sensing element, and a wearable elastic belt. This configuration amplifies chest movements caused by breathing and heartbeats, causing the FBG to stretch and modify the reflected light. By detecting these deformations, the sensor translates changes in light wavelength into signals representing both respiratory and cardiac activity. The sensor demonstrated high sensitivity, with heart rate readings closely matching those from standard ECG monitors, confirming its effectiveness for continuous, real-time cardiac monitoring. In terms of respiration monitoring, the sensor successfully captured the chest's expansion and contraction during breathing, showing its reliability across various conditions and suitability for continuous respiratory monitoring.

Continuous monitoring of respiratory activity is vital in medical contexts, such as managing anesthetized patients, monitoring infants at risk of sudden infant death syndrome (SIDS) and diagnosing sleep apnea (Shi et al. 2022). Respiratory patterns directly reflect lung and respiratory system health, with changes such as increased breathing rate indicating conditions like cardiopulmonary deterioration or respiratory failure (Ye et al. 2023). To address this, Ye and colleagues (2023) (Ye et al. 2023) developed a wearable respiration sensor integrated into a face mask. The sensor, based on La0.7Sr0.3MnO_3 (LSMO), a manganite oxide material, was deposited on a flexible mica substrate using pulsed laser deposition (PLD). This sensor detects airflow changes by measuring variations in electrical resistance due to the temperature differences between inhaled and exhaled air. Different breathing modes—normal, deep, and coughing—produced distinct resistance patterns, enabling accurate identification of respiratory activities. The sensor was tested in both wireless and wired clinical modes. In wireless mode, it transmitted data to a smartphone via Bluetooth for real-time respiratory tracking. The wired mode offered more reliable data acquisition for clinical applications (Fig. 5.1b).

Wearable sensors have proven highly effective for continuous monitoring of physiological parameters, enabling early detection and intervention for conditions like cardiovascular diseases, respiratory issues, and movement disorders. Their portability, affordability, and real-time data transmission make them ideal for both clinical and daily use. With this success, attention is now turning to chemical sensors, capable of detecting biomarkers in bodily fluids, expanding the potential of wearable technologies for non-invasive monitoring of biochemical health indicators.

5.2.3.2 Chemical Wearable Sensor

The ability to monitor chemical health parameters is the gold of the state-of-art of wearable device technology. One of the most significant advantages of these sensors is their capacity to drastically reduce the time it takes for information to travel between healthcare professionals and patients. This real-time data transmission enables quicker diagnoses and more timely interventions, improving the chances of addressing health conditions before they escalate. By integrating continuous chemical monitoring with fast communication and wearable devices offer a transformative approach to healthcare management. Body fluids-based analytes like saliva, tears, urine, sweat, and interstitial fluid are promising candidates since they are abundant sources of ions, proteins, metabolites, and amino acids that can be monitored in real time (Peng et al. 2016; Ramachandran and Liao 2022). Here, we are going to show some examples of how these fluids can be used to monitor human health:

- *Tears*: Sempionatoto et al. (2019) developed an eyeglasses-based tear biosensing system designed to monitor alcohol, glucose, and vitamins in a versatile manner (Fig. 5.2a). The system utilizes a microfluidic channel built with a super-hydrophilic polycarbonate membrane to collect tears via capillary action. After collection, the tears are directed through the microfluidic channel to flow over a screen-printed electrochemical sensor. Tear production is stimulated using a menthol tear stick applied below the eye, ensuring a controlled and continuous flow for precise measurement. The sensor is built into the nose-bridge pad of standard eyeglasses, chosen for its proximity to the eyes, which allows for efficient tear collection without obstructing the user's vision or comfort. The core

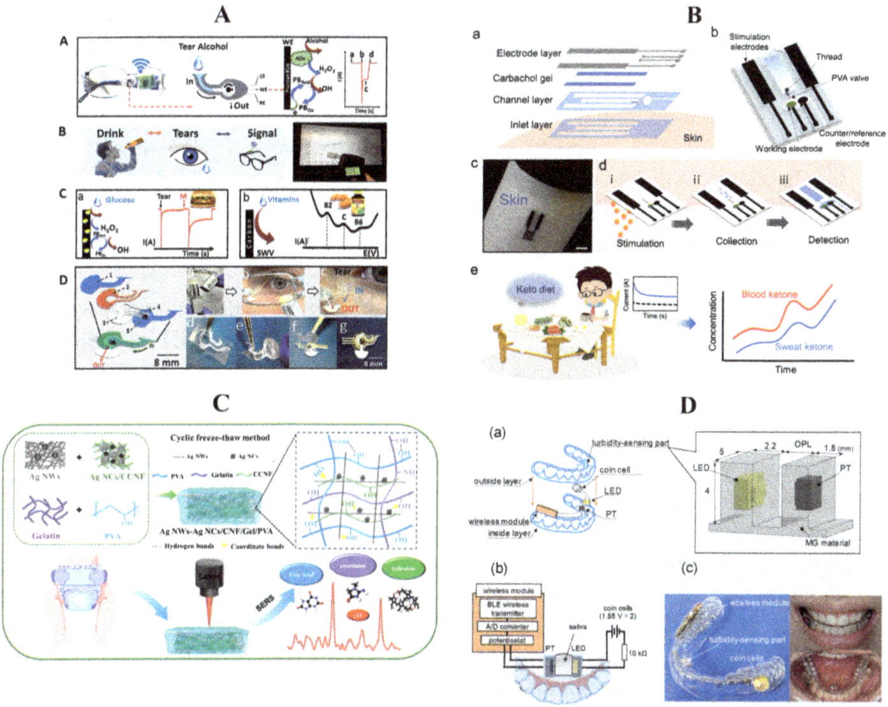

Fig. 5.2 (**a**) (a) Image and design of the fluidic device and wireless system embedded in eyeglasses. (b) Process for detecting alcohol in tears, starting with drink ingestion, tear stimulation, and measurement. (c) Illustration of glucose and vitamin detection via enzymatic reactions and Prussian blue electrode response for glucose, vitamins B2, C, and B6. (d) Components of the fluidic system; (**b**) (a) Diagram of the sweat ketone sensor showing its layered structure. (b) Assembled sensor for sweat analysis. (c) Photo of the wearable sensor positioned on the upper arm (scale: 1 cm). (d) Diagram illustrating the sweat β-HB detection stages: (e) Graphical representation of the relationship between β-HB levels in sweat and blood during a ketogenic diet; (**c**) Depiction of the construction and use of a SERS sensor integrated into a diaper for real-time urine analysis; (**d**) (a) Representation of the sensing area in the mouthguard-based turbidity sensor. (b) Setup for the turbidity measurement using the mouthguard-based turbidity sensor. (c) Image of the finalized device. (Reprinted with permission from Sempionatto et al. (2019), Shen et al. (2024), Liu et al. (2024), Ichikawa et al. (2024))

component is a screen-printed electrode modified with a Prussian blue (PB) layer, facilitating the detection of hydrogen peroxide produced during enzymatic reactions with glucose oxidase (GOx) or ethanol oxidase (AOx). The enzymes were immobilized on the sensor surface using drop-casting or mixing it with the electrode ink. Additionally, the potentiostat applied in the measurements was equipped with a Bluetooth module for real-time, wireless data transmission to a smartphone or other external devices, enabling seamless health monitoring (Sempionatto et al. 2019).

- *Sweat*: Sweat is vital for maintaining body temperature and serves as a non-invasive avenue for real-time health tracking. Shen and colleagues (2024) developed a wearable electrochemical sensor to measure β-hydroxybutyrate (β-HB), an important ketone body reflecting the body's metabolic state. The sensor features microfluidic channels made from PDMS (polydimethylsiloxane) and PET (polyethylene terephthalate), which collect sweat stimulated through iontophoresis. At the core is a screen-printed carbon electrode (SPE) modified with β-hydroxybutyrate dehydrogenase (HBD), which reacts with β-HB to produce an electrochemical signal (Fig. 5.2b). Designed for flexibility and comfort, the sensor adheres to the skin (e.g., upper arm) and transmits data wireless to external devices for real-time monitoring.

- *Urine*: Liu and coworkers (2024) developed an innovative diaper-based wearable sensor for non-invasive, real-time detection of biomarkers and pH levels in urine, specifically designed for infants and elderly individuals (Fig. 5.2c). Biomarkers are crucial for assessing kidney function, metabolic health, and liver conditions. The sensor's core component is a polyvinyl alcohol (PVA)/gelatin-based hydrogel, which absorbs urine directly from the diaper. This hydrogel features a dual-network structure, providing flexibility and strength for seamless integration into diapers. Embedded with silver nanowires (Ag NWs) and silver nanocubes (Ag NCs), the hydrogel serves as the foundation for the SERS (surface-enhanced Raman spectroscopy) system. These nanostructures create "hot spots" that amplify Raman signals, enabling enhanced detection of biomarkers in the absorbed urine. A portable Raman spectrometer reads the amplified signals, providing real-time data on biomarker concentrations and pH levels. The sensor also supports wireless data transmission, facilitating continuous health monitoring and real-time updates to caregivers or healthcare providers.
- *Saliva*: Inadequate oral hygiene is a well-established contributor to oral diseases such as periodontal disease and dental caries. Less widely known, however, is its role in systemic conditions like diabetes and cardiovascular diseases due to the intricate link between oral and overall health (Kane 2017). To address this, Ichikawa and colleagues (2024) developed a mouthguard-type sensor for non-invasive, real-time monitoring of salivary turbidity (Fig. 5.2d). The sensor, integrated into a polyethylene terephthalate-glycol (PETG) mouthguard, features a light-emitting diode (LED) and a phototransistor (PT) positioned opposite each other around the incisor area. The system measures salivary turbidity by detecting light transmittance through saliva, with reduced light transmission indicating increased turbidity. A wireless module and Bluetooth transmission enable real-time data collection while coin cells power the device. The sensor, designed for continuous use without hindering normal activities, demonstrated high repeatability and sensitivity in detecting oral hygiene conditions.
- *Interstitial fluid (ISF):* In their efforts to develop a wearable device for non-invasive glucose monitoring, Yao and coworkers (Yao et al. 2021) created a sensor capable of extracting ISF from the skin without relying on invasive methods. ISF originates from transcapillary blood exchange and is widely used in biosensing because it is a potential biofluid for biomarker analysis (Dai et al. 2023). For ISF extraction, the authors employed a process known as reverse iontophoresis (RI). This method placed a two-electrode system (working and counter elec-

trodes) on the skin, and a small electric current (2 mA/cm^2) was applied. This current mobilized ions within the skin, drawing the ISF to the surface. The process relies on ion movement to transport the fluid across the skin barrier, and the ISF was collected in the region between the electrodes. The electrodes were constructed using a nickel textile mesh as the substrate. The working electrode was a composite of graphene and carbon nanotubes (CNTs), chemically deposited on the textile mesh, followed by the immobilization of glucose oxidase. Once the ISF reached the skin's surface, it naturally diffused toward the working electrode, where the glucose in the ISF was detected. The device could measure glucose concentrations with high sensitivity and accuracy, yielding results that were highly consistent with those obtained from a commercial glucose meter.

5.2.4 Challenges and Limitations of Wearable Sensors

In the broader context of both physical and chemical wearable sensors, one of the primary challenges for widespread adoption in healthcare is the extraction of meaningful and actionable health-related information from the large volumes of data generated. This data must be transmitted securely and efficiently while maintaining privacy and allowing for real-time, uninterrupted monitoring via IoT systems (Dunn et al. 2018; Bandodkar et al. 2016). Wearable sensors are also energy-intensive, and while energy harvesting solutions are promising, consistent power generation remains an issue (Shilaih et al. 2017). Wearable chemical sensors face specific challenges, particularly in materials and construction. Unlike traditional bulky sensors, they must be small, flexible, biocompatible, and able to attach seamlessly to the body's soft, flexible surfaces. They must also withstand mechanical deformation and environmental factors like temperature and humidity (Bandodkar et al. 2016; Nasiri and Khosravani 2020). The fabrication process adds complexity, as integrating multiple components while ensuring scalability and cost-effectiveness requires advanced manufacturing techniques (Nasiri and Khosravani 2020). Another critical challenge is the selection of biomarkers. The chosen biomolecule must be present in detectable concentrations in alternative biofluids such as sweat, saliva, or tears and should correlate with health conditions. However, these biofluids are highly variable due to individual physiology, diet, and medication, making it difficult to establish a direct correlation with blood analytes (Sempionatto and Gao 2022). Moreover, maintaining sensor accuracy in fluctuating environmental conditions, including temperature and pH, is essential. Sensors with biological recognition elements can be particularly prone to instability (Bandodkar et al. 2016). Real-time calibration and sensor robustness under fluctuating conditions are significant obstacles that must be overcome (Sempionatto and Gao 2022; Nasiri and Khosravani 2020). Wearable devices offer immense potential for continuous health monitoring, providing real-time insights into physiological conditions. By overcoming these obstacles, wearable chemical sensors can be successfully commercialized, revolutionizing the way we monitor and manage health in real time. While wearable sensors offer the advantage of being non-invasive for health monitoring, they have limitations in accessing certain fluids directly. Implantable sensors, however, bypass these barriers, providing a more direct and precise method of fluid analysis and presenting new possibilities in continuous health monitoring.

5.3 Implantable Sensors

5.3.1 Definition and Overview

With the growing demand for personalized healthcare and point-of-care treatments, there is an increasing need for sensors that maintain long-term stability while ensuring excellent mechanical and biological compatibility with human tissues. Real-time monitoring is vital in this context and can be achieved through the analysis of body fluids inside the body, known as implantable sensors (Mei et al. 2022). Implantable sensors are medical devices designed to enable real-time monitoring of physiological, biochemical, or metabolic changes within the body. These sensors are implanted either temporarily or permanently, providing continuous, in situ measurements. Built with miniaturized, energy-efficient designs, they are well-suited for integration with surgical implants or prosthetics, allowing for seamless long-term use. Modern implantable sensors are engineered with advanced biocompatible materials to minimize immune responses and ensure long-term stability and reliability within the body. Their compact and robust designs facilitate prolonged monitoring without significant discomfort or interference with normal body functions, making them invaluable in both diagnostic and therapeutic applications (Mei et al. 2022; Alam et al. 2024). The invention of a compact pacemaker, as small as 2.5 cm and weighing less than 15 g, ignited a relentless drive for the development of new implantable devices (Mei et al. 2022; Yogev et al. 2023). Notable commercial examples of implantable devices include cardiac monitors like the BioMonitor 2, which continuously tracks heart rhythms and records ECGs for early arrhythmia detection, and the Reveal XT, specifically designed to detect atrial fibrillation (AF) (Omar et al. 2024). Other key examples include continuous glucose monitors for diabetic patients and pressure sensors for monitoring intracranial pressure in individuals with brain injuries (Alam et al. 2024; Huynh et al. 2023). Since then, these devices have gained increasing attention, offering enhanced features like multi-target monitoring, self-powering capabilities, and wireless data transmission through the integration of functional modules. To meet the stringent requirements of implantable devices, sensors must be miniaturized, have low power consumption, be reliable, and be biocompatible. Additionally, implantable sensors must provide high data transfer rates, low latency, high sensitivity, and long-term stability. The materials used in these devices must facilitate close contact with tissue while minimizing any potential damage during the sensing process (Ashammakhi et al. 2021). Depending on their intended purpose, sensors can remain implanted in the body for anywhere from a few months to several years, or in some cases, for the patient's entire lifetime. However, most traditional implanted sensors require a second surgery for removal once they have served their purpose. This additional procedure not only increases surgical risks but also raises hospitalization costs (Koo et al. 2021). In contrast, resorbable devices, which are designed for short-term functionality, present an ideal solution for temporary diagnostic and therapeutic applications. These implants operate effectively for a specified period before naturally degrading, eliminating the need for surgical removal. This significantly enhances patient comfort while reducing the risks and costs associated with secondary procedures (Hosseini et al. 2020).

5.3.2 Fabrication Challenges and Clinical Applications

Biodegradable materials—such as natural and synthetic polymers (Fig. 5.3), as well as hydrolyzable metals—are frequently used in temporary implants due to their ability to break down safely within the body (Hosseini et al. 2020; Li et al. 2020). The use of flexible polymers in these sensors not only simplifies fabrication but also allows them to closely mimic the structure of biological tissues, improving both performance and biocompatibility (Mobarak et al. 2023). Another key factor is the mechanical compatibility between the device material and the surrounding tissue, which is essential for maintaining stable contact. However, different organs have distinct mechanical properties. For instance, Young's modulus of muscle and skin is

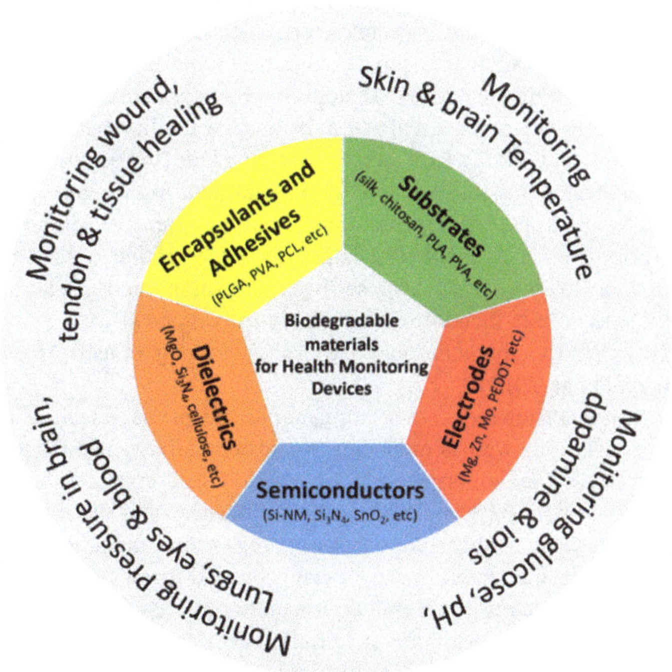

Fig. 5.3 Schematic illustration highlighting the significance of biodegradable materials in health monitoring devices. (Reprinted with permission from Hosseini et al. (2020))

approximately 10 kPa, whereas the brain's modulus is less than 1 kPa (Mei et al. 2022). This variation emphasizes the importance of selecting flexible materials tailored to the specific mechanical requirements of the target organ. In other words, material selection must be carefully aligned with the intended application and the specific tissue characteristics.

The clinical success of implantable and degradable devices is largely influenced by the body's response, particularly the immune reaction, which is shaped by the properties of the material and its degradation over time. Concerns about sensor material are well-founded, as the body's response can directly influence the degradation process. This, in turn, may impact the structure, appearance, mechanical properties, and overall function of the implant. Li and colleagues (2020) described a key mechanism that initiates the body's reaction to biomaterials: the activation of cascade systems. This process occurs at the material's surface, where changes in the conformation of recognition proteins and complex protein interactions take place (Li et al. 2020). Moreover, it is essential to assess the toxicity and potential carcinogenicity of both the original materials and their degradation byproducts (Hosseini et al. 2020).To achieve the desired flexibility and biocompatibility, there has been a growing use of naturally derived polymers for the fabrication of biodegradable devices. These include protein-based polymers such as collagen, chitosan, fibrin, silk, and gelatin, as well as plant-derived polysaccharides like alginate, cellulose, dextran, and starch (Hosseini et al. 2020; Kurowiak et al. 2023; Li et al. 2023). These natural polymers often possess bioactive properties, as their structures are similar to biological molecules. This characteristic can occasionally trigger immune responses in the body, which are used as temporary materials in implantable medical technologies. Despite their potential, applying biodegradable polymers may face challenges, like limited mechanical strength and variability between production batches (Hosseini et al. 2020). Conversely, since the late 1960s, synthetic polymers have gained increasing popularity due to their manufacturing versatility, mechanical strength, and tunable properties, which have spurred extensive research into their use across various fields. These synthetic polymers are produced through chemical reactions in laboratory or industrial settings. Common biodegradable synthetic polymers include poly(L-lactide) (PLLA), poly-dioxanone (PDO), poly(lactic-co-glycolic acid) (PLGA), polycaprolactone (PCL), among others (Hosseini et al. 2020; Kurowiak et al. 2023).

Another significant challenge for implantable sensors is selecting appropriate conductive materials, as well as managing water permeability and protecting internal components such as batteries and electrodes. These considerations influence whether additional barriers against biofluids are needed or if the materials can be used directly for sensor encapsulation. To address these challenges, a variety of biodegradable electronic materials have been proposed for developing implantable chemical sensors. These materials encompass interconnects, dielectric layers, semiconducting elements, encapsulation materials, substrates, and sophisticated data acquisition systems (Hu et al. 2024). In the field of conductive materials for implantable sensors, commonly used bio-compatible metals include platinum, gold, titanium nitride, and silicon. Bioresorbable materials like molybdenum, magnesium, tungsten, iron, germanium, and zinc (or zinc oxide) are also employed, as they naturally dissolve in biofluids through the formation of metal hydroxides or oxides. Conductive micro- and nanostructures, such as metallic flakes, nanowires, nanorods, and two-dimensional (2D) materials (e.g., MoS_2, WS_2, black phosphorus, inorganic compounds—also called MXenes—and transition metal oxides) further expand design possibilities. Carbon-based structures, like carbon nanotubes and graphene derivatives, add versatility to the range of options available for enhancing conductivity. Conducting polymers, such as poly(pyrrole) (PPy), polyaniline (PANI), poly(thiophene) (PTh), and poly(3,4-ethylenedioxythiophene) (PEDOT), are particularly valued for their electron delocalization along the polymer backbone, a result of π-orbital overlap in their aromatic structures. This delocalization creates efficient conductive pathways, allowing these materials to serve as reliable components in various implantable sensor applications due to their flexibility, conductivity, and biocompatibility. This combination of bioresorbable metals, advanced nanostructures, and conductive polymers forms a diverse toolkit for developing implantable sensors with high functionality and compatibility within the body (Sacchi et al. 2024).

In certain medical applications, biodegradable sensors must have wireless communication capabilities to reduce the risk of infection from wires penetrating the skin (Malik et al. 2020). In body-centric wireless communication networks, antennas and propagation are crucial, as they function as either transmitting or receiving antennas. If the antenna performance is subpar, it can negatively impact the overall system performance. Communication through far-field radio frequency telemetry offers advantages such as extended range and high data rates, enabling implantable medical devices (IMDs) to wirelessly monitor patient physiological data in real time (Hu et al. 2023). To ensure biocompatibility, two common approaches are typically employed. The first involves using biocompatible materials for antenna fabrication, while the second entails encapsulating the antenna with a biocompatible superstrate (Malik et al. 2020). To address this, researchers have explored the use of antennas and coils for data transmission and power supply. Wang and colleagues (2021) proposed a sensor design that integrates an antenna, battery, and microelectronic components for data and power management. To prevent direct contact with human tissue, the sensor components were encapsulated in a biocompatible aluminum box. The antenna demonstrated characteristics suitable for use as a wireless sensor, capable of monitoring pathological changes in human tissues, such as breast tumors. As a proof of concept, the authors conducted measurements using pork tissue and tissue-mimicking liquids. Additionally, an implantable antenna integrated with an immunosensor has been developed for the detection of carcinoembryonic antigen (CEA). This innovative antenna features a patch made from a gold nanoparticle-decorated graphene film, facilitating the creation of an antenna immunosensor capable of real-time monitoring of CEA levels (Chen et al. 2024). Implantable antennas have been extensively studied; however, they still encounter several challenges. These include issues related to the dispersion and variability of human tissues and organs, which lead to significant signal attenuation in vivo. Additionally, there are safety concerns and size limitations associated with the devices. As a result, implantable antennas continue to face numerous obstacles and unresolved issues (Hu et al. 2023).

Several techniques have been explored in implantable electrode fabrication, including microfabrication, direct writing/printing, and additive manufacturing methods (Qian and Liao 2023). However, current applied 3D printing technologies often fail to deliver the high resolution and sensitivity needed for implantable sensor production (Mobarak et al. 2023). To address the limitations in the fabrication of implantable sensors, alternative fabrication methods such as soft lithography, screen printing, and transfer printing have been developed for biodegradable devices (Min et al. 2023). Numerous studies have focused on advancing these implantable sensor systems, incorporating green electronics, innovative materials, and integration with IoT (Internet of Things) and AI (artificial intelligence) (Omar et al. 2024). For instance, Rawan Omar and colleagues (2024) made significant strides in implantable sensor technology using biodegradable materials like PLA, magnesium-based electrodes, and zinc nanoparticles for functional sensor layers. These sensors, integrated with IoT and wireless technologies, are fabricated using advanced techniques to create a flexible, transparent, and multifunctional nanosensor platform (Fig. 5.4). This platform, designed for cardiac monitoring, includes electrochemical biosensors for detecting lactic acid and pH, a pressure sensor, and a chemiresistor array for detecting volatile organic compounds. In vitro tests confirmed the device's biocompatibility with myocytes, while ex vivo experiments using a 3D-printed silicone heart model and cardiac tissue patches validated its biodegradability and functionality. Additionally, integrating an AI-based predictive model enhances the accuracy of health assessments by fusing sensor data, making this platform highly promising for clinical use.

Needle-based sensors have gained considerable attention for monitoring glucose and other analytes. Among these, a smartphone-controlled microneedle (MN)-based continuous glucose monitoring (CGM) system has been highlighted for its role in home-care diabetes management. This system includes a disposable glucose sensor, a reusable CGM device, and a smartphone app for real-time monitoring. The sensor is inserted into the skin to detect glucose levels in the interstitial fluid (ISF), while the CGM device wirelessly transmits data to the app for immediate feedback. Combining screen printing and laser cutting, the fabrication process enables the cost-effective production of three-electrode strips of double-sided electrochemical. To enhance performance, the sensor was modified with glucose oxidase (GOx) enzyme and incorporates conductive materials along with a biocompatible protective membrane. The sensor design features layers such as a conductive carbon nanotube/Nafion inner layer, a ferrocene-mediator/enzyme layer, and an outer biocompatible porous membrane, which collectively improve sensor functionality

Fig. 5.4 (continued) validation of the developed sensor: (a) Experimental setup. (b) Biodegradability test of the sensor applying a 3D-printed silicone heart model. (c) Pressure response of the sensor. (d) Cell growth in a 3D-printed cardiac patch. (e) Confocal microscopy image of the cells grown on the 3D-printed cardiac patch. (f) Test of the developed sensor on the 3D-printed cardiac patch. (g) pH response using the cardiac patch. (h) Lactate biodegradable biosensor response using the cardiac patch. (Reprinted with permission from Omar et al. (2024))

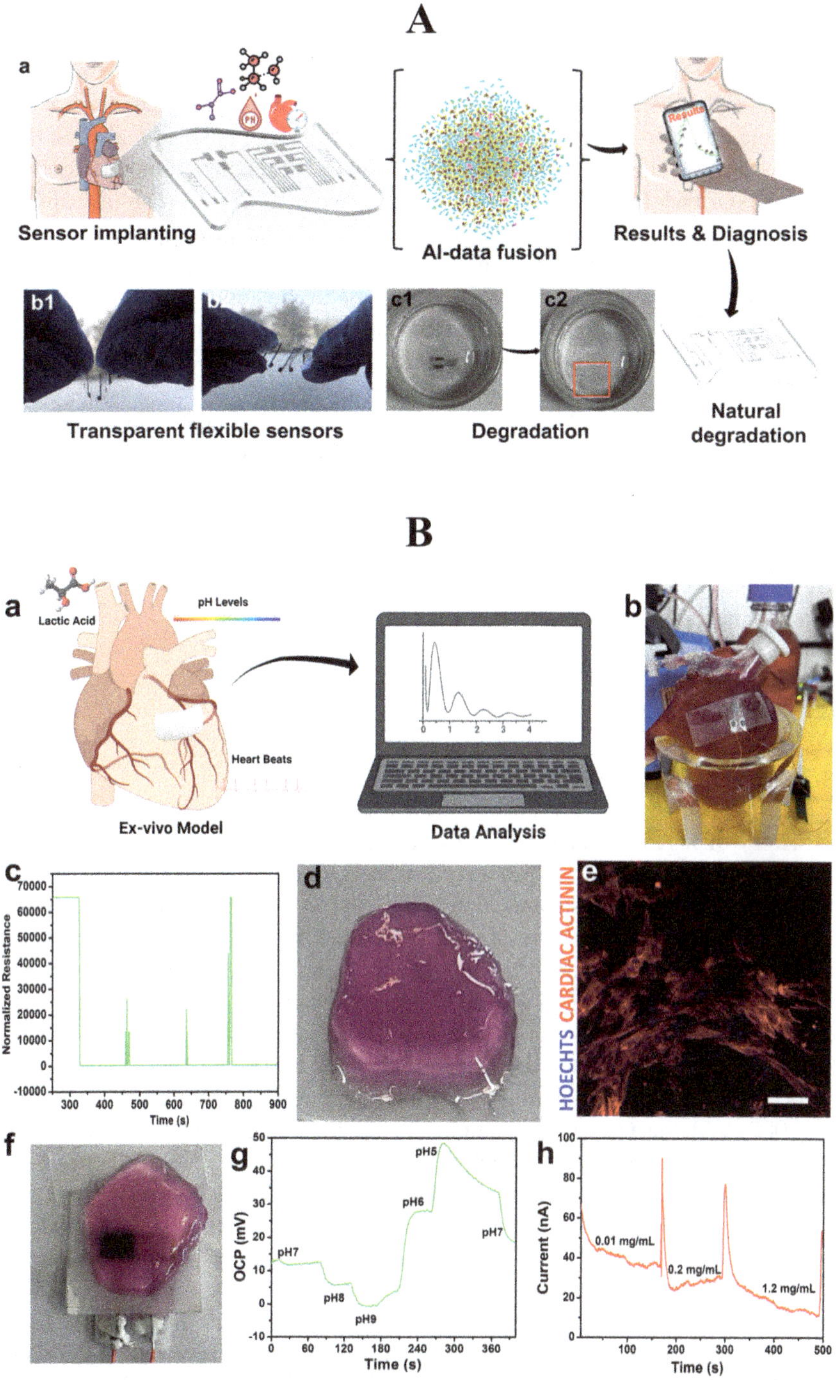

Fig. 5.4 (**a**) (a) Schematic illustration of the biodegradable nanosensor and AI-data fusion system. (b1-b2) Images demonstrating the flexibility and bendability of the polylactic acid (PLA)-based electrodes used in the cardiac monitoring sensor. (c1-c2) Sensor degradation over time. (**b**) Ex vivo

and longevity (Yang et al. 2023). Regarding the advancements of implantable devices in real-time neurochemical monitoring, Wu and colleagues developed an implantable aptamer-graphene micro-transistor probe that enables real-time detection of dopamine. This neural probe successfully monitors dopamine release in vivo following pharmacological stimulation in mice (Wu et al. 2022). A similar aptamer-functionalized field-effect transistor (FET) biosensor was developed for serotonin detection using microelectromechanical system (MEMS) technology. This process enabled the production of 150 flexible polyimide probes on a single 4-inch silicon wafer. The probes were coated with a 3 nm thin film of In_2On_3 semiconductor and functionalized with DNA aptamers, ensuring high specificity for neurotransmitter recognition, particularly for serotonin (Zhao et al. 2022).

Undoubtedly, implantable devices offer significant potential for continuous monitoring of clinically relevant analytes. However, numerous challenges remain in developing these systems. There is a growing need for bio-friendly, bioresorbable, and soft electronics that can reliably function in the body while withstanding mechanical stresses. Systems designed for implantation using minimally invasive methods hold great promise, especially in contributing to advancements in biomedicine and enhancing societal well-being.

5.4 Ingestible Sensor

5.4.1 Definition and Overview

The history of ingestible electronic devices can be traced back to the 1950s when Jacobson and Mackay (1957) pioneered using a small radio transmitter to collect data from the patient's gastrointestinal (GI) tract. Their objective was to monitor hydrogen-ion concentration, pressure, and temperature within the GI tract, leading to the development of two devices, the "pH-endosonde" and the "endo-radiosonde" (Jacobson and Mackay 1957; Mackay and Jacobson 1957). In the same year, Farrar and colleagues (Farrar et al. 1957) introduced a pressure-sensing telemeter capsule, offering valuable insights into GI tract motility. The 1980s and 1990s saw increased development of ingestible sensors, culminating in a significant milestone in the early 2000s with the advent of capsule endoscopy. A key innovation in this field was the PillCam, developed by Given Imaging (Sparling et al. 1993; O'Brien et al. 1998), which enabled non-invasive imaging of the GI tract and played a pivotal role in diagnosing small bowel diseases. Ingestible sensors are also largely applied to the measurement of physical parameters, such as respiratory rate and heart rate (Traverso et al. 2023), and GI pressure (Mitrakos et al. 2024). Figure 5.5 illustrates the components of an ingestible sensor. Like the wearable and implantable sensors discussed earlier in this chapter, ingestible biosensors also represent an advanced class of medical devices capable of monitoring and transmitting real-time physiological data (Lu et al. 2023b). What sets ingestible sensors apart is their non-invasive nature, allowing for the collection of chemical information from within the body without the challenges faced by wearable or implantable alternatives. Wearable

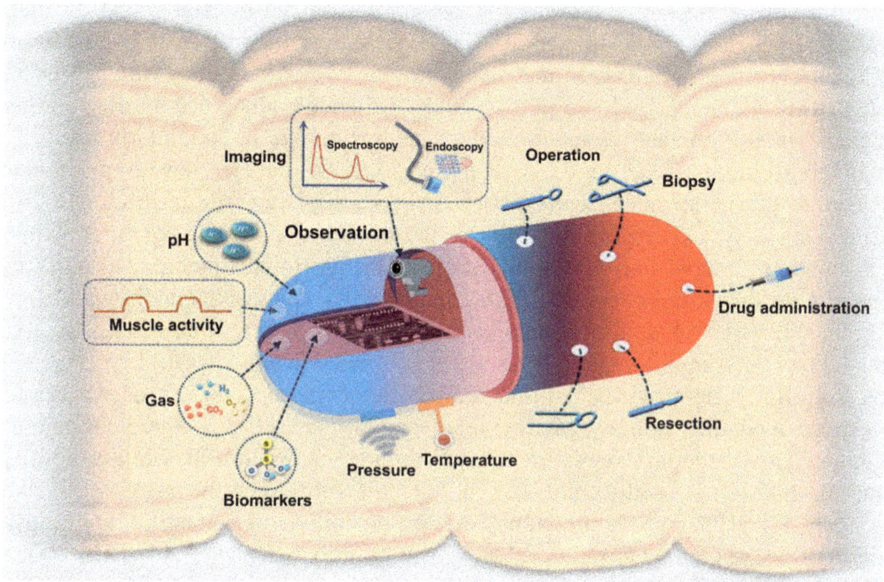

Fig. 5.5 Illustration of the components of an ingestible sensor utilized in healthcare applications. (Reprinted with permission from Lu et al. (2023b))

sensors often struggle with the quality of biological fluids they analyze, while implantable sensors, though effective, are invasive and less favorable for patient comfort (Kalantar-Zadeh et al. 2017a, 2018). Ingestible sensors, uniquely able to traverse the GI tract, can access the gut environment where the mucous membrane provides critical chemical insights (Kalantar-Zadeh et al. 2017a; Weitschies et al. 2021). Encapsulated in a pill or capsule, these devices are equipped with microelectronics—sensors and transmitters—that monitor a range of chemical components, including metabolites (Paz et al. 2022), microbial communities (Khachornsakkul et al. 2024), and biomarkers (Kalantar-Zadeh et al. 2018). This capability enables comprehensive monitoring, offering valuable insights into an individual's health.

Over time, ingestible devices have evolved beyond diagnostic applications to include smart pills designed for monitoring medication adherence (Chai et al. 2015; Liu et al. 2022) and even delivering medications (Hoffmann et al. 2024). These capsules now represent a vital technology in healthcare and nutrition, offering safe, non-invasive insights into gut health, diet, medication interactions, and environmental effects on the GI tract (Iddan et al. 2000; Swain 2003). With the increasing integration of smartphone connectivity, data collected by these devices can be easily accessed and analyzed by both patients and healthcare providers, enhancing the usability and effectiveness of these tools (Kalantar-Zadeh et al. 2017b). Significant advancements in electronics, optics, and materials science have made these devices smaller, smarter, and more power-efficient. Improvements in microprocessors and battery technology have led to more reliable, sensitive, and compact sensors, significantly boosting their practical utility and performance (Kalantar-Zadeh et al.

2017b). Several ingestible biosensors have already gained FDA approval for their varied applications in healthcare. These devices are used in fields ranging from diagnostics to chronic disease management. One example is the **Proteus Digital Health sensor**, integrated into pills to track medication adherence by transmitting signals when it contacts stomach fluids. This technology is particularly useful in managing chronic conditions like hypertension and diabetes, ensuring patients adhere to their prescribed treatments (U.S. Food and Drug Administration 2024). Another widely used ingestible device is the **PillCam**, a capsule equipped with a camera that captures high-resolution images of the gastrointestinal tract as it moves through the digestive system. It plays a crucial role in diagnosing conditions such as Crohn's disease, celiac disease, and gastrointestinal bleeding (FDA 2012; Goenka et al. 2014). For monitoring body temperature, the **e-Celsius** ingestible thermometer from BodyCap provides real-time temperature tracking. This sensor is precious in sports medicine, military environments, and clinical settings where continuous monitoring is essential (Newswire 2024). Similarly, the **SmartPill** wireless motility capsule (WMC) is used to diagnose motility disorders such as gastroparesis and constipation. This capsule measures key parameters like pH, pressure, and transit time within the GI tract, offering valuable insights into digestive health (FDA 2005). The **CorTemp®** Ingestible Core Body Temperature Sensor also provides real-time data as it moves through the GI tract, making it helpful in assessing disorders related to gastric acidity and gastroesophageal reflux disease (FDA 2015). The Atmo Gas Capsule is a sensor developed to measure the composition of intestinal gases, providing valuable insights into GI health and aiding in the diagnosis of conditions such as irritable bowel syndrome (IBS). Although the device has not yet received FDA approval for diagnostic purposes, it remains a valuable tool for clinical research, offering the potential for non-invasive monitoring in medical studies (Atmo Biosciences 2024). The growing exploration of ingestible devices highlights their potential to monitor key physiological parameters and improve patient care. In the next section, we will explore the common technologies employed in the design and construction of these innovative devices.

5.4.2 Technology for Ingestible Sensor Manufacturing

Ingestible biosensors incorporate advanced technologies to enable real-time monitoring and diagnostics within the human body. These devices rely on advancements in microelectronics to ensure their functionality, data transmission, and safety. Over the years, innovations in microelectronics have significantly broadened the capabilities of ingestible sensors, making them more versatile and clinically relevant. Typical components include sensor arrays, microcontrollers, power sources, data transmission systems, and biocompatible capsule housings (Kalantar-Zadeh et al. 2017a; Thwaites et al. 2024; Abdigazy et al. 2024) (Fig. 5.6).

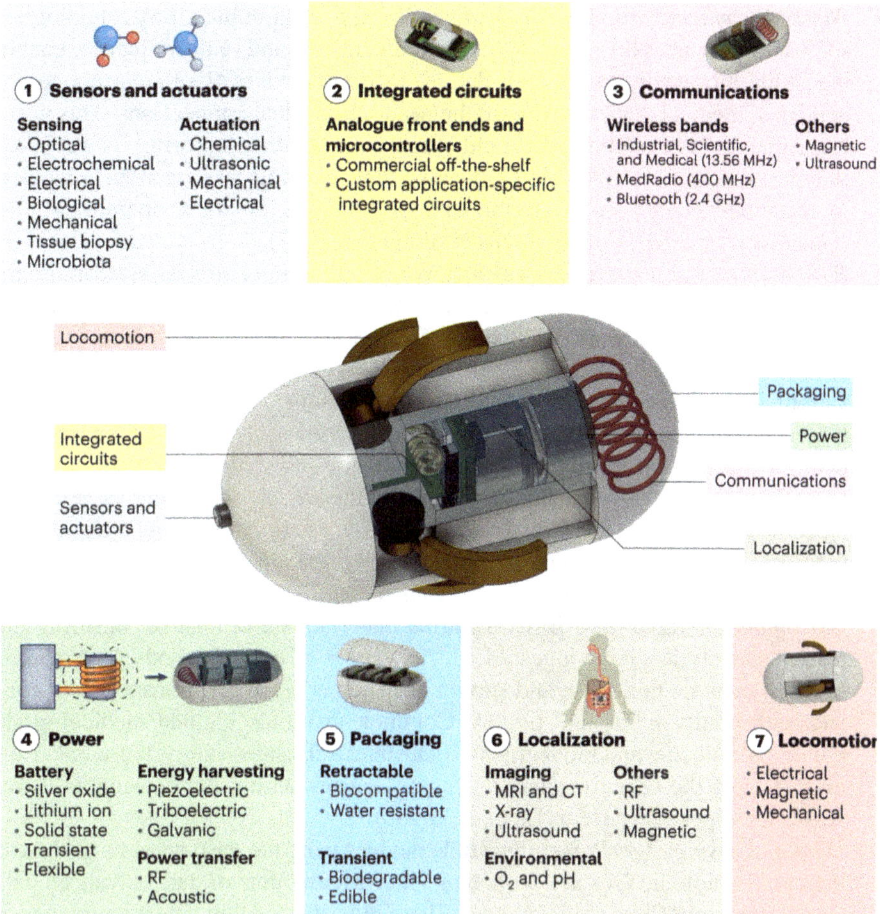

Fig. 5.6 Components include an ingestible sensor on its different applications. The transducer method on the sensors and actuators (1), the setup circuits (2), the choice of the communication (3), the application of power supplies (4), and the selection of the ideal material for the packaging drives are essential elements of the ingestible sensors. They may also have a localization (6) or any extra functionality (7). (Reprinted with permission from Abdigazy et al. (2024))

- *Microelectronics*: Microelectronics serve as the backbone of ingestible biosensors. These devices rely on miniaturized components such as microprocessors, sensors, and transmitters to gather and process data. One of the main challenges is miniaturizing components while ensuring they remain fully functional within a pill-sized device. To achieve this, advanced integrated circuits (ICs) enable real-time data acquisition and transmission, while low-power electronics help extend the device's battery life. These innovations ensure that despite the reduced size, the device can continuously monitor and collect data while passing through the GI tract (Thwaites et al. 2024; Bettinger 2015).

- *Wireless communication technologies:* Wireless communication technologies, such as Bluetooth, NFC (near-field communication), and radiofrequency, enable ingestible biosensors to transmit data to external devices (like smartphones or dedicated medical devices) without the need for physical connections. This wireless capability is essential in the field of real-time health monitoring, as it ensures seamless data transfer. Ultra-low-power wireless technologies are often employed to minimize energy consumption and prolong the device's operational life (Ghanim et al. 2023; Cao et al. 2019; Abid et al. 2017).
- *Power sources and energy harvesting:* While traditional batteries are commonly used, they must be both compact and durable enough to power the device throughout its journey inside the body. This presents limitations in terms of size, energy capacity, and potential environmental risks. To address this, recent innovations have focused on energy harvesting techniques, which enable the device to generate power from the body's internal environment, such as stomach acid, heat, or mechanical motion. These methods not only reduce reliance on traditional batteries but also extend the lifespan of the sensor, enhancing its practicality for long-term monitoring and data collection. Energy harvesting technologies represent a promising solution for increasing the operational efficiency and safety of ingestible sensors (Kalantar-Zadeh et al. 2017a; Nadeau et al. 2017).
- *Biocompatible materials:* Biocompatible materials are crucial for ensuring the patient's safety when the ingestible device passes through the body. The materials used must be non-toxic and should not elicit adverse reactions when in contact with bodily fluids and tissues. Common materials include medical-grade polymers, silicone, and biodegradable substances. Besides safety, these materials must protect the sensitive electronics and sensors within the capsule as it traverses the digestive system (Lu et al. 2023b).
- *Data processing algorithms:* Ingestible devices generate vast amounts of data as they go through the GI tract. To manage the vast amounts of data, advanced data processing algorithms are employed. These algorithms often detect anomalies or patterns, providing healthcare providers with actionable insights from the collected raw data. Machine learning and AI techniques are increasingly incorporated to enhance data analysis and improve diagnostic accuracy (Chai et al. 2015; Zhou et al. 2024).

5.4.3 Clinical Application

Ingestible biosensors serve a range of functions, from tracking medication adherence to diagnosing diseases using biomarkers. By leveraging nanotechnology, biotechnology, and wireless communication, they enable non-invasive monitoring of internal bodily processes. Integrated with digital health platforms, these sensors hold transformative potential for patient care, allowing timely interventions, reducing the need for invasive procedures, and enhancing overall healthcare outcomes. Advances in microelectronics and bioengineering have produced ingestible sensors that can measure key physiological parameters like pH, pressure, and temperature, while also performing optical imaging (e.g., PillCam, VitalSense). Research is expanding its applications to include capsule-based biopsies, surgical interventions, and better drug adherence monitoring, bringing these technologies closer to routine clinical use (Thwaites et al. 2024). Figure 5.7a summarizes the many potential targets for ingestible electronic device technology in different gastrointestinal regions. Ingestible sensors fall into three main categories: diagnostic sensors that measure biomarkers to detect diseases, digital pill systems (DPS) that monitor medication adherence by transmitting data to healthcare providers, and therapeutic ingestible pills (actuators sensors) that deliver targeted treatments inside the body.

5.4.3.1 Ingestible Diagnostic Devices

Ingestible biosensors can detect a variety of biomarkers within the gastrointestinal (GI) tract, including bacteria, metabolites, physiological gases, proteins, and neurotransmitters (Paz et al. 2022; Woo et al. 2011; Stearns et al. 2011; Lavelle and Sokol 2020; Kalantar-Zadeh et al. 2019; Dave et al. 2014; Yang et al. 2021). These biomarkers have been linked to various GI, neurodegenerative, and neurodevelopmental diseases, as demonstrated through analyses of stool samples and biopsies collected via colonoscopy (Luna et al. 2017; Fan and Pedersen 2021). Ingestible capsules equipped with optical, electrochemical, electrical, biological, and mechanical sensors provide timely, non-invasive diagnostics for a wide range of conditions, as illustrated in Fig. 5.7b (Abdigazy et al. 2024).

- *Imaging:* Capsule endoscopes utilize visible light wavelengths for optical imaging of the mucosal surface, detecting GI pathologies with high-resolution cameras and light-emitting diodes (LEDs). Companies such as Medtronic, Intro-Medic, and Jinshan manufacture various types of capsule endoscopes, and advancements in AI algorithms have enhanced the analysis of these images. Deep learning techniques have been able to classify GI disorders, including bleeding ulcers and Crohn's disease, with over 90% accuracy (Atsawarungruangkit et al. 2020; Moen et al. 2022).
- *Electrochemical sensors:* Various voltammetric techniques have been successfully integrated into ingestible devices using traditional three-electrode electrochemical sensors. A key application of this is pH sensing, which is particularly relevant for GI conditions like gastric ulcers and Crohn's disease (Machida 1981; Hauptmann et al. 1998; Maurer et al. 2015). These electrochemical sensors also provide detailed characterization of gut fluids, aiding in the diagnosis of conditions such as Crohn's disease. Additionally, the ability to measure neurotransmitters within the GI tract offers valuable insights into GI disorders and the complex relationship between the gut-brain axis, which is implicated in various neurological and digestive health conditions (Mc Caffrey et al. 2015; Salama et al. 2022; Li et al. 2022).
- *Biological sensors:* Genetically modified bacteria-based sensors have been effectively integrated into ingestible electronics. Researchers have engineered strains of *Escherichia coli* to produce luminescence in the presence of heme, providing a visual signal that ingestible devices can detect (Mimee et al. 2018).

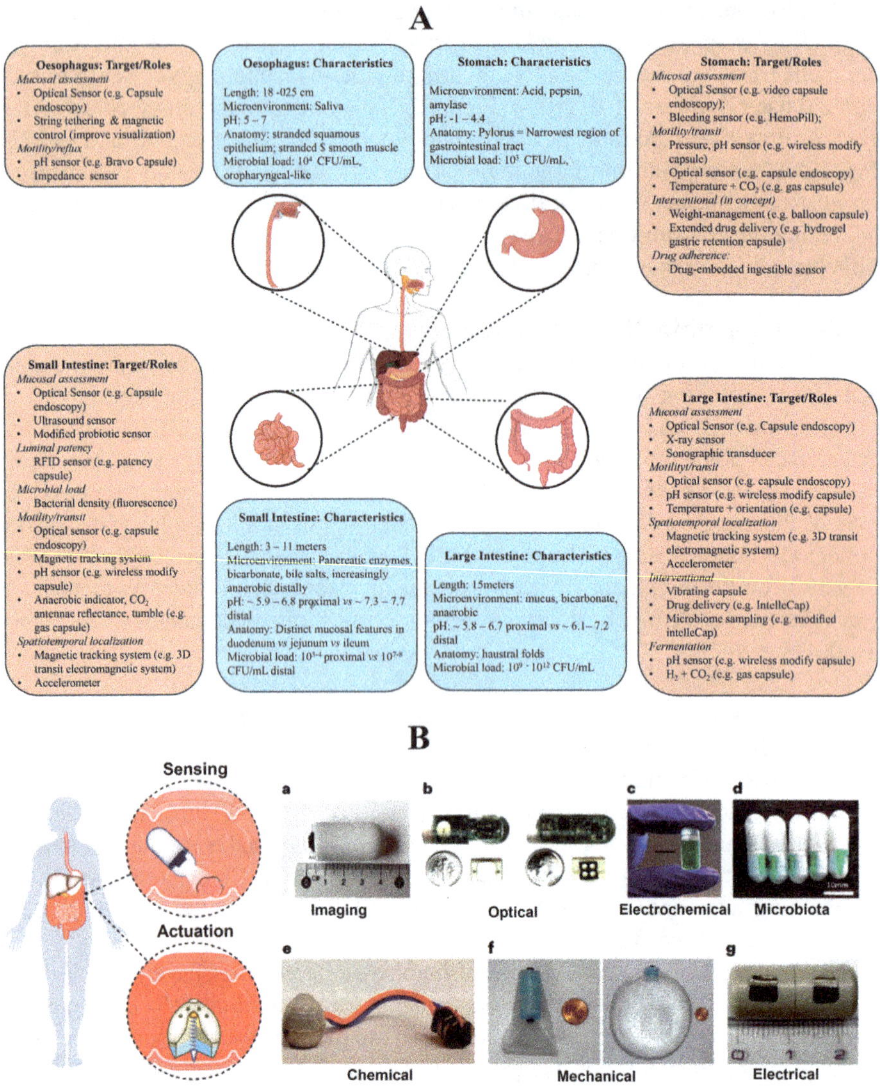

Fig. 5.7 (**a**) Overview of potential applications for ingestible electronic devices within the gastrointestinal tract. (**b**) Specific uses of ingestible pills include: (a) wireless video endoscopy for gastrointestinal imaging (Sahafi et al. 2022), (b) diagnosing gastrointestinal bleeding (Mimee et al. 2018), (c) glucose detection (Paz et al. 2022), (d) in vivo sampling of the gut microbiome (Rezaei Nejad et al. 2019), (e) drug delivery and electrical stimulation of the gastrointestinal wall (Abramson et al. 2020), (f) treatment of obesity (Freitas et al. 2021), and (g) electrical stimulation for gastrointestinal motility (Woo et al. 2011). (Reprinted with permission from Thwaites et al. (2024), Abdigazy et al. (2024))

Similarly, *E. coli* strains have been designed to respond to nitric oxide and hydrogen sulfide, two critical biomarkers associated with inflammatory bowel disease (Inda-Webb et al. 2023). These sensors, equipped with photodetectors and custom readout chips, have demonstrated effectiveness in animal models like pig intestines. However, ensuring their survival in the acidic environment of the stomach remains a challenge.

- *Mechanical sensors:* Mechanical sensing capabilities in ingestible capsules allow for the measurement of internal GI pressure, which is an important factor in detecting organ dysfunction (Guozheng et al. 2007; Liao et al. 2020). The SmartPill is currently the only FDA-approved intraluminal pressure-sensing pill, capable of measuring pH and temperature, and it has proven valuable in understanding blast injuries in military divers (Yunus et al. 2015).
- *Electrical sensing*: Ingestible capsules offer a non-invasive method for monitoring core body temperature, providing a reliable alternative to traditional measurement techniques. These capsules are particularly useful in detecting conditions such as hypothermia and hyperthermia, enabling timely interventions in both clinical and emergency settings. By continuously tracking body temperature, these devices assist healthcare professionals and individuals across various health scenarios (Beardslee et al. 2021; Hoang et al. 2020).
- *Tissue biopsy and microbiota sampling:* Collecting gastrointestinal tissue and microbiota samples is critical for diagnosing various GI diseases. Traditional methods, such as endoscopy and colonoscopy, can retrieve tissue samples but often cause patient discomfort and typically require sedation. To address the limitations of conventional methods, ingestible capsules equipped with microneedles or brushes perform tissue biopsies or fluid sampling, offering a less invasive and more comfortable alternative while maintaining diagnostic accuracy (Hoang et al. 2020; Son et al. 2020).

5.4.3.2 Digital Pill Systems (DPS)

The digital pill is a promising tool for tracking when medications have been ingested, giving information about the patient's medication adherence. The system usually consists of a sensor pill, an external adhesive monitor (wearable sensor), a wireless system, and a mobile app. After being swallowed, the stomach fluids activate the device and send an electric signal with a specific code to the wearable sensor. The patch sends data via Bluetooth to the mobile app, which uploads it to a cloud-based system for viewing on the web portal. Proteus Digital Health (Redwood City, CA, USA) developed the first FDA-approved DPS to integrate a small radiofrequency or electronic sensor into a standard gelatin capsule. When the capsule is ingested, the chloride ions in gastric fluid activate the sensor inside. This activation triggers the sensor to emit a radiofrequency signal detected by a wearable device worn by the patient. When the DTS is ingested with the medication, the dispositive can track the patient medication ingestion adherence (Liu et al. 2022; DiCarlo et al. 2012; Frias et al. 2017).

5.4.3.3 Actuators

In addition to their diagnostic and monitoring capabilities, ingestible sensors are also designed to deliver therapeutic actions, offering solutions for conditions such as chronic constipation and obesity. Various actuation mechanisms have been developed to enhance drug delivery and modulate GI functions. Ultrasonic actuation uses low-frequency ultrasound to enhance drug absorption and expedite the treatment of GI diseases. This approach has shown promise in treating conditions like colorectal cancer, where high-intensity focused ultrasound therapy offers both thermal and non-thermal therapeutic effects (Mimee et al. 2018; Inda et al. 2022). Miniature ultrasonic transducers integrated into ingestible capsules have proven effective at penetrating intestinal mucus layers, thereby improving drug delivery efficiency (Kloetzer et al. 2010). Chemical actuation allows direct drug delivery into the GI wall, offering faster pharmacokinetic responses than subcutaneous injections. This method achieves comparable plasma levels, making it a highly efficient alternative for drug administration (Gibson et al. 1993; Raj et al. 2014; Farmer et al. 2014, 2019). Many ingestible devices employ internal mechanisms like springs, gas, or osmotic pumps to control and time drug release (Ou et al. 2015). In mechanical actuation, devices such as Vibrant and VibraBot use vibrations to stimulate peristalsis, offering a non-invasive treatment for chronic constipation. These capsules generate controlled vibrations, promoting GI motility and facilitating bowel movements without the need for traditional laxatives (Liao et al. 2020; Bongers et al. 2017; Osterberg and Blaschke 2005; McKenzie and Osgood 2004; Hafezi et al. 2014). Lastly, electrical actuation applies electrical stimulation to the GI wall, modulating motility and potentially influencing appetite. This technology shows promise for treating obesity and GI dysmotility disorders by regulating digestive processes and controlling food intake through targeted electrical impulses (Söderlind et al. 2015; Becker et al. 2014; Beccani et al. 2015; Zheng et al. 2022).

5.4.4 Challenges and Perspectives

The integration of ingestible biosensors into clinical practice presents several key challenges, such as validating the accuracy of novel biomarkers, incorporating new data into established clinical workflows, and demonstrating their cost-effectiveness. Technical failures, including issues with data transmission, and safety concerns, such as compliance with international standards like ISO-62304, are critical areas that need to be addressed. Additionally, precise control and locomotion of these devices, along with detailed mapping of specific regions within the body, remain technical obstacles. Miniaturizing robotic technologies for therapeutic interventions is also a significant challenge that has yet to be fully overcome (Rao et al. 2009; Min et al. 2020; Vaccari et al. 2018; Saad and Hasler 2011; Rondonotti et al. 2005). Other limitations of ingestible devices include concerns about data integrity, security, validation, and ownership. These devices may disrupt patients' daily activities and pose environmental concerns as well (Gravina and Fortino 2020; Klugman et al. 2018; Matzeu et al. 2015). Despite these hurdles, ingestible biosensors have the potential to offer cost-effective, non-invasive diagnostic solutions that could reduce dependence on stationary equipment and hospital visits, thereby making healthcare more accessible (Thwaites et al. 2024; So et al. 2023). Future perspectives for ingestible biosensors are promising across several healthcare domains. They are expected to enhance disease diagnosis and monitoring, improve medication adherence, personalize treatments, and play a crucial role in chronic disease management. Advancements in microelectronics, material science, and integration with wearable technologies will further expand their applicability. Regulatory and ethical considerations will also play a key role in their widespread adoption, particularly in terms of data privacy and patient safety (So et al. 2023).

5.5 Conclusion

Wearable, implantable, and ingestible biosensors represent a transformative approach to health monitoring, each offering distinct advantages. Wearable sensors provide non-invasive external monitoring of physiological and chemical processes but can struggle with accuracy in dynamic environments due to external factors like movement. Implantable sensors address this by offering precise, long-term data from within the body, though they come with the downside of being invasive, often requiring surgical procedures. Ingestible sensors strike a balance by enabling internal monitoring, particularly in the gastrointestinal tract, without surgery, though they face challenges like limited power and regulatory approval. Together, these technologies offer a comprehensive approach to health monitoring, filling specific gaps and enabling more personalized, accurate, and proactive healthcare, allowing real-time decision-making for patients and providers.

References

Abdigazy A, Arfan M, Lazzi G, Sideris C, Abramson A, Khan Y (2024) End-to-end design of ingestible electronics. Nat Electron 7(2):102–118. https://doi.org/10.1038/s41928-024-01122-2

Abid A, O'Brien JM, Bensel T, Cleveland C, Booth L, Smith BR, Langer R, Traverso G (2017) Wireless power transfer to millimeter-sized gastrointestinal electronics validated in a swine model. Sci Rep 7(1):46745. https://doi.org/10.1038/srep46745

Abramson A, Dellal D, Kong YL, Zhou J, Gao Y, Collins J, Tamang S, Wainer J, McManus R, Hayward A et al (2020) Ingestible transiently anchoring electronics for microstimulation and conductive signaling. Sci Adv 6(35):0127. https://doi.org/10.1126/sciadv.aaz0127

Alam F, Ashfaq Ahmed M, Jalal AH, Siddiquee I, Adury RZ, Hossain GM, Pala N (2024) Recent progress and challenges of implantable biodegradable biosensors. Micromachines 15(4):475. https://doi.org/10.3390/mi15040475

Al-Daraghmeh MY, Stone RT (2023) A review of medical wearables: materials, power sources, sensors, and manufacturing aspects of human wearable technologies. J Med Eng Technol 47(1):67–81. https://doi.org/10.1080/03091902.2022.2097743

Ali A, Ashfaq M, Qureshi A, Muzammil U, Shaukat H, Ali S, Altabey WA, Noori M, Kouritem SA (2023) Smart detecting and versatile wearable electrical sensing mediums for healthcare. Sensors 23(14):6586. https://doi.org/10.3390/s23146586

Alipour A, Gabrielson S, Patel PB (2020) Ingestible sensors and medication adherence: focus on use in serious mental illness. Pharmacy 8(2):103. https://doi.org/10.3390/pharmacy8020103

Ashammakhi N, Hernandez AL, Unluturk BD, Quintero SA, Barros NR, Hoque Apu E, Bin Shams A, Ostrovidov S, Li J, Contag C et al (2021) Biodegradable implantable sensors: materials design, fabrication, and applications. Adv Funct Mater 31(49):2104149. https://doi.org/10.1002/adfm.202104149

Atmo Biosciences: Atmo Biosciences. Accessed 2024-10-04. https://www.atmobiosciences.com/

Atsawarungruangkit A, Elfanagely Y, Asombang AW, Rupawala A, Rich HG (2020) Understanding deep learning in capsule endoscopy: can artificial intelligence enhance clinical practice? Artif Intelli Gastrointest Endoscop 1(2):33–43. https://doi.org/10.37126/aige.v1.i2.33

Awotunde JB, Jimoh RG, Folorunso SO, Adeniyi EA, Abiodun KM, Banjo OO (2021) Privacy and security concerns in iot-based healthcare systems. In: The fusion of Internet of Things, artificial intelligence, and cloud computing in health care. Springer, pp 105–134. https://doi.org/10.1007/978-3-030-75220-0_6

Bandodkar AJ, Jeerapan I, Wang J (2016) Wearable chemical sensors: present challenges and future prospects. ACS Sens 1(5):464–482. https://doi.org/10.1021/acssensors.6b00250

Beardslee LA, Todd J, Powers CM, Fontaine DT, McCosh A, Shukla A, Casper BM (2021) Blast capsule: an embedded pressure sensing system for internal blast pressure measurement. IEEE Sensors Lett 5(11):1–4. https://doi.org/10.1109/LSENS.2021.3117890

Beccani M, Di Natali C, Aiello G, Benjamin C, Susilo E, Valdastri P (2015) A magnetic drug delivery capsule based on a coil actuation mechanism. Proc Eng 120:53–56. https://doi.org/10.1016/j.proeng.2015.08.564

Becker D, Zhang J, Heimbach T, Penland RC, Wanke C, Shimizu J, Kulmatycki K (2014) Novel orally swallowable Intellicap® device to quantify regional drug absorption in human GI tract using diltiazem as model drug. AAPS PharmSciTech 15:1490–1497. https://doi.org/10.1208/s12249-014-0172-1

Bettinger CJ (2015) Materials advances for next-generation ingestible electronic medical devices. Trends Biotechnol 33(10):575–585. https://doi.org/10.1016/j.tibtech.2015.07.0085

Bongers CC, Daanen HA, Bogerd CP, Hopman MT, Eijsvogels TM (2017) Validity, reliability, and inertia of four different temperature capsule systems. Med Sci Sports Exerc. https://doi.org/10.1249/MSS.0000000000001403

Boutry CM, Beker L, Kaizawa Y, Vassos C, Tran H, Hinckley AC, Pfattner R, Niu S, Li J, Claverie J, Wang Z, Chang J, Fox PM, Bao Z (2019) Biodegradable and flexible arterial-pulse sensor for the wireless monitoring of blood flow. Nat Biomed Eng 3(1):47–57. https://doi.org/10.1038/s41551-018-0336-5

Buckley N, Davey P, Jensen L, Baptist K, Jansen B, Campbell A, Downs J (2022) Can wearable inertial measurement units be used to measure sleep biomechanics? establishing initial feasibility and validity. Biomimetics 8(1):2. https://doi.org/10.3390/biomimetics8010002

Cao Z, Chen P, Ma Z, Li S, Gao X, Wu R-X, Pan L, Shi Y (2019) Near-field communication sensors. Sensors 19(18):3947. https://doi.org/10.3390/s19183947

Casson AJ, Galvez AV, Jarchi D (2016) Gyroscope vs. accelerometer measurements of motion from wrist PPG during physical exercise. ICT Exp 2(4):175–179. https://doi.org/10.1016/j.icte.2016.11.003

Chai PR, Castillo-Mancilla J, Buffkin E, Darling C, Rosen RK, Horvath KJ, Boudreaux ED, Robbins GK, Hibberd PL, Boyer EW (2015) Utilizing an ingestible biosensor to assess real-time medication adherence. J Med Toxicol 11:439–444. https://doi.org/10.1007/s13181-015-0494-8

Chen Y, Xuan X, Li M, Jiang D, Li H (2024) Implantable antenna immunosensor based on au-decorated graphene film for wireless cea detection. Sensors Actuators B Chem 136037. https://doi.org/10.1016/j.snb.2024.136037

Choi EJ, Drago NP, Humphrey NJ, Houten JV, Ahn J, Lee J, Kim I-D, Ogata AF, Penner RM (2023) Transduced sensors and biosensors. Mater Today 62(February):129–150. https://doi.org/10.1016/j.mattod.2022.11.021

Choudry M, Stead TS, Mangal RK, Ganti L (2022) The history and evolution of the stethoscope. Cureus 14(8):8–11. https://doi.org/10.7759/cureus.28171

Dai Y, Nolan J, Madsen E, Fratus M, Lee J, Zhang J, Lim J, Hong S, Alam MA, Linnes JC et al (2023) Wearable sensor patch with hydrogel microneedles for in situ analysis of interstitial fluid. ACS Appl Mater Interfaces 15(49):56760–56773. https://doi.org/10.1021/acsami.3c12740

Dave LA, Montoya CA, Rutherfurd SM, Moughan PJ (2014) Gastrointestinal endogenous proteins as a source of bioactive peptides-an in silico study. PLoS One 9(6):98922. https://doi.org/10.1371/journal.pone.0098922

DiCarlo L, Moon G, Intondi A, Duck R, Frank J, Hafazi H, Behzadi Y, Robertson T, Costello B, Savage G et al (2012) A digital health solution for using and managing medications: wirelessly observed therapy. IEEE Pulse 3(5):23–26. https://doi.org/10.1109/MPUL.2012.2205777

Donoso FA, Arriagada SD (2020) René théophile hyacinthe laënnec (1781-1826). Two hundred years of the stethoscope. A brief overview. Arch Argent Pediatr 118(5):444–8

Dunn J, Runge R, Snyder M (2018) Wearables and the medical revolution. Pers Med 15(5):429–448. https://doi.org/10.2217/pme-2018-0044

Eidi A (2023) Design and evaluation of an implantable mems based biosensor for blood analysis and real-time measurement. Microsyst Technol 29(6):857–864

Erdem A, Eksin E, Senturk H, Yildiz E, Maral M (2024) Trends in analytical chemistry recent developments in wearable biosensors for healthcare and biomedical applications. Trends Anal Chem 171(December 2023):117510. https://doi.org/10.1016/j.trac.2023.117510

Fan Y, Pedersen O (2021) Gut microbiota in human metabolic health and disease. Nat Rev Microbiol 19(1):55–71. https://doi.org/10.1038/s41579-020-0433-9

Farmer AD, Mohammed SD, Dukes GE, Scott SM, Hobson AR (2014) Caecal pH is a biomarker of excessive colonic fermentation. World J Gastroenterol 20(17):5000. https://doi.org/10.3748/wjg.v20.i17.5000

Farmer AD, Ruffle JK, Hobson AR (2019) Linaclotide increases cecal ph, accelerates colonic transit, and increases colonic motility in irritable bowel syndrome with constipation. Neurogastroenterol Motil 31(2):13492. https://doi.org/10.1111/nmo.13492

Farrar JT, Zworykin VK, Baum J (1957) Pressure-sensitive telemetering capsule for study of gastrointestinal motility. Science 126(3280):975–976. https://doi.org/10.1126/science.126.3280.975

Faust O, Lei N, Chew E, Ciaccio EJ, Acharya UR (2020) A smart service platform for cost efficient cardiac health monitoring. Int J Environ Res Public Health 17(17):1–18. https://doi.org/10.3390/ijerph17176313

FDA (2005) 510(k) Summary - K053547. Accessed 2024-10-07. https://www.accessdata.fda.gov/cdrh_docs/pdf5/k053547.pdf

FDA (2012) 510(k) Summary - K123666. Accessed 2024-10-07. https://www.accessdata.fda.gov/cdrh_docs/pdf12/K123666.pdf

FDA (2015) MAUDE adverse event report: HQ, Inc. CorTemp® ingestible core body temperature sensor. Accessed 2024-10-07. https://www.accessdata.fda.gov/scripts/cdrh/cfdocs/cfmaude/detail.cfm?mdrfoi_id=4904973&pc=FLL

Freitas R, Rodrigues J, Puga H, Correia J (2021) Design, simulation, and fabrication of an ingestible capsule with gastric balloon for obesity treatment. Biomed Phys Eng Exp 7(5):055024

Frias J, Virdi N, Raja P, Kim Y, Savage G, Osterberg L (2017) Effectiveness of digital medicines to improve clinical outcomes in patients with uncontrolled hypertension and type 2 diabetes: prospective, open-label, cluster-randomized pilot clinical trial. J Med Internet Res 19(7):246. https://doi.org/10.2196/jmir.7833

Gambhir SS, Ge TJ, Vermesh O, Spitler R, Gold GE (2021) Continuous health monitoring: an opportunity for precision health. Sci Transl Med 13(597):1–6. https://doi.org/10.1126/scitranslmed.abe5383

Ghanim R, Kaushik A, Park J, Abramson A (2023) Communication protocols integrating wearables, ingestibles, and implantables for closed-loop therapies. Device 1(3). https://doi.org/10.1016/j.device.2023.100092

Gharbi A, Kallel AY, Kanoun O, Cheikhrouhou-Koubaa W, Contag CH, Antoniac I, Derbel N, Ashammakhi N (2023) A biodegradable bioactive glass-based hydration sensor for biomedical applications. Micromachines 14(1). https://doi.org/10.3390/mi14010226

Gibson G, Macfarlane G, Cummings J (1993) Sulphate reducing bacteria and hydrogen metabolism in the human large intestine. Gut 34(4):437. https://doi.org/10.1136/gut.34.4.437

Goenka MK, Majumder S, Goenka U (2014) Capsule endoscopy: present status and future expectation. World J Gastroenterol 20(29):10024. https://doi.org/10.3748/wjg.v20.i29.10024

Gravina R, Fortino G (2020) Wearable body sensor networks: state-of-the-art and research directions. IEEE Sensors J 21(11):12511–12522. https://doi.org/10.1109/JSEN.2020.3044447

Grodzinsky E, Sund Levander M (2020) History of the thermometer. Understand Fever Body Temp:23–35. https://doi.org/10.1007/978-3-030-21886-7_3

Guozheng Y, Biao H, Peng Z (2007) Design of battery-less and real-time telemetry system for gastrointestinal applications. In: 2007 IEEE international conference on control and automation. IEEE, pp 245–249. https://doi.org/10.1109/ICCA.2007.4376356

Gupta S, Kakkar V, Ahmed S, Khanday FA, Sharma S, Singh S, Yoon B (2021) Modeling of on-chip biosensor for the in vivo diagnosis of hypertension in wireless body area networks. IEEE Access 9:95072–95082

Hafezi H, Robertson TL, Moon GD, Au-Yeung K-Y, Zdeblick MJ, Savage GM (2014) An ingestible sensor for measuring medication adherence. IEEE Trans Biomed Eng 62(1):99–109. https://doi.org/10.1109/TBME.2014.2341272

Handler DT, Hauge L, Spognardi A, Dragoni N (2017) Security and privacy issues in healthcare monitoring systems: a case study. In: HEALTHINF 2017 - 10th international conference on health informatics, proceedings; part of 10th international joint conference on biomedical engineering systems and technologies, BIOSTEC 2017 5, pp 383–388. https://doi.org/10.5220/0006224603830388

Hauptmann H, Fuchs F, Ewe R (1998) Gastrointestinal pH profiles in patients with inflammatory bowel disease. Aliment Pharmacol Ther 12(7):673–678. https://doi.org/10.1046/j.1365-2036.1998.00358.x

Heikenfeld J, Jajack A, Rogers J, Gutruf P, Tian L, Pan T, Li R, Khine M, Kim J, Wang J, Kim J (2018) Wearable sensors: modalities, challenges, and prospects. Lab Chip 18(2):217–248. https://doi.org/10.1039/c7lc00914c

Hoang MC, Le VH, Nguyen KT, Nguyen VD, Kim J, Choi E, Bang S, Kang B, Park J-O, Kim C-S (2020) A robotic biopsy endoscope with magnetic 5-dof locomotion and a retractable biopsy punch. Micromachines 11(1):98. https://doi.org/10.3390/mi11010098

Hoffmann SV, O'Shea JP, Galvin P, Jannin V, Griffin BT (2024) State-of-the-art and future perspectives in ingestible remotely controlled smart capsules for drug delivery: a genegut review. Eur J Pharm Sci 106911. https://doi.org/10.1016/j.ejps.2024.106911

Hosseini ES, Dervin S, Ganguly P, Dahiya R (2020) Biodegradable materials for sustainable health monitoring devices. ACS Appl Bio Mater 4(1):163–194. https://doi.org/10.1021/acsabm.0c01139

Hu X-Y, Yin W-L, Du F, Zhang C, Xiao P, Li G (2023) Biomedical applications and challenges of in-body implantable antenna for implantable medical devices: a review. AEU Int J Electron Commun 155053. https://doi.org/10.1016/j.aeue.2023.155053

Hu C, Wang L, Liu S, Sheng X, Yin L (2024) Recent development of implantable chemical sensors utilizing flexible and biodegradable materials for biomedical applications. ACS Nano 18(5):3969–3995. https://doi.org/10.1021/acsnano.3c11832

Hulanicki A, Glab S, Ingman F (1991) Chemical sensors: definitions and classification. Pure Appl Chem 63(9):1247–1250. https://doi.org/10.1351/pac199163091247

Huynh MA, Nguyen T-H, Nguyen T-K, Nguyen N-T (2023) Convergence of implantable bioelectronics and brain–computer interfaces. ACS Appl Electron Mater 5(11):5777–5793. https://doi.org/10.1021/acsaelm.3c00879

Ichikawa K, Iitani K, Kawase G, Toma K, Arakawa T, Dao DV, Mitsubayashi K (2024) Real-time salivary turbidity measurement with a mouthguard-type wearable sensor. Sensors Actuators A Phys 376:115660. https://doi.org/10.1016/j.sna.2024.115660

Iddan G, Meron G, Glukhovsky A, Swain P (2000) Wireless capsule endoscopy. Nature 405(6785):417–417. https://doi.org/10.1038/35013140

Inda M, Jimenez M, Liu Q, Phan N, Ahn J, Steiger C, Wentworth A, Riaz A, Zirtiloglu T, Wong K et al (2022) Ingestible capsule for detecting labile inflammatory biomarkers in situ. bioRxiv, 2022–02. https://doi.org/10.1101/2022.02.16.480562

Inda-Webb M, Jimenez M, Liu Q, Phan N, Ahn J, Steiger C, Wentworth A, Riaz A, Zirtiloglu T, Wong K et al (2023) Sub-1.4 cm3 capsule for detecting labile inflammatory biomarkers in situ. Nature 620(7973):386–392. https://doi.org/10.1038/s41586-023-06369-x

Jacobson B, Mackay RS (1957) A pH-endoradiosonde. Lancet 269(6981):1224. https://doi.org/10.1016/S0140-6736(57)91792-0

Kalantar-Zadeh K, Ha N, Ou JZ, Berean KJ (2017a) Ingestible Sensors. ACS Sens 2(4):468–483. https://doi.org/10.1021/acssensors.7b00045

Kalantar-Zadeh K, Ha N, Ou JZ, Berean KJ (2017b) Ingestible sensors. ACS Sens 2(4):468–483. https://doi.org/10.1021/acssensors.7b00045

Kalantar-Zadeh K, Berean KJ, Ha N, Chrimes AF, Xu K, Grando D, Ou JZ, Pillai N, Campbell JL, Brkljača R et al (2018) A human pilot trial of ingestible electronic capsules capable of sensing different gases in the gut. Nat Electron 1(1):79–87

Kalantar-Zadeh K, Berean KJ, Burgell RE, Muir JG, Gibson PR (2019) Intestinal gases: influence on gut disorders and the role of dietary manipulations. Nat Rev Gastroenterol Hepatol 16(12):733–747. https://doi.org/10.1038/s41575-019-0193-z

Kane SF (2017) The effects of oral health on systemic health. Gen Dent 65(6):30–34

Khachornsakkul K, Del-Rio-Ruiz R, Zeng W, Santos D, Silva DRR, Widmer G, Sonkusale S (2024) Ingestible sensor based on fluorescent gold nanopaticles and molecularly imprinted polymers on thread for localized quantification of escherichia coli in the gastrointestinal tract. Adv Funct Mater 2408056. https://doi.org/10.1002/adfm.202408056

Kloetzer L, Chey WD, McCallum R, Koch K, Wo J, Sitrin M, Katz L, Lackner J, Parkman HP, Wilding G et al (2010) Motility of the antroduodenum in healthy and gastroparetics characterized by wireless motility capsule. Neurogastroenterol Motil 22(5):527–117. https://doi.org/10.1111/j.1365-2982.2010.01468.x

Klugman CM, Dunn LB, Schwartz J, Cohen IG (2018) The ethics of smart pills and self-acting devices: autonomy, truth-telling, and trust at the dawn of digital medicine. Am J Bioeth 18(9):38–47. https://doi.org/10.1080/15265161.2018.1498933

Ko Y-F, Kuo P-H, Wang C-F, Chen Y-J, Chuang P-C, Li S-Z, Chen B-W, Yang F-C, Lo Y-C, Yang Y et al (2022) Quantification analysis of sleep based on smartwatch sensors for parkinson's disease. Biosensors 12(2):74

Koo JH, Song J-K, Kim D-H, Son D (2021) Soft implantable bioelectronics. ACS Mater Lett 3(11):1528–1540. https://doi.org/10.1021/acsmaterialslett.1c00438

Kurowiak J, Klekiel T, Będziński R (2023) Biodegradable polymers in biomedical applications: a review—developments, perspectives and future challenges. Int J Mol Sci 24(23):16952. https://doi.org/10.3390/ijms242316952

Launius RD (2023) Inventing Apollo: Spaceflight biomedical sensors. Accessed 2024-10-02. https://airandspace.si.edu/stories/editorial/inventing-apollo-spaceflight-biomedical-sensors

Lavelle A, Sokol H (2020) Gut microbiota-derived metabolites as key actors in inflammatory bowel disease. Nat Rev Gastroenterol Hepatol 17(4):223–237. https://doi.org/10.1038/s41575-019-0258-z

Li C, Guo C, Fitzpatrick V, Ibrahim A, Zwierstra MJ, Hanna P, Lechtig A, Nazarian A, Lin SJ, Kaplan DL (2020) Design of biodegradable, implantable devices towards clinical translation. Nat Rev Mater 5(1):61–81. https://doi.org/10.1038/s41578-019-0150-z

Li J, Liu Y, Yuan L, Zhang B, Bishop ES, Wang K, Tang J, Zheng Y-Q, Xu W, Niu S et al (2022) A tissue-like neurotransmitter sensor for the brain and gut. Nature 606(7912):94–101. https://doi.org/10.1038/s41586-022-04615-2

Li Z, Li C, Deng Y (2023) Bioresorbable pressure sensor and its applications in abnormal respiratory event identification. ACS Appl Electron Mater 5(3):1761–1769. https://doi.org/10.1021/acsaelm.2c01782

Li X, Li Y, Wei H, Wang C, Liu B (2024a) A review of wearable optical fiber sensors for rehabilitation monitoring. Sensors 24(11):3602. https://doi.org/10.3390/s24113602

Li Z, Tang X, Zhao T, Chen K, Zhang T (2024b) Highly sensitive skin-like wearable optical sensor for human physiological signals monitoring. Opt Fiber Technol 82:103652. https://doi.org/10.1016/j.yofte.2023.103652

Liao C-H, Cheng C-T, Chen C-C, Jow U-M, Chen C-H, Lai Y-L, Chen Y-C, Ho D-R (2020) An ingestible electronics for continuous and real-time intraabdominal pressure monitoring. J Personal Med 11(1):12. https://doi.org/10.3390/jpm11010012

Liu H, Wang Y, Huang Y, Xiong D, Shen J, Siqueiros L, Bala V, Savage GM, Guerrero M, Corado K et al (2022) Ingestible sensor system for measuring, monitoring and enhancing adherence to antiretroviral therapy: an open-label, usual care-controlled, randomised trial. EBioMedicine 86. https://doi.org/10.1016/j.ebiom.2022.104330

Liu R, Nie Q, Wang Y, Wu Y, Tu Y, Xie C, Xiao X, You R, Lu Y (2024) Diaper-based wearable SERS sensing system with a silver nano dual-structure composite hydrogel for the detection of biomarkers and pH in urine. Chem Eng J 498:155207. https://doi.org/10.1016/j.cej.2024.155207

Lu T, Ji S, Jin W, Yang Q, Luo Q, Ren T-L (2023a) Wearable, ingestible and implantable biosensors: reform. Sensors 23(2991):27. https://doi.org/10.3390/s23062991

Lu T, Ji S, Jin W, Yang Q, Luo Q, Ren T-L (2023b) Biocompatible and long-term monitoring strategies of wearable, ingestible and implantable biosensors: reform the next generation healthcare. Sensors 23(6):2991. https://doi.org/10.3390/s23062991

Luna RA, Oezguen N, Balderas M, Venkatachalam A, Runge JK, Ver-salovic J, Veenstra-VanderWeele J, Anderson GM, Savidge T, Williams KC (2017) Distinct microbiome-neuroimmune signatures correlate with functional abdominal pain in children with autism spectrum disorder. Cell Mol Gastroenterol Hepatol 3(2):218–230. https://doi.org/10.1016/j.jcmgh.2016.11.008

Machida T (1981) A study of intragastric pH in patients with peptic ulcer-with special reference to the clinical significance of basal pH value. Gastroenterol Jpn 16:447–458. https://doi.org/10.1007/BF02774516

Mackay RS, Jacobson B (1957) Endradiosonde letter to the editor. Nature 179(4572):1239–1240. https://doi.org/10.1038/1791239a0

Makino M, Yoshimoto R, Ono M, Itoko T, Katsuki T, Koseki A, Kudo M, Haida K, Kuroda J, Yanagiya R, Saitoh E, Hoshinaga K, Yuzawa Y, Suzuki A (2019) Artificial intelligence predicts the progression of diabetic kidney disease using big data machine learning. Sci Rep 9(1):1–9. https://doi.org/10.1038/s41598-019-48263-5

Malasinghe LP, Ramzan N, Dahal K (2019) Remote patient monitoring: a comprehensive study. J Ambient Intell Humaniz Comput 10(1):57–76. https://doi.org/10.1007/s12652-017-0598-x

Malik NA, Sant P, Ajmal T, Ur-Rehman M (2020) Implantable antennas for bio-medical applications. IEEE J Electromagnet RF Microw Med Biol 5(1):84–96. https://doi.org/10.1109/JERM.2020.3026588

Mathesul S, Swain D, Satapathy SK, Rambhad A, Acharya B, Gero-giannis VC, Kanavos A (2023) COVID-19 detection from chest X-ray images based on deep learning techniques. Algorithms 16(10):1–19. https://doi.org/10.3390/a16100494

Matzeu G, Florea L, Diamond D (2015) Advances in wearable chemical sensor design for monitoring biological fluids. Sensors Actuators B Chem 211:403–418. https://doi.org/10.1016/j.snb.2015.01.077

Maurer JM, Schellekens RC, Van Rieke HM, Wanke C, Iordanov V, Stellaard F, Wutzke KD, Dijkstra G, Zee M, Woerdenbag HJ et al (2015) Gastrointestinal pH and transit time profiling in healthy volunteers using the Intellicap system confirms Ileo-colonic release of Colopulse tablets. PLoS One 10(7):0129076. https://doi.org/10.1371/journal.pone.0129076

Mc Caffrey C, Twomey K, Ogurtsov V (2015) Development of a wireless swallowable capsule with potentiostatic electrochemical sensor for gastrointestinal track investigation. Sensors Actuators B Chem 218:8–15. https://doi.org/10.1016/j.snb.2015.04.063

McCallum C, Rooksby J, Gray CM et al (2018) Evaluating the impact of physical activity apps and wearables: interdisciplinary review. JMIR Mhealth Uhealth 6(3):9054. https://doi.org/10.2196/mhealth.9054

McKenzie J, Osgood D (2004) Validation of a new telemetric core temperature monitor. J Therm Biol 29(7–8):605–611. https://doi.org/10.1016/j.jtherbio.2004.08.020

Mei X, Ye D, Zhang F, Di C-A (2022) Implantable application of polymer-based biosensors. J Polym Sci 60(3):328–347. https://doi.org/10.1002/pol.20210543

Mimee M, Nadeau P, Hayward A, Carim S, Flanagan S, Jerger L, Collins J, McDonnell S, Swartwout R, Citorik RJ et al (2018) An ingestible bacterial-electronic system to monitor gastrointestinal health. Science 360(6391):915–918. https://doi.org/10.1126/science.aas9315

Min J, Yang Y, Wu Z, Gao W (2020) Robotics in the gut. Adv Therap 3(4):1900125. https://doi.org/10.1002/adtp.201900125

Min J, Jung Y, Ahn J, Lee JG, Lee J, Ko SH (2023) Recent advances in biodegradable green electronic materials and sensor applications. Adv Mater 35(52):2211273. https://doi.org/10.1002/adma.202211273

Mitrakos V, Cummins G, Tauber FJ, Cox BF, Pavuluri SK, Wood GS, Potter MA, Clutton E, Cochran S, Speck T et al (2024) Pressurecap: an endoscopic sensor capsule for real-time gastrointestinal pressure monitoring. Device 2(5). https://doi.org/10.1016/j.device.2024.100325

Mobarak MH, Islam MA, Hossain N, Al Mahmud MZ, Rayhan MT, Nishi NJ, Chowdhury MA (2023) Recent advances of additive manufacturing in implant fabrication–a review. Appl Surf Sci Adv 18:100462. https://doi.org/10.1016/j.apsadv.2023.100462

Moen S, Vuik FE, Kuipers EJ, Spaander MC (2022) Artificial intelligence in colon capsule endoscopy—a systematic review. Diagnostics 12(8):1994. https://doi.org/10.3390/diagnostics12081994

Mukherjee S, Suleman S, Pilloton R, Narang J, Rani K (2022) State of the art in smart portable, wearable, ingestible and implantable devices for health status monitoring and disease management. Sensors 22(11). https://doi.org/10.3390/s22114228

Nadeau P, El-Damak D, Glettig D, Kong YL, Mo S, Cleveland C, Booth L, Roxhed N, Langer R, Chandrakasan AP et al (2017) Prolonged energy harvesting for ingestible devices. Nat Biomed Eng 1(3):0022. https://doi.org/10.1038/s41551-016-0022

Nasiri S, Khosravani MR (2020) Progress and challenges in fabrication of wearable sensors for health monitoring. Sensors Actuators A Phys 312:112105. https://doi.org/10.1016/j.sna.2020.112105

Newswire P (2024) Full Press Release. Accessed 2024-10-07. https://mma.prnewswire.com/media/2238451/Full_Press_Release.pdf

Noura Z, Shah I, Aziz S, Ahmed A, Jung DW, Brahim L, ElMostafa R (2022) Wearable healthcare monitoring based on a microfluidic electrochemical integrated device for sensing glucose in natural sweat. Sensors 22(22). https://doi.org/10.3390/s22228971

O'Brien C, Hoyt RW, Buller MJ, Castellani JW, Young AJ (1998) Telemetry pill measurement of core temperature in humans during active heating and cooling. Med Sci Sports Exerc 30:468–472. https://doi.org/10.1097/00005768-199803000-00020

Omar R, Saliba W, Khatib M, Zheng Y, Pieters C, Oved H, Silberman E, Zohar O, Hu Z, Kloper V et al (2024) Biodegradable, biocompatible, and implantable multifunctional sensing platform for cardiac monitoring. ACS Sens 9(1):126–138. https://doi.org/10.1021/acssensors.3c01755

Ometov A, Shubina V, Klus L, Skibińska J, Saafi S, Pascacio P, Fluera-toru L, Gaibor DQ, Chukhno N, Chukhno O, Ali A, Channa A, Svertoka E, Qaim WB, Casanova-Marqués R, Holcer S, Torres-Sospedra J, Caste-leyn S, Ruggeri G, Araniti G, Burget R, Hosek J, Lohan ES (2021) A survey on wearable technology: history, state-of-the-art and current challenges. Comput Netw 193:108074. https://doi.org/10.1016/J.COMNET.2021.108074

Organization, W.H (2023) Cardiovascular Diseases (CVDs). Accessed 2024- 10-02. https://www.who.int/news-room/fact-sheets/detail/cardiovascular-diseases-(cvds)

Osterberg L, Blaschke T (2005) Adherence to medication. N Engl J Med 353(5):487–497. https://doi.org/10.1056/NEJMra050100

Ou JZ, Yao CK, Rotbart A, Muir JG, Gibson PR, Kalantar-zadeh K (2015) Human intestinal gas measurement systems: in vitro fermentation and gas capsules. Trends Biotechnol 33(4):208–213. https://doi.org/10.1016/j.tibtech.2015.02.002

Pan J, Zhang Z, Jiang C, Zhang L, Tong L (2020) A multifunctional skin-like wearable optical sensor based on an optical micro-/nanofibre. Nanoscale 12(33):17538–17544. https://doi.org/10.1039/d0nr03446k

Park E, Kim KJ, Kwon SJ (2016) Understanding the emergence of wearable devices as next-generation tools for health communication. Inf Technol People 29(4):717–732. https://doi.org/10.1108/ITP-04-2015-0096

Paz E, Maganti NH, Trifonov A, Jeerapan I, Mahato K, Yin L, Sonsa-Ard T, Ma N, Jung W, Burns R et al (2022) A self-powered ingestible wireless biosensing system for real-time in situ monitoring of gastrointestinal tract metabolites. Nat Commun 13(1):7405. https://doi.org/10.1038/s41467-022-35074-y

Peng H (2023) Wearable electronics. Natl Sci Rev 10(1):193. https://doi.org/10.1093/nsr/nwac193

Peng R, Sonner Z, Hauke A, Wilder E, Kasting J, Gaillard T, Swaille D, Sherman F, Mao X, Hagen J et al (2016) A new oil/membrane approach for integrated sweat sampling and sensing: sample volumes reduced from μL's to nL's and reduction of analyte contamination from skin. Lab Chip 16(22):4415–4423. https://doi.org/10.1039/c6lc01013j

Potts RO, Tamada JA, Tierney MJ (2002) Glucose monitoring by reverse iontophoresis. Diabetes Metab Res Rev 18(SUPPL. 1):49–53. https://doi.org/10.1002/dmrr.210

Promphet N, Ummartyotin S, Ngeontae W, Puthongkham P, Rodthongkum N (2021) Non-invasive wearable chemical sensors in real-life applications. Anal Chim Acta 1179:338643. https://doi.org/10.1016/j.aca.2021.338643

Qian X, Liao C (2023) Engineering liquid metal-based implantable electrodes toward brain-machine interfaces. Health Sci Rev 100118. https://doi.org/10.1016/j.hsr.2023.100118

Raj R, Ussavarungsi K, Nugent K (2014) Accelerometer-based devices can be used to monitor sedation/agitation in the intensive care unit. J Crit Care 29(5):748–752. https://doi.org/10.1016/j.jcrc.2014.05.014

Ramachandran B, Liao Y-C (2022) Microfluidic wearable electrochemical sweat sensors for health monitoring. Biomicrofluidics 16(5). https://doi.org/10.1063/5.0116648

Rao SS, Kuo B, McCallum RW, Chey WD, DiBaise JK, Hasler WL, Koch KL, Lackner JM, Miller C, Saad R et al (2009) Investigation of colonic and whole-gut transit with wireless motility capsule and radiopaque markers in constipation. Clin Gastroenterol Hepatol 7(5):537–544. https://doi.org/10.1016/j.cgh.2009.01.017

Rezaei Nejad H, Oliveira BC, Sadeqi A, Dehkharghani A, Kondova I, Langermans JA, Guasto JS, Tzipori S, Widmer G, Sonkusale SR (2019) Ingestible osmotic pill for in vivo sampling of gut microbiomes. Adv Intell Syst 1(5):1900053. https://doi.org/10.1002/aisy.201900053

Rondonotti E, Herrerias JM, Pennazio M, Caunedo A, Mascarenhas-Saraiva M, Franchis R (2005) Complications, limitations, and failures of capsule endoscopy: a review of 733 cases. Gastrointest Endosc 62(5):712–716. https://doi.org/10.1016/j.gie.2005.05.002

Saad RJ, Hasler WL (2011) A technical review and clinical assessment of the wireless motility capsule. Gastroenterol Hepatol 7(12):795

Sacchi M, Sauter-Starace F, Mailley P, Texier I (2024) Resorbable conductive materials for optimally interfacing medical devices with the living. Front Bioeng Biotechnol 12(1294238):1294238. https://doi.org/10.3389/fbioe.2024

Sahafi A, Wang Y, Rasmussen C, Bollen P, Baatrup G, Blanes-Vidal V, Herp J, Nadimi E (2022) Edge artificial intelligence wireless video capsule endoscopy. Sci Rep 12(1):13723. https://doi.org/10.1038/s41598-022-17502-7

Salama R, Arshavsky-Graham S, Sella-Tavor O, Massad-Ivanir N, Segal E (2022) Design considerations of aptasensors for continuous monitoring of biomarkers in digestive tract fluids. Talanta 239:123124. https://doi.org/10.1016/j.talanta.2021.123124

Sempionatto JR, Gao W (2022) Wearable chemosensors. In: Wearable physical, chemical and biological sensors. Elsevier, pp 219–234. https://doi.org/10.1016/B978-0-12-821661-3.00010-0

Sempionatto JR, Brazaca LC, García-Carmona L, Bolat G, Campbell AS, Martin A, Tang G, Shah R, Mishra RK, Kim J et al (2019) Eyeglasses-based tear biosensing system: non-invasive detection of alcohol, vitamins andglucose. Biosens Bioelectron 137:161–170. https://doi.org/10.1016/j.bios.2019.04.058

Seneviratne S, Hu Y, Nguyen T, Lan G, Khalifa S, Thilakarathna K, Hassan M, Seneviratne A (2017) A survey of wearable devices and challenges. IEEE Commun Surv Tutor 19(4):2573–2620. https://doi.org/10.1109/COMST.2017.2731979

Shah A (2023) Best heart rate monitor watches to help you keep track of your fitness. Accessed 2024-10-02. https://www.standard.co.uk/shopping/esbest/health-fitness/best-heart-rate-monitor-watches-b1110190.html

Shao M, Yuan Y, Wang M, Liu Y, Qiao X (2024) Optical fiber sensor for wearable and accurate human respiratory monitoring. Biomed Opt Express 15(7):4132–4146. https://doi.org/10.1364/boe.524355

Shen Y, Chen G, Chen Y, Gao Y, Hou C, Liao K, Ma B, Liu H (2024) Wearable microfluidic electrochemical sensor integrated with iontophoresis for non-invasive sweat ketone monitoring. Sensors Actuators B Chem 421:136518. https://doi.org/10.1016/j.snb.2024.136518

Shi C, Tang Z, Zhang H, Liu Y (2022) Development of an FBG-based wearable sensor for simultaneous respiration and heartbeat measurement. IEEE Trans Instrum Meas 72(1–9). https://doi.org/10.1109/TIM.2022.3228276

Shilaih M, Clerck VD, Falco L, Kübler F, Leeners B (2017) Pulse rate measurement during sleep using wearable sensors, and its correlation with the menstrual cycle phases, a prospective observational study. Sci Rep 7(1):1294. https://doi.org/10.1038/s41598-017-01433-9

Shrivastava S, Trung TQ, Lee NE (2020) Recent progress, challenges, and prospects of fully integrated mobile and wearable point-of-care testing systems for self-testing. Chem Soc Rev 49(6):1812–1866. https://doi.org/10.1039/c9cs00319c

So D, Yao CK, Gill PA, Thwaites PA, Ardalan ZS, McSweeney CS, Denman SE, Chrimes AF, Muir JG, Berean KJ et al (2023) Detection of changes in regional colonic fermentation in response to supplementing a low fodmap diet with dietary fibres by hydrogen concentrations, but not by luminal ph. Aliment Pharmacol Ther 58(4):417–428. https://doi.org/10.1111/apt.17629

Söderlind E, Abrahamsson B, Erlandsson F, Wanke C, Iordanov V, Cor-swant C (2015) Validation of the Intellicap® system as a tool to evaluate extended release profiles in human GI tract using metoprolol as model drug. J Control Release 217:300–307. https://doi.org/10.1016/j.jconrel.2015.09.024

Son D, Gilbert H, Sitti M (2020) Magnetically actuated soft capsule endoscope for fine-needle biopsy. Soft Robot 7(1):10–21. https://doi.org/10.1089/soro.2018.0171

Song Z, Zhou S, Qin Y, Xia X, Sun Y, Han G, Shu T, Hu L, Zhang Q (2023) Flexible and wearable biosensors for monitoring health conditions. Biosensors 13(630):1–32. https://doi.org/10.3390/bios13060630

Sparling P, Snow T, Millard-Stafford M (1993) Monitoring core temperature during exercise: ingestible sensor vs. rectal thermistor. Aviat Space Environ Med 64(8):760–763

Stearns JC, Lynch MD, Senadheera DB, Tenenbaum HC, Goldberg MB, Cvitkovitch DG, Croitoru K, Moreno-Hagelsieb G, Neufeld JD (2011) Bacterial biogeography of the human digestive tract. Sci Rep 1(1):170. https://doi.org/10.1038/srep00170

Suarez-Ibarrola R, Sigle A, Eklund M, Eberli D, Miernik A, Benndorf M, Bamberg F, Gratzke C (2022) Artificial intelligence in magnetic resonance imaging–based prostate cancer diagnosis: where do we stand in 2021? Eur Urol Focus 8(2):409–417. https://doi.org/10.1016/j.euf.2021.03.020

Surantha N, Atmaja P, David Wicaksono M (2021) A review of wearable Internet-of-Things device for healthcare. Proc Comput Sci 179(2020):939–943. https://doi.org/10.1016/j.procs.2021.01.083

Swain P (2003) Wireless capsule endoscopy. Gut 52(suppl 4):48–50. https://doi.org/10.1136/gut.52.suppl_4.iv48

Team P (2023) How heart rate monitors changed endurance sports. Accessed 2024-10-02. https://www.polar.com/blog/how-heart-rate-monitors-changed-endurance-sports/

Thwaites PA, Yao CK, Halmos EP, Muir JG, Burgell RE, Berean KJ, Kalantar-zadeh K, Gibson PR (2024) Current status and future directions of ingestible electronic devices in gastroenterology. Aliment Pharmacol Ther 59(4):459–474. https://doi.org/10.1111/apt.17844

Traverso G, Finomore V, Mahoney J, Kupec J, Stansbury R, Bacher D, Pless B, Schuetz S, Hayward A, Langer R, Rezai A (2023) First-in-human trial of an ingestible vitals-monitoring pill. Device 1(5):100125. https://doi.org/10.1016/j.device.2023.100125

U.S. Food and Drug Administration FDA approves pill with sensor that digitally tracks if patients have ingested their medication. Accessed 2024-10-04. https://www.fda.gov/news-events/press-announcements/fda-approves-pill-sensor-digitally-tracks-if-patients-have-ingested-their-medication

Vaccari M, Tudor T, Perteghella A (2018) Costs associated with the management of waste from healthcare facilities: an analysis at national and site level. Waste Manag Res 36(1):39–47. https://doi.org/10.1177/0734242X17739968

Wang X, Liu Z, Zhang T (2017) Flexible sensing electronics for wearable/attachable health monitoring. Small 13(25):1602790. https://doi.org/10.1002/smll.201602790

Wang Y, Cang S, Yu H (2019) A survey on wearable sensor modality centred human activity recognition in health care. Expert Syst Appl 137:167–190. https://doi.org/10.1016/j.eswa.2019.04.057

Wang W, Xuan X-W, Zhao W-Y, Nie H-K (2021) An implantable antenna sensor for medical applications. IEEE Sensors J 21(13):14035–14042. https://doi.org/10.1109/JSEN.2021.3068957

Weitschies W, Müller L, Grimm M, Koziolek M (2021) Ingestible devices for studying the gastrointestinal physiology and their application in oral biopharmaceutics. Adv Drug Deliv Rev 176:113853. https://doi.org/10.1016/j.addr.2021.113853

Woo SH, Kim TW, Mohy-Ud-Din Z, Park IY, Cho J-H (2011) Small intestinal model for electrically propelled capsule endoscopy. Biomed Eng Online 10:1–20. https://doi.org/10.1186/1475-925X-10-108

Wu G, Zhang N, Matarasso A, Heck I, Li H, Lu W, Phaup JG, Schneider MJ, Wu Y, Weng Z et al (2022) Implantable aptamer-graphene micro-transistors for real-time monitoring of neurochemical release in vivo. Nano Lett 22(9):3668–3677. https://doi.org/10.1021/acs.nanolett.2c00289

Xu YD, Lin M, Xu ZY, Kang H, Li ZT, Luo ZZ, Lin SY (2022) Holter electrocardiogram research trends and hotspots: bibliometrics and visual analysis. Eur Rev Med Pharmacol Sci 26(17):6027–6039. https://doi.org/10.26355/eurrev_202209_29617

Xu F, Meng L, Lin H, Xu W, Guo H, Peng F (2024) Systematic review of leadless pacemaker. Acta Cardiol 79(3):284–294. https://doi.org/10.1080/00015385.2023.2276537

Yang X, Lou J, Shan W, Ding J, Jin Z, Hu Y, Du Q, Liao Q, Xie R, Xu J (2021) Pathophysiologic role of neurotransmitters in digestive diseases. Front Physiol 12:567650. https://doi.org/10.3389/fphys.2021.567650

Yang J, Gong X, Chen S, Zheng Y, Peng L, Liu B, Chen Z, Xie X, Yi C, Jiang L (2023) Development of smartphone-controlled and microneedle-based wearable continuous glucose monitoring system for home-care diabetes management. ACS Sens 8(3):1241–1251. https://doi.org/10.1021/acssensors.2c02635

Yao Y, Chen J, Guo Y, Lv T, Chen Z, Li N, Cao S, Chen B, Chen T (2021) Integration of interstitial fluid extraction and glucose detection in one device for wearable non-invasive blood glucose sensors. Biosens Bioelectron 179:113078. https://doi.org/10.1016/j.bios.2021.113078

Ye L, Wu F, Xu R, Zhang D, Lu J, Wang C, Dong A, Xu S, Xue L, Fan Z et al (2023) Face mask integrated with flexible and wearable manganite oxide respiration sensor. Nano Energy 112:108460. https://doi.org/10.1016/j.nanoen.2023.108460

Yogev D, Goldberg T, Arami A, Tejman-Yarden S, Winkler TE, Maoz BM (2023) Current state of the art and future directions for implantable sensors in medical technology: clinical needs and engineering challenges. APL Bioeng 7(3). https://doi.org/10.1063/5.0152290

Yun S-O, Lee JH, Lee J, Kim C-Y (2019) A flexible wireless sensor patch for real-time monitoring of heart rate and body temperature. IEICE Trans Inf Syst 102(5):1115–1118

Yunus NAM, Halin IA, Sulaiman NB, Ismail NF, Sheng OK (2015) Valuation on mems pressure sensors and device applications. World Acad Sci Eng Technol Int J Electr Comput Energ Electron Commun Eng 9:834–842

Zhao C, Man T, Cao Y, Weiss PS, Monbouquette HG, Andrews AM (2022) Flexible and implantable polyimide aptamer-field-effect transistor biosensors. ACS Sens 7(12):3644–3653. https://doi.org/10.1021/acssensors.2c01909

Zheng L, Guo S, Kawanishi M (2022) Magnetically controlled multifunctional capsule robot for dual-drug delivery. IEEE Syst J 16(4):6413–6424. https://doi.org/10.1109/JSYST.2022.3145869

Zhou Z, Tian D, Yang Y, Cui H, Li Y, Ren S, Han T, Gao Z (2024) Machine learning assisted biosensing technology: an emerging powerful tool for improving the intelligence of food safety detection. Curr Res Food Sci 100679. https://doi.org/10.1016/j.crfs.2024.100679

Zokaeinikoo M, Kazemian P, Mitra P, Kumara S (2021) AIDCOV: an interpretable artificial intelligence model for detection of COVID-19 from chest radiography images. ACM Trans Manag Inf Syst 12(4):1–20. https://doi.org/10.1145/3466690

Electrochemical Biosensors for Therapeutic Drug Monitoring: Next Generation of Disease Management

6

Martins Lateef Yusuf, Renzhe Lu, Haozhe Zhou, Hui Li, Shaoguang Li, and Fan Xia

Abstract

The precision of drug therapy has garnered significant attention for enhancing disease treatment at the individual patient level. Monitoring therapeutic drugs—such as anticonvulsants, antimicrobials, cardioactives, antidepressants, and antineoplastics—is essential to ensure safe and effective administration. Therapeutic drug monitoring (TDM) focuses on understanding the pharmacokinetics and pharmacodynamics of these drugs, as these factors critically influence drug activity at the target site. Mismanagement of these parameters often leads to underdosing or overdosing, which can be detrimental to patient outcomes. TDM plays a crucial role in risk assessment, improving drug efficacy, and providing actionable insights for optimized disease management. In this chapter, we explore the integration of nanobioelectronics into TDM, emphasizing the incorporation of bioreceptors and advanced nanomaterials. The unique properties of functional nanomaterials and nanoscale fabrication techniques enable the customization of sensors to meet specific monitoring requirements. Nanobioelectronics offer innovative solutions for real-time, high-sensitivity drug monitoring, enhancing the precision of therapeutic interventions. We discuss current strategies for the application of nanobioelectronics in TDM, highlighting advances in nanoscale material engineering and sensor design. Additionally, we

M. L. Yusuf · R. Lu · H. Zhou
State Key Laboratory of Biogeology and Environmental Geology, Engineering Research Center of Nano-Geomaterials of the Ministry of Education, Faculty of Materials Science and Chemistry, China University of Geosciences, Wuhan, China

H. Li · S. Li (✉) · F. Xia
State Key Laboratory of Geomicrobiology and Environmental Changes, Engineering Research Center of Nano-Geomaterials of Ministry of Education, Faculty of Materials Science and Chemistry, China University of Geosciences, Wuhan, China
e-mail: lisg@cug.edu.cn

K. Mahato, P. Chandra (eds.), *Nano-bioelectronics for Precision Health Monitoring*, https://doi.org/10.1007/978-981-95-2117-3_6

address the challenges associated with these approaches, such as scalability, reliability, and regulatory considerations. Finally, this chapter examines future perspectives, emphasizing the potential of nanobioelectronics to transform TDM practices and contribute to precision healthcare through enhanced drug monitoring and individualized treatment plans.

Keywords

Therapeutic drug monitoring · Electrochemical biosensor · Nanobioelectronics · Nanomaterials · Interface

6.1 Introduction to Therapeutic Drug Monitoring

Therapeutic drug monitoring (TDM) involves the continuous and precise observation of drug behavior within different systems of the body (Roberts et al. 2011). Therapeutic drugs, which are typically prescribed by professional healthcare providers to patients in need of medical intervention, require meticulous management to prevent adverse health outcomes (Vande Casteele et al. 2017). Improper management of these drugs can exacerbate a patient's condition rather than ameliorate it. Therefore, TDM is a crucial clinical practice that involves direct concentration measurement of drugs at their site of action (Stott and Hope 2017; Rutherford et al. 2023), which is essential for accurately determining a patient's drug exposure (Zentner et al. 2018; Heysell et al. 2020). The response to a drug can vary significantly between individuals due to factors such as body weight, blood composition, age, genetics, organ function, and overall health condition (Lauschke and Ingelman-Sundberg 2016). These factors can affect the dosage required for therapeutic efficacy, leading to considerable pharmacokinetic variations (Stott and Hope 2017). Moreover, variations within an individual over time, influenced by changes in health status or concurrent medications, can also impact drug efficacy and safety (Yamada et al. 2024; Arnold et al. 2015). Consequently, continuous monitoring becomes essential to tailor drug regimens to individual needs.

Given the overwhelming volume of patients that healthcare professionals must attend to, managing individual patient responses to drugs can be challenging. This is where the integration of bioelectronics has proven invaluable. The continuous monitoring capabilities of bioelectronic sensors enable healthcare providers to maintain drugs within their therapeutic windows, preventing underdosing or overdosing. For example, antimicrobial resistance has become a major issue in healthcare management (Wicha et al. 2021). Antonio et al. demonstrated the importance of de-escalation of antibiotic therapy in noncritically ill patients as a strategy to reduce the risk of overdose and resistance (Gutiérrez-Pizarraya et al. 2017). Their study showed that de-escalation decreased mortality in immunocompetent patients with severe sepsis and septic shock. Additionally, the timing of antibiotic administration has been shown to significantly affect patient mortality rates (Liu et al. 2017a). Similarly, in cancer treatment, Moghaddas et al. conducted studies on drug–drug interactions in hospitalized patients with solid tumors, focusing on interactions between chemotherapeutic and nonchemotherapeutic drugs (Moghaddas et al. 2021). They found that moderate to major drug–drug interactions frequently occur among these patients, which can complicate treatment and affect patient outcomes. This highlights the purpose of precise therapeutic drug management in patient care to avoid adverse interactions and ensure the efficacy of treatment protocols.

Therapeutic drug monitoring extends beyond merely measuring drug concentrations in biological matrices. It involves integrating valuable information to make informed decisions that enhance patient care. This process encompasses an understanding of pharmacokinetics (how the drug is absorbed, distributed, metabolized, and excreted), pharmacodynamics (the drug's effects on the body), sampling time, drug history, and the patient's clinical condition (Wicha et al. 2021). By synthesizing this information, healthcare providers can optimize therapeutic outcomes, minimize adverse effects, and ensure patient safety. Several conventional approaches have been employed for TDM, including liquid chromatography (Turković et al. 2022), high-performance liquid chromatography (HPLC) (Pereira et al. 2021), mass spectroscopy (Abouzid et al. 2024; Rosado et al. 2018), and spectrofluorometric (Ibrahim et al. 2018; Alqahtani et al. 2023; Magdy et al. 2022; Ibrahim et al. 2017; El-Deen et al. 2023; Magdy et al. 2021). While these methods have proven effective, they present numerous challenges. The processes involved are often cumbersome, requiring skilled professionals to operate complex instruments. There is a significant time lag between sample collection and obtaining results, which can delay critical clinical decisions. Additionally, these techniques often involve the use of organic solvents, contributing to environmental pollution, and the high operational costs can be prohibitive.

6.2 Electrochemical Biosensors

Electrochemical approaches have shown considerable promise for the development of novel bioelectronic technologies for continuous therapeutic drug monitoring (Qian et al. 2021). Electrochemical sensors offer several advantages, including portability, rapid response times, high sensitivity, and cost-effectiveness. These devices can be designed for self-use by patients, facilitating continuous monitoring without the need for professional intervention (Liu et al. 2017b; Gao et al. 2024). The electrochemical approach aligns with green chemistry principles as it typically does not involve the use of organic or hazardous materials, thereby contributing to environmental sustainability. Electrochemical sensors operate by measuring the current or voltage changes that occur when a drug undergoes a redox process or interacts with immobilized recognition elements on the electrode surface (Fig. 6.1) (Bonazza et al. 2018; Novak Jovanović et al. 2017). These changes are proportional to the concentration of the drug in the sample, allowing for precise quantification (Temerk et al. 2015).

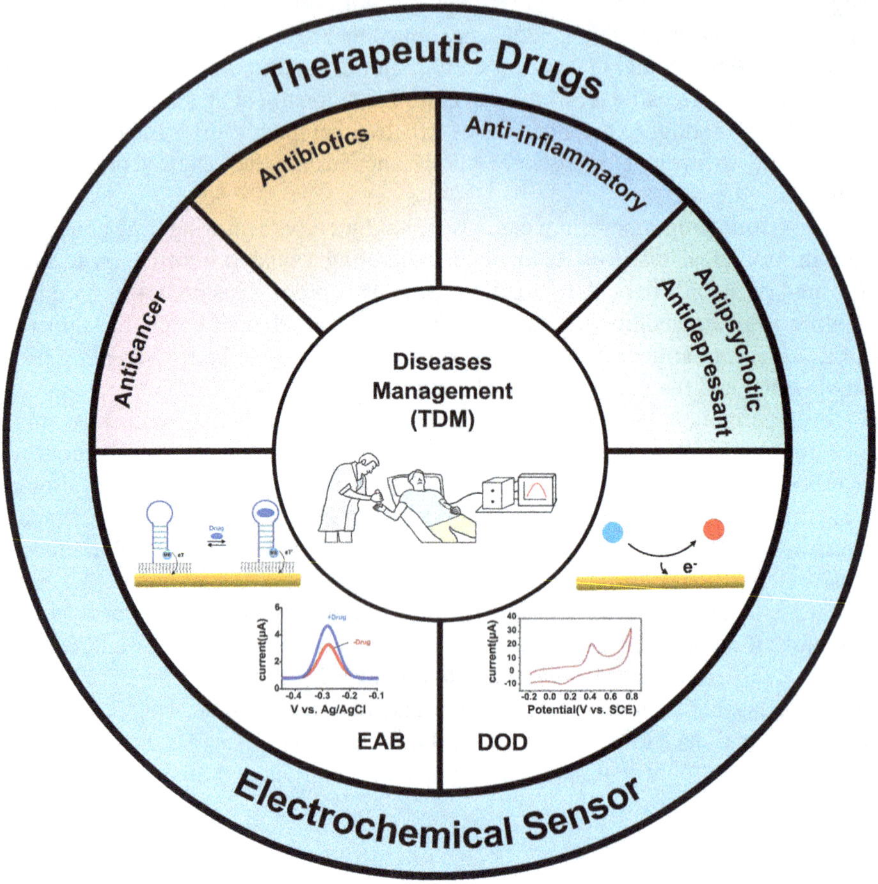

Fig. 6.1 Disease management can be enhanced by monitoring therapeutic drugs using electrochemical aptamer biosensors (EAB) or through the direct oxidation of the drug (DOD) on the electrode surface

The development of these sensors has been enhanced by advances in materials science, particularly with the use of reporters such as methylene blue (MB) (Zhu et al. 2023), anthraquinone (AQ) (Zhu et al. 2020), ferrocene (FC), nanomaterials (Mahmoud et al. 2020), etc. These materials enhance the sensitivity and selectivity of electrochemical sensors, making them capable of detecting low concentrations of drugs in complex biological matrices. One significant advantage of bioelectronics for therapeutic drug monitoring is their ability to perform effectively in various biological matrices and their potential drift correction and calibration within these matrices, such as blood (Leung et al. 2024), interstitial fluid (ISF) (Mishra et al. 2020; Teymourian et al. 2019), and even within living tissues (Chamorro-Garcia et al. 2023; Li et al. 2019). This adaptability is crucial for achieving accurate and reliable drug monitoring, considering the complexity and dynamic nature of biological environments. The integration of the electrochemical approach into therapeutic drug monitoring, particularly through wearable devices (Teymourian et al. 2020) and the use of nanomaterials (Qian et al. 2021), has been reviewed. This chapter focuses on the recent developments in the field of electrochemical sensors, particularly those employing bioelectronics for continuous therapeutic drug monitoring.

To put more insight, we provide detailed information on the biorecognition elements and probes used to enhance sensors for therapeutic drug monitoring. We also classify these sensors based on the types of drugs they monitor, including anticancer drugs, antibiotics, anti-inflammatory drugs, and others (Fig. 6.2). Finally, we discuss the challenges associated with the application of bioelectronic electrochemical approaches for TDM and suggest ways to overcome these challenges for further enhancement.

6.2.1 Recognition Elements and Reporters

In fabricating bioelectronic sensors capable of continuous monitoring of therapeutic drugs in complex matrices, it is essential to utilize components that are sensitive, stable, and specific to the drug target. A critical component in these sensor designs is the biorecognition element. Various biorecognition elements, such as DNA/RNA, aptamers, proteins, and antibodies, have been employed in sensor design. Among these, aptamers have garnered significant attention for therapeutic drug monitoring due to their unique properties. Aptamers are short single-stranded or double-stranded oligonucleotides capable of folding into secondary and tertiary structures upon binding to their target drug. They are typically selected through the systematic evolution of ligands by exponential enrichment (SELEX) and can be postmodified to enhance their properties. Aptamers offer excellent stability when probing drug samples in complex matrices, making them ideal for use in bioelectronic sensors. Electrochemical biosensors typically rely on redox-active reporters for signaling output, which can enhance the oxidation signal during sensing and provide information on sensor performance during drug monitoring. These reporters facilitate electron transfer with the electrode when the recognition element undergoes conformational changes in the presence of the target. Plaxco and team explored several reporters, surveying over a dozen to determine not only their electron transfer efficiency but also their stability in complex matrices such as blood (Kang et al. 2016). Methylene blue demonstrated the best performance with excellent stability, which explains its frequent use in the fabrication of EAB sensors. While anthraquinone, Nile blue, and ferrocene exhibited relatively good properties, they could not match the performance of methylene blue. In a recent study, an EAB sensor was designed to monitor doxorubicin (DOX) using an aptamer with two nonoverlapping reporters to improve sensor calibration (Zhu et al. 2023). The reporters were methylene blue (MB) and ferrocene (Fc). When the aptamer probe binds to the DOX target, the induced structural change results in an MB signal due to fast electron transfer, while the Fc signal changes due to alterations in its local environment (Fig. 6.3a). This approach substantially enhanced sensor performance, achieving a

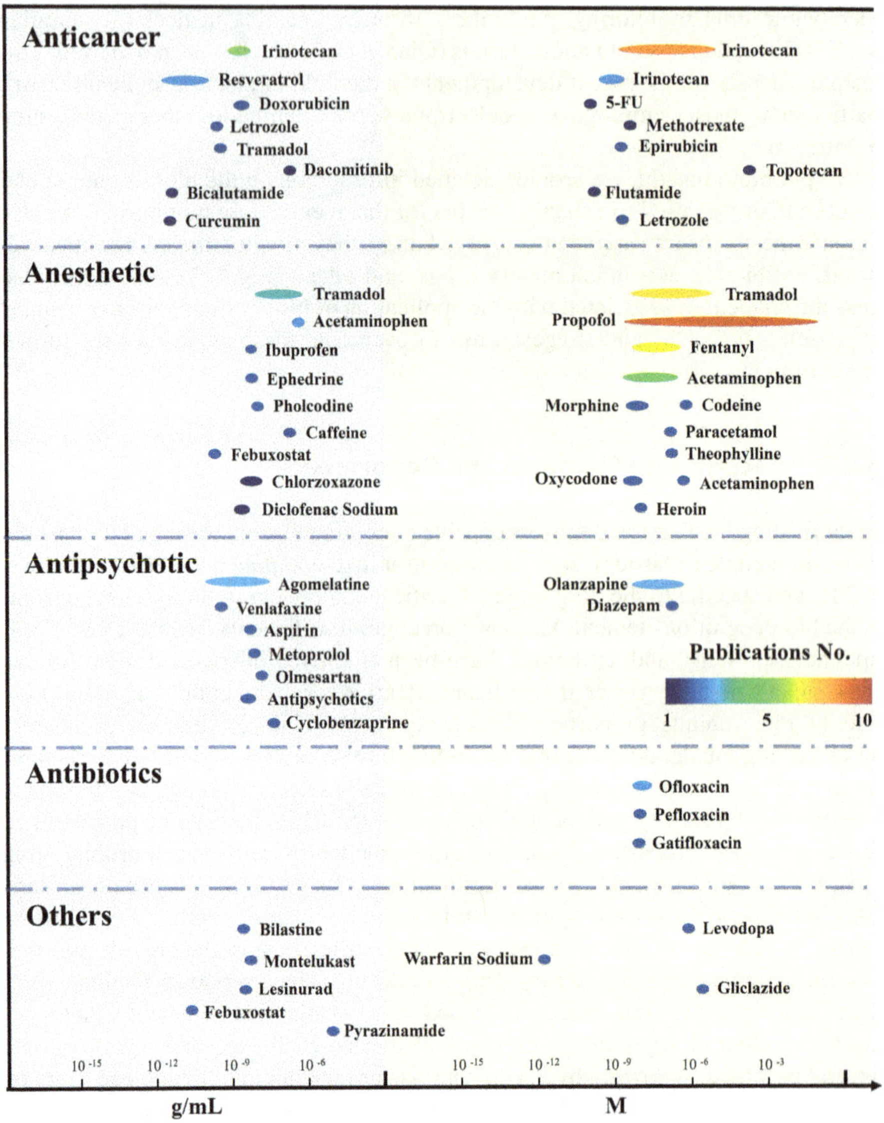

Fig. 6.2 Therapeutic drugs encompass various classes such as anticancer, antibiotics, anti-inflammatory, antipsychotic, and several other less common categories. Over the past decades, electrochemical bioelectronic sensors have been developed to monitor these drugs, as reflected in numerous publications

recovery range of doxorubicin concentrations from 1.0×10^{-6} M to 2.5×10^{-5} M in buffer.

Similar approach was employed to address the degradation of EAB sensors when monitoring therapeutic drugs in complex matrices. In this case, anthraquinone (AQ)

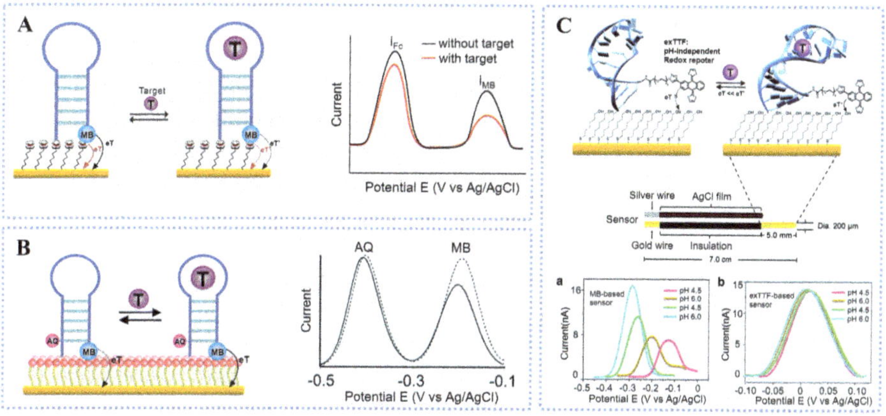

Fig. 6.3 Recognition elements enhanced with redox reporters. (**a**) Schematic illustration of double redox reporters (methylene blue (MB) and ferrocene (Fc) with different electron transfer responses. (Reproduced with permission from Zhu et al. (2023). Copyright 2023 American Chemical Society). (**b**) Schematic illustration of target binding alters the kinetics with which electrons exchange from two reporters, MB and anthraquinone (AQ). (Reproduced with permission from Li et al. (2019). Copyright 2022 The Royal Society of Chemistry). (**c**) Schematic illustration of an exTTF derivative as a PH-independent redox reporter. (Reproduced with permission from Li et al. (2022). Copyright 2022 The Royal Society of Chemistry)

was used alongside MB (Li et al. 2019). The MB was placed on the distal terminus of the DNA aptamer, which is affected by conformational changes in the presence of the drug target, while the AQ served as a reference reporter positioned in a manner that was not affected by these changes (Fig. 6.3b).

The chemical and physical properties of the matrix in which a sensor is used for therapeutic drug monitoring are critical factors in sensor design. Parameters such as pH and temperature must be considered to ensure optimal sensor performance. One effective strategy for addressing these variations is by modifying the sensor reporter. For example, in the fabrication of sensors for cocaine and kanamycin detection, a pH-independent redox reporter was coupled to DNA aptamer to overcome pH variability across different matrices (Li et al. 2022). Specifically, a π-extended tetrathiafulvalene (exTTF) was used as a redox reporter (Fig. 6.3c). This choice is advantageous because exTTF's redox process does not involve protons, making it unaffected by pH changes. The sensor demonstrated good stability in biological systems, with only a 5% signal fluctuation compared to 80% observed with methylene blue (MB). The high binding affinity of the exTTF redox reporter across varying pH conditions underscores its value in sensor fabrication. Despite optimizing the sensor through redox reporter modulation, it is also crucial to consider the aptamer itself and the sensor surface modification. These elements play significant roles in drift correction and calibration, which are essential for achieving overall sensor performance and effective therapeutic drug monitoring. The aptamer's binding characteristics and the surface properties of the sensor can significantly impact the reliability and accuracy of the measurements.

6.2.2 Surface Modification

To enhance sensor performance, both the biorecognition element (aptamer) and the sensor surface must be meticulously engineered. Surface modifications can improve the sensor's stability and selectivity, and such modifications can also enhance specificity and reduce nonspecific interactions. Together, these optimizations contribute to the development of high-performance sensors capable of accurate and reliable therapeutic drug monitoring in complex biological matrices. Considering the self-assembly mode of constructing an aptamer-based sensor for drug monitoring, caution must be taken to ensure the sensor does not deplete easily, especially when probed in a complex matrix. In addressing the challenge of sensor depletion, particularly in the context of probing whole blood samples, a continuous real-time sensing of molecular analysis has been reported (Fig. 6.4a) (Li et al. 2017a). The approach centered on the fabrication of a biomimetic surface treatment aimed at mitigating drift issues encountered when probing whole blood, without the need for a physical barrier, thus ensuring stable sensor performance.

The mechanism of the sensor design is based on the use of phosphatidylcholine (PC), chosen for its antifouling properties against eukaryotic cellular membranes, facilitation of electron transfer, and resistance to blood cells when applied as a coating on gold surfaces. The self-assembled monolayers (SAM) formation of the E-AB sensor when employing PC was compared with mercaptohexanol (MCH), using a 28-base DNA in a buffer solution. Both compounds demonstrated activity toward the target in the buffer medium. However, when tested in whole blood samples, the peak observed with the PC-treated sensor was notably higher, attributed to the impact of charged monolayers on sensor performance. This innovative approach underscores the efficacy of PC-terminated SAM in mimicking the properties of cell membranes, thereby enhancing sensor performance, particularly in whole blood environments (Zhang et al. 2022).

In one of the recent studies, an alternative approach to achieve stable E-DNA sensors via SAM mobilization has been explored. Here, the use of hydrophobic moieties as monolayers for the E-DNA sensor, comparing them to the traditional molecule MCH, has been investigated (Zhang et al. 2022). Various hydrophobic membranes were examined, including 1-hexylmercaptan (HEM), polyhedral oligomeric silsesquioxane (POSS) with n-propyl mercaptan functional groups,

Fig. 6.4 (continued) GmbH & Co. KGaA, Weinheim). (**b**) Schematic illustration of SAMs as a blocking layer that desorbs the nonspecific DNA on the surface and prevents the fouling biocomponent. (Reproduced with permission from Li et al. (2021). Copyright 2021 American Chemical Society). (**c**) Schematic illustration of hydrophobic SAMs influenced by the ending groups and their molecular length. (Reproduced with permission from Zhang et al. (2022). Copyright 2021 American Chemical Society). (**d**) Schematic illustration of gel layer limits the approach of high-molecular-weight components, such as enzymes and blood cells, to the electrode surface. (Reproduced with permission from Li et al. (2023). Copyright 2023 American Chemical Society). (**e**) Schematic illustration of "xenonucleic" acids (XNAs) reduced enzymatic degradation of oligonucleotides by employing 2′-methoxy-ribose backbone (OMe RNA) for sensor fabrication. (Reproduced with permission from Leung et al. (2024). Copyright 2024 Wiley-VCH GmbH)

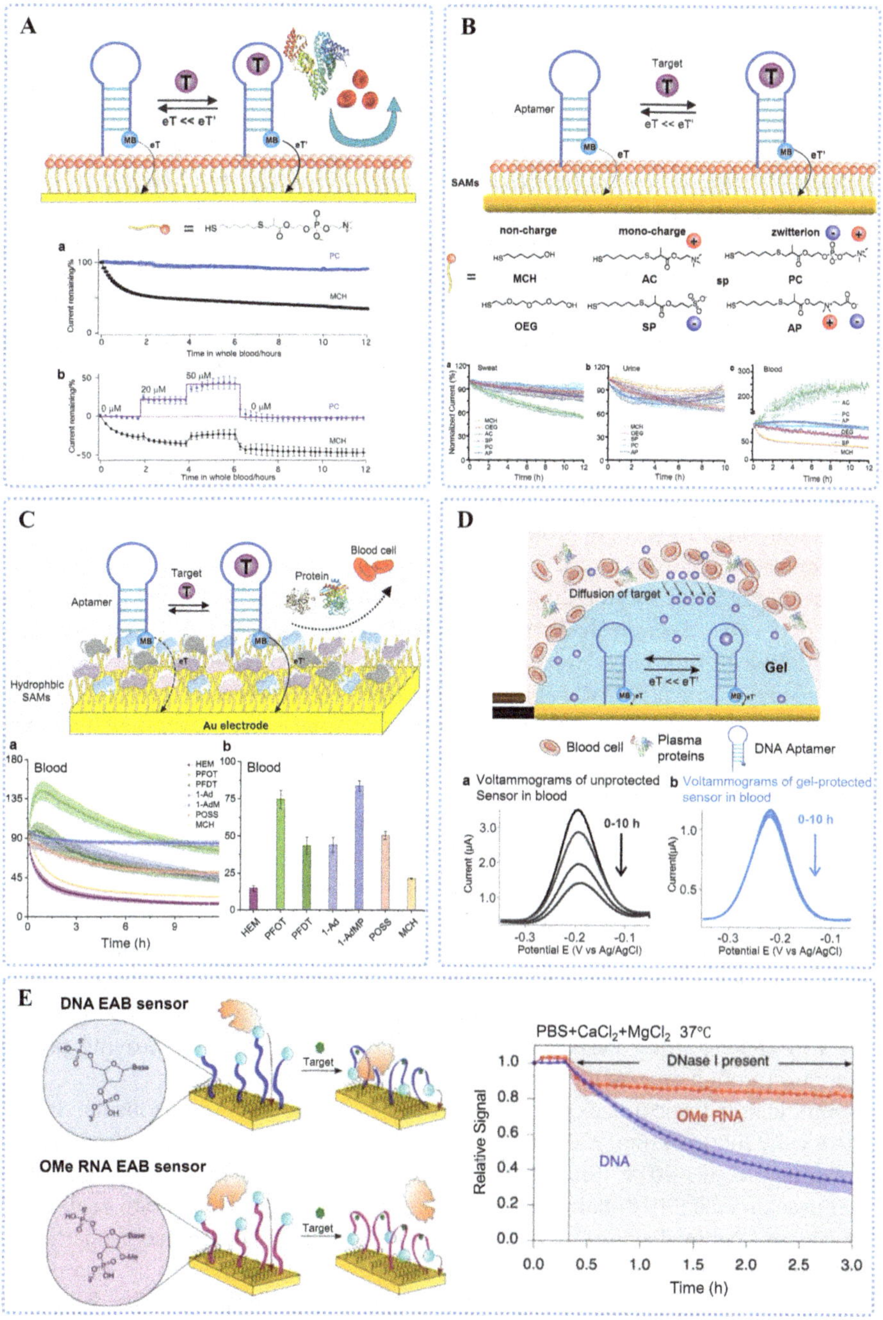

Fig. 6.4 Surface and recognition element modification. (**a**) Schematic illustration of biomimetic monolayers incorporating phosphatidylcholine head-groups used to largely eliminate the signal drift. (Reproduced with permission from Li et al. (2017a). Copyright 2017 Wiley-VCH Verlag

perfluorinated alkane derivatives 1H,1H,2H,2H-perfluorooctanethiol (PFOT), and 1H,1H,1H,2H-perfluorodecanethiol (PFDT), as well as two adamantine derivatives: 1-adamantylmercaptan (1-Ad) and (3S,5S,7S)-adamantan-1-yl 3-((2-mercaptoethyl) thio)-2-methylpropanoate (1-AdMP) with thiol and longer alkyl thiol modifications, respectively (Fig. 6.4b). To evaluate the antifouling properties of these molecules in complex biological samples such as urine, sweat, and blood, with the aim of facilitating the monitoring of doxorubicin (DOX), it was observed that the length and hydrophobic moieties of these molecules influence the wettability and antifouling properties of the self-assembled monolayer (SAM), consequently affecting the sensor performance. These approaches have shown that surface modification via immobilization of the biorecognition element could improve the sensor performance toward drug monitoring (Fig. 6.4c).

While the surface modification improves sensor performance, and stability in complex matrices, the stability of the sensor is still important for the overall sensor performance. For instance, the sensor for doxorubicin monitoring in the earlier studies had about a 40% current drop in blood. To address this issue, there has been a demonstration of hydrogel protection for DNA aptamer-based sensors to improve performance in complex matrices (Li et al. 2023). Coating the sensor with hydrogel prevents degradation by shielding the sensor from interfering elements like proteins and blood cells. The mechanism is based on preventing the diffusion of high-molecular-weight fouling elements, ensuring stability (Fig. 6.4d).

More recently, a pragmatic approach was taken by Leung et al. to optimize sensor stability when monitoring drugs in complex matrices. They leveraged the topology of the biorecognition element itself by using an unsaturated oligomer of xenonucleic acid 2′-methoxyRNA (Ome RNA) (Leung et al. 2024). This sensor was used to monitor tobramycin in blood with high affinity, improving the in vivo detection limit and overcoming degradation by maintaining stability in blood compared to the unmodified DNA aptamer. The signal degradation of the xenonucleic acid after 5 hours was 7%, compared to 48% in the DNA aptamer (Fig. 6.4e). These approaches show promise in optimizing biorecognition elements for bioelectronic sensors, enabling more effective and stable monitoring of therapeutic drugs in complex samples. Continued advancements in the design and application of these sensors will enhance their performance and reliability, ultimately improving patient outcomes through precise therapeutic drug monitoring.

Significant advancements have also been made in employing various nanomaterials for the modification of electrode surfaces used in therapeutic drug monitoring (Hatamluyi et al. 2019; Shawky and El-Tohamy 2021; Alossaimi et al. 2023; Muthusankar et al. 2018; Salandari-Jolge et al. 2020; Hojjati-Najafabadi et al. 2021; Sohrabi and Amini-Fazl 2024; Alavi-Tabari et al. 2018). The unique properties of nanomaterials, such as high surface area, excellent electroconductivity, and porosity, make them ideal candidates for enhancing sensor performance. The deposition of these nanomaterials on the surface electrode has been shown to significantly improve the sensitivity and functionality of bioelectronic sensors. For instance, Hassannezhad et al. demonstrated the application of a nanocomposite, g-C_3N_4/Fe_3O_4, for detecting tramadol (Hassannezhad et al. 2019). This sensor achieved a detection limit of 0.1 μM within a concentration range of 14.0–120 μM. The deposition of the nanocomposite material on the electrode surface facilitated electroconductivity and enhanced the oxidation of tramadol, thereby improving the sensor's performance. In another study, the electrochemical properties of Fe_3O_4 nanoparticles were utilized to form a composite with carbon nanotubes for detecting imatinib (Ghapanvari et al. 2020). This composite material leveraged the high surface area and electroconductivity of both components, resulting in a sensor capable of precise and reliable drug monitoring.

These properties make nanomaterials a valuable addition to the design of sensors capable of continuous monitoring of therapeutic drugs. The combination of nanomaterials with biological probes could result in sensors with enhanced stability, sensitivity, and specificity, suitable for real-time monitoring in complex biological matrices.

6.2.3 Calibration-Free Approach to Correct the Signaling Drift for TDM

Drift correction and calibration are essential for the clinical relevance of sensors, particularly in therapeutic drug monitoring. The fabrication of calibration-free sensors is critical, as it ensures accurate signal responses and reduces the time required to obtain pharmacokinetic information. One approach to achieve this is the dual-frequency calibration-free method (Li et al. 2017b). This method relies on the binding-induced change of the aptamer, which affects the electron transfer kinetics of the attached methylene blue redox reporter. The signal generated is proportional to the minimum current and independent of the target drug concentration, allowing the sensor output to be normalized without the need for individual calibration. The sensor was used to monitor doxorubicin and kanamycin, achieving accuracy within ±20% of the spiked target concentration (Fig. 6.5a). This method shows potential for use in electrochemical aptamer-based sensors for therapeutic drug monitoring. However, the requirement to obtain two different signals per data point limits the time resolution of the dual-frequency method. To address the time resolution issue, a more recent approach proposed the use of dual aptamers for overcoming drift correction in drug monitoring. In this method, the methylene blue redox reporter is intercalated at two different positions on identical aptamer sequences that respond differently to target molecules while exhibiting common drifts (Tsai et al. 2023). The difference in the sensor signals provides an avenue for drift correction (Fig. 6.5b). Despite demonstrating the cancellation of drift in complex biological matrices, this technique faces setbacks, including a complex screening process for aptamers. Additionally, it was reported that 10 minutes is required to capture the drift rate when the sensor is exposed to an unknown sample medium, indicating low time resolution.

Beyond using kinetic parameters (dual frequency) and biorecognition elements (dual aptamer), dual reporters have also been explored and have been shown to provide better time resolution for electrochemical aptamer sensors for drug

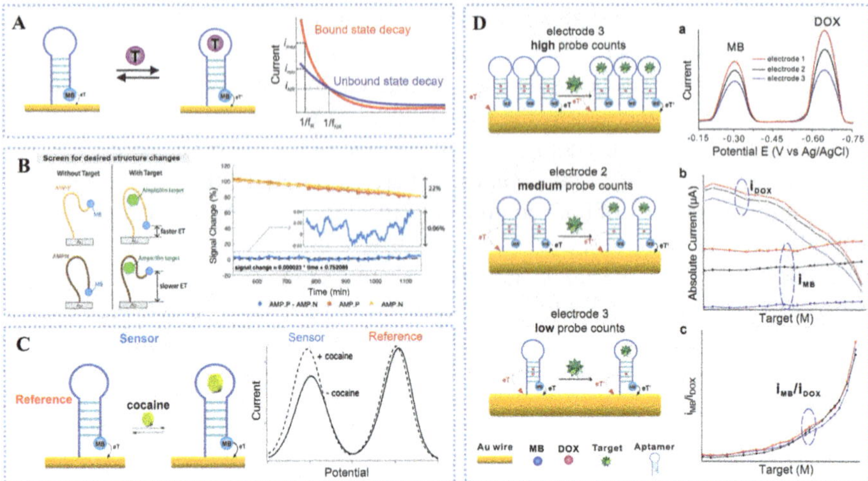

Fig. 6.5 Calibration-free sensor design. (**a**) Schematic illustration of the rapid electrochemical sensor. (Reproduced with permission from Li et al. (2017b). Copyright 2017 American Chemical Society). (**b**) Schematic illustration of drift cancellation by a dual aptamer. (Reproduced with permission from Tsai et al. (2023). Copyright 2023 American Chemical Society). (**c**) Schematic illustration of dual reporter drift correction. (Reproduced with permission from Li et al. (2016). Copyright 2016 American Chemical Society). (**d**) Schematic illustration of Calibration-free sensor using MB and DOX. (Reproduced with permission from Zhu et al. (2020). Copyright 2020 American Chemical Society)

monitoring. The concept behind the dual reporter mechanism involves incorporating two distinct redox elements into the recognition probe. In another study conducted by our group, methylene blue was intercalated to aptamer for target response and anthraquinone (AQ) for reference (Li et al. 2016). These two reporters exhibit distinct signals at different voltages, allowing for simultaneous monitoring of their current responses (Fig. 6.5c). The results demonstrated a significant reduction in baseline drift, with the electrochemical signal drift reduced by up to 50% when the sensor was probed in whole blood samples. This innovative approach not only addressed the steady drift observed during blood flow but also mitigated the dramatic signal jump encountered when transitioning to a fresh sample.

Further studies have also demonstrated the dual reporter approach by intercalating doxorubicin into the aptamer alongside methylene blue. The ratio of their signals was reproducible, allowing for the correction of signal variation without the need for a separate calibration process. The dual reporter approach has shown promise, offering a valuable option for calibration-free sensors without the need for second signal measurements or excessive process time and complexity (Fig. 6.5d) (Zhu et al. 2020). This approach provides a robust method for improving the reliability and accuracy of therapeutic drug monitoring, ensuring timely and precise drug administration in clinical settings.

6.3 The Electrochemical Biosensors for Anticancer Drug Analysis

Cancer remains one of the leading causes of mortality worldwide, prompting extensive research and development efforts to combat this devastating disease. One significant area of focus is the development of anticancer drugs, also known as antineoplastic agents, which are critical components of chemotherapy regimens. These drugs target and kill cancer cells, but their administration must be carefully monitored to avoid severe side effects and ensure therapeutic efficacy. The continuous monitoring of therapeutic drugs at the point of care using bioelectronics sensors has become an imperative aspect of modern cancer treatment. For instance, irinotecan (CPT-11), an anticancer drug belonging to the class of camptothecins, undergoes metabolic conversion to its bioactive metabolite, 7-ethyl-10-hydroxy camptothecin (SN-38) (Tartaggia et al. 2018). Monitoring the concentrations of CPT-11 and its metabolite SN-38 in patients undergoing chemotherapy is crucial to maintaining therapeutic levels and avoiding toxicity.

The voltammetry response of CPT-11 and SN-38 in organic solvents on the surface of electrodes demonstrates the possibility of direct redox activity of the drug (Bonazza et al. 2018; Novak Jovanović et al. 2017). However, the direct voltammetry approach faces several limitations. One significant limitation is interference, which can affect the sensitivity of drug monitoring. Additionally, this method requires the use of organic solvents, which may not be suitable for point-of-care bioelectronic sensors intended for therapeutic drug monitoring. These limitations have motivated researchers to design bioelectronic sensors that are highly sensitive to drug monitoring in complex matrices. These advanced bioelectronics sensors aim to provide accurate and reliable measurements in biological environments, overcoming the challenges of interference and the impracticality of organic solvents.

One notable approach involves enzymatic sensors that leverage the inhibitory properties of CPT-11 on acetylcholinesterase (AChE). AChE, an enzyme found at neuronal muscular junctions and in the central nervous system, catalyzes the hydrolysis of acetylcholine into choline and acetic acid, producing a measurable current. In the presence of CPT-11, the enzymatic activity of AChE is inhibited, leading to a reduction in current production. This change in current can be quantified, allowing for the determination of CPT-11 concentration in the patient's plasma. Alvau et al. developed an enzymatic sensor that quantifies and monitors CPT-11 within a concentration range of 10–10,000 ng/mL, which typically encompasses the levels found in patients undergoing chemotherapy (Alvau et al. 2018). This sensor exploits the fact that the presence of CPT-11 impairs the enzymatic conversion of acetylcholine, resulting in slower production of hydrogen peroxide and consequently lower current intensity (Fig. 6.6a). This method has shown promise in laboratory settings with a detection limit of 1.6 ng/mL. However, due to the complexity of the enzymatic reaction steps and other interfering factors in human plasma, further studies are required to optimize this sensor for practical clinical use.

To develop a more practical bioelectronics sensor for point-of-care monitoring of the antineoplastic drug, Arroyo-Currás and the team designed an aptamer-based

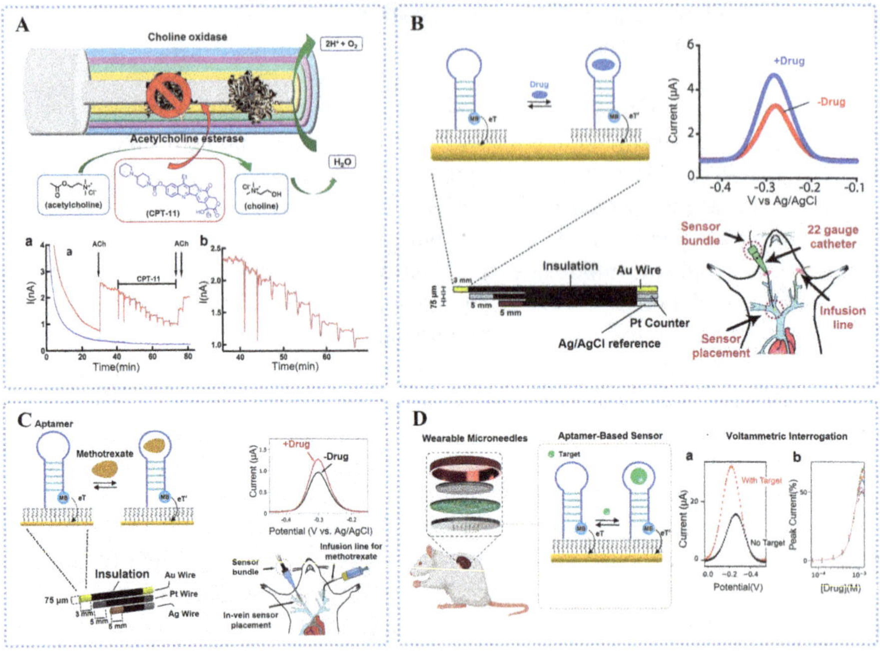

Fig. 6.6 Electrochemical bioelectronics sensor design for anticancer drug monitoring. (**a**) Schematic diagram of Ach sensor for CPT-11. (Reproduced with permission from Alvau et al. (2018). Copyright 2018 American Chemical Society). (**b**) Schematic illustration of a G-quadruplex aptamer-based sensor for irinotecan cancer drug. (Reproduced with permission from Idili et al. (2019). Copyright 2019 The Royal Society of Chemistry). (**c**) Schematic illustration of a ssDNA aptamer-based sensor for Methotrexate. (Reproduced with permission from Chamorro-Garcia et al. (2023). Copyright 2023 American Chemical Society). (**d**) Schematic illustration of wearable microneedle aptamer-based sensors for continuous, real-time therapeutic drug monitoring. (Reproduced with permission from Wu et al. (2022). Copyright 2022 Analytical Chemistry)

electrochemical sensor, considering that aptamers are short, single-stranded DNA or RNA molecules that can fold into specific three-dimensional shapes to bind tightly to their target molecules (Idili et al. 2019). A 40-base DNA sequence that forms a G-quadruplex structure in the presence of CPT-11, flanked by a 12-base pair stem was immobilized on the surface of the gold electrode. This structural change alters the redox state of a methylene blue (MB) reporter (Fig. 6.6b). This affects electron transfers and generates a measurable signal. When tested in buffer solutions, the sensor demonstrated significant signal gain. However, its performance in blood was initially poor due to protein interference, which reduced the drug concentration and signal gain. To improve performance in blood, truncating the DNA aptamer, and also introducing mismatches, improves the signal output. This approach enabled in situ monitoring of CPT-11 in the veins of live rats, with drug concentrations tracked over several hours. The sensor successfully measured plasma CPT-11 at micromolar concentrations, providing valuable pharmacokinetic data in a living organism.

A similar aptamer-based approach was employed by Chamorro-Garcia et al. to study methotrexate, another widely used anticancer drug (Chamorro-Garcia et al. 2023). Methotrexate-specific aptamers were immobilized on a gold wire electrode and the binding of methotrexate-induced conformational changes that altered the electrochemical signal. This sensor achieved clinically relevant sensitivity, with a dissociation constant of 1.05 ± 0.09 mM in blood, indicating good performance in comparison to clinical therapy. The sensor's time-resolved, real-time monitoring capabilities were demonstrated in live rats, where it was implanted in the jugular vein. Methotrexate was infused intravenously at a dose of 300 mg/kg over 180 seconds, resulting in a detectable rise in plasma levels (Fig. 6.6c). This approach provides real-time monitoring of drug pharmacokinetics, which is essential for optimizing chemotherapy regimens.

To further enhance the practicality of point-of-care bioelectric sensors for cancer patients, researchers have explored noninvasive methods of drug monitoring. Interstitial fluid (ISF), a fluid that surrounds cells and tissues, represents an ideal matrix for this purpose. Wu et al. developed a microneedle sensor that immobilizes an aptamer for monitoring CPT-11 in ISF (Fig. 6.6d) (Wu et al. 2022). The aptamer undergoes a conformational change upon binding to CPT-11, altering the electron transfer of the methylene blue reporter and leading to a measurable current change. The sensor demonstrated a strong response to both CPT-11 and its metabolite SN-38, with higher signal gain for the drug. This approach allows for continuous, noninvasive monitoring of anticancer drugs, offering significant advantages in clinical settings by reducing the need for frequent blood sampling and improving patient comfort.

Doxorubicin (DOX) is another widely used and effective anticancer drug, but when administered to patients, it can lead to side effects by attacking normal cells if not well-dosed. Moreover, DOX is often co-administered with other drugs, making continuous monitoring essential to understand drug interactions. In this context, electrochemical aptamer-based sensors have been developed for continuous DOX monitoring by our group and others. Recently, an impedimetric electrochemical approach was developed to monitor DOX (Ou et al. 2024). An aptamer was combined with a rigid tetrahedral DNA nanostructure and immobilized on a gold electrode. In the presence of DOX, a conformational change occurs, leading to the formation of a dense structure on the electrode surface. This formation impedes the ferri/ferrocyanide ions from reaching the electrode surface. Based on the charge transfer resistance, the sensor monitored DOX concentration in the range of 10.0–100.0 nM, with a detection limit of 3.0 nM.

These advanced sensor technologies represent significant progress in the continuous monitoring of anticancer drugs. They offer the potential for personalized treatment by providing real-time data on drug concentrations, enabling healthcare providers to adjust dosages promptly and optimize therapeutic outcomes. The development and implementation of these sensors in clinical practice could greatly enhance the management of chemotherapy, improving both the efficacy and safety of cancer treatments.

6.4 The Electrochemical Biosensors for Antibiotics Analysis

Antibiotics are a critical class of drugs for the treatment of antimicrobial infections. Administering these drugs should be tailored to the specific type of microbial infection to avoid exposing patients to broad-spectrum antibiotics, which can often lead to antibiotic resistance, a growing global health issue. Consequently, monitoring these drugs in patients is essential. One method to study antibiotic susceptibility involves a simple electrochemical approach.

Specifically, the retention of tobramycin (TOB) in *Pseudomonas aeruginosa*, a Gram-negative aerobic bacterium, has been studied (Clarindo Lopes et al. 2022). The therapeutic mechanism of TOB against this bacterium involves inhibiting protein biosynthesis, leading to the formation of complex polypeptides. When the bacterium is resistant, this process is hindered, providing a means to determine TOB retention time. In this study, the oxidation of TOB was demonstrated by immobilizing the drug on the surface of an electrode. TOB oxidation, due to the presence of amine (NH_2) groups, was affected by pH, with an increase in pH slightly enhancing TOB oxidation. The susceptibility of TOB to *P. aeruginosa* was evaluated by comparing the electrochemical signals from different bacterial strains: the susceptible PA01 and the resistant PA094 and PA085 strains. Results indicated that PA085 exhibited a higher degree of resistance to TOB, as evidenced by a higher current after incubation with TOB compared to PA094. In contrast, PA01 showed a significant decrease in current after incubation, indicating susceptibility to TOB (Fig. 6.7a). This approach demonstrates that antibiotic drugs can not only be monitored but also be used to distinguish between resistant and susceptible bacterial strains, aiding in the efficient management of antibiotic resistance. However, this approach requires improvement, as the current system design does not support in-person use due to the need for sample preparation and incubation time.

A versatile sensor specifically targeted toward kanamycin is imperative for rapid drug monitoring and timely feedback, which can improve the pharmacokinetics of this drug. An example of such innovation is the aptamer-based electrochemical sensor used for detecting cancer drugs, which relies on the conformational change of the aptamer in the presence of the drug target, affecting the electron transfer of the redox reporter. Arroyo-Currás et al. developed an aptamer sensor for real-time, continuous monitoring of kanamycin in the body of living rats (Arroyo-Currás et al. 2018). This sensor was enhanced by using kinetic differential measurement (KDM), which exploits the square wave frequency dependence resulting from electron transfer when the aptamer undergoes a conformational change in the presence of the target. The sensor showed a response to kanamycin, measuring concentrations ranging from 34 to 400 μM. The concentration of kanamycin determined by the sensor increased with the dosage, ranging from 5 mg/kg to 30 mg/kg over time (Fig. 6.7b). This sensor also provided insights into the pharmacokinetics of the antibiotic, demonstrating its effectiveness in a living body. The specificity of the aptamer toward the target drug makes this approach reliable, as it is not affected by endogenous factors.

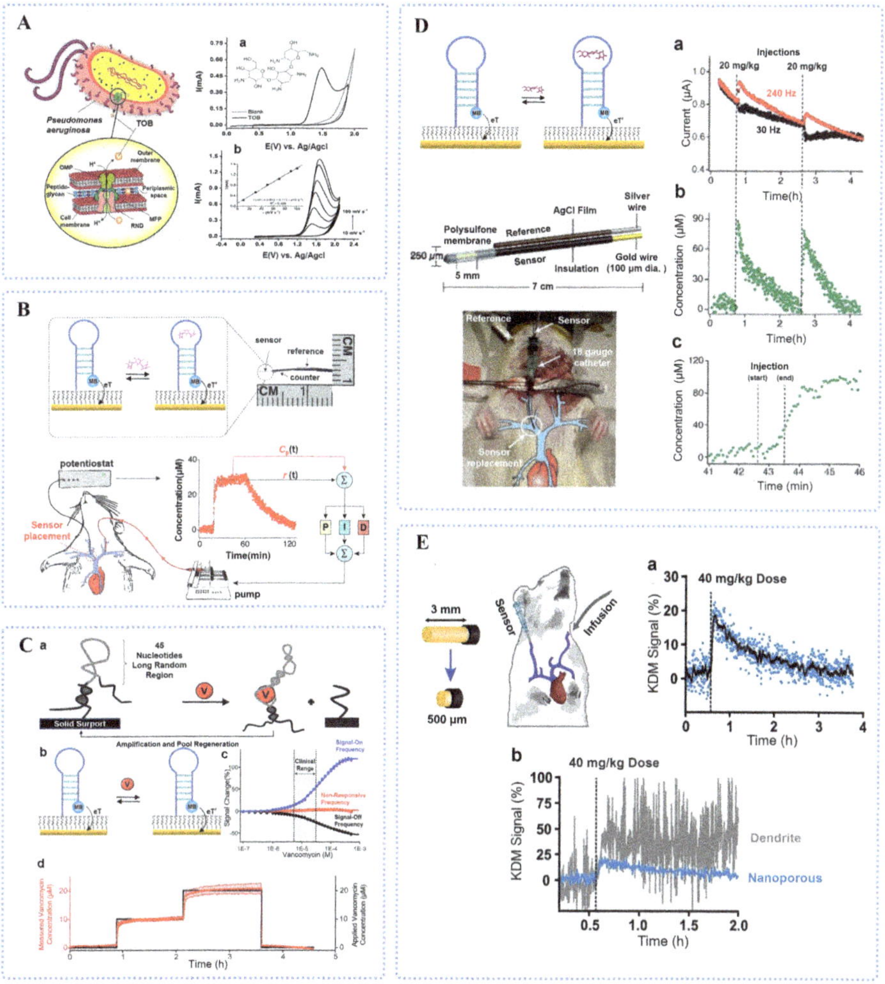

Fig. 6.7 Electrochemical bioelectronics sensor design for antibiotics drugs monitoring. (**a**) Schematic diagram of a *P. aeruginosa* bacterial cell and its efflux pump-mediated resistance mechanism. (Reproduced with permission from Clarindo Lopes et al. (2022). Copyright 2022 American Chemical Society). (**b**) Schematic illustration of closed-loop, feedback-controlled drug delivery. (Reproduced with permission from Arroyo-Currás et al. (2018). Copyright 2018 American Chemical Society). (**c**) Schematic illustration of a ssDNA aptamer-based sensor for vancomycin. (Reproduced with permission from Dauphin-Ducharme et al. (2019). Copyright 2019 American Chemical Society). (**d**) Schematic illustration of catheter electrochemical aptamer-based sensor probe in the right jugular vein of an anesthetized rat and infuse vancomycin via the left jugular vein. (Reproduced with permission from Arroyo-Currás et al. (2017). Copyright 2017 Proceedings of the National Academy of Science). (**e**) Schematic illustration of real-time continuous monitoring of tobramycin in a living body. (Reproduced with permission from Downs et al. (2021). Copyright 2021 American Chemical Society)

Building on these developments, Dauphin-Ducharme and Arroyo-Currás, along with their respective teams, utilized feedback-controlled systems for in vivo measurement of antibiotic drug concentrations (Dauphin-Ducharme et al. 2019; Arroyo-Currás et al. 2017). Dauphin-Ducharme monitored TOB using an aminoglycoside-detecting sensor inserted into the jugular vein of live rats. The dosing rate was computed using a proportional-integral-derivative (PID) controller, based on the drug concentration measured by the sensor and varying reference concentrations. This system was tested by measuring a 20-minute baseline before initiating PID-controlled infusion. The feedback-controlled system achieved sustained drug levels within ±5 μM of the specified concentration set point, allowing precise mapping of the plasma concentration of the drug. The high precision control over plasma drug levels allowed for consistent drug concentration despite significant metabolic variations until elimination. However, continuous drug delivery by infusion limits the feedback control, and time lag in the feedback control process remains an issue requiring further improvement (Fig. 6.7c, d). One aspect of development in therapeutic drug monitoring involves the matrix in which the drug is monitored.

Aside from TOB, vancomycin is another antibiotic used primarily to treat severe bacterial infections, particularly those caused by Gram-positive bacteria resistant to other antibiotics, such as methicillin-resistant *Staphylococcus aureus* (MRSA). Continuous monitoring of vancomycin levels in patients is critical to ensure therapeutic efficacy and minimize toxicity. A nanoporous electrochemical sensor capable of alloying and dealloying has been demonstrated for vancomycin detection (Downs et al. 2021). This sensor design offers significant potential for reducing sensor size for real-time monitoring. The alloying and dealloying properties increased the sensor's surface area by 100-fold, enhancing the detection signal and improving the signal-to-noise ratio. Notably, the sensor's performance was unaffected by material diffusion on the surface; exposure to 50 μM vancomycin resulted in a 90% signal within 60 seconds (Fig. 6.7e). These advancements in sensor technology for antibiotic monitoring are valuable in clinical settings, providing precise, real-time data that can improve patient management and effectively combat antibiotic resistance.

6.5 The Electrochemical Biosensors for Anti-inflammatory Drug Analysis

Analgesics are powerful therapeutic drugs used for pain relief and anesthesia during surgical operations. Unfortunately, these drugs are often subject to abuse, particularly opioids, which are commonly misused. Therefore, it is essential to develop sensors capable of continuous, close monitoring for therapeutic efficacy and to control abuse. Innovative approaches are needed to create sensors that facilitate easy and reliable detection. Electrochemical sensors have demonstrated the capacity to detect various drug classes, including opioids. Fentanyl is one of the widely used opioids, and several electrochemical sensors have been developed for its continuous monitoring. These sensors are portable, sensitive, and suitable for self-use

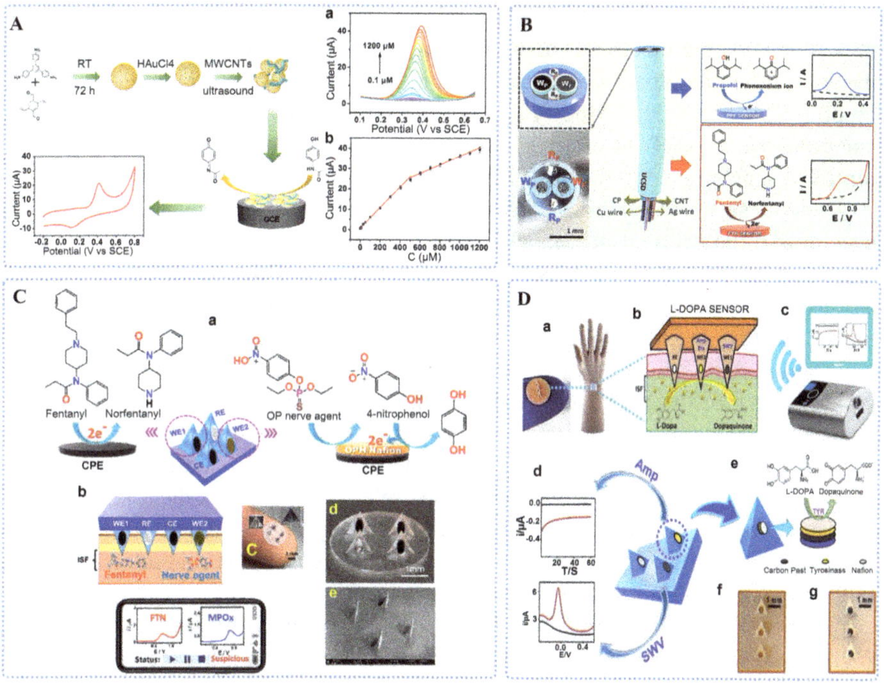

Fig. 6.8 Electrochemical bioelectronics sensor design for anti-inflammatory drug monitoring. (**a**) Schematic illustration of COF@AuNPs@MWCNTs Sensor for oxidation of AP. (Reproduced with permission from Marenco et al. (2024). Copyright 2024 American Chemical Society). (**b**) Schematic illustration of integrated dual microcatheter sensor for PPF/FTN detection. (Reproduced with permission from Moonla et al. (2020). Copyright 2020 Elsevier). (**c**) Schematic illustration of the microneedle sensor array for simultaneous minimally invasive dual-threat detection, along with wireless data transmission to the mobile device. (Reproduced with permission from Mishra et al. (2020). Copyright 2020 American Chemical Society). (**d**) Schematic illustration of three electrodes used for L-dopa sensing using SWV and amperometry. (Reproduced with permission from Goud et al. (2019). Copyright 2019 American Chemical Society)

applications. For instance, Marenco et al. demonstrated the use of carboxyl-functionalized carbon nanofiber modified on SPCE electrodes for the direct oxidation of fentanyl (Marenco et al. 2024). The oxidation process involves a two-electron-coupled two-proton dealkylation of its piperidine nitrogen, followed by hydrolysis, producing norfentanyl and phenyl acetaldehyde (Fig. 6.8a). This sensor exhibited sensitivity toward fentanyl with a detection limit of 75 nM. Despite this sensitivity in actual samples like urine, the challenge of electrode poisoning leads to current degradation, which could hinder the continuous usage of these sensors, making them impractical.

One way to overcome these limitations is by designing electrode surface modifications capable of preventing fouling and ensuring electrocatalytic performance. In this regard, a microcatheter-based dual-analyte sensor was designed for propofol and fentanyl (Moonla et al. 2020). Unlike conventional electrodes, the microcatheter contained Teflon-based microtubes patched with modified carbon paste as a working electrode, demonstrating sensitivity and good performance in plasma and human blood with detection limits of 2.18 nM for fentanyl and 4.3 μM for propofol (Fig. 6.8b). Modifying the surface of electrodes with highly porous nanocomposites, such as gold nanoparticles and organic frameworks, has also improved sensor performance in opioid monitoring.

While these approaches have shown effectiveness in designing sensors for analgesic drug monitoring, it is crucial to develop rapid and noninvasive sensors, often depending on the matrix such as interstitial fluid (ISF). This would allow point-of-care use and ensure timely responses. ISF has been demonstrated to be a suitable matrix for drug monitoring, as shown by antibiotic and other drug designs. Grant et al. reported the detection of fentanyl from ISF using a microneedle filled with a carbon paste transducer (Mishra et al. 2020). This sensor could distinguish between episodes of fentanyl overdose and nerve agent poisoning by organophosphates. The fentanyl detection mechanism is based on the two-electron oxidation of the amine group, while organophosphate detection relies on the anodic oxidation of p-nitrophenol, a product of the organophosphonate hydrolase enzymatic reaction on the carbon paste-filled microneedle (Fig. 6.8c). The sensor was characterized over a wide range of 10–200 μM and showed good sensitivity.

Another study demonstrated easy monitoring of fentanyl by fabricating an electrochemical sensor on a glove. Barfidokht et al. described a glove-based sensor for fentanyl detection, which relies on its direct electrocatalytic oxidation on fingertip-printed electrodes (Barfidokht et al. 2019). The working electrode at the index fingertip was functionalized with a nanocomposite composed of carbon nanotubes and an ionic liquid (4-(3-butyl-1-imidazolio)-1-butanesulfonate) to enhance electrochemical performance. The resulting glove opioid sensor displayed high selectivity toward fentanyl detection in the powder state, even in the presence of common cutting agents (e.g., acetaminophen, caffeine, and glucose), showing promise for on-site detection by minimizing potential false positive/negative results. The sensor performed well over a concentration range of 10–100 μM with a detection limit of 10 μM.

Acetaminophen, unlike fentanyl, is an analgesic that is widely accessible due to its over-the-counter status in many areas and its affordability, making it a common choice for patients suffering from pain and fever. Designing sensors that allow patients to self-monitor their acetaminophen intake could prevent overdosing, which is a frequent issue. While aptamer-based sensors hold potential for future development, electrochemical sensors have already been designed for this purpose. One such electrochemical sensor employs FeS_2/CoS_2 nanohybrids to monitor acetaminophen. The sensor operates through the oxidation of the drug, achieving a detection limit of 0.88 μM within a wide concentration range of 1–27 μM (Zhao and Zheng 2024).

In another study of a similar approach, a flexible microsensor based on gold dendrite nanoclusters on flexible carbon fiber was utilized. This microelectrode facilitated the oxidation of acetaminophen, achieving a detection limit of 5 nM within a broad range of 5.0–500.0 nM (Hefayathullah et al. 2024). Liu et al.

immobilized a gold nanoparticle-supported covalent organic framework on the electrode surface, coupled with carbon nanotubes, for monitoring the electrocatalytic activity of acetaminophen (Liu et al. 2024). The composite showed good adsorption capacity for acetaminophen with a wide detection range of 0.1–200 μM and a detection limit of 80 nM. These advancements in electrochemical sensor technology demonstrate the potential for effective self-monitoring of acetaminophen, helping to prevent overdosing and ensuring safer use of this common analgesic.

The development of electrochemical sensors for monitoring analgesics is crucial for both therapeutic efficacy and controlling misuse. These sensors offer portability, sensitivity, and suitability for self-use applications. Various innovative approaches have been explored to enhance the performance and reliability of these sensors.

6.6 The Electrochemical Biosensors for Other Drug Analysis

Electrochemical approaches have also been employed to monitor other therapeutic drugs, including antidepressants and antipsychotics. Although the development of aptamer-based electrochemical sensors for this class of drugs is still in its infancy, there has been significant progress in fabricating sensors for small molecules associated with the diseases treated by these drugs, such as dopamine and serotonin (Liu et al. 2022). However, for monitoring drug efficacy, the focus has been on nanomaterials. For instance, a composite sensor comprising sulfur-doped quantum dots and Fe_2O_3 immobilized on a glassy carbon electrode (GCE) has been developed to monitor olanzapine (Muthusankar et al. 2018). The intrinsic enzymatic activity of Fe_2O_3 combined with the high conductivity of the quantum dots enhances the catalytic performance of the sensor. The sensing mechanism involves the oxidation of olanzapine facilitated by hydrogen bond formation between the diazepine ring of olanzapine and the carbonyl functional group of the quantum dots. This sensor exhibits a linear response in the range of 0.1–35 μM and achieves a detection limit of 0.006 μM. Another notable example is the use of a microfluidic chip based on molecular imprinting technology for the continuous monitoring of warfarin (Liu et al. 2017b). This sensor was used to track the pharmacokinetics of the drug in the plasma of live rabbits. The drug, upon entering the microfluidic chip, alters the current due to the oxidation of warfarin, facilitated by the catalytic properties of the microfluidic system. This approach enables warfarin detection within a concentration range of 5×10^{-6} to 4.0×10^{-4} M, with a detection limit of 2.0×10^{-6} M.

Advancements have also been made in the minimally invasive monitoring of drugs through interstitial fluid (ISF). A noteworthy example is the continuous monitoring of levodopa, a drug used to elevate dopamine levels in patients with Parkinson's disease. Accurate dosing of levodopa is critical, as underdosing can exacerbate Parkinson's symptoms, anxiety, and depression while overdosing can lead to dyskinesia, psychosis, and orthostatic hypotension. An orthogonal biocatalytical microneedle sensor array has been designed for this purpose, using two working electrode systems (Goud et al. 2019). Carbon paste is packed into the hollow structure of the microneedles. The first electrode employs nonenzymatic square wave detection of levodopa, while the neighboring microneedle uses a tyrosinase-based biocatalytic detection through chronoamperometry of the corresponding dopaquinone product (Fig. 6.8d). Both electrodes exhibit high sensitivity and selectivity toward levodopa, facilitating a closed-loop feedback delivery system that adjusts the L-Dopa dose based on real-time blood levels.

6.7 The Challenges for the Development of Electrochemical Biosensors for TDM

The electrochemical biosensor has proven highly important due to its sensitivity, selectivity, ease of use, and real-time monitoring capabilities, among other advantages for therapeutic drug monitoring. Despite these advances, several challenges remain in the development and application of electrochemical sensors for therapeutic drug monitoring. These challenges and potential solutions can be summarized as follows:

1. The properties of aptamers still require further enhancement. This could be achieved through post-SELEX modifications, such as altering their topology, merging them with nanomaterials, or combining them with other types of biorecognition elements like enzymes or proteins. These enhancements could pave the way for promising recognition elements with significantly improved sensitivity and selectivity toward specific drugs.
2. Electrode fouling can lead to current degradation and reduced sensor performance over time, limiting the practical continuous use of such sensors. Although various approaches have been developed for electrode surface modifications that prevent fouling and maintain electrocatalytic performance, further research is needed to achieve practical sensors for use in human body matrices.
3. Ensuring accurate sensor calibration and addressing drift correction are crucial for reliable drug monitoring. Combining aptamers with nanomaterials could enhance the calibration and stability of electrochemical sensors for therapeutic drug monitoring.
4. Few studies have focused on therapeutic drug monitoring in interstitial fluid (ISF), which allows for minimal invasiveness compared to blood in in vivo studies. Designing electrochemical sensors for use in matrices that permit less invasive approaches, such as ISF, is essential for patients already dealing with significant pain from health issues.
5. Current sensor designs often focus on monitoring a specific type of therapeutic drug. However, certain treatments, such as cancer therapy, involve the administration of multiple drugs. Developing sensors capable of simultaneously monitoring multiple drugs can provide valuable feedback to improve treatment efficacy and prevent underdosing or overdosing in patients.

Addressing these challenges would optimize electrochemical sensors to provide more practical, accurate, and less invasive solutions for therapeutic drug monitoring.

6.8 Conclusion and Outlook

Managing numerous diseases requires the development of sensors capable of continuous monitoring of therapeutic drugs. While several methods, such as spectrofluorometric, HPLC, GCM, and mass spectrometry, have been reported for drug monitoring, future advancements in disease management necessitate the development of bioelectronics sensors. These sensors should enable real-time, continuous, point-of-care monitoring that is less invasive and cost-effective. Such capabilities allow healthcare practitioners to obtain immediate feedback on drug interactions, thereby helping patients manage their medication and improving drug administration and efficacy. Electrochemical approaches, unlike the aforementioned methods, offer the potential to develop bioelectronics sensors with these desired attributes. One approach to fabricating electrochemical-based bioelectronics sensors involves the use of nanomaterials and biorecognition elements, such as DNA aptamers. DNA aptamers provide sensitivity and selectivity for these sensors due to their ability to interact specifically with target drugs, undergoing conformational changes upon binding. Since electrochemical sensing relies on redox behavior, redox reporters are attached to DNA aptamers to report electron transfer events when the biorecognition element interacts with the drug of interest, resulting in drug concentration measurement.

This approach has been further refined through various methods, such as surface modifications, biorecognition element truncation, and the use of multiple redox reporters, to improve sensor performance and drift correction in complex matrices. Sensors developed using this approach have been employed to monitor a range of drugs, including anticancer agents, antibiotics, anti-inflammatory drugs, antipsychotics, and many others. Despite advances in developing bioelectronics sensors for therapeutic drug monitoring, further studies are needed to enhance these sensors. Future research should focus on incorporating nanomaterials, such as quantum dots, organic framework nanoparticles, and nanocomposites, with biorecognition elements (aptamers) to further enhance performance in complex matrices and enable the simultaneous monitoring of multiple drugs. Additionally, there is a need to develop bioelectronics sensors with high sensitivity to therapeutic drugs in less invasive sample matrices, such as interstitial fluid (ISF), urine, and saliva. In this chapter, we have discussed how bioelectronics sensors are designed. These sensors are designed to advance the performance and efficacy of therapeutic drugs, providing an effective approach to health monitoring and disease management. It is hoped that readers will be inspired by the ongoing development of bioelectronics sensors, which hold the promise of significantly improving patient care and health outcomes.

Acknowledgments This work was supported by the National Natural Science Foundation of China (22274144, 22474130, 22122410, 22090050), the National Key Research and Development Program of China (2021YFA1200403), the Joint NSFC-ISF Research Grant Program (Grant No: 22161142020), and the Natural Science Foundation of Shenzhen (JCYJ20220530162406014).

References

Abouzid M, Kosicka-Noworzyń K, Karaźniewicz-Łada M, Rao P, Modi N, Xie YL, Heysell SK, Główka A, Kagan L (2024) Development and validation of a Uplc-Ms/Ms method for therapeutic drug monitoring, pharmacokinetic and stability studies of first-line antituberculosis drugs in urine. Molecules 29:337

Alavi-Tabari SAR, Khalilzadeh MA, Karimi-Maleh H, Zareyee D (2018) An amplified platform nanostructure sensor for the analysis of Epirubicin in the presence of Topotecan as two important chemotherapy drugs for breast cancer therapy. New J Chem 42:3828–3832

Alossaimi MA, Elmansi H, Alajaji M, Altharawi A, Altamimi ASA, Magdy G (2023) A novel quantum dots-based fluorescent sensor for determination of the anticancer Dacomitinib: application to dosage forms. Molecules 28:2351

Alqahtani A, Alqahtani T, Gahtani RM, Ramzy S (2023) Green Spectrofluorimetric quantification of Aspirin, Olmesartan, and Metoprolol in spiked human plasma. Sci Rep 13:20182

Alvau MD, Tartaggia S, Meneghello A, Casetta B, Calia G, Serra PA, Polo F, Toffoli G (2018) Enzyme-based electrochemical biosensor for therapeutic drug monitoring of anticancer drug Irinotecan. Anal Chem 90:6012–6019

Arnold LM, Sarzi-Puttini P, Arsenault P, Khan T, Bhadra Brown P, Clair A, Scavone JM, Driscoll J, Landen J, Pauer L (2015) Efficacy and safety of Pregabalin in patients with fibromyalgia and comorbid depression taking concurrent antidepressant medication: a randomized, placebo-controlled study. J Rheumatol 42:1237–1244

Arroyo-Currás N, Somerson J, Vieira PA, Ploense KL, Kippin TE, Plaxco KW (2017) Real-time measurement of small molecules directly in awake, ambulatory animals. Proc Natl Acad Sci 114:645–650

Arroyo-Currás N, Ortega G, Copp DA, Ploense KL, Plaxco ZA, Kippin TE, Hespanha JP, Plaxco KW (2018) High-precision control of plasma drug levels using feedback-controlled dosing. ACS Pharmacol Transl Sci 1:110–118

Barfidokht A, Mishra RK, Seenivasan R, Liu S, Hubble LJ, Wang J, Hall DA (2019) Wearable electrochemical glove-based sensor for rapid and on-site detection of fentanyl. Sensors Actuators B Chem 296:126422

Bonazza G, Tartaggia S, Toffoli G, Polo F, Daniele S (2018) Voltammetric behaviour of the anticancer drug Irinotecan and its metabolites in acetonitrile. Implications for electrochemical therapeutic drug monitoring. Electrochim Acta 289:483–493

Chamorro-Garcia A, Gerson J, Flatebo C, Fetter L, Downs AM, Emmons N, Ennis HL, Milosavic N, Yang K, Stojanovic M, Ricci F, Kippin TE, Plaxco KW (2023) Real-time, seconds-resolved measurements of plasma methotrexate in situ in the living body. ACS Sens 8:150–157

Clarindo Lopes L, Lima D, Hayat M, Li Y, Kumar A, Kuss S (2022) Electrochemical quantification of Tobramycin retention in Pseudomonas Aeruginosa as antimicrobial susceptibility indicator. Anal Chem 94:12553–12558

Dauphin-Ducharme P, Yang K, Arroyo-Currás N, Ploense KL, Zhang Y, Gerson J, Kurnik M, Kippin TE, Stojanovic MN, Plaxco KW (2019) Electrochemical Aptamer-based sensors for improved therapeutic drug monitoring and high-precision, feedback-controlled drug delivery. ACS Sens 4:2832–2837

Downs AM, Gerson J, Hossain MN, Ploense K, Pham M, Kraatz HB, Kippin T, Plaxco KW (2021) Nanoporous gold for the miniaturization of in vivo electrochemical Aptamer-based sensors. ACS Sens 6:2299–2306

El-Deen AK, Radwan AS, Belal F, Magdy G (2023) Spider diagram with greenness evaluation metrics for assessing the new synchronous Spectrofluorimetric determination of Bicalutamide and resveratrol in human plasma. Luminescence 38:1996–2006

Gao Y, Guo Y, He P, Liu Z, Chen Y (2024) Enhanced sensitivity and selectivity of an electrochemical sensor for real-time Propofol monitoring in anesthesia. Alex Eng J 87:47–55

Ghapanvari M, Madrakian T, Afkhami A, Ghoorchian A (2020) A modified carbon paste electrode based on Fe3o4@Multi-Walled carbon nanotubes@Polyacrylonitrile nanofibers for determination of Imatinib anticancer drug. J Appl Electrochem 50:281–294

Goud KY, Moonla C, Mishra RK, Yu C, Narayan R, Litvan I, Wang J (2019) Wearable electrochemical microneedle sensor for continuous monitoring of Levodopa: toward Parkinson management. ACS Sens 4:2196–2204

Gutiérrez-Pizarraya A, Leone M, Garnacho-Montero J, Martin C, Martin-Loeches I (2017) Collaborative approach of individual participant data of prospective studies of de-escalation in non-immunosuppressed critically ill patients with sepsis. Expert Rev Clin Pharmacol 10:457–465

Hassannezhad M, Hosseini M, Ganjali MR, Arvand M (2019) A graphitic carbon nitride (G-C3n4/Fe3o4) nanocomposite: an efficient electrode material for the electrochemical determination of tramadol in human biological fluids. Anal Methods 11:2064–2071

Hatamluyi B, Es'haghi Z, Modarres Zahed F, Darroudi M (2019) A novel electrochemical sensor based on Gqds-Pani/Zno-Ncs modified glassy carbon electrode for simultaneous determination of Irinotecan and 5-fluorouracil in biological samples. Sensors Actuators B Chem 286:540–549

Hefayathullah M, Arivazhagan M, Maduraiveeran G (2024) Gold Nanodendrites enriched with gold nanoclusters on carbon-fiber as flexible acetaminophen sensor for clinical diagnostics. Mater Chem Phys 320:129449

Heysell S, Rao P, Zhdanova S, Ogarkov O, Orlova E, Ebers A, Stroup S, Mirawdaly S, Aartsen D, Koshkina O, Suzdalnitsky A, Moiseeva E, Dillingham R (2020) Urine Colorimetry for Levofloxacin pharmacokinetics and personalized dosing in people with drug-resistant tuberculosis. Int J Mycobacteriol 9:411

Hojjati-Najafabadi A, Salmanpour S, Sen F, Asrami PN, Mahdavian M, Khalilzadeh MA (2021) A tramadol drug electrochemical sensor amplified by biosynthesized Au nanoparticle using Mentha aquatic extract and ionic liquid. Top Catal 65:587–594

Ibrahim F, El-Deen AK, Shimizu K (2017) Application of Quinone-based fluorophore and native fluorescence for the Spectrofluorimetric determination of Agomelatine in dosage form: identification of acidic and alkaline- induced degradation products by Lc–Ms/Tof. Luminescence 33:225–231

Ibrahim F, Wahba MEK, Magdy G (2018) Analytical method development and validation of spectrofluorimetric and spectrophotometric determination of some antimicrobial drugs in their pharmaceuticals. Spectrochim Acta A Mol Biomol Spectrosc 188:525–536

Idili A, Arroyo-Currás N, Ploense KL, Csordas AT, Kuwahara M, Kippin TE, Plaxco KW (2019) Seconds-resolved pharmacokinetic measurements of the chemotherapeutic Irinotecan in situ in the living body. Chem Sci 10:8164–8170

Kang D, Ricci F, White RJ, Plaxco KW (2016) Survey of redox-active moieties for application in multiplexed electrochemical biosensors. Anal Chem 88:10452–10458

Lauschke V, Ingelman-Sundberg M (2016) The importance of patient-specific factors for hepatic drug response and toxicity. Int J Mol Sci 17:1714

Leung KK, Gerson J, Emmons N, Heemstra JM, Kippin TE, Plaxco KW (2024) The use of Xenonucleic acids significantly reduces the in vivo drift of electrochemical Aptamer-based sensors. Angew Chem Int Ed Engl 63:e202316678

Li H, Arroyo-Currás N, Kang D, Ricci F, Plaxco KW (2016) Dual-reporter drift correction to enhance the performance of electrochemical Aptamer-based sensors in whole blood. J Am Chem Soc 138:15809–15812

Li H, Dauphin-Ducharme P, Arroyo-Currás N, Tran CH, Vieira PA, Li S, Shin C, Somerson J, Kippin TE, Plaxco KW (2017a) A biomimetic phosphatidylcholine-terminated monolayer greatly improves the in vivo performance of electrochemical Aptamer-based sensors. Angew Chem Int Ed 56:7492–7495

Li H, Dauphin-Ducharme P, Ortega G, Plaxco KW (2017b) Calibration-free electrochemical biosensors supporting accurate molecular measurements directly in undiluted whole blood. J Am Chem Soc 139:11207–11213

Li H, Li S, Dai J, Li C, Zhu M, Li H, Lou X, Xia F, Plaxco KW (2019) High frequency, calibration-free molecular measurements in situ in the living body. Chem Sci 10:10843–10848

Li S, Wang Y, Zhang Z, Wang Y, Li H, Xia F (2021) Exploring End-Group effect of Alkanethiol self-assembled monolayers on electrochemical Aptamer-based sensors in biological fluids. Anal Chem 93:5849–5855

Li S, Ferrer-Ruiz A, Dai J, Ramos-Soriano J, Du X, Zhu M, Zhang W, Wang Y, Herranz MÁ, Jing L, Zhang Z, Li H, Xia F, Martín N (2022) A Ph-independent electrochemical Aptamer-based biosensor supports quantitative, real-time measurement in vivo. Chem Sci 13:8813–8820

Li S, Dai J, Zhu M, Arroyo-Curras N, Li H, Wang Y, Wang Q, Lou X, Kippin TE, Wang S, Plaxco KW, Li H, Xia F (2023) Implantable hydrogel-protective DNA Aptamer-based sensor supports accurate, continuous electrochemical analysis of drugs at multiple sites in living rats. ACS Nano 17:18525–18538

Liu VX, Fielding-Singh V, Greene JD, Baker JM, Iwashyna TJ, Bhattacharya J, Escobar GJ (2017a) The timing of early antibiotics and hospital mortality in sepsis. Am J Respir Crit Care Med 196:856–863

Liu J, Zhang Y, Jiang M, Tian L, Sun S, Zhao N, Zhao F, Li Y (2017b) Electrochemical microfluidic chip based on molecular imprinting technique applied for therapeutic drug monitoring. Biosens Bioelectron 91:714–720

Liu Y, Liu Z, Tian Y (2022) Real-time tracking of electrical signals and an accurate quantification of chemical signals with long-term stability in the live brain. Acc Chem Res 55:2821–2832

Liu Y, Yan X, Xing Y, Zhao P, Zhu Y, Li L, Liu N, Zhang Z (2024) Dispersed Au nanoparticles anchored on covalent organic frameworks/carbon nanotubes via self-reduction for electrochemical sensing of Acetaminophen. ACS Appl Nano Mater 7:4980–4988

Magdy G, Belal F, Abdel-Megied AM, Abdel Hakiem AF (2021) Two different synchronous Spectrofluorimetric approaches for simultaneous determination of Febuxostat and Ibuprofen. R Soci Open Sci 8:210354

Magdy G, Belal F, El-Deen AK (2022) Green synchronous Spectrofluorimetric method for the simultaneous determination of Agomelatine and Venlafaxine in human plasma at part per billion levels. Sci Rep 12:22559

Mahmoud AM, Mahnashi MH, Alkahtani SA, El-Wekil MM (2020) Nitrogen and sulfur co-doped graphene quantum dots/Nanocellulose Nanohybrid for electrochemical sensing of anti-Schizophrenic drug Olanzapine in pharmaceuticals and human biological fluids. Int J Biol Macromol 165:2030–2037

Marenco AJ, Pillai RG, Harris KD, Chan NWC, Jemere AB (2024) Electrochemical determination of fentanyl using carbon nanofiber-modified electrodes. ACS Omega 9:17592

Mishra RK, Goud KY, Li Z, Moonla C, Mohamed MA, Tehrani F, Teymourian H, Wang J (2020) Continuous opioid monitoring along with nerve agents on a wearable microneedle sensor array. J Am Chem Soc 142:5991–5995

Moghaddas A, Adib-Majlesi M, Sabzghabaee AM, Hajigholami A, Riechelmann R (2021) Potential drug-drug interactions in hospitalized cancer patients: a report from the middle-east. J Oncol Pharm Pract 27:46–53

Moonla C, Goud KY, Teymourian H, Tangkuaram T, Ingrande J, Suresh P, Wang J (2020) An integrated microcatheter-based dual-Analyte sensor system for simultaneous, real-time measurement of Propofol and fentanyl. Talanta 218:121205

Muthusankar G, Sangili A, Chen S-M, Karkuzhali R, Sethupathi M, Gopu G, Karthick S, Devi RK, Sengottuvelan N (2018) In situ assembly of sulfur-doped carbon quantum dots surrounded iron(Iii) oxide nanocomposite; a novel electrocatalyst for highly sensitive detection of antipsychotic drug Olanzapine. J Mol Liq 268:471–480

Novak Jovanović I, Komorsky-Lovrić Š, Lucić Vrdoljak A, Popović AR, Neuberg M (2017) Voltammetric characterisation of anticancer drug Irinotecan. Electroanalysis 30:336–344

Ou D, Yan H, Chen Z (2024) An impedance labeling free electrochemical Aptamer sensor based on tetrahedral DNA nanostructures for doxorubicin determination. Microchim Acta 191:94

Pereira FJ, Rodríguez-Cordero A, López R, Robles LC, Aller AJ (2021) Development and validation of an Rp-Hplc-Pda method for determination of paracetamol, caffeine and tramadol hydrochloride in pharmaceutical formulations. Pharmaceuticals 14:466

Qian L, Durairaj S, Prins S, Chen A (2021) Nanomaterial-based electrochemical sensors and biosensors for the detection of pharmaceutical compounds. Biosens Bioelectron 175:112836

Roberts JA, Norris R, Paterson DL, Martin JH (2011) Therapeutic drug monitoring of antimicrobials. Br J Clin Pharmacol 73:27–36

Rosado T, Oppolzer D, Cruz B, Barroso M, Varela S, Oliveira V, Leitão C, Gallardo E (2018) Development and validation of a gas chromatography/Tandem mass spectrometry method for simultaneous quantitation of several antipsychotics in human plasma and oral fluid. Rapid Commun Mass Spectrom 32:2081–2095

Rutherford SH, Hutchison CDM, Greetham GM, Parker AW, Nordon A, Baker MJ, Hunt NT (2023) Optical screening and classification of drug binding to proteins in human blood serum. Anal Chem 95:17037–17045

Salandari-Jolge N, Ensafi AA, Rezaei B (2020) A novel three-dimensional network of Cucr2o4/Cuo nanofibers for Voltammetric determination of anticancer drug methotrexate. Anal Bioanal Chem 412:2443–2453

Shawky AM, El-Tohamy MF (2021) Highly functionalized modified metal oxides polymeric sensors for potentiometric determination of Letrozole in commercial oral tablets and biosamples. Polymers 13:1384

Sohrabi N, Amini-Fazl MS, Mohammadi R (2024) Adsorption and detection Tramadol in aqueous solution by fluorescent- magnetic molecularly imprinting polymers. J Taiwan Inst Chem Eng 155:105314

Stott KE, Hope WW (2017) Therapeutic drug monitoring for invasive Mould infections and disease: pharmacokinetic and Pharmacodynamic considerations. J Antimicrob Chemother 72:i12–i18

Tartaggia S, Alvau MD, Meneghello A, Casetta B, Polo F, Toffoli G (2018) Practical Fluorimetric assay for the detection of anticancer drug Sn-38 in human plasma. J Pharm Biomed Anal 159:73–81

Temerk YM, Ibrahim HSM, Schuhmann W (2015) Square wave cathodic adsorptive stripping Voltammetric determination of the anticancer drugs Flutamide and Irinotecan in biological fluids using renewable pencil graphite electrodes. Electroanalysis 28:372–379

Teymourian H, Moonla C, Tehrani F, Vargas E, Aghavali R, Barfidokht A, Tangkuaram T, Mercier PP, Dassau E, Wang J (2019) Microneedle-based detection of Ketone bodies along with glucose and lactate: toward real-time continuous interstitial fluid monitoring of diabetic Ketosis and Ketoacidosis. Anal Chem 92:2291–2300

Teymourian H, Parrilla M, Sempionatto JR, Montiel NF, Barfidokht A, Van Echelpoel R, De Wael K, Wang J (2020) Wearable electrochemical sensors for the monitoring and screening of drugs. ACS Sens 5:2679–2700

Tsai Y-C, Weng W-Y, Yeh Y-T, Chien J-C (2023) Dual-Aptamer drift canceling techniques to improve long-term stability of real-time structure-switching Aptasensors. ACS Sens 8:3380–3388

Turković L, Bočkor L, Ekpenyong O, Silovski T, Lovrić M, Crnković S, Nigović B, Sertić M (2022) Development and validation of a novel Lc-Ms/Ms method for the simultaneous determination of Abemaciclib, Palbociclib, Ribociclib, Anastrozole, Letrozole, and Fulvestrant in plasma samples: a prerequisite for personalized breast cancer treatment. Pharmaceuticals 15:614

Vande Casteele N, Herfarth H, Katz J, Falck-Ytter Y, Singh S (2017) American Gastroenterological Association Institute technical review on the role of therapeutic drug monitoring in the management of inflammatory bowel diseases. Gastroenterology 153:835–857.e6

Wicha SG, Märtson AG, Nielsen EI, Koch BCP, Friberg LE, Alffenaar JW, Minichmayr IK (2021) From therapeutic drug monitoring to model-informed precision dosing for antibiotics. Clin Pharmacol Therap 109:928–941

Wu Y, Tehrani F, Teymourian H, Mack J, Shaver A, Reynoso M, Kavner J, Huang N, Furmidge A, Duvvuri A, Nie Y, Laffel LM, Doyle FJ, Patti M-E, Dassau E, Wang J, Arroyo-Currás N (2022) Microneedle Aptamer-based sensors for continuous, real-time therapeutic drug monitoring. Anal Chem 94:8335–8345

Yamada J, Fukui T, Yatani A, Mimura C, Fukuda K, Hazama D, Katsurada N, Nagano T, Yamamoto M, Tachihara M (2024) Impact of concurrent medications on the outcome of immunotherapy in non-small cell lung carcinoma. Thoracic Cancer 15:1228–1236

Zentner I, Modongo C, Zetola NM, Pasipanodya JG, Srivastava S, Heysell SK, Mpagama S, Schlect HP, Gumbo T, Bisson GP, Vinnard C (2018) Urine Colorimetry for therapeutic drug monitoring of pyrazinamide during tuberculosis treatment. Int J Infect Dis 68:18–23

Zhang Z, Wang Y, Mei Z, Wang Y, Li H, Li S, Xia F (2022) Incorporating hydrophobic Moieties into self-assembled monolayers to enable electrochemical Aptamer-based sensors deployed directly in a complex matrix. ACS Sens 7:2615–2624

Zhao Z, Zheng L (2024) Cobalt-generated reactive oxygen species promoted Fes2/Cos2 for electrochemical detection of acetaminophen with the widest linear range and simultaneous detection of Dopamine. J Electroanal Chem 956:118101

Zhu M, Li S, Li H, Li H, Xia F (2020) Employing an intercalated Redox Reporter in electrochemical Aptamer-based biosensors to enable calibration-free molecular measurements in undiluted serum. Anal Chem 92:12437–12441

Zhu M, Kuang Z, Xu F, Li S, Li H, Xia F (2023) Employing a Redox Reporter-modified self-assembly monolayer in electrochemical Aptamer-based sensors to enable calibration-free measurements. ACS Appl Bio Mater 6:1586–1593

Biosensors for Cardiovascular Health Monitoring

7

Wanxue Zhang, Ziyin Mei, Xinyue Guo, Shaoguang Li, Hui Li, and Fan Xia

Abstract

Cardiovascular disease (CVD) remains a leading global cause of mortality, posing a significant challenge to human health. Early detection of CVD through effective monitoring of cardiac biomarkers is critical for improving outcomes. However, conventional diagnostic approaches rely heavily on centralized laboratory procedures and large, expensive equipment, which are time-intensive and have notable limitations. Recent advancements in nanotechnology and bioelectronics have driven the development of biosensors as transformative tools for point-of-care analysis of cardiovascular biomarkers. These devices offer distinct advantages, including low cost, portability, and rapid response, making them highly suitable for decentralized healthcare settings. In this chapter, we provide a comprehensive overview of state-of-the-art biosensor technologies for CVD diagnostics. We begin with a classification of common CVD types and an introduction to key biomarkers currently used in clinical practice. The discussion highlights notable biosensor platforms designed for biomarker analysis, emphasizing their functionality and potential applications. Finally, this chapter explores the prospects of biosensors in CVD diagnostics, addressing the challenges associated with their widespread adoption. By overcoming hurdles such as sensitivity, specificity, and regulatory approval, biosensors could revolutionize CVD diagnostics, enabling accessible and timely detection of this critical disease class.

Keywords

Cardiovascular disease · Biosensor · Disease biomarker · Immunoassay · Aptamer

W. Zhang · Z. Mei · X. Guo · S. Li · H. Li (✉) · F. Xia
State Key Laboratory of Geomicrobiology and Environmental Changes, Engineering Research Center of Nano-Geomaterials of Ministry of Education, Faculty of Materials Science and Chemistry, China University of Geosciences, Wuhan, China
e-mail: lihui-chem@cug.edu.cn

K. Mahato, P. Chandra (eds.), *Nano-bioelectronics for Precision Health Monitoring*, https://doi.org/10.1007/978-981-95-2117-3_7

7.1 Introduction

Cardiovascular health is one of the crucial determinants of human life quality, directly affecting both lifespan and living standards. According to the World Health Organization, the global mortality rate due to cardiovascular disease (CVD) reached 31% in 2019 (Organization WH 2021). Many factors can induce CVD (Mozaffarian et al. 2016; Benjamin et al. 2019), including smoking, excessive alcohol consumption, stress, hyperlipidemia, hypertension, lack of exercise, and poor diet. Unfortunately, in recent years, due to changes in human lifestyles and the aging population, the incidence of CVD has been on the rise, placing additional pressure on healthcare systems worldwide (Timmis et al. 2020). Therefore, effective management and monitoring of cardiovascular diseases are of utmost importance. However, there are still challenges in current cardiovascular health monitoring (Ginsburg et al. 2005; Fihn et al. 2012). Firstly, the symptoms of early-stage CVD are not obvious, and the levels of relevant biomarkers in patients are low, making detection difficult. Additionally, traditional diagnostic methods for CVD rely on professional instruments and require healthcare personnel to operate, making continuous and real-time monitoring challenging. Biosensor technology, with its high specificity and sensitivity, and capability for real-time continuous monitoring, shows great potential in the early diagnosis and management of cardiovascular diseases (Luong et al. 2008). It can specifically identify disease-related biomarkers and convert biological information into electrical signals to detect biomarker concentrations. Meanwhile, with the continuous development of micro–nanotechnology, materials science, and information technology, biosensors are gradually evolving toward miniaturization, wearability, intelligence, and integration, making their application in home health monitoring, remote medical care, and personalized medicine possible (Guk et al. 2019). This chapter summarizes the common types of cardiovascular diseases and their related biomarkers, including troponin, C-reactive protein, natriuretic peptides, creatine kinase MB, myoglobin, and copeptin, focusing on their physiological principles and biosensor detection methods. It will compare the working range and sensitivity of different sensors for biomarker detection. Finally, it will look forward to the future development trends and challenges of biosensor technology in the field of cardiovascular health monitoring.

7.2 The Common Classifications of CVD and Biomarkers

CVDs are diverse, with common types including coronary artery disease, heart failure, arrhythmias, valvular heart disease, and stroke (Fig. 7.1) (Tsao et al. 2022; Shi et al. 2020). Traditional diagnostic methods for cardiovascular diseases include electrocardiograms (ECG), echocardiography, and imaging techniques such as CT and MRI. However, these methods have various limitations. For example, while right ventricular enlargement can indicate pressure or volume overload, imaging may not clarify the underlying cause (Voelkel et al. 2006), and early myocardial infarction patients may not show obvious abnormalities on ECG readings (Lee et al.

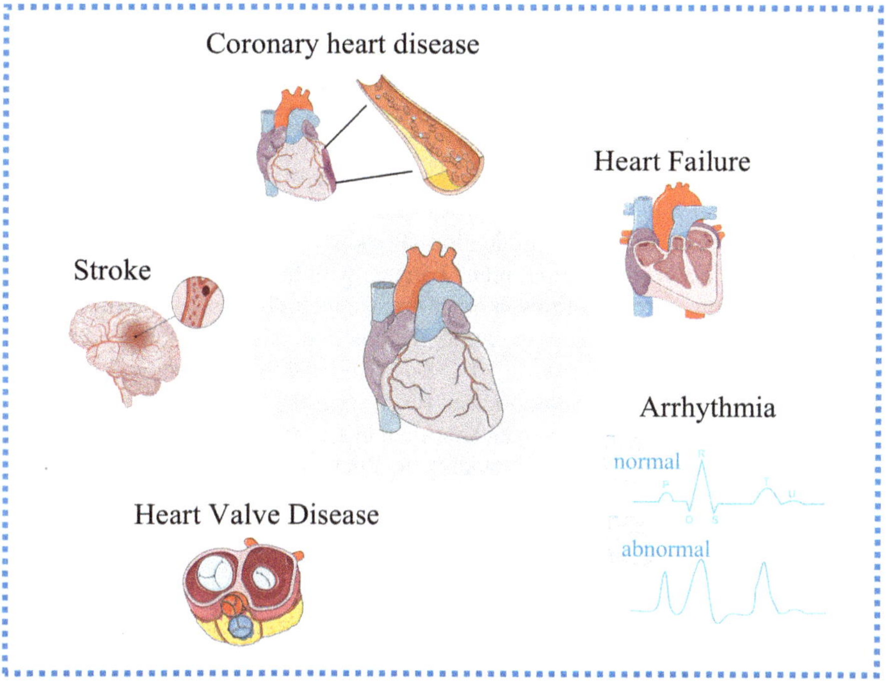

Fig. 7.1 Common types of cardiovascular disease

1989). As of the mid-twentieth century, the analysis of various CVD biomarkers has been employed as a significant, complementary diagnostic tool in clinical settings (Alpert et al. 2000). These biomarkers can be used to assess normal or pathological biological processes and responses to therapeutic interventions. Specifically, cardiac troponins, creatine kinase-MB, and myoglobin are associated with myocardial injury; BNP and NT-proBNP reflect cardiac function levels; C-reactive protein, an inflammation marker, can be employed as a complementary biomarker; and copeptin is an emerging cardiac biomarker that can be used in combination with other markers to improve diagnostic efficiency.

7.2.1 The Common Classifications of CVD

Coronary heart disease (CHD) is a common cardiovascular disease characterized by the formation of atherosclerotic plaques on the inner walls of the coronary arteries, leading to the narrowing or blockage of the vessels (Hansson 2005; Libby and Theroux 2005). This results in inadequate blood supply, causing myocardial ischemia, hypoxia, or even myocardial necrosis (Yasue et al. 2008). The main clinical manifestations include angina pectoris, silent myocardial ischemia, myocardial infarction, ischemic cardiomyopathy, and sudden death. Among these, acute myocardial infarction is a sudden event caused by the abrupt rupture of a plaque, leading to a sudden obstruction of blood flow and causing irreversible damage to the heart tissue (Boersma et al. 2003). It is the most urgent and severe form of coronary heart disease, posing a significant threat to the patient's life if not treated promptly. Heart failure (HF) is a complex clinical syndrome caused by structural and/or functional abnormalities of the heart, characterized by various symptoms and/or signs. Due to the nonspecific nature of its symptoms, the diagnosis of heart failure requires elevated levels of natriuretic peptides and/or objective evidence of pulmonary or systemic congestion on medical examination (Bozkurt et al. 2021). Typically, heart failure does not occur in isolation; many patients with heart failure also have functional and structural abnormalities in other organs, including but not limited to diabetes, dyslipidemia, lung disease, renal dysfunction, and respiratory disorders (Braunstein et al. 2003; Mcalister et al. 2004). This makes the treatment and management of heart failure quite complicated (Daubert 2004; Bleasdale and Frenneaux 2004). Arrhythmia, which refers to an abnormal heart rhythm, can interfere with the heart's normal pumping function, leading to fluctuations in blood pressure and reduced blood flow. The heart functions as a complex pump, efficiently coordinating the contraction of the atria and ventricles to circulate blood throughout the body, thereby maintaining essential life activities and various bodily functions (Keating and Sanguinetti 2001). The brain is particularly sensitive to changes in blood pressure and blood flow. Once the contraction of the ventricle is briefly interrupted for a few seconds, blood pressure will drop, leading to a loss of consciousness. If the interruption lasts for more than a few minutes, it may result in permanent brain damage or even death. Heart valves function like a "switching system" between the atria and ventricles, ensuring the correct direction of blood flow during each cardiac cycle (Taghizadeh et al. 2020). Heart valve diseases include valve insufficiency and valve stenosis (Crousillat and Wood 2019; Topilsky 2018). Valve insufficiency occurs when the valve leaflets do not close completely, causing blood to leak or flow backward into the originating chamber. Valve stenosis results from stiffening and reduced elasticity of the valve, limiting its opening and thereby reducing the flow of blood through the valve. Heart valve diseases can be congenital or develop due to various risk factors over time.

Stroke (also known as cerebral apoplexy) is an acute medical event caused by an interruption in the blood supply to the brain, leading to changes or damage in brain tissue. It can be classified into ischemic and hemorrhagic stroke. Ischemic stroke is the most common type, accounting for approximately 70% of all cases (Simats and Liesz 2022). It occurs due to thrombosis (a blood clot in the cerebral blood vessels), embolism (a clot formed elsewhere in the vascular system that travels to the brain and causes an obstruction), or a significant drop in blood pressure that results in insufficient or interrupted blood supply to the brain (Girnar and Mahajan 2021). Hemorrhagic stroke includes intracerebral hemorrhage and subarachnoid hemorrhage. It occurs when a blood vessel within the brain or between the brain and the skull ruptures, causing blood to flow into the brain parenchyma or into the subarachnoid space outside the skull (Feigin et al. 2009).

7.2.2 The Common Classifications of Biomarkers

1. Troponin

Cardiac troponin (cTn) is a complex protein found in myocardial cells, which plays a crucial role in regulating calcium-mediated muscle contraction in both skeletal and cardiac muscles (Takeda et al. 2003). The troponin complex consists of three subunits: troponin I (cTnI), troponin T (cTnT), and troponin C (cTnC). While cTnC is present in both myocardial and skeletal muscle, cTnI and cTnT are solely present in cardiac muscle. Thus, the latter two are recognized as golden-standard biomarker for myocardial injury (Thygesen et al. 2018). cTn can be further subdivided into conventional cTn and hypersensitive cTn. High-sensitivity cardiac troponin T (hs-cTnT) is more associated with heart failure caused by renal insufficiency and is a reliable biomarker for evaluating systemic cardiovascular disease risk and all-cause mortality (Dong et al. 2024). On the other hand, high-sensitivity cardiac troponin I (hs-cTnI) is more sensitive and reliable in diagnosing and prognosing coronary artery disease, particularly myocardial infarction (AMI). However, certain interference factors must be considered when detecting cTn. For instance, cTnI may bind with autoantibodies, delaying clearance and potentially causing false positives (Li et al. 2021). Additionally, hemolysis may lead to false-negative results in hs-cTnT testing (Jing et al. 2021).

2. Creatine Kinase MB

Creatine Kinase-MB (CK-MB), a subtype of creatine kinase, is primarily distributed in the myocardium (Zhong et al. 2023), with very small amounts present in skeletal muscle. It was introduced in 1965 as a biochemical marker for assessing myocardial injury, making it one of the earliest markers in this field (Jaffe 1991). In healthy individuals, serum levels of CK are maintained within a very low range, but when myocardial injury occurs, CK levels in the serum become abnormal 3–4 hours after the injury, peaking at 18–24 hours (Morrow et al. 2007a). Although the discovery of cTn has quickly replaced CK-MB as the gold standard, CK-MB's rapid response and good sensitivity still provide it with clinical value in the diagnosis and prognosis of cardiovascular diseases (Saenger and Jaffe 2008). It is often used in conjunction with other cardiac markers to improve diagnostic accuracy and reliability (Dong et al. 2024).

3. Myoglobin

Myoglobin (Myo) is a small molecular weight protein primarily found in cardiac and skeletal muscle. Its function is to bind, transport, and store oxygen. Structurally, myoglobin consists of a polypeptide chain backbone composed of amino acids and a heme prosthetic group (Hendgen-Cotta et al. 2010). During myocardial injury or necrosis, Myo is rapidly released into the bloodstream, rising within 2–3 hours, peaking at around 6–9 hours, and returning to baseline within 24 hours (Negandary

2020). This early release characteristic is likely related to Myo's small size and cytoplasmic location. As an early detection biomarker for acute myocardial infarction (AMI), myoglobin has a high negative predictive value (Lewandrowski et al. 2002). However, because it is also present in skeletal muscle, its concentration can increase in conditions such as renal failure, trauma, and inflammatory myopathies, leading to poor specificity in the clinical detection of cardiovascular diseases (Hu et al. 2023). Therefore, Myo is recognized as a complementary biomarker to others.

4. B-Type Natriuretic Peptide (BNP) and NT-proBNP

BNP is a neurohormone synthesized and secreted by ventricular myocytes. Within myocytes, the precursor protein proBNP is cleaved at the C-terminal end to generate and release biologically active BNP and the inactive NT-proBNP (N-terminal pro-B-type natriuretic peptide). Due to its high specificity ($\approx$98%) in differentiating left ventricular dysfunction (Dong et al. 2024), BNP and NT-proBNP are widely recognized as powerful biomarkers for HF. According to the literature, plasma BNP levels in heart failure patients are greater than 100 pg/mL (Maisel et al. 2002). Persistent elevation of BNP concentration in patients with myocardial infarction may indicate ventricular remodeling, with its concentration closely related to various cardiac function indices. Therefore, BNP and NT-proBNP are also recommended as biomarkers for ACS, particularly for early risk stratification and prognosis assessment of patients (Amsterdam et al. 2014). Moreover, changes in BNP levels have been observed in stroke-heart syndrome, with NT-proBNP closely associated with ischemic stroke and included in atrial fibrillation stroke risk assessments (Hijazi et al. 2016). Regarding the specific role of BNP in pulmonary arterial hypertension and right ventricular dysfunction, current research is still in the exploratory stage. Although numerous studies have indicated the indicative role of BNP in right ventricular dysfunction, using a multiparameter risk assessment method is considered more reliable for a comprehensive evaluation of the patient's condition (Yap et al. 2004; Tulevski et al. 2001).

5. C-Reactive Protein

C-reactive protein (CRP) is a nonspecific acute phase biomarker, composed of five identical polypeptide chains connected by noncovalent bonds (Black et al. 2004). Under normal physiological conditions, CRP exists as a trace protein in plasma, maintaining a very low concentration (approximately 0.8 mg/L). However, during acute events such as myocardial infection or tissue injury, CRP plasma levels significantly increase, becoming a sensitive indicator of the body's inflammatory response. Its expression is precisely regulated by various cytokines, such as interleukin-6 (IL-6), interleukin-1 (IL-1), and tumor necrosis factor-alpha (TNF-α) (Battistoni et al. 2012). In cardiovascular diseases, CRP serves as a risk predictor and prognostic indicator. Coronary heart disease can be classified into three levels of risks with regarding to the concentrations of CRP: low risk (<1 μg/mL), moderate risk (1–3 μg/mL), and high risk (>3 μg/mL) (Morrow et al. 2007b). Thus, high-sensitivity CRP (hs-CRP) has a more significant relationship with the risk of cardiovascular events and can serve as an important risk predictor (Ridker 2004). Additionally, multiple prospective studies have shown that hs-CRP holds substantial prognostic significance for patients with heart failure (Prokopidis et al. 2024; Lakhani et al. 2021).

6. Copeptin

Copeptin, an arginine vasopressin (AVP)-associated glycopeptide, is the C-terminal part of proAVP (Morgenthaler et al. 2008). AVP has clinical value in maintaining vascular tone and fluid balance, but measuring mature AVP is challenging. Consequently, copeptin has emerged as a surrogate marker for AVP. With a longer half-life in the blood, copeptin is easier to quantify and has thus garnered increasing clinical attention. A series of observational studies have found that copeptin levels significantly rise in the early stages of acute myocardial infarction (AMI). Although its specificity is low (e.g., elevated levels in chronic obstructive pulmonary disease) (Reichlin et al. 2009a), its clinical value when used in conjunction with cTn is substantial. The combined detection of copeptin with conventional and high-sensitivity cardiac troponin (cTn) is an excellent marker for ruling out AMI, and its high negative predictive value improves diagnostic efficiency (Möckel et al. 2015; Roffi et al. 2016) (Table 7.1).

7.3 Biosensors for CVD Biomarkers Analysis

Biosensors are analytical devices used for monitoring and quantifying specific target molecules in biological systems. They consist of three core components: a bioreceptor, which acts as a recognition element that can specifically bind to the target molecule; a transducer, which converts the biochemical signal into a measurable physical signal (such as electrochemical or optical signals); and a signal processing unit, which receives and processes these signals, converting them into data that can be directly analyzed or further processed. For the detection of cardiovascular diseases, the primary bioreceptors used are antibodies and aptamers. Antibodies, being the most commonly used and accounting for 70% (Fig. 7.2), utilize molecular affinity to target specific analytes. However, their purification process is complex, often requiring multiple steps, making it not cost-effective (Derin and Inci 2022). Aptamers, on the other hand, are single-stranded oligonucleotides that exhibit high selectivity and affinity for their targets. They can be obtained through multiple rounds of screening and counterscreening using the Systematic Evolution of Ligands by Exponential Enrichment (SELEX) technology. Although this process is time-consuming, the actual synthesis time for aptamers is short, and they are easily chemically modifiable (Crivianu-Gaita and Thompson 2016). To date, these two elements have been widely incorporated into various types of sensor designs to develop high-performance sensors for cardiovascular diseases, including electrochemical, colorimetric, surface-enhanced Raman spectroscopy, fluorescence assay,

Table 7.1 Several common cardiac biomarkers

Cardiac biomarker	CVD indicator	Initial elevation	Cut-off concentration and range	References
Troponin	MI, HF	4–6 h	cTnI 0.1 ng/mL cTnT 14 ng/mL	Hjort et al. (2018), Giménez et al. (2016), Qureshi et al. (2012)
CK-MB	Re-infarction, MI (early detection)	4–6 h	10 ng/mL	Qureshi et al. (2012), Friess and Stark (2009)
MYO	MI (early detection)	1 h	70–200 ng/mL	Mccord et al. (2001), Mair et al. (1992)
BNP	ACS, HF	/	9–142 ng/mL	Di Angelantonio et al. (2009, Thygesen et al. (2012)
NT-proBNP	ACS, HF	/	30–750 ng/mL	Di Angelantonio et al. (2009), Thygesen et al. (2012)
CRP	Inflammation	4–6 h	Low CVDs risk (<1 μg/mL); medium CVDs risk (1–3 μg/mL); and high CVDs risk (>3 μg/mL)	Yang et al. (2009)
Copeptin	MI, HF	/	0.04–0.12 ng/mL	Potocki et al. (2010), Reichlin et al. (2009b)

and chemiluminescent sensors. The following focuses on the detection of six cardiovascular disease biomarkers using biosensors.

7.3.1 Biosensors for Troponin

Due to the high specificity of cTnI and cTnT in the diagnosis of cardiovascular diseases, various types of biosensors have been designed for the detection of these two proteins. Electrochemical biosensors immobilize recognition probes on the electrode surface and quantify target molecules by detecting the changes in current, potential, or impedance before and after the target-probe binding. Li et al. successfully constructed a linear relationship between the peak current and target concentration by recording the voltammetric signals of the sensor in response to the target, achieving highly sensitive detection of cTnI with a detection limit as low as 0.92 fg/mL (Fig. 7.3a) (Li et al. 2024). Specifically, the study immobilized primary capture antibodies on a glassy carbon electrode and used a Fe(III)-2-aminoterephthalic acid (Fe-ATA) metal-organic framework (MOF) as a nanocarrier to load secondary aptamers and dopamine (DA) redox indicator molecules through amide reactions.

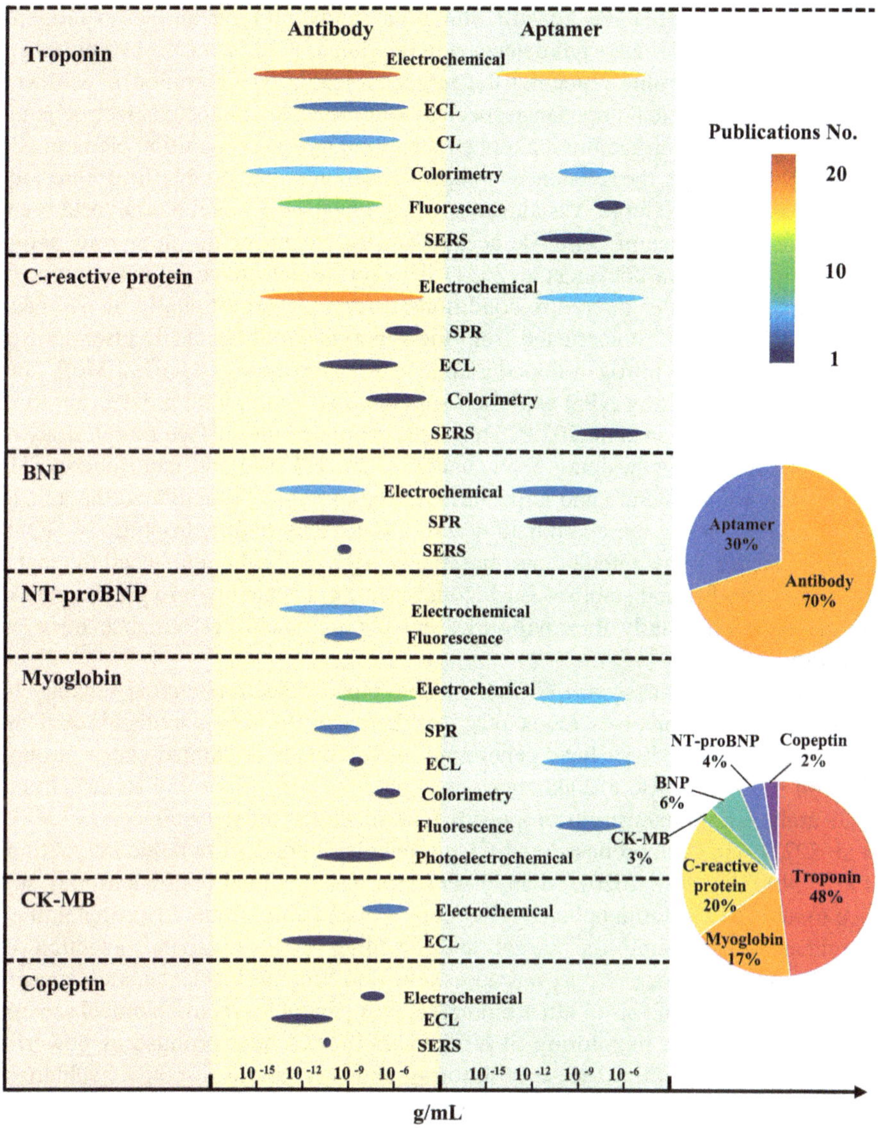

Fig. 7.2 This figure summarizes various sensor technologies based on cardiovascular disease biomarker analysis (with troponin studies only covering major publications from the last four years). Each technology has a specific label, with the color and size representing the number of reported publications and the detection range achieved by each sensor platform, respectively. From top to bottom, the first pie chart illustrates the statistics of recognition elements, with antibodies being the predominant biomolecules. The second pie chart shows the proportion of all biomarkers, with troponin being the most studied

This approach enabled the construction of a sandwich-type antibody–antigen–aptamer sensor, achieving cTnI detection in the range of 10 fg/mL to 100 ng/mL.

The impedance value reflecting the target molecule concentration is achieved through electrochemical impedance spectroscopy (EIS), an unlabeled electrochemical method. When cardiac biomarkers are captured by the recognition elements on the electrode surface, the resistance or capacitance within the double-layer interface is modulated. For example, Vasudevan et al. synthesized a novel nanohybrid combining the advantages of cellulose acetate and MoS_2, providing numerous active sites for aptamers (Vasudevan et al. 2021). This sensor detected cTnI in the range of 10 fM to 1 nM under nonredox conditions using EIS. Additionally, it has been reported that a sensor constructed from the synergistic effects of 3D laser-scribed graphene (3D LSG), nitrogen-doped graphene quantum dots (N-GQDs), MoS_2, and silver nanoparticles (Ag NPs) exhibited superior sensitivity, with a detection limit of 100 aM (Vasudevan et al. 2023). This study regulated the surface morphology of MoS_2 precursors, synthesizing MoS_2 nanoflowers (H3 NF) and nanospheres (H3 NS). The porous structure and large surface area of 3D LSG facilitated the attachment of MoS_2, while the combination with low-toxicity and hydrophilic N-GQDs through van der Waals interactions, and finally Ag NPs, further enhanced the material's dispersibility and stability. Each component of this nanohybrid plays a role in biosensing, significantly improving the sensor's analytical performance for cTnI detection.

Electrochemiluminescence (ECL) is also a type of electrochemical sensing. Its working principle involves redox reactions between the target molecule and the probe on the electrode surface, generating intermediates or excited states through electron transfer. These excited states emit photons when returning to the ground state, and the light intensity is proportional to the target molecule concentration (Li et al. 2023). Yin et al. proposed an ECL strategy based on hydrophobic interactions (Fig. 7.3b) (Yin et al. 2020), where cholesterol was modified on the antibody surface to attract ECL luminophores ($Ru(dcbpy)_3^{2+}$) and coreactants (tripropylamine), significantly enhancing the signal and enabling high-sensitivity detection of cTnI. This strategy is reusable, has adjustable detection performance, and is easily extendable to the analysis of other biomarkers. Implantable electrochemical sensors capable of real-time monitoring of biomarkers in vivo hold promise as powerful tools for cardiovascular disease monitoring and therapeutic response evaluation. The inverted molecular pendulum model provides a universal design approach for implantable sensors to directly monitor physiologically relevant proteins in biological fluids (Das et al. 2021). As shown in Fig. 7.3c, double-stranded DNA serves as the pendulum rod, a specific recognition antibody acts as the spherical pendulum

Fig. 7.3 (continued) from Elsevier). (**b**) Target-induced self-enrichment based on a hydrophobic interaction, significantly enhances the ECL signal for ultrasensitive detection of cTnI. (Reprinted with permission from Yin et al. (2020). Copyright 2024 with permission from American Chemical Society). (**c**) In vivo animal heart biomarker monitoring based on the inverted molecular pendulum model. (Reprinted with permission from Das et al. (2021). Copyright 2024 with permission from Springer Nature)

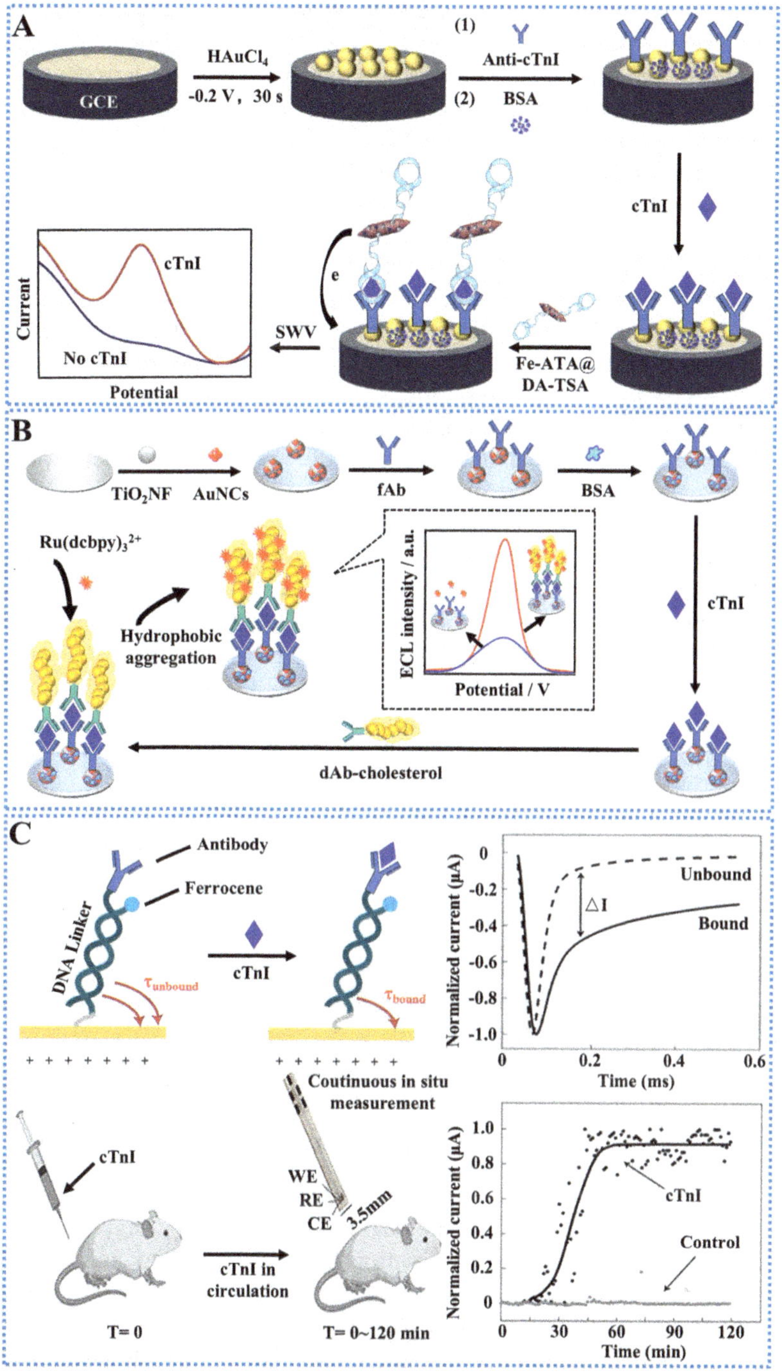

Fig. 7.3 (**a**) Principle of the Fe-ATA@DA-TSA sandwich-type electrochemical biosensor for cTnI detection. (Reprinted with permission from Li et al. (2024). Copyright 2024 with permission

bob to capture the target analyte, and ferrocene attached to the double strand serves as the indicator molecule. Changes in the pendulum's oscillation mode before and after capturing the analyte alter the electron transfer kinetics of ferrocene, which is measured through chronoamperometry. This achieved in situ detection of cTnI in live rats, with a detection limit of 1 pg/mL.

In addition to electrochemical methods, optical techniques are also common for detecting cardiovascular disease biomarkers, including fluorescence assays, chemiluminescence, colorimetry, and surface-enhanced Raman spectroscopy (SERS). Fluorescent sensors use fluorescence signals to detect and quantify target molecules, and their performance relies on the design of the fluorescent probes. They are often combined with nanotechnology to improve detection performance. For instance, when combined with quantum dots, the sensors exhibit significantly enhanced fluorescence brightness and stability compared to traditional organic dyes. In the study by Liu et al., a fluorescent probe composed of composite magnetic nanoparticles and quantum dots, with flower-shaped magnetic fluorescent nanoparticles as the shell and surface-modified carboxyl molecules coupled with cTnI antibodies, was developed (Liu et al. 2024). In fluorescence mode, the detection limit for cTnI was 0.0097 ng/mL. Recently, a sandwich immunosensor based on lanthanide-ion-doped upconversion nanoparticles (UCNPs) fluorescent probe was developed (Fig. 7.4a) for the detection of cTnI in serum (Tseng et al. 2024). They used a resonant waveguide grating (RWG) substrate coated with a carboxymethyl dextran layer, where the primary antibody captures the target molecule, binding UCNPs to form a sandwich structure. The RWG substrate can excite the resonance-enhanced upconversion luminescence (UCL) of UCNPs, greatly improving the sensor's sensitivity (LOD of 0.24 fg/mL).

Unlike fluorescence detection, colorimetric sensors require a chromogenic marker, where the color intensity indicates the target concentration. Zeng et al. developed an enhanced colorimetric aptamer sensing platform that successfully detected cTnI in serum samples within 65 minutes (Zeng et al. 2023). The specific principle is shown in Fig. 7.4c. A tetrahedral structure with three vertices modified with cTnI aptamers is constructed to achieve specific recognition. This structure is combined with Fe_3O_4-Au magnetic nanoparticles as capture probes. Simultaneously, it is loaded onto MIL-101(Fe) nanoparticles labeled with HRP to form signal probes. The two probes and cTnI are incubated together. After magnetic separation, the residual signal probe in the supernatant catalyzes the TMB-H_2O_2 solution, resulting in a visible color change. Quantitative analysis can also be performed using a UV spectrophotometer.

Colorimetry can be combined with other signal detection modes. For example, in the electrochemical-colorimetric dual-mode sensing strategy (Zhang et al. 2020), Zhang et al. used aptamer-functionalized Fe^{3+}-polydopamine (Apt@Fe^{3+}-PDA) as a reaction-triggered beacon. Under acidic conditions, the beacon decomposes, releasing a large amount of Fe^{3+}, which then converts to the dual-attribute product Prussian Blue (PB). On the one hand, cTnI can be quantified by measuring the electrochemical response of PB. On the other hand, different amounts of PB produce color changes when $K_3[Fe(CN)_6]$ is added, enabling visual detection of

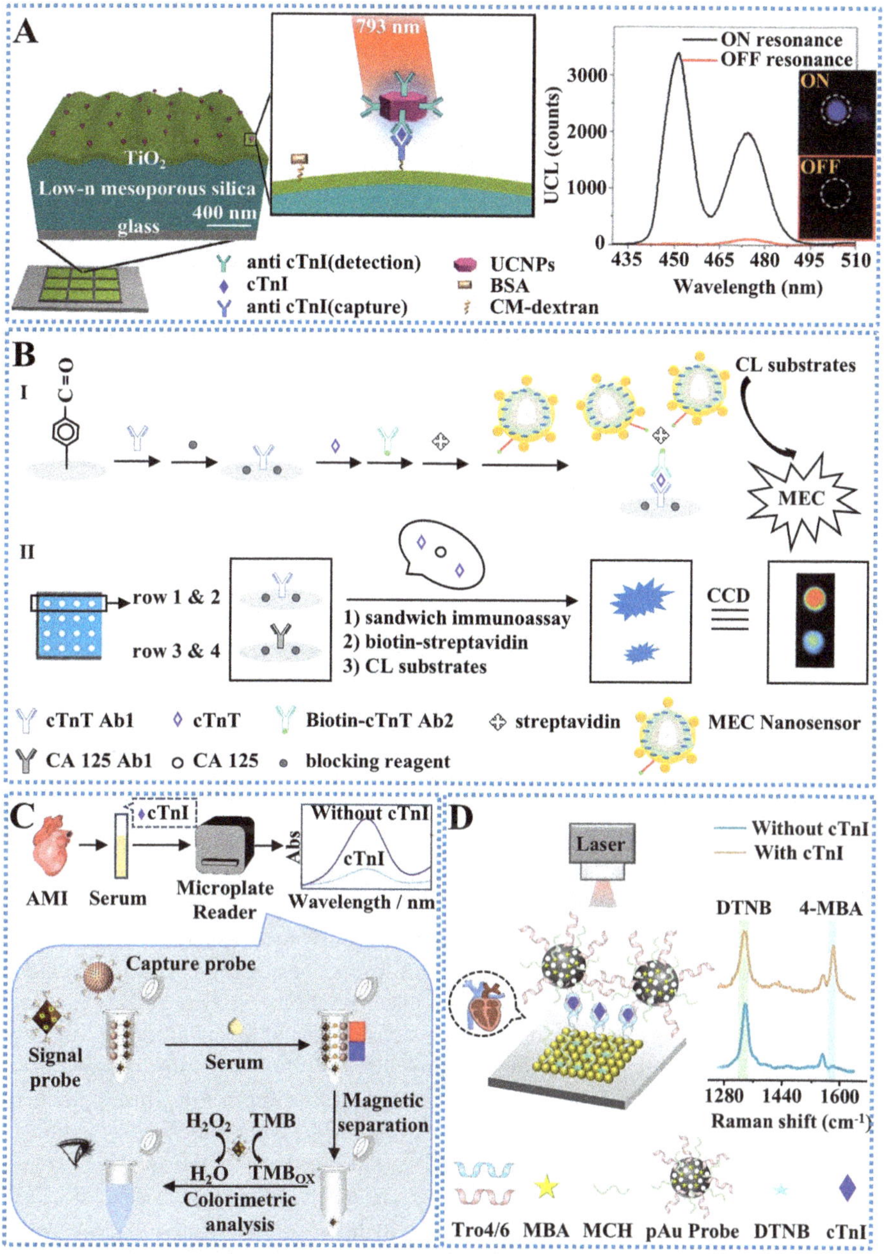

Fig. 7.4 (**a**) Design of the RWG-UCNPs sandwich immunosensor for detecting cTnI. (Reprinted with permission from Liu et al. (2024). Copyright 2024 with permission from American Chemical Society). (**b**) Sensing procedure for metal-enhanced chemiluminescence and the preparation and imaging of the protein immunoarray. (Reprinted with permission from Zong et al. (2021). Copyright 2024 with permission from Elsevier). (**c**) Anisotropic aptamer-mediated colorimetric detection platform. (Reprinted with permission from Zeng et al. (2023). Copyright 2024 with permission from Elsevier). (**d**) Schematic diagram of the ratiometric aptasensor based on SERS. (Reprinted with permission from Wang et al. (2024). Copyright 2024 with permission from Elsevier)

cTnI. Under optimal experimental conditions, the sensing strategy maintained a linear response in the range of 1.0×10^{-2} to 1.0×10^{-3} ng/mL, with detection limits of 7.4 pg/mL and 3.2 pg/mL for the colorimetric and electrochemical modes, respectively.

Chemical luminescence (CL) technology has gained widespread attention due to its characteristics such as not requiring an external light source for excitation, absence of background interference, and fast response speed. The luminol-H_2O_2 reaction is a classic luminescent reaction known for its high stability and nontoxicity. Catalysts are often incorporated into sensors to enhance the CL signal. For example, Ouyang et al. prepared a Co single-atom catalyst (Co SACs) supported on a metal-organic framework, which can promote the decomposition of H_2O_2 to generate abundant singlet oxygen, significantly increasing the CL release from the luminol-H_2O_2 reaction by approximately 1349 times (Ouyang et al. 2020). The immunosensor using Co SACs achieved a detection limit of 3.3 pg/mL. Metal nanoparticles are also widely used in CL sensing platforms due to their strong catalytic performance. Zong et al. reported an ultrasensitive nanosensor based on functional DNA dendrimers conjugated with silver nanoparticles (Fig. 7.4b) (Zong et al. 2021). This sensor comprises three modules: a functional DNA dendrimer scaffold, a G-quadruplex/hemin DNAzyme signal reporting element, and AgNPs as a chemiluminescence enhancer. It achieved a 9-fold CL intensity amplification and a 6.4-fold metal enhancement, with a response range for cardiac troponin T spanning over 5 orders of magnitude and a detection limit as low as 5×10^{-5} ng/L.

Surface-enhanced Raman scattering (SERS) is an ultrasensitive molecular spectroscopy technique that has garnered extensive attention in analytical and imaging fields due to its excellent sensitivity, label-free recognition characteristics, and rapid response capability. It primarily relies on nanomaterials as plasmonic signal amplifiers to enhance Raman signal detection efficiency. For highly sensitive detection of cTnI, researchers have explored various SERS tags. Wang et al. constructed a ratiometric aptasensor based on surface-enhanced Raman scattering (SERS) for detecting cTnI (Wang et al. 2024). The principle, as shown in Fig. 7.4d, involves immobilizing aptamers on a gold nanoparticle array using a self-assembly strategy to form the capture substrate. They introduced 5,5′-dithiobis-(2-nitrobenzoic acid) (DTNB) as an internal reference signal tag for signal self-calibration. Simultaneously, another aptamer and 4-mercaptobenzoic acid (4-MBA) were anchored on a porous gold surface to form the SERS signal tag. In the presence of the target cTnI, the capture substrate, cTnI, and the signal tag form a typical sandwich structure. The ratio of the SERS peak intensities of $I_{4\text{-}MBA}$ to I_{DTNB} can be used for quantitative analysis, allowing detection of cTnI in the range of 0.001–100 ng/mL. Wen et al. developed a reusable microfluidic biosensing platform for detecting cTnI in the serum of cardiovascular disease patients using peptide-coated gold nanostar "antennas" (GNS) to construct a GNS-SERS barcode (Wen et al. 2020). The limit of detection (LOD) achieved was 0.0055 ng/mL.

7.3.2 Biosensors for C-Reactive Protein

Although CRP has low specificity in the diagnosis of CVD, it plays a crucial role in the pathophysiology of CVD and is widely recognized as an effective marker for the early identification of cardiovascular diseases and therapeutic intervention. As a typical predictor of inflammatory response, CRP helps clinicians develop preventive measures to reduce mortality caused by cardiovascular diseases. Various types of biosensing methods have been developed to detect CRP concentration levels in CVD. Gupta et al. proposed a nonenzymatic colorimetric immunoassay platform (Fig. 7.5a) using nanozymes as alternatives to natural enzymes (Gupta et al. 2021). Citrate-functionalized ultrafine bimetallic PtRu alloy nanoparticles (CA@PtRu ANPs) were generated from platinum and ruthenium metal precursors in a 1:1 ratio as enzyme mimics, which significantly amplified the color signal. Amine-functionalized silica-coated magnetic beads (APTES/SiO_2@Fe_3O_4) were used as CRP recognition and capture probes. When combined with CRP in the presence of TMB and H_2O_2, a blue visual signal was produced, with a detection range of 0.01–180 μg/mL.

Surface plasmon resonance (SPR) sensors are a mature label-free technology based on the excitation and resonance of surface plasmon waves at the interface of a metal and an adjacent biological medium. By monitoring changes in the resonance angle, molecular interactions occurring on the metal surface can be detected in real time. Thao et al. used a silver thin film as the SPR coating material, achieving a 2.21-fold enhancement of the fluorescence signal by coating the silver film with Alexa Fluor 488-labeled anti-C-reactive protein antibodies (CRP@Alexa Fluor 488) for immunoassay, with a detection limit of 0.001 mg/L (Thao et al. 2021). A fiber-based SPR aptamer sensor was developed for POCT detection of cTnI and CRP (Hua et al. 2022). Aptamers were immobilized on the gold film-coated fiber surface using gold-thiol bonds. At optimal aptamer concentrations, the detection limits for CRP and cTnI were 0.204 μg/mL and 57.5 ng/mL, respectively.

The combination of electrochemical techniques and nanomaterials has significantly contributed to the high-sensitivity detection of CRP (Fig. 7.2). For example, Hong et al. reported an electrochemiluminescence lateral flow immunosensor (ECL-LFI) (Fig. 7.5b) (Hong et al. 2021). The authors inserted an ECL electrode into the test strip, using $Ru(bpy)_3^{2+}$-labeled gold nanoparticles (AuNPs) as the targeting probes for CRP. When TPA was pre-added to the sample solution, the targeting probe, CRP, and antibodies immobilized on the test line formed a sandwich immunocomplex, triggering an ECL signal through the $Ru(bpy)_3^{2+}$/tripropylamine system. ECL-LFI could detect CRP in serum within 15 minutes, with a detection range of 0.01–1000 ng/mL and a detection limit of 4.6 pg/mL. Validation with serum samples showed good linear correlation with results from clinical chemistry analyzers ($R = 0.9896$). Huang et al. used rhombic dodecahedral carbonized ZIF67 composite nanomaterials loaded with AuNPs as a sensing platform (Fig. 7.5c) (Huang et al. 2022). Aptamers anchored to the nanomaterials were used for specific recognition. When CRP was present, the aptamer captured CRP and formed a sandwich structure with horseradish peroxidase-labeled antibodies. The rapid electron

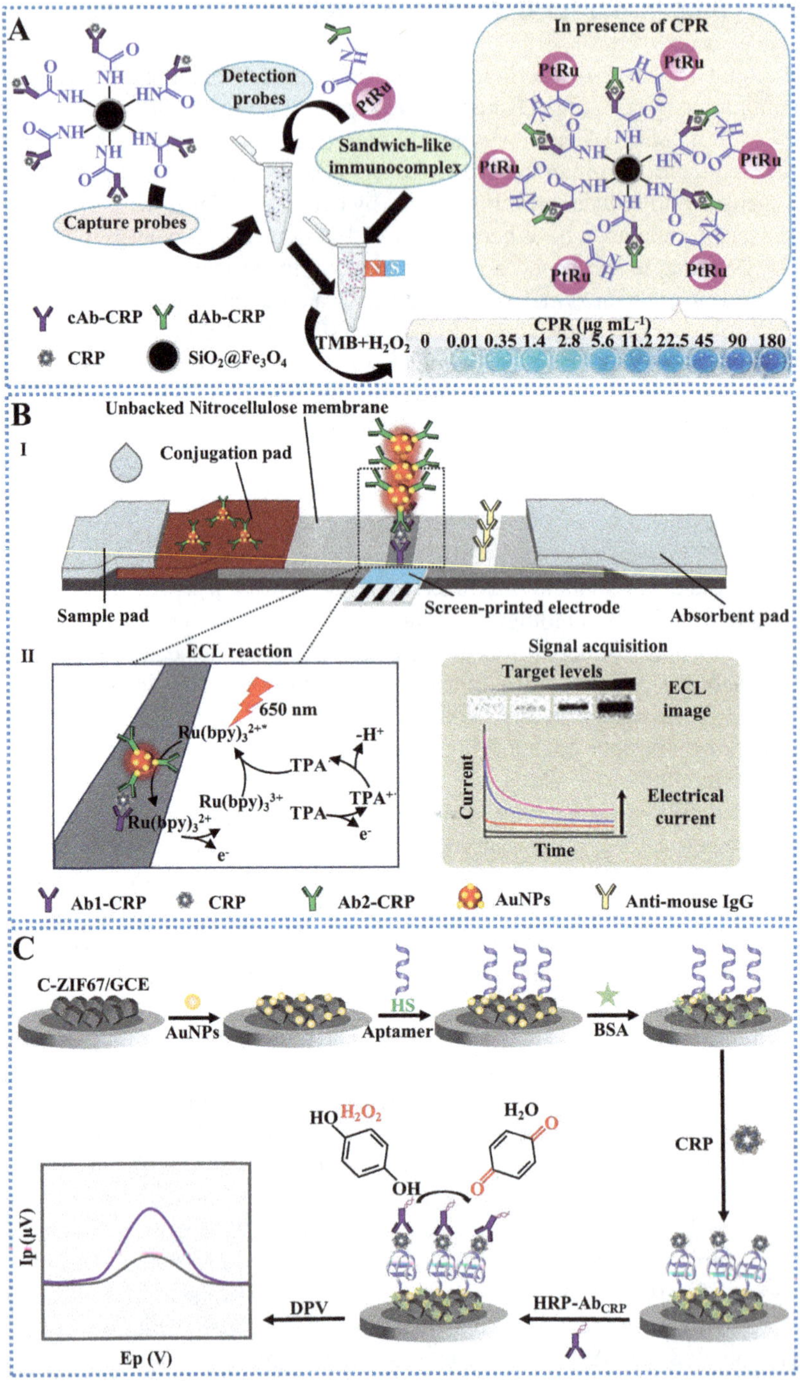

Fig. 7.5 (**a**) Colorimetric sensing platform based on nanozyme catalysis. (Reprinted with permission from Gupta et al. (2021). Copyright 2024 with permission from Springer Nature). (**b**) Principle of ECL-LFI, with the electrode fixed below the T-line of the backing-free NC membrane.

transfer capability of the nanomaterials, combined with enzyme-catalyzed reactions, significantly enhanced the electrochemical signal, achieving ultrasensitivity (detection limit of 0.44 pg/mL) and rapid detection of CRP (the read time is 5 minutes).

7.3.3 Biosensors for BNP and NT-proBNP

As shown in Fig. 7.2, the number of BNP sensors is limited, most of which use immunoassays. Gong and his team optimized the size of gold nanoparticles (GNPs) and the concentration of antibodies in the reaction system by adjusting various conditions during the coupling process of GNPs with antibodies and improved the construction of test strips. They successfully developed an enhanced lateral flow immunoassay (LFIA) method (Fig. 7.6a) (Gong et al. 2017). This method uses the classic sandwich immunoassay and can sensitively detect BNP in just 10–15 minutes, with a detection limit of 0.1 ng/mL. Zhang et al. proposed a novel AlGaN/GaN sensor device with a Kelvin connection structure (Zhang et al. 2021). Compared with traditional gold electrodes, it can effectively suppress background interference and improve the signal-to-noise ratio. The detection limit for BNP was improved from 0.47 ng/mL to 1.29 pg/mL.

In recent years, aptamers have also been applied to BNP sensing analysis, becoming an efficient tool for BNP detection. Zhao et al. constructed an SPR sensing platform with a dual recognition strategy using both aptamers and antibodies (Fig. 7.6b) (Zhao et al. 2019). Aptamer-functionalized gold nanoparticles (GNP-Apt) and antibody-modified magnetic plasmonic nanoparticles (MNPs-Ab) worked together to doubly capture BNP in complex matrices. The resulting MNP-Ab/BNP/GNP-Apt nanocomposite can be rapidly separated from complex samples using a magnet, preventing analyte degradation during its half-life. The constructed SPR biosensor showed high selectivity, with a detection range for BNP from 100 fg/mL to 10 ng/mL and a detection limit as low as 28.2 fg/mL.

Recently, an electrochemical aptamer sensor utilizing entropy-driven reactions as a "switch" to regulate CRISPR/Cas12a activity was reported (Wang et al. 2023). When BNP binds to the aptamer on the electrode surface, it releases the trigger for the entropy-driven reaction, activating CRISPR/Cas12a to cleave the ssDNA on the electrode surface, ultimately generating an electrochemical signal proportional to the BNP concentration. The sensor demonstrated its capability for BNP detection in both buffer and human serum samples. Currently, reports on NT-proBNP biosensors are scarce and primarily involve antigen–antibody immunoassays. Squire et al. demonstrated an immunosensor based on biogenic silica photonic crystal materials to enhance fluorescence imaging, capable of detecting low levels of NT-proBNP

Fig. 7.5 (continued) (Reprinted with permission from Hong et al. (2021). Copyright 2024 with permission from American Chemical Society). (**c**) Electrochemical sensor based on an aptamer–antibody sandwich strategy. (Reprinted with permission from Huang et al. (2022). Copyright 2024 with permission from Elsevier)

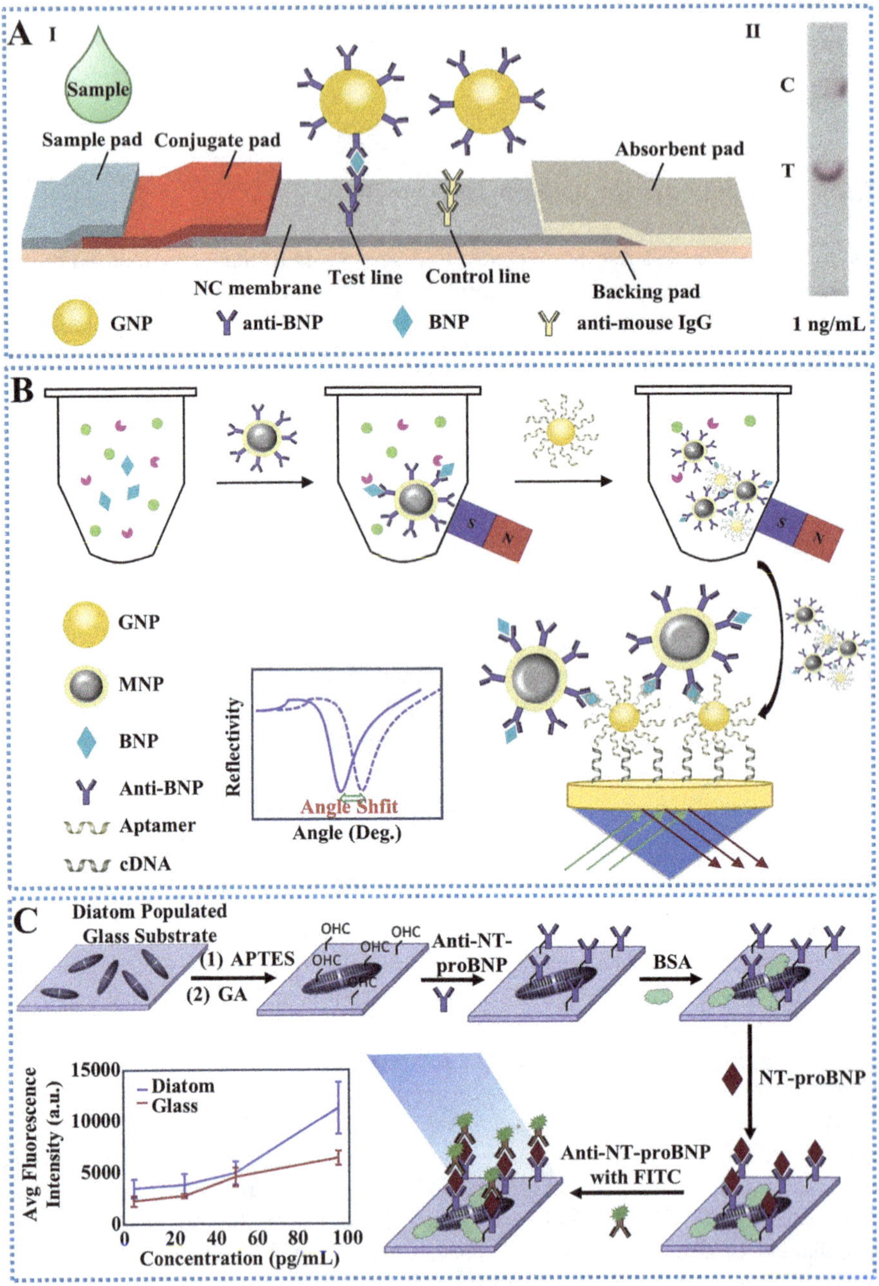

Fig. 7.6 (**a**) Optimized LFIA for detecting BNP. (Reprinted with permission from Gong et al. (2017). Copyright 2024 with permission from Dove Medical Press). (**b**) BNP dual recognition strategy for SPR signal amplification using magnetic plasmonic nanomaterials. (Reprinted with permission from Zhao et al. (2019). Copyright 2024 with permission from Elsevier). (**c**) Fluorescence assay enhanced by photonic crystals for NT-proBNP detection. (Reprinted with permission from Squire et al. (2019). Copyright 2024 with permission from Elsevier)

from 0 to 100 pg/mL (Squire et al. 2019). The specific principle, as shown in Fig. 7.6c, involves isolating biosilica shells from diatoms and depositing them on a glass slide to form a monolayer film. The surface-fixed antibody captures NT-proBNP and a secondary antibody labeled with a fluorophore, enabling the detection of clinically relevant levels of NT-proBNP in plasma. Machine learning algorithms were used to train the fluorescence images for distinguishing between high and low concentrations of NT-proBNP, achieving a specificity of 93%, accuracy of 78%, and sensitivity of 65%.

7.3.4 Biosensors for Myoglobin

Myoglobin (Myo) serves as a complementary biomarker for cardiac diseases. During myocardial injury or necrosis, myoglobin is rapidly released into the bloodstream, rising within 2–3 hours, peaking at around 6–9 hours, and returning to baseline within 24 hours. Thus, it holds unique value as a biomarker of detection platform research.

Zhang et al. developed an electrochemical immunosensor using m-SiO2@PDA@PtPd nanocrystals and PtNi nanodendrite composites, demonstrating the ability to detect Myo within the concentration range of 0.001–1000 ng/mL, with a detection limit as low as 0.21 pg/mL, suitable for human serum sample analysis (Squire et al. 2019). Wang et al. showcased a separation-free ECL immunosensor exhibiting good analytical performance for Myo and cTnI/T (Fig. 7.7a) (Wang et al. 2020). Specifically, target protein-specific antibodies were linked to specific DNA sequences (DNA3-ab1 and DNA4-ab2) to form recognition probes, while DNA5 labeled with a ruthenium complex served as the ECL signal reporting molecule (DNA5-Ru1). On the surface of a glassy carbon electrode, a capture probe DNA1 was covalently immobilized, and a competitive single-stranded DNA2 (ss-DNA2) was introduced to hybridize with DNA1, forming the initial biosensor structure. When the sensor was exposed to a solution containing the target protein, recognition probes, DNA5-Ru1, and tripropylamine (as a coreactant), the binding of the target protein to the antibody triggered a dynamic DNA competition hybridization reaction, effectively localizing DNA5-Ru1 near the electrode surface, thereby completing the ECL measurement. This method achieved a detection limit of 0.5 ng/mL for Myo.

Combined detection of Myo and other cardiovascular disease biomarkers holds significance for improving diagnostic accuracy and optimizing patient management. He et al. achieved simultaneous detection of cTnI and Myo using a potential decomposition strategy (He et al. 2022). In the experiment, molecularly imprinted polymers (MIPs) were used to mimic the function of antibodies for specific capture of ECL probes, including $Ru(bpy)_3^{2+}$-modified mesoporous carbon nanospheres (MOCs) for cTnI and MoS_2 QDs modified with Myo aptamer. Dual ECL signal output was realized through a single potential scan, achieving detection limits of 0.0184 ng/mL for cTnI and 0.0492 ng/mL for Myo.

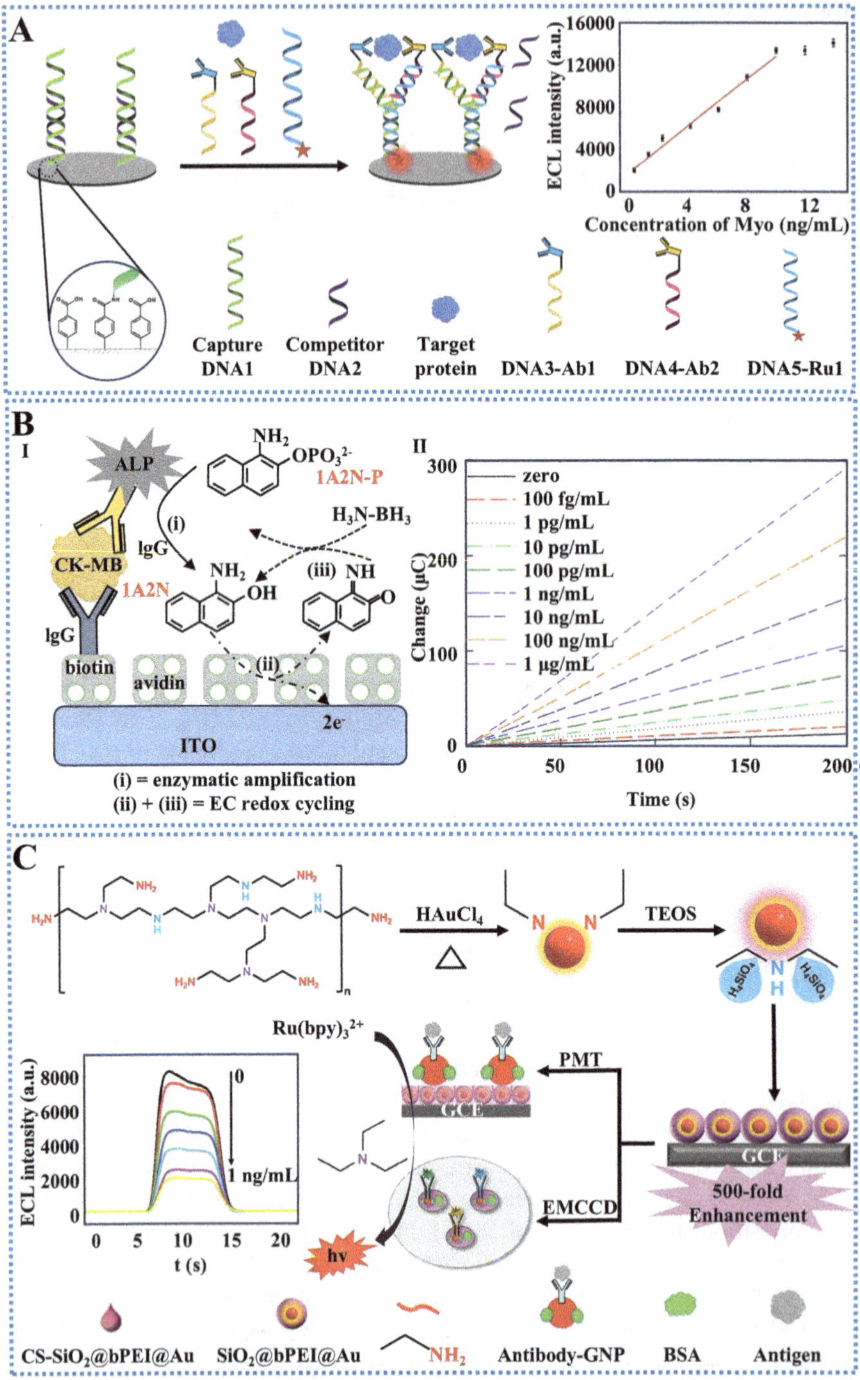

Fig. 7.7 (**a**) Schematic diagram of the target-induced proximity hybridization ECL immunosensor for Myo analysis. (Reprinted with permission from Wang et al. (2020). Copyright 2024 with permission from American Chemical Society). (**b**) Low-background electrochemical immunosensor

7.3.5 Biosensors for CK-MB

Currently, the detection methods for CK-MB mainly focus on immunoreactions and electrochemical strategies using creatine phosphatase (Pcrea) as the recognition substrate and the aptamer technology has not been involved. Moreira et al. utilized the catalytic reaction between Pcrea self-assembled on the electrode surface and CK-MB to generate an electrochemical response signal, achieving mediator-free electrochemical sensing of CK-MB with a detection limit of 0.11 μg/mL (Moreira et al. 2014). Similarly, the enzyme-catalyzed SPR sensor aimed to enhance detection sensitivity by using conductive polymers such as polyethyleneimine (PEI) and poly(vinyl sulfonic acid) (PVS), as well as gold nanoparticles (AuNP) stabilized with cysteamine, which were layer-by-layer (LbL) deposited as a self-assembly substrate for Pcrea. The optimized sensor achieved a dynamic range of 5.0–100.0 ng/mL, with a detection limit of 0.209 ng/mL. Seo et al. reported an electrochemical immunosensor for ultrasensitive detection of CK-MB (Seo et al. 2017). As shown in Fig. 7.7b, this study selected stable, solid 1-amino-2-naphthyl phosphate (1A2N-P) as the substrate for alkaline phosphatase (ALP) and compared four different reductants. Ammonia-borane (H_3N-BH_3) was chosen as the reductant for the electrochemical cyclic oxidation-reduction reaction due to its lowest background current, achieving CK-MB detection with a detection limit of 80 fg/mL, which is much lower than the detection limit without H_3N-BH_3 (40 pg/mL).

7.3.6 Biosensors for Copeptin

As an emerging cardiovascular disease biomarker, copeptin is also of great significant for early diagnostic of this class of disease, while the development of biosensors for this biomarker are relatively limited due to the lack of corresponding specific aptamers (Fig. 7.2). The state-of-art detection methods for copeptin are mainly relying on immunoassays. For instance, Liang et al. developed an enhanced ECL immunoassay strategy using gold core-silica shell nanomaterials loaded with branched polyethyleneimine (bPEI) (SiO_2@bPEI@Au) for the simultaneous detection of cTn, human-type fatty acid binding protein (hFABP), and copeptin (Fig. 7.7c) (Liang et al. 2021). Specifically, the authors synthesized SiO_2@bPEI@Au nanocomposites through a simple one-pot method and sequentially modified SiO_2@bPEI@Au and Ab-GNP onto a glassy carbon electrode (GCE). As an efficient coreactant for $Ru(bpy)_3^{2+}$, bPEI effectively enhanced the luminescence efficiency, amplifying the ECL signal by nearly 500 times. Using the photomultiplier tube (PMT) method, the detection limits for cTnI, hFABP, and copeptin were 0.56, 0.9,

Fig. 7.7 (continued) for CK-MB detection. (Reprinted with permission from Seo et al. (2017). Copyright 2024 with permission from American Chemical Society). (**c**) Schematic illustration of the ECL sensing platform and electrode array construction based on SiO2@bPEI@Au nanostructures. (Reprinted with permission from Liang et al. (2021). Copyright 2024 with permission from Springer Nature)

and 0.40 fg/mL, respectively. Based on this strategy, an electrode array was used to simultaneously detect cTnI, hFABP, and copeptin using the electron multiplying charge-coupled device (EMCCD) method, achieving detection limits at the pg/mL level, sufficient to meet most analytical needs.

In another example, Tu et al. proposed a paper-based sensing platform capable of simultaneously detecting cTn, hFABP, and copeptin on a single test line (Tu et al. 2022). By reading the SERS spectra and characteristic peaks of different GeNPs on the test line, the platform could differentiate the three biomarkers. This method enabled detection within clinically relevant ranges in human serum samples, ensuring sensitivity while maintaining simplicity, cost-effectiveness, and user-friendliness.

7.4 Summary and Future Prospects

This chapter delves into the major advancements in the monitoring of cardiovascular diseases using biosensors. Cardiac biomarker sensing and monitoring technology have shown tremendous potential in tracking early symptoms and providing rapid diagnosis of cardiovascular diseases, making it a significant area of research in the biomedical field.

The sensitivity of currently developed biosensors for cardiac biomarkers has barely met clinical needs despite the ultrasensitive assays. Challenges remain in the robustness of the sensing interface, real-time monitoring in complex environments, and the transition from laboratory to commercialization. For example, in biosensors applied to actual patient samples or in vivo, issues such as nonspecific adsorption of biomacromolecules like proteins, interference from impurities in the samples, and aging of sensor materials often lead to signal distortion. To address these challenges, researchers have proposed solutions that focus on antifouling layers, material modification, and the incorporation of nanotechnology.

Additionally, the development of multimodal sensor systems for cardiac biomarkers holds promise as a new tool for assessing cardiovascular health, providing more comprehensive and in-depth health information, and facilitating reliable diagnosis and efficient management of cardiovascular diseases at different stages. On the one hand, multimodal sensor systems can simultaneously monitor multiple disease-related biomarkers and physiological parameters, helping clinicians more comprehensively assess a patient's cardiovascular health, thereby improving diagnostic accuracy. On the other hand, multimodal sensor systems facilitate the development of personalized treatment plans for cardiovascular diseases by capturing subtle changes in the early pathological processes, thereby promoting effective management of cardiovascular diseases at different stages.

In recent years, artificial intelligence (AI) has garnered attention from various sectors due to its potential for automation, which can replace manual operations with higher reliability and accuracy. In the future, the integration of AI might open new avenues for cardiovascular health monitoring. For instance, developing complex machine learning algorithms embedded in microchips and integrated into portable electronic devices could enable automated and precise analysis of biomarkers, thereby greatly aiding the prevention and diagnosis of cardiovascular diseases.

Overall, although significant progress has been made in cardiovascular monitoring technology, there is still ample room for innovation and improvement. Future scientific research should focus on key performance aspects of biosensing technology, including but not limited to enhancing sensor sensitivity to capture subtle physiological changes during pathological processes, optimizing portability for various application scenarios, improving cost-effectiveness to promote technology dissemination, and strengthening biocompatibility to ensure safety during use.

Acknowledgments This work was supported by the National Natural Science Foundation of China (22122410, 22474130, 22090050, 22274144), the National Key Research and Development Program of China (2021YFA1200403), the Joint NSFC-ISF Research Grant Program (Grant No: 22161142020), and the Natural Science Foundation of Shenzhen (JCYJ20220530162406014).

References

Alpert JS, Thygesen K, Antman E et al (2000) Myocardial infarction redefined--a consensus document of The Joint European Society of Cardiology/American College of Cardiology Committee for the redefinition of myocardial infarction. J Am Coll Cardiol 36:959–969

Amsterdam EA, Wenger NK, Brindis RG et al (2014) 2014 AHA/ACC guideline for the management of patients with non-ST-elevation acute coronary syndromes: a report of the American College of Cardiology/American Heart Association Task Force on practice guidelines. J Am Coll Cardiol 64:e139–e228

Battistoni A, Rubattu S, Volpe M (2012) Circulating biomarkers with preventive, diagnostic and prognostic implications in cardiovascular diseases. Int J Cardiol 157:160–168

Benjamin EJ, Muntner P, Alonso A et al (2019) Heart disease and stroke statistics-2019 update: a report from the American Heart Association. Circulation 139:e56–e528

Black S, Kushner I, Samols D (2004) C-reactive protein. J Biol Chem 279:48487–48490

Bleasdale RA, Frenneaux MP (2004) Cardiac resynchronisation therapy: when the drugs don't work. Heart 90 Suppl 6:vi2–4

Boersma E, Mercado N, Poldermans D et al (2003) Acute myocardial infarction. Lancet 361:847–858

Bozkurt B, Coats AJS, Tsutsui H et al (2021) Universal definition and classification of heart failure: a report of the Heart Failure Society of America, Heart Failure Association of the European Society of Cardiology, Japanese Heart Failure Society and Writing Committee of the Universal Definition of Heart Failure: Endorsed by the Canadian Heart Failure Society, Heart Failure Association of India, Cardiac Society of Australia and New Zealand, and Chinese Heart Failure Association. Eur J Heart Fail 23:352–380

Braunstein JB, Anderson GF, Gerstenblith G et al (2003) Noncardiac comorbidity increases preventable hospitalizations and mortality among Medicare beneficiaries with chronic heart failure. J Am Coll Cardiol 42:1226–1233

Crivianu-Gaita V, Thompson M (2016) Aptamers, antibody scFv, and antibody Fab' fragments: an overview and comparison of three of the most versatile biosensor biorecognition elements. Biosens Bioelectron 85:32–45

Crousillat DR, Wood MJ (2019) Valvular heart disease and heart failure in women. Heart Fail Clin 15:77–85

Das J, Gomis S, Chen JB et al (2021) Reagentless biomolecular analysis using a molecular pendulum. Nat Chem 13:428-+

Daubert JC (2004) Introduction to atrial fibrillation and heart failure: a mutually noxious association. Europace 5(Suppl 1):S1–S4

Derin E, Inci F (2022) Advances in biosensor technologies for acute kidney injury. ACS Sens 7:358–385

Di Angelantonio E, Chowdhury R, Sarwar N et al (2009) B-type natriuretic peptides and cardiovascular risk: systematic review and meta-analysis of 40 prospective studies. Circulation 120:2177–2187

Dong T, Zhu WA, Yang ZC et al (2024) Advances in heart failure monitoring: biosensors targeting molecular markers in peripheral bio-fluids. Biosens Bioelectron 255:116090

Feigin VL, Lawes CM, Bennett DA et al (2009) Worldwide stroke incidence and early case fatality reported in 56 population-based studies: a systematic review. Lancet Neurol 8:355–369

Fihn SD, Gardin JM, Abrams J et al (2012) 2012 ACCF/AHA/ACP/AATS/PCNA/SCAI/STS guideline for the diagnosis and management of patients with stable ischemic heart disease: a report of the American College of Cardiology Foundation/American Heart Association Task Force on practice guidelines, and the American College of Physicians, American Association for Thoracic Surgery, Preventive Cardiovascular Nurses Association, Society for Cardiovascular Angiography and Interventions, and Society of Thoracic Surgeons. J Am Coll Cardiol 60:e44–e164

Friess U, Stark M (2009) Cardiac markers: a clear cause for point-of-care testing. Anal Bioanal Chem 393:1453–1462

Giménez MR, Twerenbold R, Boeddinghaus J et al (2016) Clinical effect of sex-specific cutoff values of high-sensitivity cardiac Troponin T in suspected myocardial infarction. JAMA Cardiol 1:912–920

Ginsburg GS, Donahue MP, Newby LK (2005) Prospects for personalized cardiovascular medicine: the impact of genomics. J Am Coll Cardiol 46:1615–1627

Girnar GA, Mahajan HS (2021) Cerebral ischemic stroke and different approaches for treatment of stroke. Future J Pharm Sci 7:134

Gong Y, Hu J, Choi JR et al (2017) Improved LFIAs for highly sensitive detection of BNP at point-of-care. Int J Nanomedicine 12:4455–4465

Guk K, Han G, Lim J et al (2019) Evolution of wearable devices with real-time disease monitoring for personalized healthcare. Nanomaterials 9:813

Gupta PK, Son SE, Seong GH (2021) Functionalized ultra-fine bimetallic PtRu alloy nanoparticle with high peroxidase-mimicking activity for rapid and sensitive colorimetric quantification of C-reactive protein. Microchim Acta 188:119

Hansson GK (2005) Inflammation, atherosclerosis, and coronary artery disease. N Engl J Med 352:1685–1695

He S, Zhang P, Sun J et al (2022) Integrating potential-resolved electrochemiluminescence with molecularly imprinting immunoassay for simultaneous detection of dual acute myocardial infarction markers. Biosens Bioelectron 201:113962

Hendgen-Cotta UB, Kelm M, Rassaf T (2010) A highlight of myoglobin diversity: the nitrite reductase activity during myocardial ischemia-reperfusion. Nitric Oxide-Biol Ch 22:75–82

Hijazi Z, Lindbäck J, Alexander JH et al (2016) The ABC (age, biomarkers, clinical history) stroke risk score: a biomarker-based risk score for predicting stroke in atrial fibrillation. Eur Heart J 37:1582–1590

Hjort M, Lindahl B, Baron T et al (2018) Prognosis in relation to high-sensitivity cardiac troponin T levels in patients with myocardial infarction and non-obstructive coronary arteries. Am Heart J 200:60–66

Hong DG, Kim K, Jo EJ et al (2021) Electrochemiluminescence-incorporated lateral flow immunosensors using $Ru(bpy)_3^{2+}$-labeled gold nanoparticles for the full-range detection of physiological C-reactive protein levels. Anal Chem 93:7925–7932

Hu BB, Li LY, Zhao DY et al (2023) Research progress of electrochemistry in detecting markers of cardiovascular disease. Microchem J 195:109489

Hua Y, Wang RD, Li DC (2022) A fiber-based SPR Aptasensor for the in vitro detection of inflammation biomarkers. Micromachines 13:1036

Huang SQ, Liu ZP, Yan Y et al (2022) Triple signal-enhancing electrochemical aptasensor based on rhomboid dodecahedra carbonized-ZIF_{67} for ultrasensitive CRP detection. Biosens Bioelectron 207:114129

Jaffe A (1991) Biochemical detection of acute myocardial infarction. Acute myocardial infarction. Elsevier, New York, pp 110–127

Jing W, Wang Y, Chen C et al (2021) Gradient-based rapid digital immunoassay for high-sensitivity cardiac Troponin T (hs-cTnT) detection in 1 μL plasma. ACS Sens 6:399–407

Keating MT, Sanguinetti MC (2001) Molecular and cellular mechanisms of cardiac arrhythmias. Cell 104:569–580

Lakhani I, Wong MV, Hung JKF et al (2021) Diagnostic and prognostic value of serum C-reactive protein in heart failure with preserved ejection fraction: a systematic review and meta-analysis. Heart Fail Rev 26:1141–1150

Lee TH, Weisberg MC, Brand DA et al (1989) Candidates for thrombolysis among emergency room patients with acute chest pain - potential true-positive and false-positive rates. Ann Intern Med 110:957–962

Lewandrowski K, Chen A, Januzzi J (2002) Cardiac markers for myocardial infarction. A brief review. Am J Clin Pathol 118(Suppl):S93–99

Li S, She F, Lv T et al (2021) The prognostic role of high-sensitivity cardiac troponin T over time in ischemic and non-ischemic heart failure. Postepy Kardiol Interwencyjnej 17:54–59

Li YP, Gao X, Fang YS et al (2023) Nanomaterials-driven innovative electrochemiluminescence aptasensors in reporting food pollutants. Coord Chem Rev 485:215136

Li C, Yang HZ, Song ZP et al (2024) Rational utilization of Fe(III)-2-amino terephthalic acid MOF as a bifunctional nanocarrier of dopamine and aptamer for constructing ultrasensitive sandwich biosensor of cardiac Troponin I. Sens Actuators B Chem 411:135746

Liang X, Mi L, Yu ZY et al (2021) Multifunctional polyethyleneimine for synthesis of core-shell nanostructures and electrochemiluminescent detection of three AMI biomarkers. Sci China Chem 64:2230–2238

Libby P, Theroux P (2005) Pathophysiology of coronary artery disease. Circulation 111:3481–3488

Liu G, Chen MR, Wang JL et al (2024) Flower-shaped magnetic fluorescent nanoparticles based lateral flow biosensor for highly sensitive detection of cardiac troponin I. Sens Actuators B Chem 414:414

Luong JH, Male KB, Glennon JD (2008) Biosensor technology: technology push versus market pull. Biotechnol Adv 26:492–500

Mair J, Artnerdworzak E, Lechleitner P et al (1992) Early diagnosis of acute myocardial-infarction by a newly developed rapid immunoturbidimetric assay for myoglobin. Br Heart J 68:462–468

Maisel AS, Krishnaswamy P, Nowak RM et al (2002) Rapid measurement of B-type natriuretic peptide in the emergency diagnosis of heart failure. N Engl J Med 347:161–167

Mcalister FA, Ezekowitz J, Tonelli M et al (2004) Renal insufficiency and heart failure: prognostic and therapeutic implications from a prospective cohort study. Circulation 109:1004–1009

Mccord J, Nowak RM, Mccullough PA et al (2001) Ninety-minute exclusion of acute myocardial infarction by use of quantitative point-of-care testing of myoglobin and troponin I. Circulation 104:1483–1488

Möckel M, Searle J, Hamm C et al (2015) Early discharge using single cardiac troponin and copeptin testing in patients with suspected acute coronary syndrome (ACS): a randomized, controlled clinical process study. Eur Heart J 36:369-+

Moreira FTC, RaF D, Noronha JP et al (2014) Novel sensory surface for creatine kinase electrochemical detection. Biosens Bioelectron 56:217–222

Morgenthaler NG, Struck J, Jochberger S et al (2008) Copeptin: clinical use of a new biomarker. Trends Endocrinol Metab 19:43–49

Morrow DA, Cannon CP, Jesse RL et al (2007a) National Academy of Clinical Biochemistry Laboratory medicine practice guidelines: clinical characteristics and utilization of biochemical markers in acute coronary syndromes. Clin Chem 53:552–574

Morrow DA, Cannon CP, Jesse RL et al (2007b) National Academy of Clinical Biochemistry Laboratory medicine practice guidelines: clinical characteristics and utilization of biochemical markers in acute coronary syndromes. Circulation 115:e356–e375

Mozaffarian D, Benjamin EJ, Go AS et al (2016) Heart disease and stroke statistics-2016 update: a report from the American Heart Association. Circulation 133:e38–e360
Negandary M (2020) Aptamers in nanostructure-based electrochemical biosensors for cardiac biomarkers and cancer biomarkers: a review. Biosens Bioelectron 152:112018
Organization WH (2021) Cardiovascular diseases (CVDs). In: https://www.who.int/news-room/fact-sheets/detail/cardiovascular-diseases-(cvds)
Ouyang H, Zhang LX, Jiang S et al (2020) Co single-atom catalysts boost Chemiluminescence. Chem Eur J 26:7583–7588
Potocki M, Breidthardt T, Mueller A et al (2010) Copeptin and risk stratification in patients with acute dyspnea. Crit Care 14:R213
Prokopidis K, Irlik K, Ishiguchi H et al (2024) Natriuretic peptides and C-reactive protein in in heart failure and malnutrition: a systematic review and meta-analysis. ESC Heart Fail 11:3052
Qureshi A, Gurbuz Y, Niazi JH (2012) Biosensors for cardiac biomarkers detection: a review. Sens Actuators B Chem 171:62–76
Reichlin T, Hochholzer W, Stelzig C et al (2009a) Incremental value of Copeptin for rapid rule out of acute myocardial infarction. J Am Coll Cardiol 54:60–68
Reichlin T, Hochholzer W, Stelzig C et al (2009b) Incremental value of copeptin for rapid rule out of acute myocardial infarction. J Am Coll Cardiol 54:60–68
Ridker PM (2004) High-sensitivity C-reactive protein, inflammation, and cardiovascular risk: from concept to clinical practice to clinical benefit. Am Heart J 148:S19–S26
Roffi M, Patrono C, Collet JP et al (2016) 2015 ESC guidelines for the management of acute coronary syndromes in patients presenting without persistent ST-segment elevation Task Force for the management of acute coronary syndromes in patients presenting without persistent ST-segment elevation of the European Society of Cardiology (ESC). Eur Heart J 37:267–+
Saenger AK, Jaffe AS (2008) Requiem for a heavyweight the demise of Creatine Kinase-MB. Circulation 118:2200–2206
Seo J, Ha H, Park S et al (2017) Immunosensor employing stable, Solid 1-Amino-2-naphthyl phosphate and ammonia-borane toward ultrasensitive and simple point-of-care testing. ACS Sens 2:1240–1246
Shi CH, Xie HT, Ma YF et al (2020) Nanoscale technologies in highly sensitive diagnosis of cardiovascular diseases. Front Bioeng Biotechnol 8:531
Simats A, Liesz A (2022) Systemic inflammation after stroke: implications for post-stroke comorbidities. EMBO Mol Med 14:e16269
Squire KJ, Zhao Y, Tan AL et al (2019) Photonic crystal-enhanced fluorescence imaging immunoassay for cardiovascular disease biomarker screening with machine learning analysis. Sens Actuators B Chem 290:118–124
Taghizadeh B, Ghavami L, Derakhshankhah H et al (2020) Biomaterials in valvular heart diseases. Front Bioeng Biotechnol 8:529244
Takeda S, Yamashita A, Maeda K et al (2003) Structure of the core domain of human cardiac troponin in the Ca^{2+}-saturated form. Nature 424:35–41
Thao NT, Hoang TX, Phan TB et al (2021) Metal-enhanced sensing platform for the highly sensitive detection of C-reactive protein antibody and rhodamine B with applications in cardiovascular diseases and food safety. Dalton Trans 50:6962–6974
Thygesen K, Mair J, Mueller C et al (2012) Recommendations for the use of natriuretic peptides in acute cardiac care: a position statement from the Study Group on biomarkers in cardiology of the ESC Working Group on acute cardiac care. Eur Heart J 33:2001–2006
Thygesen K, Alpert JS, Jaffe AS et al (2018) Fourth universal definition of myocardial infarction (2018). Glob Heart 13:305–338
Timmis A, Townsend N, Gale CP et al (2020) European Society of Cardiology: cardiovascular disease statistics 2019. Eur Heart J 41:12–85
Topilsky Y (2018) Tricuspid valve regurgitation: epidemiology and pathophysiology. Minerva Cardioangiol 66:673–679
Tsao CW, Aday AW, Almarzooq ZI et al (2022) Heart disease and stroke statistics-2022 update: a report from the American Heart Association. Circulation 145:E153–E639

Tseng YT, Chiu YC, Pham V et al (2024) Ultrasensitive Upconversion nanoparticle immunoassay for human serum cardiac Troponin I detection achieved with resonant waveguide grating. ACS Sens 9:455–463

Tu DD, Holderby A, Guo H et al (2022) Spectrally multiplexed assay using gap enhanced nanoparticle for detection of a myocardial infarction biomarker panel. Anal Chim Acta 1198:339562

Tulevski I, Hirsch A, Sanson BJ et al (2001) Increased brain natriuretic peptide as a marker for right ventricular dysfunction in acute pulmonary embolism. Thromb Haemost 86:1193–1196

Vasudevan M, Tai MJY, Perumal V et al (2021) Cellulose acetate-MoS_2 nanopetal hybrid: a highly sensitive and selective electrochemical aptasensor of Troponin I for the early diagnosis of Acute Myocardial Infarction. J Taiwan Inst Chem 118:245–253

Vasudevan M, Perumal V, Raja PB et al (2023) A quadruplet 3-D laser scribed graphene/MoS_2, functionalised N_2-doped graphene quantum dots and lignin-based Ag-nanoparticles for biosensing. Int J Biol Macromol 253:126620

Voelkel NF, Quaife RA, Leinwand LA et al (2006) Right ventricular function and failure - report of a National Heart, lung, and Blood Institute working group on cellular and molecular mechanisms of right heart failure. Circulation 114:1883–1891

Wang B, Shi SW, Yang XL et al (2020) Separation-free electrogenerated Chemiluminescence immunoassay incorporating target assistant proximity hybridization and dynamically competitive hybridization of a DNA signal probe. Anal Chem 92:884–891

Wang JY, Wang XL, Li B et al (2023) Entropy-driven reactions for controlling CRISPR/Cas12a and constructing an electrochemical biosensor for cardiac biomarkers detection. Microchim Acta 190:440

Wang T, Tan HS, Zhao LX et al (2024) A novel ratiometric aptasensor based on SERS for accurate quantification of cardiac troponin I. Sens Actuators B Chem 412:412

Wen XN, Ou YC, Zarick HF et al (2020) PRADA: portable reusable accurate diagnostics with nanostar antennas for multiplexed biomarker screening. Bioeng Transl Med 5:e10165

Yang YN, Lin HI, Wang JH et al (2009) An integrated microfluidic system for C-reactive protein measurement. Biosens Bioelectron 24:3091–3096

Yap LB, Ashrafian H, Mukerjee D et al (2004) The natriuretic peptides and their role in disorders of right heart dysfunction and pulmonary hypertension. Clin Biochem 37:847–856

Yasue H, Nakagawa H, Itoh T et al (2008) Coronary artery spasm - Clinicat features, diagnosis, pathogenesis, and treatment. J Cardiol 51:2–17

Yin XR, Yang P, Zhang HM et al (2020) Hydrophobic-driven Electrochemiluminescence enhancement via target-induced self-enrichment for ultrasensitive bioassay. Anal Chem 92:15120–15128

Zeng LZ, Lin CJ, Liu PQ et al (2023) Anisotropic aptamer-modified DNA tetrahedra/MOF nanoprobes for enhanced colorimetric aptasensing of cardiac troponin I. Chem Eng J 474:145525

Zhang GH, Zhang L, Yu Y et al (2020) Dual-mode of electrochemical-colorimetric imprinted sensing strategy based on self-sacrifice beacon for diversified determination of cardiac troponin I in serum. Biosens Bioelectron 167:112502

Zhang HY, Gan Y, Yang S et al (2021) Low limit of detection of the AlGaN/GaN-based sensor by the Kelvin connection detection technique. Microsyst Nanoeng 7:51

Zhao JL, Liang DL, Gao SW et al (2019) Analyte-resolved magnetoplasmonic nanocomposite to enhance SPR signals and dual recognition strategy for detection of BNP in serum samples. Biosens Bioelectron 141:111440

Zhong S, Chen L, Shi X et al (2023) Recent advances in electrochemical aptasensors for detecting cardiac biomarkers: a review. Microchem J 193:109063

Zong C, Wang RK, Jiang F et al (2021) Metal enhanced chemiluminescence nanosensor for ultrasensitive bioassay based on silver nanoparticles modified functional DNA dendrimer. Anal Chim Acta 1165:1165

8 Bioelectronic Devices for Diabetes: Monitoring, Management, and Precision Health

Kuldeep Mahato and Shekhar Kummari

Abstract

The integration of bioelectronics into diabetes management marks a transformative era in precision healthcare, offering innovative solutions to overcome the limitations of traditional glucose-centric approaches. This chapter explores the evolution of bioelectronic technologies, emphasizing their role in enabling real-time, continuous monitoring of glucose and other biomarkers critical for comprehensive diabetes care. While glucose remains the cornerstone of diabetes management, the monitoring of additional biomarkers, such as ketones, lactate, insulin, and inflammatory markers, provides a holistic understanding of metabolic health, supporting early detection of complications and personalized interventions. The chapter discusses the advancements in biosensor design, including implantable, wearable, and non-invasive technologies that enhance accuracy, usability, and patient adherence. It also discusses the integration of biosensing data with telemedicine and remote patient monitoring platforms, enabling proactive, data-driven interventions and reducing the need for in-person care. Furthermore, the role of artificial intelligence in refining predictive algorithms and optimizing insulin delivery is explored, highlighting its potential to revolutionize closed-loop systems. Key challenges such as sensor accuracy, calibration, data security, and accessibility are critically examined alongside innovative solutions, including anti-fouling coatings, flexible electronics, and multimodal biosensing. Ethical considerations and regulatory frameworks are also addressed to ensure equitable access and patient-centered care. This chapter underscores the

K. Mahato (✉)
Department of Biotechnology, Motilal Nehru National Institute of Technology Allahabad, Prayagraj, Uttar Pradesh, India

S. Kummari
Department of Electrical and Computer Engineering, University of California San Diego, La Jolla, CA, USA

K. Mahato, P. Chandra (eds.), *Nano-bioelectronics for Precision Health Monitoring*, https://doi.org/10.1007/978-981-95-2117-3_8

promise of bioelectronics in driving a paradigm shift towards personalized, precise, and integrative diabetes management, paving the way for improved outcomes and enhanced quality of life for individuals with diabetes.

Keywords

Bioelectronics · Diabetes management · Continuous glucose monitoring (CGM) · Personalized medicine · Multimodal biosensing · Wearable sensors

8.1 Diabetes: An Introduction

Diabetes mellitus, a metabolic disorder characterized by chronic hyperglycemia, arises from defects in insulin secretion, insulin action, or both (Antar et al. 2023). Insulin is a hormone produced by the beta cells of the pancreas, and its primary function is to regulate blood glucose levels. When insulin production is inadequate, or the body cannot efficiently utilize the insulin produced, glucose accumulates in the bloodstream, leading to hyperglycemia (Rahman et al. 2021; Karpova et al. 2019). Over time, this condition can result in various complications affecting the cardiovascular, renal, and nervous systems, among others. Diabetes is typically categorized into three major types: Type 1 diabetes mellitus (T1DM) (Roep et al. 2021), Type 2 diabetes mellitus (T2DM) (DeFronzo et al. 2015), and gestational diabetes mellitus (GDM) (Modzelewski et al. 2022). The T1DM is primarily an autoimmune disorder, wherein the immune system attacks and destroys the insulin-producing beta cells in the pancreas. As a result, individuals with T1DM are unable to produce insulin and thus require lifelong insulin therapy for survival. T1DM typically manifests during childhood or adolescence, but it can occur at any age. It accounts for approximately 5–10% of all diabetes cases globally (Green et al. 2021; Ward et al. 2022). Management of T1DM is particularly challenging due to the need for continuous insulin administration and the risk of both hypoglycemia (low blood sugar) and hyperglycemia. The T2DM is the most common form of diabetes, accounting for 90–95% of all diabetes cases (Zheng et al. 2018). Unlike T1DM, T2DM is characterized by insulin resistance, where the body's cells fail to respond to insulin adequately. Initially, the pancreas compensates by producing more insulin, but over time, this compensation becomes insufficient, leading to hyperglycemia. T2DM is often associated with obesity, sedentary lifestyle, poor diet, and genetic predisposition (Eaton and Eaton 2017; Ingelsson and Mark 2018). It tends to develop later in life, but with increasing rates of obesity, T2DM is becoming more prevalent in younger populations. Management of T2DM typically involves lifestyle modifications, oral hypoglycemic agents, and, in some cases, insulin therapy. The GDM occurs during pregnancy and affects approximately 2–10% of pregnant women worldwide (Hod et al. 2019). It results from hormonal changes during pregnancy that induce insulin resistance. While GDM typically resolves after childbirth, it increases the risk of developing T2DM later in life for both the mother and the child. GDM also poses short-term risks, such as macrosomia (large birth weight) and neonatal hypoglycemia, necessitating careful monitoring and management during pregnancy (Murray and Reynolds 2020; Mitanchez et al. 2015).

The global prevalence of diabetes is rapidly rising, making it a significant public health concern. According to the International Diabetes Federation (IDF), an estimated 537 million adults were living with diabetes in 2021, and this number is projected to increase to 643 million by 2030 (Williams et al. 2020; Sun et al. 2022). The rising prevalence is driven by factors such as population aging, urbanization, and the increasing prevalence of obesity and sedentary lifestyles. Low- and middle-income countries are disproportionately affected, as they experience the highest rates of diabetes-related morbidity and mortality. The socioeconomic impact of diabetes is substantial. The cost of diabetes care includes direct medical expenses, such as hospitalization, medication, and outpatient care, as well as indirect costs, including lost productivity, absenteeism, and premature mortality (Shrestha et al. 2019; Kennedy-Martin et al. 2017; Dall et al. 2019). In 2021, global healthcare expenditures related to diabetes were estimated at $966 billion, representing a 316% increase over the previous 15 years (JHEOR 2022). Furthermore, diabetes-related complications, such as cardiovascular disease (Canto et al. 2019), nephropathy (Umanath and Lewis 2018), and retinopathy (Kropp et al. 2023), contribute to long-term disability, further increasing the economic burden. The substantial healthcare costs associated with diabetes, coupled with its impact on quality of life, underscore the need for effective management strategies. Traditional diabetes management primarily focuses on controlling blood glucose levels through a combination of lifestyle modifications, pharmacotherapy, and, in the case of T1DM, insulin therapy. However, these strategies have limitations, particularly in maintaining long-term glycemic control and preventing complications (Perkins et al. 2021). As a result, there is a growing interest in innovative technologies, particularly biosensing and bioelectronic devices, to improve diabetes management (Dua et al. 2024; Güemes Gonzalez et al. 2020).

Historically, diabetes management has relied on self-monitoring of blood glucose (SMBG) using finger-stick blood samples and glucometers (Oparaji 2024). While this method provides valuable information on blood glucose levels, it requires frequent manual testing, which can be inconvenient, painful, and prone to user error (Yamada 2011). Moreover, SMBG only offers intermittent snapshots of glucose levels, leaving gaps in glycemic data that may result in unrecognized hyperglycemic or hypoglycemic episodes, particularly during sleep or physical activity (Erbach et al. 2016). Another challenge of traditional diabetes management is the delayed response to fluctuations in blood glucose levels. Insulin therapy, for instance, is often reactive rather than proactive (Nikpour et al. 2022). Patients adjust insulin dosages based on past glucose readings or subjective assessments of symptoms, which can lead to inaccurate dosing and an increased risk of hypo- or hyperglycemia. Additionally, insulin administration is not always optimized for individual variability in insulin sensitivity, which can vary based on factors such as stress, illness, or physical activity. Oral hypoglycemic agents used in T2DM management also present challenges (Chaudhury et al. 2017). While medications like metformin and sulfonylureas help regulate blood glucose levels, they may have side effects, including gastrointestinal disturbances or the risk of hypoglycemia (Nabrdalik et al. 2022). Moreover, patient adherence to medication regimens is often suboptimal due to the chronic nature of diabetes and the complexities of managing multiple medications. These limitations highlight the need for more dynamic and precise management tools that provide real-time data, reduce the burden of frequent testing, and support personalized treatment decisions. This is where biosensing and bioelectronic technologies come into play (Liu et al. 2021; Psoma and Kanthou 2023).

8.2 Bioelectronics for Diabetes Monitoring

Bioelectronic devices, including continuous glucose monitoring (CGM) systems, represent a significant advancement in diabetes care, offering a more comprehensive and real-time approach to blood glucose and diabetes management (Fig. 8.1) (Liu and Zhang 2024). Unlike traditional SMBG, CGM devices continuously track glucose levels throughout the day and night, providing a more complete picture of glycemic trends and patterns. CGM devices typically consist of a small sensor inserted under the skin that measures interstitial glucose levels, a transmitter that sends data to a display device, and a receiver, such as a smartphone or dedicated monitor, that displays real-time glucose data (Johnston et al. 2021). One of the primary advantages of CGM is its ability to provide trend information, such as the direction and rate of glucose level changes. This enables users and healthcare providers to make more informed treatment decisions and take proactive measures to prevent hypo- or hyperglycemia. For example, if glucose levels are trending downward, a patient can consume carbohydrates to prevent hypoglycemia before it occurs. Conversely, if glucose levels are rising, insulin can be administered to prevent hyperglycemia using the bioeletronic pumps. The bioeletronic devices reduce the burden of frequent finger-stick based non-invasive testing, which can improve patient adherence and quality of life. Furthermore, these systems offer features such

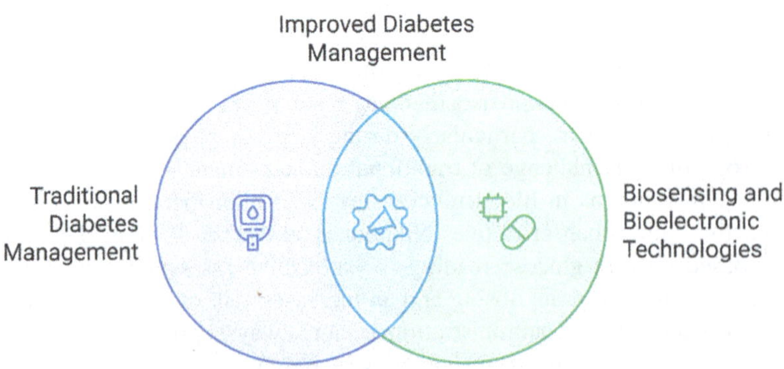

Fig. 8.1 Schematic illustration of an integrated strategy for enhanced diabetes management, combining conventional methods with biosensing-enabled bioelectronic technologies for real-time monitoring and personalized interventions

as alarms and alerts that notify users of impending glycemic excursions, enabling prompt intervention. For instance, studies have shown that CGM use is associated with improved glycemic control, reduced A1c levels, and a lower risk of hypoglycemia, particularly in patients with T1DM (Teo et al. 2022). The progression of diabetes can lead to various complications, including shifts in cardiovascular parameters. Bioelectronics, which offers the measurement of cardiovascular parameters such as oxygen saturation and heart rate variability, provides efficient management of diabetes. Diabetic wounds are another complication that results from poorly regulated blood glucose levels. Recent advancements in the bioelectronic bandages and patches for monitoring the biomarkers associated with wound healing and progression. Additionally, the devices coupled with drug delivery modules using hydrogels have also been developed towards facilitating quicker assisted healing.

8.3 Bioelectronics for Glucose Monitoring

Bioelectronics refers to the integration of biological components with electronic systems to create devices that can interact with the body at a molecular or cellular level (Kim et al. 2018a). These devices often combine biosensing technologies with electronics to detect glucose levels and other biomarkers, deliver therapeutic interventions, and provide feedback to both patients and healthcare providers. Over the years, glucose monitoring has undergone significant advancements, transforming from rudimentary urine tests to modern continuous glucose monitoring (CGM) systems that provide real-time data (Didyuk et al. 2020). Understanding the evolution of glucose monitoring technologies highlights the need for more precise and less invasive solutions in diabetes management, laying the foundation for the adoption of biosensing technologies in precision healthcare. In the early twentieth century, glucose monitoring relied heavily on urine testing, which measured glucose levels indirectly by detecting its presence in the urine. This method was neither timely nor accurate, as glucose only appeared in urine when blood glucose levels were already significantly elevated, making it unsuitable for tight glycemic control. The advent of blood glucose monitoring (BGM) in the 1970s, particularly the development of the first portable glucometer, revolutionized diabetes management (Hirsch 2018). By providing direct, immediate glucose measurements from blood samples, glucometers enabled patients to monitor their glucose levels more frequently and make informed decisions about insulin administration and diet (Weinstock et al. 2020). Despite this advancement, blood glucose monitoring using glucometers remains invasive and often uncomfortable, requiring regular finger pricking to obtain blood samples. This method only provides discrete glucose readings, offering limited insight into glucose fluctuations throughout the day. Such intermittent monitoring increases the risk of missing critical glucose excursions, such as postprandial hyperglycemia or nocturnal hypoglycemia (Palumbo 2001). Moreover, the discomfort associated with finger pricking may lead to poor patient adherence, reducing the effectiveness of diabetes management. The shortcomings of traditional blood glucose monitoring led to the development of continuous glucose monitoring (CGM) systems, which provide real-time glucose readings throughout the day and night. The first generation of CGM systems, introduced in the early 2000s, used subcutaneous sensors to measure glucose levels in interstitial fluid, the fluid surrounding cells (Hovorka 2006). While this method offered more frequent glucose measurements than traditional glucometers, early CGM devices required frequent calibration with finger-stick blood glucose measurements and were often bulky and uncomfortable to wear. Over the past two decades, CGM technology has improved significantly, with modern devices offering longer sensor lifetimes, greater accuracy, and user-friendly interfaces. These advancements have made CGM more accessible and appealing to patients, leading to increased adoption, particularly among individuals with Type 1 diabetes (T1DM). Furthermore, the integration of CGM with insulin pumps and mobile health platforms has created a more dynamic and personalized approach to diabetes care (Marks et al. 2022). Despite these advances, CGM still faces limitations, such as sensor lag, high costs, and the need for invasive sensor insertion. These challenges underscore the importance of continued innovation in biosensing technologies, particularly in the pursuit of non-invasive, real-time glucose monitoring solutions that can be integrated into wearable and bioelectronic platforms (Jain et al. 2024).

Biosensors are analytical devices that convert a biological response into an electrical signal, enabling the detection of various analytes, such as glucose, in biological fluids (Xu et al. 2022). Glucose biosensors have become the cornerstone of diabetes management, and their development has largely focused on improving accuracy, sensitivity, and ease of use (Teymourian et al. 2020). Broadly, glucose biosensors are categorized into enzymatic and non-enzymatic systems, each with distinct mechanisms of glucose detection (Govindaraj et al. 2023). Enzymatic Glucose Biosensors: Enzymatic biosensors are the most widely used glucose sensing devices, particularly in commercial glucometers and CGM systems (Yoon et al. 2018). These sensors rely on the catalytic activity of glucose oxidase (GOx), an enzyme that oxidizes glucose to gluconic acid and hydrogen peroxide (H_2O_2) in the presence of oxygen (Valdes and Moussy 2000). The generation of H_2O_2 serves as the key detectable species, as it can be electrochemically oxidized at an electrode surface, producing an electrical current proportional to the glucose concentration in the sample (Wu et al. 2015). The resulting current is measured and correlated to the glucose concentration in the sample. Enzymatic glucose biosensors are highly selective for glucose, as glucose oxidase specifically catalyzes the oxidation of glucose. However, the performance of these biosensors can be affected by various factors, such as oxygen concentration, pH, and temperature, which may introduce variability in the signal. To enhance the performance of enzymatic biosensors, researchers have explored various strategies, such as immobilizing GOx on nanomaterial-based electrodes (e.g., carbon nanotubes, gold nanoparticles) to improve electron transfer efficiency, sensitivity, and stability (Bharath et al. 2015; Sabury et al. 2015; Palanisamy et al. 2014). Despite these advancements, enzymatic biosensors are susceptible to fouling and degradation over time, limiting their long-term use, particularly in continuous monitoring applications. Another approach, based on non-enzymatic glucose biosensors bypass the use of enzymes and rely on the direct electro-oxidation of glucose at the electrode surface (Park et al. 2006). These biosensors typically utilize nanomaterials, such as metal oxides (Zhu et al. 2016) (e.g., CuO, NiO) or noble metals (Hoa et al. 2015; Dilmac and Guler 2020) (e.g., Pt, Au), which possess catalytic properties that facilitate the oxidation of glucose. Non-enzymatic biosensors offer several advantages over enzymatic systems, including greater stability, resistance to fouling, and independence from oxygen concentration (Naikoo et al. 2021). The basic principle of non-enzymatic glucose sensing involves the direct oxidation of glucose at a high-potential electrode surface, typically facilitated by a nanostructured catalyst. The oxidation of glucose generates an electrical current that is proportional to the glucose concentration (Wang et al. 2013). While non-enzymatic glucose biosensors offer the potential for long-term use and greater stability, they often suffer from lower selectivity compared to enzymatic sensors, as other electroactive species in the sample (e.g., ascorbic acid, uric acid) may interfere with glucose detection. To address this challenge, researchers have focused on developing selective nanomaterials and electrode coatings that enhance the specificity of non-enzymatic biosensors. Both enzymatic and non-enzymatic glucose biosensors have played a pivotal role in advancing glucose monitoring technologies (Hassan et al. 2021). However, further research is needed to address the limitations of each system and to develop biosensors that are more reliable, accurate, and user-friendly, particularly in the context of continuous glucose monitoring.

A major area of interest in the development of glucose biosensors is the pursuit of non-invasive monitoring methods that eliminate the need for blood sampling or subcutaneous sensor insertion (Soni and Jha 2017; Rachim and Chung 2019). Non-invasive glucose monitoring involves the detection of glucose in alternative body fluids, such as sweat (Xu et al. 2021), saliva (Soni and Jha 2015), or interstitial fluid (Yao et al. 2021), using wearable biosensors (Kim et al. 2018b). These approaches have the potential to significantly improve patient adherence and comfort, making glucose monitoring more accessible and convenient for individuals with diabetes. As sweat contains various metabolites, including glucose, it has been employed to detect using biosensors integrated into wearable devices, such as smartwatches, skin patches, or textile-based sensors (Zhao et al. 2019; Manasa et al. 2022; Li et al. 2023). Sweat-based glucose monitoring offers the advantage of non-invasive sample collection, as sweat can be easily collected from the skin's surface without the need for invasive procedures. However, the concentration of glucose in sweat is typically much lower than in blood, and sweat composition can be influenced by factors such as hydration status, temperature, and physical activity (Baker et al. 2019). To overcome these challenges, researchers have developed highly sensitive biosensors that can detect glucose at low concentrations in sweat. For example, enzymatic glucose sensors immobilized on flexible, stretchable substrates have been integrated into skin patches that can continuously monitor glucose levels in sweat. These devices often include additional sensors to measure sweat rate or temperature, allowing for real-time calibration and improved accuracy (Zhou et al. 2021). Despite the promise of sweat-based glucose monitoring, there are still challenges related to variability in sweat production and the need for reliable calibration against blood glucose levels. Additionally, sweat-based sensors may not be suitable for individuals who do not produce sufficient sweat, such as those with sedentary lifestyles or certain medical conditions. Saliva is another alternative biofluid that has been explored for non-invasive glucose monitoring. Saliva contains glucose, albeit at lower concentrations than blood, and its composition is less variable than sweat, making it an attractive medium for biosensing. Saliva-based glucose sensors typically involve the immobilization of glucose oxidase on an electrode surface, similar to blood-based glucose sensors (Giaretta 2024). One of the primary challenges of saliva-based glucose monitoring is the low concentration of glucose in saliva, which requires highly sensitive biosensors capable of detecting minute changes in glucose levels (Jurysta 2009). Additionally, saliva composition can be affected by factors such as food intake, oral hygiene, and hydration status, which may introduce variability in glucose readings. Recent advances in microfluidic technology and nanomaterial-based biosensors have enabled the development of saliva-based glucose sensors with improved sensitivity and specificity (Panwar et al. 2024). However, like sweat-based sensors, saliva-based glucose monitoring still faces challenges related to calibration and long-term stability. Tear fluid, which bathes the surface of the eye, is another biofluid that contains glucose and has been explored for non-invasive glucose monitoring. The concentration of glucose in tears is lower than in blood, but it correlates with blood glucose levels, making it a potential target for continuous monitoring. Tear-based glucose sensors have been integrated into smart contact lenses, which can detect glucose levels in real time and transmit the data to a mobile device (Sempionatto et al. 2019). One of the most notable examples of tear-based glucose monitoring is Google's development of a smart contact lens for diabetes management (eyeSTYLE 2025). The lens contains a miniaturized glucose sensor and wireless communication system, enabling continuous monitoring of glucose levels in tear fluid. While this approach offers a novel and potentially non-invasive solution for glucose monitoring, there are challenges related to the comfort and safety of wearing contact lenses for extended periods, as well as the need for reliable calibration against blood glucose levels. In addition, interstitial fluid, the fluid surrounding cells in tissues, has a glucose concentration that closely mirrors blood glucose levels, making it a suitable target for glucose monitoring. CGM systems currently rely on subcutaneous sensors to measure glucose in interstitial fluid, but researchers are exploring non-invasive methods for accessing interstitial fluid, such as microneedle-based sensors. Microneedles are tiny, minimally invasive needles that penetrate the outermost layer of the skin without reaching nerve endings, making them painless and suitable for continuous glucose monitoring (Takeuchi and Kim 2018). Microneedle-based glucose sensors can detect glucose in interstitial fluid and transmit the data wirelessly to a monitoring device. These sensors offer the potential for long-term, continuous glucose monitoring with minimal discomfort and interference (Yang et al. 2023). Thus, wearable sensors for non-invasive glucose monitoring represent a promising direction for improving diabetes management. While challenges remain, such as low analyte concentrations and the need for calibration, advances in biosensor design and nanotechnology are bringing these solutions closer to clinical application.

In addition to wearable sensors, several emerging technologies are being developed to improve the accuracy, convenience, and comfort of glucose monitoring. These technologies include implantable sensors, flexible electronics, and glucose-sensitive on skin tattoos, each offering unique advantages for continuous glucose monitoring. Implantable glucose sensors are designed to be inserted beneath the skin, where they can continuously monitor glucose levels in interstitial fluid (Renard 2008). These sensors offer the advantage of long-term use without the need for frequent sensor replacement, as they are protected from external environmental factors such as sweat and contamination. Implantable glucose sensors are typically based on enzymatic glucose detection, using glucose oxidase to catalyze the oxidation of glucose and generate an electrical signal. One of the most well-known examples of an implantable glucose sensor is the Eversense system, which uses an implanted sensor that lasts over 90 days (Dehennis et al. 2015). The sensor wirelessly transmits glucose data to a wearable transmitter, which in turn communicates with a mobile app. This system offers several advantages, including fewer sensor replacements and reduced maintenance compared to traditional CGM systems. However, implantable sensors require a minor surgical procedure for insertion and removal, which may limit their appeal to some patients. Another type of glucose monitoring is done by flexible electronics. Flexible electronics are a rapidly growing field in biosensing, offering the potential for comfortable, wearable devices that conform to the skin's surface (Yang and Gao 2019). These devices are typically composed of stretchable, flexible materials such as polymers or thin films, which house the glucose sensor and associated electronics. Flexible glucose sensors can be integrated into a variety of wearable platforms, such as skin patches, wristbands, or smart clothing, allowing for continuous, non-invasive glucose monitoring (Jin et al. 2022). These devices often incorporate additional sensors to measure physiological parameters such as heart rate, temperature, and activity levels, enabling a more comprehensive approach to diabetes management. The use of flexible electronics in glucose monitoring is still in the early stages of development, but initial studies have demonstrated promising results in terms of sensor performance, accuracy, and patient comfort. Furthermore, glucose-sensitive tattoos are another novel and futuristic approach to glucose monitoring, utilizing tattoo-like patches that change color in response to changes in glucose levels. These tattoos typically consist of a glucose-sensitive dye or polymer that reacts with glucose and produces a visible color change. The color change can be detected using a smartphone camera or other optical devices, providing a non-invasive, real-time glucose reading (Bandodkar et al. 2015). Glucose-sensitive tattoos offer the potential for completely non-invasive glucose monitoring, with the added benefit of being visually discrete and easy to use. However, this technology is still in the experimental stages, and further research is needed to improve the sensitivity, stability, and accuracy of the sensors.

8.4 Insulin Delivery Bioelectronic Systems Towards Diabetes Management

The management of diabetes, particularly Type 1 diabetes mellitus (T1DM), revolves around maintaining optimal blood glucose levels through precise insulin administration. Insulin, a hormone essential for glucose uptake by cells, is either deficient or ineffective in people with diabetes, requiring exogenous insulin therapy for survival and glycemic control (Leto and Saltiel 2012). The journey of insulin delivery has evolved from early, rudimentary methods of manual injections to sophisticated bioelectronic systems designed to automate insulin delivery and achieve tight glycemic control (Bally et al. 2018). While insulin injections were life-saving, they posed significant challenges in terms of precision and timing. Manual insulin injections require patients to carefully calculate insulin doses based on blood glucose levels, carbohydrate intake, and physical activity, often leading to errors and increased risk of hypo- or hyperglycemia. Additionally, the fixed dosing schedules of traditional insulin regimens, often involving long-acting insulin analogs combined with rapid-acting insulin for meals, do not account for the body's dynamic insulin requirements throughout the day. In response to these challenges, insulin pump therapy was introduced in the 1980s as an alternative to multiple daily injections (MDIs) (Torrance et al. 2003). Insulin pumps deliver continuous subcutaneous insulin infusion (CSII) via a small catheter inserted under the skin. Patients can program the pump to deliver basal insulin throughout the day and bolus insulin at mealtimes, providing greater flexibility and control over blood glucose levels. The ability to adjust insulin delivery based on individual needs, including changes in activity levels or stress, represents a significant advancement in diabetes management. However, insulin pumps still require manual input from the patient to adjust insulin doses based on blood glucose readings, which can be cumbersome and prone to human error. This has led to the development of automated insulin delivery systems, often referred to as closed-loop or artificial pancreas systems, which aim to take human guesswork out of diabetes management by integrating real-time glucose monitoring with automated insulin delivery (Davis et al. 2023).

The concept of the artificial pancreas involves a closed-loop system that mimics the function of the human pancreas by continuously monitoring blood glucose levels and automatically delivering the appropriate amount of insulin. These systems integrate continuous glucose monitoring (CGM) with insulin pumps and advanced algorithms, enabling real-time adjustments in insulin delivery without patient intervention. In a closed-loop system, glucose data from a CGM sensor is transmitted to a control algorithm, which calculates the optimal insulin dose based on glucose trends, insulin sensitivity, and other factors (Bekiari et al. 2018). The algorithm then communicates with the insulin pump to deliver the calculated dose of insulin. By continuously adjusting insulin delivery in response to real-time glucose fluctuations, closed-loop systems offer the potential for more stable glycemic control and a reduced risk of hypo- and hyperglycemia. The first commercially available closed-loop systems, such as the Medtronic MiniMed 670G, were hybrid systems that required manual input from the user for bolus insulin doses at mealtimes (Knebel and Neumiller 2019). These systems provided automated basal insulin delivery but still relied on the patient to determine bolus doses based on carbohydrate intake. Despite this limitation, hybrid closed-loop systems represented a significant step forward in diabetes management, as they provided more consistent glycemic control than manual insulin pumps or MDI. Clinical trials of hybrid closed-loop systems demonstrated improved glycemic outcomes, including lower HbA1c levels and reduced time spent in hyperglycemia (Abraham et al. 2025). Additionally, these systems were associated with a lower risk of hypoglycemia, particularly during the night, when blood glucose levels tend to fluctuate and are harder to manage manually. However, patients still had to remain actively involved in their care, particularly in calculating bolus doses for meals.

The goal of diabetes technology is to create a fully closed-loop system that automates both basal and bolus insulin delivery, eliminating the need for manual input. Fully closed-loop systems, also known as "bionic pancreas" systems, aim to provide autonomous glycemic control by continuously adjusting insulin delivery based on real-time glucose data and advanced predictive algorithms (Russel et al. 2014). One example of a fully closed-loop system is the iLet Bionic Pancreas, which is currently undergoing clinical trials (Oser et al. 2025). The iLet system combines a CGM sensor with an insulin pump and a sophisticated control algorithm that automatically delivers insulin based on glucose readings and trends. Unlike hybrid systems, the iLet does not require the user to manually input carbohydrate intake or bolus doses, offering true autonomy in diabetes management. Preliminary studies of fully closed-loop systems have shown promising results, with patients achieving near-normal glycemic control and spending more time within the target glucose range compared to traditional methods. These systems are particularly beneficial for reducing the burden of diabetes management, as they eliminate the need for frequent blood glucose checks, carbohydrate counting, and insulin dose adjustments. However, the success of fully closed-loop systems depends on the accuracy of CGM sensors and the reliability of the control algorithm in predicting and responding to glucose fluctuations.

The integration of bioelectronic systems with insulin delivery devices has been instrumental in advancing real-time data collection and insulin optimization. Bioelectronics refers to the fusion of biological components, such as sensors or tissue, with electronic systems that can process and transmit data (Dirr et al. 2020). In diabetes management, bioelectronic systems enable continuous monitoring of glucose levels and facilitate more precise, personalized insulin delivery. The CGMs play a pivotal role in bioelectronic systems by providing continuous, real-time glucose data that informs insulin dosing decisions. CGM sensors typically measure glucose in interstitial fluid, which correlates with blood glucose levels but may lag behind during rapid changes in glucose concentrations, such as after meals. Modern CGM systems, such as the Dexcom G6 and the Abbott FreeStyle Libre, have improved accuracy, sensor longevity, and ease of use, making them essential components of closed-loop insulin delivery systems (Castorino et al. 2020; Jafri et al. 2020). The real-time data provided by CGM allows bioelectronic systems to detect trends in glucose levels, such as rising or falling glucose, and predict future glucose values based on past trends. This predictive capability enables automated insulin delivery systems to make proactive adjustments to insulin dosing, such as reducing insulin delivery during exercise to prevent hypoglycemia or increasing insulin delivery in anticipation of postprandial hyperglycemia.

A key component of bioelectronic insulin delivery systems is the control algorithm, which processes real-time glucose data and calculates the optimal insulin dose. Traditional algorithms, such as proportional-integral-derivative (PID) controllers, use feedback from glucose readings to make adjustments to insulin delivery. While these algorithms are effective in managing basal insulin needs, they may struggle to account for the complex and dynamic factors that influence glucose levels, such as meals, exercise, stress, and illness (Padula and Visioli 2012). To address this challenge, researchers are increasingly turning to machine learning algorithms, which can learn from large datasets of glucose readings, insulin doses, and patient behaviors to improve insulin delivery decisions. Machine learning algorithms can identify patterns in glucose data and predict future glucose fluctuations with greater accuracy than traditional models. For example, machine learning models can be trained to recognize the impact of different types of meals on glucose levels, allowing the system to adjust insulin delivery based on the specific carbohydrate load of a meal (Chan et al. 2023). Incorporating machine learning into closed-loop insulin delivery systems represents a significant advancement in personalized diabetes care, as these algorithms can adapt to the individual needs of each patient. By continuously learning from real-time data, machine learning algorithms can fine-tune insulin dosing to optimize glycemic control while minimizing the risk of hypo- or hyperglycemia.

While glucose levels are the primary target for bioelectronic insulin delivery systems, other physiological parameters can also influence glucose metabolism and insulin sensitivity. For example, physical activity, heart rate, stress levels, and sleep patterns can all impact blood glucose levels (Sevil et al. 2021). Integrating additional biosensors into bioelectronic systems allows for a more comprehensive approach to diabetes management. For instance, wearable devices that monitor physical activity and heart rate can provide valuable data on the patient's energy expenditure and insulin sensitivity (Prieto-Avalos et al. 2022). Bioelectronic systems can use this information to adjust insulin delivery during periods of increased physical activity, such as exercise, when insulin requirements typically decrease. Similarly, stress monitoring through biosensors that measure cortisol levels or heart rate variability can help predict glucose fluctuations associated with emotional stress, allowing for more accurate insulin dosing. The integration of multimodal biosensing into bioelectronic insulin delivery systems represents a shift toward more holistic diabetes management. By considering the many factors that influence glucose levels, these systems can provide more personalized and precise insulin therapy, ultimately improving patient outcomes and quality of life.

Despite the significant advancements in bioelectronic insulin delivery systems, several challenges remain that must be addressed to ensure the widespread adoption and success of these technologies. CGM sensors are a critical component of bioelectronic insulin delivery systems, as they provide the real-time glucose data needed for automated insulin dosing. However, the accuracy and reliability of CGM sensors can be affected by factors such as sensor lag, calibration errors, and interference from other electroactive species. While modern CGM systems have improved accuracy, particularly in the hypoglycemic range, further advancements are needed to ensure reliable glucose measurements over extended periods. The success of closed-loop systems depends on the performance of the control algorithm, which must accurately calculate insulin doses based on real-time glucose data. While machine learning algorithms offer the potential for more personalized and predictive insulin delivery, they require large datasets and continuous learning to adapt to individual patient needs. Ensuring the safety and efficacy of these algorithms in diverse patient populations is a key challenge. While bioelectronic insulin delivery systems offer significant benefits in terms of glycemic control, patient acceptance and adherence are critical to their success (Waataja et al. 2022). Some patients may find the devices cumbersome or uncomfortable to wear, particularly if they require frequent sensor replacements or involve bulky components. Additionally, the cognitive burden of learning to use these systems, particularly for older adults or individuals with limited technological literacy, may pose a barrier to adoption. The cost of bioelectronic insulin delivery systems, including CGM sensors, insulin pumps, and control algorithms, can be prohibitively expensive for many patients, particularly in low-resource settings. Ensuring equitable access to these advanced technologies is essential for reducing disparities in diabetes care and improving outcomes for all patients.

8.5 Bioelectronics for Monitoring Non-Glucose Biomarkers Towards Diabetes Management

Diabetes is a multifaceted metabolic disorder that, while primarily characterized by impaired glucose regulation, also involves complex biochemical and physiological imbalances. The traditional focus of diabetes management has been on controlling blood glucose levels through insulin administration or oral hypoglycemic agents (Rizza et al. 1980). However, it is increasingly evident that a glucose-centric approach to diabetes management is insufficient for comprehensive disease control. Diabetes, particularly Type 2 diabetes mellitus (T2DM), is associated with a range of metabolic derangements, including insulin resistance, dyslipidemia, chronic inflammation, and endothelial dysfunction, which contribute to long-term complications such as cardiovascular disease, nephropathy, and neuropathy (Lima et al. 2022). Non-glucose biomarkers provide valuable insights into these underlying pathophysiological processes and offer the potential for a more holistic and precise approach to diabetes management. By tracking these biomarkers alongside glucose, healthcare providers can gain a more complete understanding of a patient's metabolic health, enabling early detection of complications, personalized treatment adjustments, and better prevention strategies. The growing interest in non-glucose biomarkers in diabetes management reflects the shift toward precision medicine, where treatment decisions are tailored to the individual's unique biochemical profile, rather than relying solely on glucose levels. This section explores several key non-glucose biomarkers, their relevance in diabetes management, and how emerging biosensing technologies can facilitate their monitoring in clinical practice.

The metabolic complexity of diabetes has led to the identification of several non-glucose biomarkers that are valuable for assessing the overall metabolic state of individuals with diabetes (Jin and Ma 2021). These biomarkers can provide insights into insulin sensitivity, lipid metabolism, inflammation, and oxidative stress, factors that play a crucial role in the progression of diabetes and its associated complications.

Insulin and C-Peptide Levels:

Insulin is the primary hormone responsible for regulating blood glucose levels by facilitating glucose uptake into cells. In Type 1 diabetes mellitus (T1DM), insulin production is severely reduced or absent due to autoimmune destruction of pancreatic beta cells. In contrast, individuals with T2DM often exhibit hyperinsulinemia in the early stages of the disease, as the pancreas compensates for insulin resistance by producing more insulin. Over time, however, beta-cell function declines, leading to inadequate insulin production. Measuring insulin levels in individuals with T2DM can provide valuable information about the severity of insulin resistance and beta-cell function (LeRoith 2002). C-peptide, a byproduct of insulin production, is often used as a proxy for endogenous insulin secretion, particularly in individuals receiving exogenous insulin therapy. Elevated C-peptide levels indicate preserved beta-cell function, while low C-peptide levels suggest beta-cell exhaustion. Monitoring insulin and C-peptide levels can help guide treatment decisions, such as the initiation of insulin therapy or the use of medications that preserve beta-cell function.

Ketone Bodies:

Ketone bodies, including beta-hydroxybutyrate (BHB), acetoacetate, and acetone, are produced by the liver during periods of fasting, prolonged exercise, or insulin deficiency (Puchalska and Crawford 2021). In the absence of sufficient insulin, the body cannot effectively utilize glucose for energy and instead breaks down fat stores, resulting in the production of ketones. While ketone production is a normal metabolic response to fasting or carbohydrate restriction, excessive ketone accumulation can lead to diabetic ketoacidosis (DKA), a life-threatening complication of T1DM. Ketone monitoring is particularly important in individuals with T1DM, as it helps prevent the development of DKA by identifying elevated ketone levels early. Traditionally, ketones are measured in urine or blood samples using point-of-care tests (Chong and Reineke 2016). However, advances in biosensing technology have led to the development of wearable devices capable of non-invasive ketone detection, offering a more convenient and continuous approach to ketone monitoring (Moonla et al. 2024; Shen et al. 2024).

Lactate:

Lactate is a byproduct of anaerobic glycolysis, the metabolic pathway that generates energy when oxygen is limited. Elevated lactate levels, or lactic acidosis, can occur in individuals with diabetes, particularly during episodes of hypoxia, shock, or sepsis. Lactic acidosis is also a potential side effect of certain diabetes medications, such as metformin, particularly in individuals with renal impairment or other contraindications. Monitoring lactate levels in individuals with diabetes can provide early warning of metabolic derangements and guide adjustments in medication regimens. For example, lactate measurements can help clinicians identify patients at risk of lactic acidosis and make decisions about whether to discontinue metformin or other medications associated with lactate accumulation. Emerging biosensing technologies are enabling continuous lactate monitoring in real time, providing a more dynamic picture of a patient's metabolic state during periods of stress or illness (Jia et al. 2013; Xu et al. 2023).

Lipid Profiles (Triglycerides, LDL, HDL, and Total Cholesterol):

Dyslipidemia, characterized by elevated levels of triglycerides and low-density lipoprotein (LDL) cholesterol and reduced levels of high-density lipoprotein (HDL) cholesterol, is a common feature of T2DM and contributes to the increased risk of cardiovascular disease in individuals with diabetes. Insulin resistance impairs the body's ability to regulate lipid metabolism, leading to an accumulation of circulating lipids and the development of atherogenic lipoprotein particles. Monitoring lipid profiles is essential for cardiovascular risk assessment and management in individuals with diabetes. Regular lipid testing helps identify patients who may benefit from lipid-lowering therapies, such as statins, and lifestyle interventions to reduce the risk of atherosclerosis and other cardiovascular complications. Biosensors capable of real-time lipid monitoring are being developed, offering the potential for more frequent and convenient lipid assessments in clinical practice (Ye et al. 2015).

Inflammatory Markers (C-Reactive Protein, Interleukin-6, Tumor Necrosis Factor-Alpha):

Chronic inflammation plays a central role in the pathogenesis of insulin resistance, beta-cell dysfunction, and the development of diabetes complications. Several pro-inflammatory cytokines, including interleukin-6 (IL-6), tumor necrosis factor-alpha (TNF-alpha), and C-reactive protein (CRP), are elevated in individuals with diabetes and are associated with poor glycemic control and an increased risk of cardiovascular disease. Monitoring inflammatory markers provides valuable insights into the inflammatory state of individuals with diabetes and can help guide interventions aimed at reducing inflammation, such as lifestyle modifications, anti-inflammatory medications, or dietary changes. Advances in biosensing technology have made it possible to measure inflammatory markers in real time using wearable or implantable sensors, offering the potential for continuous monitoring of inflammation in diabetes care (Gouveia et al. 2025; Lee et al. 2020).

Oxidative Stress Markers (Malondialdehyde, Advanced Glycation End Products, and 8-OHdG):

Oxidative stress, resulting from an imbalance between the production of reactive oxygen species (ROS) and the body's antioxidant defenses, is a key contributor to the development of diabetes complications, particularly cardiovascular disease, nephropathy, and neuropathy. Hyperglycemia leads to the production of ROS, which damage cellular structures, including lipids, proteins, and DNA. Oxidative stress markers, such as malondialdehyde (MDA), advanced glycation end products (AGEs), and 8-hydroxy-2′-deoxyguanosine (8-OHdG), provide insights into the extent of oxidative damage in individuals with diabetes. Monitoring oxidative stress markers can help identify patients at risk of complications and guide antioxidant therapies aimed at reducing oxidative damage. Biosensors capable of detecting oxidative stress markers in real time are currently under development, offering the potential for more precise monitoring of oxidative stress in diabetes care (Bencivenga et al. 2023; Aravapally et al. 2025).

8.6 Multimodal Modules for Diabetes Monitoring

Multimodal biosensing module involves the simultaneous detection of multiple biomarkers using a single device, providing a more comprehensive picture of an individual's physiological state. In the context of diabetes management, multimodal biosensing offers the potential to monitor not only glucose but also a range of non-glucose biomarkers that reflect insulin sensitivity, inflammation, oxidative stress, and shift in lipid metabolism (Chu et al. 2022). This approach enables clinicians to tailor treatment strategies based on a more detailed understanding of the patient's metabolic health. The integration of multiple biosensors into a single wearable or implantable device is a growing area. These devices can continuously monitor several biomarkers in real time, providing dynamic insights into the complex interplay of metabolic pathways in diabetes. For example, a multimodal sensor might measure glucose, lactate, and ketones simultaneously, allowing for a more accurate assessment of the patient's metabolic state during exercise, fasting, or illness. Similarly, the combination of glucose monitoring with lipid or inflammatory marker detection could help clinicians identify patients at high risk of cardiovascular complications and adjust treatment accordingly. Wearable and implantable biosensors represent a key innovation in multimodal biosensing for diabetes care (Coates et al. 2016). These devices are designed to be worn on the skin, embedded in textiles, or implanted subcutaneously, where they continuously monitor multiple biomarkers in interstitial fluid, sweat, or other biological fluids. The data collected by these sensors is transmitted wirelessly to a mobile device or cloud-based platform, where it can be analyzed and used to inform treatment decisions. Recent advances in nanotechnology, materials science, and microfluidics have enabled the development of highly sensitive, flexible, and biocompatible sensors capable of detecting multiple biomarkers in real time (Gideon et al. 2024). These sensors often incorporate nanomaterials, such as graphene, carbon nanotubes, or gold nanoparticles, which enhance the sensitivity and specificity of the device (Shekher et al. 2023; Kummari et al. 2020). The use of microfluidic channels allows for the precise control of fluid flow, enabling the detection of small changes in biomarker concentrations over time. Non-invasive biosensors, which detect biomarkers in sweat, saliva, or tears, are an attractive option for continuous monitoring of multiple biomarkers in individuals with diabetes. Saliva-based sensors offer another non-invasive approach to multimodal monitoring. Saliva contains a variety of biomarkers, including glucose, ketones, and inflammatory markers, making it a useful medium for diabetes monitoring. Saliva-based sensors have the advantage of being less affected by external factors, such as hydration status or temperature, compared to sweat-based sensors. However, the lower concentrations of biomarkers in saliva present a challenge for sensor sensitivity, requiring the development of highly sensitive detection methods. In addition, implantable biosensors offer the potential for long-term, continuous monitoring of multiple biomarkers in interstitial fluid, providing a more stable and reliable data stream compared to non-invasive sensors (Gray et al. 2018). These devices are typically inserted subcutaneously and can remain in place for weeks or months, reducing the need for frequent sensor replacement. Implantable multimodal sensors are particularly well-suited for monitoring glucose, ketones, and insulin levels in individuals with diabetes, offering a more comprehensive approach to metabolic control. For example, the Eversense CGM system, which is an implantable glucose sensor, is being explored for its potential to incorporate additional biosensors for ketones or lactate (Guiseppi-Elie et al. 2005; Ridder et al. 2024). By combining multiple sensors in a single device, implantable multimodal systems can provide a more detailed understanding of the patient's metabolic state and enable more precise adjustments in insulin dosing or other therapies.

8.7 Challenges and Innovations in Bioelectronic Devices for Diabetes Monitoring

8.7.1 Technological Challenges in Biosensor Accuracy, Calibration, and Longevity

The development of biosensing devices for diabetes management, particularly glucose monitoring systems, has made significant strides over the past few decades. However, there remain several persistent technological challenges that must be addressed to optimize their performance, enhance usability, and ensure long-term reliability. These challenges span several domains, including sensor accuracy, calibration, and device longevity.

Sensor Accuracy Accuracy is paramount in glucose monitoring, as imprecise measurements can lead to inappropriate therapeutic decisions, such as incorrect insulin dosing, which can result in life-threatening hypo- or hyperglycemic episodes. The CGM systems, while revolutionary in providing real-time data, face inherent limitations in their accuracy compared to traditional BGM. CGM devices measure glucose levels in interstitial fluid rather than blood, and there is often a time lag, ranging from 5 to 15 minutes, between the glucose concentration in the blood and the interstitial fluid. This delay is particularly problematic during periods of rapid glucose fluctuations, such as after meals or during physical activity. Moreover, sensor inaccuracies can arise from various physiological and environmental factors, including temperature, hydration, and sweat composition, which may interfere with the sensor's ability to produce accurate glucose readings. To address these issues, researchers are working on improving the accuracy of biosensors through various means, including:

- *Advanced Calibration Algorithms*: Calibration is essential for ensuring sensor accuracy. Traditional CGM devices require periodic finger-stick calibrations using a BGM device to adjust for any discrepancies between the interstitial and blood glucose levels. Newer sensors, such as the Dexcom G6 and Abbott FreeStyle Libre 2, have been developed with factory calibration, which eliminates the need for manual calibration by the user (Davis et al. 2021; Nakagawa et al. 2022). However, achieving truly calibration-free devices remains a challenge, as sensors must adapt to individual physiological variations.
- *Improved Sensor Materials*: The use of advanced materials, such as nanomaterials, can enhance the sensitivity and selectivity of glucose biosensors (Nichols et al. 2013). For example, sensors based on graphene, carbon nanotubes, or gold nanoparticles have been shown to improve electron transfer kinetics, leading to more accurate and reliable glucose detection (Carbone et al. 2015; Branagan and Breslin 2019). These materials also enable the miniaturization of sensors, making them less invasive and more comfortable for the user.

Calibration and Sensor Drift One of the significant challenges in CGM technology is the need for calibration to maintain accuracy. Calibration is the process of adjusting the CGM system based on known glucose values (usually obtained from finger-stick measurements) to correct for any discrepancies between the CGM sensor and blood glucose levels. In many early CGM systems, users had to calibrate the device multiple times a day, which added to the burden of diabetes management and reduced the appeal of these devices.

Sensor drift, a phenomenon where the sensor's accuracy degrades over time, compounds the issue of calibration. Over the course of days or weeks, CGM sensors can accumulate biofouling (the buildup of proteins, cells, or other biological substances on the sensor's surface), which impairs their ability to measure glucose levels accurately. Biofouling can lead to signal attenuation, delayed response times, and a loss of sensor sensitivity, necessitating frequent recalibration or sensor replacement (Tong et al. 2025). Innovations aimed at addressing these challenges include:

- *Factory-Calibration Systems*: Some modern CGM systems, such as the Abbott FreeStyle Libre, are designed to be factory-calibrated, meaning that they do not require users to perform finger-stick calibrations during their wear period (Szadkowska et al. 2021). These systems rely on improved sensor design and more accurate manufacturing processes to deliver consistent results without the

need for user intervention. However, maintaining accuracy over extended periods (10–14 days) without calibration remains an ongoing challenge.
- *Anti-Fouling Coatings*: One approach to minimizing sensor drift and extending sensor longevity is the use of anti-fouling coatings on the sensor surface. These coatings are designed to repel biological substances that can cause biofouling, thereby preserving the sensor's functionality. Materials such as polyethylene glycol (PEG), zwitterionic polymers, and hydrophilic coatings are being explored for their anti-fouling properties in biosensor applications (Lin and Li 2020).

Sensor Longevity and Replacement Frequency The longevity of glucose sensors is another key challenge in diabetes biosensing. Most CGM sensors are designed to last between 7 and 14 days before they need to be replaced. While this represents a significant improvement over early CGM systems, which required daily sensor replacement, it still imposes a burden on users. Each sensor replacement involves removing the old sensor, inserting a new one, and restarting the monitoring system. For many users, this can be a source of discomfort and inconvenience, particularly for those with needle phobia or sensitive skin. Research efforts are focused on extending the lifetime of CGM sensors, with the ultimate goal of developing sensors that can last for several months or even years without replacement. Strategies for extending sensor longevity include:

- *Encapsulation Technologies*: One approach to prolonging sensor lifespan is the use of encapsulation materials that protect the sensor from the harsh biological environment in the body. These materials can prevent biofouling, reduce immune system response, and protect the sensor from degradation. For example, encapsulating the sensor in a hydrogel matrix can create a more biocompatible interface between the sensor and the body, reducing inflammation and extending sensor life (Andersson et al. 2009).
- *Self-Cleaning Sensors*: Another innovation is the development of self-cleaning sensors that can remove biofouling automatically. These sensors use microfluidic channels or mechanical vibration to dislodge biological materials from the sensor surface, thereby maintaining accuracy and prolonging sensor functionality (Wang et al. 2024).

8.7.2 Patient Adherence and Usability Concerns in Biosensor Devices

For any biosensing device to be effective, it must be designed with the user in mind. Usability and patient adherence are critical factors that determine the success of a biosensing device, particularly in the context of chronic diseases like diabetes, where daily monitoring and management are required. While advancements in biosensor technology have improved the accuracy and functionality of these devices, challenges remain in ensuring that they are easy to use, comfortable, and conducive to long-term adherence.

Invasiveness and Comfort One of the most significant barriers to patient adherence is the invasiveness of current biosensing devices. CGM systems, for example, require users to insert a sensor subcutaneously, which can be uncomfortable or painful for some individuals, particularly for those with needle phobia or sensitive skin. Additionally, the adhesive used to attach the sensor to the skin can cause irritation or allergic reactions, leading to discomfort and reduced adherence. To address these concerns, researchers are developing less invasive or non-invasive biosensing technologies, such as:

- *Microneedle Sensors*: Microneedle-based sensors are designed to penetrate the outermost layer of the skin without reaching the nerve endings, making them virtually painless. These sensors can measure glucose levels in interstitial fluid and transmit the data wirelessly to a mobile device (Ribet et al. 2018). Microneedles offer the potential for more comfortable and user-friendly glucose monitoring, encouraging greater adherence.
- *Non-invasive Sensors*: Non-invasive glucose monitoring technologies, such as optical sensors or sweat-based biosensors, aim to eliminate the need for needles or subcutaneous sensor insertion altogether (Vashist 2012). For example, optical sensors use light to measure glucose levels through the skin, while sweat-based sensors detect glucose in perspiration (Delbeck et al. 2019; Zafar et al. 2022). While these technologies are still in the early stages of development, they hold promise for improving patient comfort and adherence.

Complexity of Use Another barrier to adherence is the complexity of using biosensing devices, particularly for older adults or individuals with limited technological literacy. Some biosensors require multiple steps to set up, calibrate, and operate, which can be overwhelming for users. Additionally, devices that require frequent maintenance, such as sensor replacement or recalibration, can reduce adherence over time.

To improve usability, device manufacturers are focusing on simplifying the user experience, with innovations such as:

- *Intuitive Interfaces*: Developing user-friendly interfaces that are easy to navigate, even for individuals with limited technological experience, is essential for improving adherence. Devices should provide clear instructions, minimal steps for operation, and visual or auditory alerts to guide the user through the monitoring process.
- *Automated Systems*: Automation is key to reducing the burden of diabetes management. Automated insulin delivery systems, for example, eliminate the need for manual insulin dosing by integrating CGM data with insulin pumps (Vettoretti and Facchinetti 2019). These systems adjust insulin delivery based on real-time glucose readings, simplifying the user's daily routine and improving adherence.

Device Aesthetics and Discretion For many individuals, the visibility of biosensing devices can be a concern, particularly for those who wish to manage their diabetes discretely. Large or bulky devices may draw unwanted attention or interfere with daily activities, leading to lower adherence. To address this, device manufacturers are designing smaller, more discrete sensors that can be worn comfortably under clothing or embedded in accessories such as wristbands or watches.

The development of flexible, stretchable electronics also holds promise for creating biosensors that conform to the body's contours, making them more comfortable and less noticeable. These innovations in device design can enhance both comfort and aesthetics, leading to improved adherence.

8.7.3 Innovations to Address Usability, Accuracy, and Long-Term Stability

To overcome the challenges associated with accuracy, usability, and long-term stability, ongoing innovations in biosensing technology are aimed at enhancing the performance, comfort, and convenience of these devices. Several key areas of innovation include the development of advanced materials, integration with artificial intelligence (AI), and the use of multimodal biosensors for comprehensive monitoring (Vettoretti et al. 2020).

Advanced Sensor Materials The development of new materials with enhanced properties, such as improved sensitivity, selectivity, and biocompatibility, is a major focus of research in diabetes biosensing. Nanomaterials, such as graphene, gold nanoparticles, and carbon nanotubes, have shown great promise in improving the performance of biosensors by facilitating faster electron transfer and reducing signal interference from other substances in biological fluids (Rahul et al. 2023; Kummari et al. 2019; Kafley et al. 2023). In addition to nanomaterials, researchers are exploring the use of biocompatible polymers and hydrogels that can encapsulate biosensors, protecting them from biofouling and immune responses. These materials can extend sensor longevity, reduce the need for frequent calibration, and improve overall sensor stability (Chan et al. 2022).

Integration with Artificial Intelligence (AI) The integration of AI into biosensing devices offers the potential to improve both accuracy and usability. AI algorithms can analyze large datasets of glucose readings and other biomarker data to identify patterns and trends, enabling more personalized and predictive diabetes management (Ahmed et al. 2023). For example, AI-driven algorithms can predict glucose fluctuations based on meal timing, physical activity, and other factors, allowing biosensors to adjust insulin delivery proactively. AI also enhances the usability of biosensing devices by automating many of the tasks that users would otherwise have to perform manually. For example, AI algorithms can automatically calibrate biosensors, reducing the need for user intervention and improving the overall accuracy of the device (Jin et al. 2023).

Multiplexed Biosensor Multiplex biosensors, which are capable of detecting multiple biomarkers simultaneously, represent the next frontier in diabetes biosensing (Tai et al. 2024). By measuring glucose alongside other relevant biomarkers, such as ketones, lactate, or inflammatory markers, multimodal biosensors provide a more comprehensive picture of an individual's metabolic state (Moonla et al. 2024; Chu et al. 2022). This holistic approach enables more precise adjustments in therapy and reduces the risk of complications. Multiplex biosensors are being developed using microfluidic technology, which allows for the simultaneous detection of multiple analytes in a single device. These sensors can be integrated into wearable platforms, such as skin patches or wristbands, providing continuous monitoring of multiple biomarkers in real time.

8.8 Conclusions and Outlook

The field of diabetes management is undergoing a profound transformation, driven by the rapid advancements in bioelectronics and the broader shift towards precision healthcare and personalized medicine. Precision healthcare refers to the tailoring of medical treatment to the individual characteristics of each patient, taking into account genetic, environmental, and lifestyle factors. The integration of bioelectronics into diabetes care represents a transformative shift toward precision healthcare and personalized medicine, enabling real-time, continuous monitoring of glucose levels and other biomarkers crucial for comprehensive disease management. This approach leverages advanced data analytics, AI, and machine learning to tailor interventions to the individual characteristics of each patient, optimizing glycemic control and reducing the risk of complications. Precision medicine in diabetes aims to create individualized treatment plans based on metabolic profiles, insulin sensitivity, lifestyle habits, and risk factors. Bioelectronics facilitates this through real-time monitoring and dynamic adjustment of insulin therapy. For example, closed-loop systems or "artificial pancreas" technologies combine CGM with insulin pumps to automate insulin delivery, reducing the burden on patients. Furthermore, future bioelectronic devices will integrate multi-biomarker monitoring, and tracking parameters such as ketones, lactate, and inflammatory markers to provide a more holistic view of metabolic health. Additionally, genetic and epigenetic information could enhance treatment personalization, allowing algorithms to adjust therapy based on an individual's genetic predispositions and lifestyle-influenced epigenetic changes. Bioelectronic devices, combined with telemedicine platforms, allow for continuous, remote patient monitoring, enabling proactive, data-driven interventions. Real-time data empowers healthcare providers to detect patterns in glucose levels and intervene early to prevent complications. Patient engagement is enhanced through user-friendly apps that provide personalized feedback, improving adherence to treatment plans. However, interoperability between devices and data platforms remains a challenge, necessitating standardized protocols to ensure seamless data integration and actionable insights for clinicians. Next-generation bioelectronic devices are becoming more patient-friendly through innovations in implantable, non-invasive, and wearable technologies. Implantable glucose monitors, such as the Eversense CGM, offer long-term functionality with minimal maintenance, while advances in biocompatible materials aim to extend sensor lifespans to a year or more. Non-invasive glucose monitoring methods, including optical, sweat-based, and saliva-based sensors, promise to eliminate the need for needle-based systems, enhancing patient comfort and adherence. Wearable devices, equipped with multimodal sensors, are also emerging as versatile tools for tracking additional biomarkers, such as lactate and ketones, alongside glucose. As bioelectronic devices become more integrated into diabetes care, addressing regulatory and ethical issues is crucial. Regulatory agencies like the FDA and EMA must balance safety evaluations with the need for timely innovation. Data privacy and security are significant concerns, given the sensitive health data transmitted by bioelectronic devices. End-to-end encryption, multi-factor authentication, and patient control over data access are essential to protecting patient privacy. Ethical considerations also include ensuring equitable access to these technologies, particularly for underserved populations, and establishing clear guidelines on data ownership and usage to prevent exploitation. Bioelectronics is poised to revolutionize diabetes care by enabling precise, personalized, and proactive management. Through real-time monitoring of glucose and other biomarkers, bioelectronic devices empower patients and clinicians with actionable insights to optimize therapy. Telemedicine integration and advancements in non-invasive and wearable technologies will reduce the burden on patients, making diabetes management more seamless and less intrusive. However, achieving this vision requires addressing challenges related to device accuracy, accessibility, and ethical standards. Regulatory frameworks must evolve to accommodate rapid innovation, while policies should ensure affordability and equitable access to these life-changing technologies. In conclusion, continued innovation, collaboration, and ethical stewardship will enable bioelectronics to realize its full potential, transforming diabetes management into a more precise, personalized, and effective field, significantly improving outcomes for millions of patients worldwide.

The future of bioelectronics in diabetes management holds immense promise for transforming the way we monitor and treat this chronic condition. As bioelectronic devices become more advanced, they will enable more precise, personalized, and proactive care, reducing the burden on patients and healthcare providers alike. Through continuous, real-time monitoring of glucose levels and other biomarkers, bioelectronic devices will empower patients to take control of their diabetes and achieve better health outcomes. The integration of bioelectronics with telemedicine, remote patient monitoring, and artificial intelligence will further enhance the ability to provide personalized care, allowing healthcare providers to make data-driven decisions and intervene earlier to prevent complications. Meanwhile, the development of non-invasive and implantable sensors will improve patient comfort and adherence, making diabetes management more seamless and less burdensome. However, as these technologies evolve, it is crucial to address the regulatory, ethical, and societal challenges they present. Ensuring equitable access to bioelectronic devices, protecting patient data, and upholding ethical standards in their development and use are essential to realizing the full potential of bioelectronics in diabetes care.

References

Abraham MB et al (2025) Glycemic and psychosocial outcomes of advanced hybrid closed-loop therapy in youth with high HbA1c: a randomized clinical trial. Diabetes Care. American Diabetes Association. https://diabetesjournals.org/care/article/48/1/67/157201/Glycemic-and-Psychosocial-Outcomes-of-Advanced. Accessed 2025-04-23

Ahmed A, Aziz S, Qidwai U, Abd-Alrazaq A, Sheikh J (2023) Performance of artificial intelligence models in estimating blood glucose level among diabetic patients using non-invasive wearable device data. Comput Methods Programs Biomed Update 3:100094. https://doi.org/10.1016/j.cmpbup.2023.100094

Andersson O, Larsson A, Ekblad T, Liedberg B (2009) Gradient hydrogel matrix for microarray and biosensor applications: an imaging SPR study. Biomacromolecules 10(1):142–148. https://doi.org/10.1021/bm801029b

Antar SA, Ashour NA, Sharaky M, Khattab M, Ashour NA, Zaid RT, Roh EJ, Elkamhawy A, Al-Karmalawy AA (2023) Diabetes mellitus: classification, mediators, and complications; a gate to identify potential targets for the development of new effective treatments. Biomed Pharmacother 168:115734. https://doi.org/10.1016/j.biopha.2023.115734

Aravapally PSN, Chandrasekar N, Verma A, Shah RP (2025) Strategic approaches to assess and quantify the oxidative stress biomarkers in complex biological systems. Bioanalysis 17(8):1–14. https://doi.org/10.1080/17576180.2025.2486929

Baker LB et al (2019) Physiology of sweat gland function: the roles of sweating and sweat composition in human health. Temperature. https://www.tandfonline.com/doi/full/10.1080/23328940.2019.1632145. Accessed 2025-04-23

Bally L et al (2018) Closed-loop insulin delivery for glycemic control in noncritical care. N Engl J Med. https://www.nejm.org/doi/full/10.1056/NEJMoa1805233?casa_token=86Yq6tVac14AAAAA%3ARh_vPm8ayeG3Y6eM5P5wf6vO1MaenBnB-zD3VhEiBTPWKkSy7V3Q4Qjivr_LtFWfDiMOkEtY7Qigqr0. Accessed 2025-04-23

Bandodkar AJ, Jia W, Yardımcı C, Wang X, Ramirez J, Wang J (2015) Tattoo-based noninvasive glucose monitoring: a proof-of-concept study. Anal Chem 87(1):394–398. https://doi.org/10.1021/ac504300n

Bekiari E, Kitsios K, Thabit H, Tauschmann M, Athanasiadou E, Karagiannis T, Haidich A-B, Hovorka R, Tsapas A (2018) Artificial pancreas treatment for outpatients with type 1 diabetes: systematic review and meta-analysis. BMJ 361:k1310. https://doi.org/10.1136/bmj.k1310

Bencivenga D, Arcadio F, Piccirillo A, Annunziata M, Della Ragione F, Cennamo N, Borriello A, Zeni L, Guida L (2023) Plasmonic optical fiber biosensor development for point-of-care detection of malondialdehyde as a biomarker of oxidative stress. Free Radic Biol Med 199:177–188. https://doi.org/10.1016/j.freeradbiomed.2023.02.020

Bharath G, Madhu R, Chen S-M, Veeramani V, Balamurugan A, Mangalaraj D, Viswanathan C, Ponpandian N (2015) Enzymatic electrochemical glucose biosensors by Mesoporous 1D hydroxyapatite-on-2D reduced graphene oxide. J Mater Chem B 3(7):1360–1370. https://doi.org/10.1039/C4TB01651C

Branagan D, Breslin CB (2019) Electrochemical detection of glucose at physiological pH using gold nanoparticles deposited on carbon nanotubes. Sens Actuators B Chem 282:490–499. https://doi.org/10.1016/j.snb.2018.11.089

Canto ED, Ceriello A, Rydén L, Ferrini M, Hansen TB, Schnell O, Standl E, Beulens JWJ (2019) Diabetes as a cardiovascular risk factor: an overview of global trends of macro and micro vascular complications. Eur J Prev Cardiol. https://journals.sagepub.com/doi/full/10.1177/2047487319878371?casa_token=q7CekWjB4nAAAAAA%3ATv3ez_NDCSh8Yl_tlj1bfzUtQA-oCDkhYoK6kS98anP0cxWSxyfL9gcTQcZuzqKoasmvn7iTvAFOmQ. Accessed 2025-04-22

Carbone M, Gorton L, Antiochia R (2015) An overview of the latest graphene-based sensors for glucose detection: the effects of graphene defects. Electroanalysis 27(1):16–31. https://doi.org/10.1002/elan.201400409

Castorino K, Polsky S, O'Malley G, Levister C, Nelson K, Farfan C, Brackett S, Puhr S, Levy CJ (2020) Performance of the Dexcom G6 continuous glucose monitoring system in pregnant women with diabetes. Diabetes Technol Ther 22(12):943–947. https://doi.org/10.1089/dia.2020.0085

Chan D, Chien J-C, Axpe E, Blankemeier L, Baker SW, Swaminathan S, Piunova VA, Zubarev DY, Maikawa CL, Grosskopf AK, Mann JL, Soh HT, Appel EA (2022) Combinatorial polyacrylamide hydrogels for preventing biofouling on implantable biosensors. Adv Mater 34(24):2109764. https://doi.org/10.1002/adma.202109764

Chan NB, Li W, Aung T, Bazuaye E, Montero RM (2023) Machine learning–based time in patterns for blood glucose fluctuation pattern recognition in type 1 diabetes management: development and validation study. JMIR AI 2(1):e45450. https://doi.org/10.2196/45450

Chaudhury A, Duvoor C, Reddy Dendi VS, Kraleti S, Chada A, Ravilla R, Marco A, Shekhawat NS, Montales MT, Kuriakose K, Sasapu A, Beebe A, Patil N, Musham CK, Lohani GP, Mirza W (2017) Clinical review of antidiabetic drugs: implications for type 2 diabetes mellitus management. Front Endocrinol 8:6. https://doi.org/10.3389/fendo.2017.00006

Chong SK, Reineke EL (2016) Point-of-care glucose and ketone monitoring. Top Companion Anim Med 31(1):18–26. https://doi.org/10.1053/j.tcam.2016.05.005

Chu SS, Nguyen HA, Zhang J, Tabassum S, Cao H (2022) Towards multiplexed and multimodal biosensor platforms in real-time monitoring of metabolic disorders. Sensors 22(14):5200. https://doi.org/10.3390/s22145200

Coates J, Chipperfield A, Clough G (2016) Wearable multimodal skin sensing for the diabetic foot. Electronics 5(3):45. https://doi.org/10.3390/electronics5030045

Dall TM, Yang W, Gillespie K, Mocarski M, Byrne E, Cintina I, Beronja K, Semilla AP, Iacobucci W, Hogan PF (2019) The economic burden of elevated blood glucose levels in 2017: diagnosed and undiagnosed diabetes, gestational diabetes mellitus, and prediabetes. Diabetes Care 42(9):1661–1668. https://doi.org/10.2337/dc18-1226

Davis GM et al (2021) Accuracy of Dexcom G6 continuous glucose monitoring in non–critically ill hospitalized patients with diabetes. Diabetes Care. American Diabetes Association. https://diabetesjournals.org/care/article/44/7/1641/138804/Accuracy-of-Dexcom-G6-Continuous-Glucose. Accessed 2025-04-30

Davis GM et al (2023) Automated insulin delivery with remote real-time continuous glucose monitoring for hospitalized patients with diabetes: a multicenter, single-arm, feasibility trial. Diabetes Technol Ther. https://www.liebertpub.com/doi/full/10.1089/dia.2023.0304. Accessed 2025-04-23

DeFronzo RA, Ferrannini E, Groop L, Henry RR, Herman WH, Holst JJ, Hu FB, Kahn CR, Raz I, Shulman GI, Simonson DC, Testa MA, Weiss R (2015) Type 2 diabetes mellitus. Nat Rev Dis Primer 1(1):1–22. https://doi.org/10.1038/nrdp.2015.19

Dehennis A, Mortellaro MA, Ioacara S (2015) Multisite study of an implanted continuous glucose sensor over 90 days in patients with diabetes mellitus. J Diabetes Sci Technol 9(5):951–956. https://doi.org/10.1177/1932296815596760

Delbeck S, Vahlsing T, Leonhardt S, Steiner G, Heise HM (2019) Non-invasive monitoring of blood glucose using optical methods for skin spectroscopy—opportunities and recent advances. Anal Bioanal Chem 411(1):63–77. https://doi.org/10.1007/s00216-018-1395-x

Didyuk O, Econom N, Guardia A, Livingston K, Klueh U (2020) Continuous glucose monitoring devices: past, present, and future focus on the history and evolution of technological innovation. J Diabetes Sci Technol 15(3):676–683. https://doi.org/10.1177/1932296819899394

Dilmac Y, Guler M (2020) Fabrication of non-enzymatic glucose sensor dependent upon au nanoparticles deposited on carboxylated graphene oxide. J Electroanal Chem 864:114091. https://doi.org/10.1016/j.jelechem.2020.114091

Dirr EW, Urdaneta ME, Patel Y, Johnson RD, Campbell-Thompson M, Otto KJ (2020) Designing a bioelectronic treatment for type 1 diabetes: targeted parasympathetic modulation of insulin secretion. Bioelectron Med 3(2):17–31. https://doi.org/10.2217/bem-2020-0006

Dua A et al (2024) Advancements of glucose monitoring biosensor: current state, generations of technological progress, and innovation dynamics: Ingenta connect fast track article. Curr Pharm Biotechnol. https://www.ingentaconnect.com/content/ben/cpb/pre-prints/content-bms-cpb-2024-52. Accessed 2025-04-22

Eaton SB, Eaton SB (2017) Physical inactivity, obesity, and type 2 diabetes: an evolutionary perspective. Res Q Exerc Sport 88(1) Get Access. https://www.tandfonline.com/doi/pdf/10.1080/02701367.2016.1268519?casa_token=EbFSfno8ghYAAAAA:52LaMc5dgonwat9z2uHxKXLkYqqKqenpNvMmy8P4vaVw_IbgL2YhO1nRQVihhT6m3OEgywXvmd8EHQ. Accessed 2025-04-22

Erbach M, Freckmann G, Hinzmann R, Kulzer B, Ziegler R, Heinemann L, Schnell O (2016) Interferences and limitations in blood glucose self-testing. J Diabetes Sci Technol 10(5):1161–1168. https://doi.org/10.1177/1932296816641433

eyeSTYLE, Editor. Google's "Smart" contact lenses: how does they transform diabetes management and eye care - eyeSTYLE blog Lens.com. https://www.lens.com/eyestyle/uncategorized/googles-smart-contact-lenses/. Accessed 2025-04-23

Giaretta (2024) Glucose sensing in saliva. Adv Sensor Res - Wiley Online Library. https://advanced.onlinelibrary.wiley.com/doi/full/10.1002/adsr.202400065. Accessed 2025-04-23

Gideon O, Samuel HS, Okino IA (2024) Biocompatible materials for next-generation biosensors. Discov Chem 1(1):34. https://doi.org/10.1007/s44371-024-00042-6

Gouveia G, Saateh A, Swietlikowska A, Scarpellini C, Tsang E, Altug H, Merkx M, Dillen A, Leirs K, Spasic D, Lammertyn J, Gothelf KV, Bonedeau E, Porzberg N, Smeets RL, Koenen HJPM, Prins MWJ, de Jonge MI (2025) Continuous biosensing to monitor acute systemic inflammation, a diagnostic need for therapeutic guidance. ACS Sens 10(1):4–14. https://doi.org/10.1021/acssensors.4c02569

Govindaraj M, Srivastava A, Muthukumaran MK, Tsai P-C, Lin Y-C, Raja BK, Rajendran J, Ponnusamy VK, Arockia Selvi J (2023) Current advancements and prospects of enzymatic and non-enzymatic electrochemical glucose sensors. Int J Biol Macromol 253:126680. https://doi.org/10.1016/j.ijbiomac.2023.126680

Gray M, Meehan J, Ward C, Langdon SP, Kunkler IH, Murray A, Argyle D (2018) Implantable biosensors and their contribution to the future of precision medicine. Vet J 239:21–29. https://doi.org/10.1016/j.tvjl.2018.07.011

Green A, Hede SM, Patterson CC, Wild SH, Imperatore G, Roglic G, Beran D (2021) Type 1 diabetes in 2017: global estimates of incident and prevalent cases in children and adults. Diabetologia 64(12):2741–2750. https://doi.org/10.1007/s00125-021-05571-8

Güemes Gonzalez A, Etienne-Cummings R, Georgiou P (2020) Closed-loop bioelectronic medicine for diabetes management. Bioelectron Med 6(1):11. https://doi.org/10.1186/s42234-020-00046-4

Guiseppi-Elie A, Brahim S, Slaughter G, Ward KR (2005) Design of a subcutaneous implantable biochip for monitoring of glucose and lactate. IEEE Sensors J 5(3):345–355. https://doi.org/10.1109/JSEN.2005.846173

Hassan MH et al (2021) Recent advances in enzymatic and non-enzymatic electrochemical glucose sensing. Sensors. https://www.mdpi.com/1424-8220/21/14/4672. Accessed 2025-04-23

Hirsch IB (2018) Introduction: history of glucose monitoring | Compendia. American Diabetes Association. https://diabetesjournals.org/compendia/article/2018/1/1/144616/Introduction-History-of-Glucose-Monitoring. Accessed 2025-04-23

Hoa LT, Sun KG, Hur SH (2015) Highly sensitive non-enzymatic glucose sensor based on Pt nanoparticle decorated graphene oxide hydrogel. Sens Actuators B Chem 210:618–623. https://doi.org/10.1016/j.snb.2015.01.020

Hod M, Kapur A, McIntyre HD (2019) Evidence in support of the International Association of Diabetes in Pregnancy Study Groups' criteria for diagnosing gestational diabetes mellitus worldwide in 2019. Am J Obstet Gynecol 221(2):109–116. https://doi.org/10.1016/j.ajog.2019.01.206

Hovorka R (2006) Continuous glucose monitoring and closed-loop systems. Diabet Med 23(1):1–12. https://doi.org/10.1111/j.1464-5491.2005.01672.x

Ingelsson E, Mark IMC (2018) Human genetics of obesity and type 2 diabetes mellitus. Circ Genom Precis Med. https://www.ahajournals.org/doi/full/10.1161/CIRCGEN.118.002090. Accessed 2025-04-22

Jafri RZ, Balliro CA, El-Khatib F, Maheno MM, Hillard MA, O'Donovan A, Selagamsetty R, Zheng H, Damiano ER, Russell SJ (2020) A three-way accuracy comparison of the Dexcom G5, Abbott Freestyle Libre Pro, and Senseonics Eversense continuous glucose monitoring devices in a home-use study of subjects with type 1 diabetes. Diabetes Technol Ther 22(11):846–852. https://doi.org/10.1089/dia.2019.0449

Jain P et al (2024) Non-invasive glucose measurement technologies: recent advancements and future challenges. IEEE J Magaz | IEEE Xplore. https://ieeexplore.ieee.org/abstract/document/10501933. Accessed 2025-04-23

JHEOR (2022) Global increase in diabetes prevalence imposes a substantial health and economic burden. J Health Econ Outcomes Res. https://jheor.org/post/1265-global-increase-in-diabetes-prevalence-imposes-a-substantial-health-and-economic-burden. Accessed 2025-04-22

Jia W, Bandodkar AJ, Valdés-Ramírez G, Windmiller JR, Yang Z, Ramírez J, Chan G, Wang J (2013) Electrochemical tattoo biosensors for real-time noninvasive lactate monitoring in human perspiration. Anal Chem 85(14):6553–6560. https://doi.org/10.1021/ac401573r

Jin Q, Ma RCW (2021) Metabolomics in diabetes and diabetic complications: insights from epidemiological studies. Cells 10(11):2832. https://doi.org/10.3390/cells10112832

Jin X, Li G, Xu T, Su L, Yan D, Zhang X (2022) Fully integrated flexible biosensor for wearable continuous glucose monitoring. Biosens Bioelectron 196:113760. https://doi.org/10.1016/j.bios.2021.113760

Jin X, Cai A, Xu T, Zhang X (2023) Artificial intelligence biosensors for continuous glucose monitoring. Interdiscip Mater 2(2):290–307. https://doi.org/10.1002/idm2.12069

Johnston L, Wang G, Hu K, Qian C, Liu G (2021) Advances in biosensors for continuous glucose monitoring towards wearables. Front Bioeng Biotechnol 9:733810. https://doi.org/10.3389/fbioe.2021.733810

Jurysta (2009) Salivary glucose concentration and excretion in normal and diabetic subjects. BioMed Res Int - Wiley Online Library. https://onlinelibrary.wiley.com/doi/full/10.1155/2009/430426. Accessed 2025-04-23

Kafley C, Kummari S, Lakavath K, Kotagiri YG (2023) Ti3C2Tx-rGO-Chitosan-based Microcatheter sensor for real-time continuous monitoring of Propofol: toward improved anesthetic management. Microchim Acta 190(10):388. https://doi.org/10.1007/s00604-023-05969-8

Karpova EV, Shcherbacheva EV, Galushin AA, Vokhmyanina DV, Karyakina EE, Karyakin AA (2019) Noninvasive diabetes monitoring through continuous analysis of sweat using flow-through glucose biosensor. Anal Chem 91(6):3778–3783. https://doi.org/10.1021/acs.analchem.8b05928

Kennedy-Martin T, Boye KS, Peng X (2017) Cost of medication adherence and persistence in type 2 diabetes mellitus: a literature review. Patient Prefer Adherence 11:1103–1117. https://doi.org/10.2147/PPA.S136639

Kim J et al (2018a) Wearable bioelectronics: enzyme-based body-worn electronic devices. Acc Chem Res. https://pubs.acs.org/doi/full/10.1021/acs.accounts.8b00451?casa_token=UioP_

sJQt5oAAAAA%3Aj6KUK6XXBij6YFl_ixK25AlNcB192fes9VjwYzg9IX0PgLKTSHy-Vt4fVKZK6KhD8hCHBQlOEGixF-14. Accessed 2025-04-23

Kim J, Campbell AS, Wang J (2018b) Wearable non-invasive epidermal glucose sensors: a review. Talanta 177:163–170. https://doi.org/10.1016/j.talanta.2017.08.077

Knebel T, Neumiller JJ (2019) Medtronic MiniMed 670G hybrid closed-loop system. Clin Diabetes. American Diabetes Association. https://diabetesjournals.org/clinical/article/37/1/94/32611/Medtronic-MiniMed-670G-Hybrid-Closed-Loop-System. Accessed 2025-04-23

Kropp M et al (2023) Diabetic retinopathy as the leading cause of blindness and early predictor of cascading complications—risks and mitigation. EPMA J. https://link.springer.com/article/10.1007/s13167-023-00314-8. Accessed 2025-04-22

Kummari S, Kumar VS, Satyanarayana M, Gobi KV (2019) Direct electrochemical determination of methotrexate using functionalized carbon nanotube paste electrode as biosensor for in-vitro analysis of urine and dilute serum samples. Microchem J 148:626–633

Kummari S, Sunil Kumar V, Vengatajalabathy Gobi K (2020) Facile electrochemically reduced graphene oxide-multi-walled carbon nanotube nanocomposite as sensitive probe for *in-vitro* determination of nitrofurantoin in biological fluids. Electroanalysis 32(11):2452–2462. https://doi.org/10.1002/elan.202060157

Lee D, Park K, Seo J (2020) Recent advances in anti-inflammatory strategies for implantable biosensors and medical implants. Biochip J 14(1):48–62. https://doi.org/10.1007/s13206-020-4105-7

LeRoith D (2002) β-Cell dysfunction and insulin resistance in type 2 diabetes: role of metabolic and genetic abnormalities. Am J Med 113(6):3–11. https://doi.org/10.1016/S0002-9343(02)01276-7

Leto D, Saltiel AR (2012) Regulation of glucose transport by insulin: traffic control of GLUT4. Nat Rev Mol Cell Biol 13(6):383–396. https://doi.org/10.1038/nrm3351

Li S et al (2023) Advanced textile-based wearable biosensors for healthcare monitoring. Biosensors. https://www.mdpi.com/2079-6374/13/10/909. Accessed 2025-04-23

Lima JEBF, Moreira NCS, Sakamoto-Hojo ET (2022) Mechanisms underlying the pathophysiology of type 2 diabetes: from risk factors to oxidative stress, metabolic dysfunction, and hyperglycemia. Mutat Res Toxicol Environ Mutagen 874–875:503437. https://doi.org/10.1016/j.mrgentox.2021.503437

Lin P-H, Li B-R (2020) Antifouling strategies in advanced electrochemical sensors and biosensors. Analyst 145(4):1110–1120. https://doi.org/10.1039/C9AN02017A

Liu X, Zhang J (2024) Continuous glucose monitoring: a transformative approach to the detection of prediabetes. J Multidiscip Healthc 17:5513–5519. https://doi.org/10.2147/JMDH.S493128

Liu Y, Yu Q, Luo X, Yang L, Cui Y (2021) Continuous monitoring of diabetes with an integrated microneedle biosensing device through 3D printing. Microsyst Nanoeng 7(1):1–12. https://doi.org/10.1038/s41378-021-00302-w

Manasa G, Mascarenhas RJ, Shetti NP, Malode SJ, Mishra A, Basu S, Aminabhavi TM (2022) Skin patchable sensor surveillance for continuous glucose monitoring. ACS Appl Bio Mater 5(3):945–970. https://doi.org/10.1021/acsabm.1c01289

Marks BE, Williams KM, Sherwood JS, Putman MS (2022) Practical aspects of diabetes technology use: continuous glucose monitors, insulin pumps, and automated insulin delivery systems. J Clin Transl Endocrinol 27:100282. https://doi.org/10.1016/j.jcte.2021.100282

Mitanchez D, Yzydorczyk C, Siddeek B, Boubred F, Benahmed M, Simeoni U (2015) The offspring of the diabetic mother – short- and long-term implications. Best Pract Res Clin Obstet Gynaecol 29(2):256–269. https://doi.org/10.1016/j.bpobgyn.2014.08.004

Modzelewski R, Stefanowicz-Rutkowska MM, Matuszewski W, Bandurska-Stankiewicz EM (2022) Gestational diabetes mellitus—recent literature review. J Clin Med 11(19):5736. https://doi.org/10.3390/jcm11195736

Moonla C, Reynoso M, Casanova A, Chang A-Y, Djassemi O, Balaje A, Abbas A, Li Z, Mahato K, Wang J (2024) Continuous Ketone monitoring via wearable microneedle patch platform. ACS Sens 9(2):1004–1013. https://doi.org/10.1021/acssensors.3c02677

Murray SR, Reynolds RM (2020) Short- and long-term outcomes of gestational diabetes and its treatment on fetal development. Prenat Diagn. Wiley Online Library. https://obgyn.onlinelibrary.wiley.com/doi/full/10.1002/pd.5768. Accessed 2025-04-22

Nabrdalik K, Skonieczna-Żydecka K, Irlik K, Hendel M, Kwiendacz H, Łoniewski I, Januszkiewicz K, Gumprecht J, Lip GYH (2022) Gastrointestinal adverse events of metformin treatment in patients with type 2 diabetes mellitus: a systematic review, meta-analysis and meta-regression of randomized controlled trials. Front Endocrinol:13. https://doi.org/10.3389/fendo.2022.975912

Naikoo GA, Salim H, Hassan IU, Awan T, Arshad F, Pedram MZ, Ahmed W, Qurashi A (2021) Recent advances in non-enzymatic glucose sensors based on metal and metal oxide nanostructures for diabetes management-a review. Front Chem:9. https://doi.org/10.3389/fchem.2021.748957

Nakagawa Y, Hirota Y, Yamamoto A, Takayoshi T, Takeuchi T, Hamaguchi T, Matsuoka A, Sakaguchi K, Ogawa W (2022) Accuracy of a professional continuous glucose monitoring device in individuals with type 2 diabetes mellitus. Kobe J Med Sci 68(1):E5–E10

Nichols SP et al (2013) Biocompatible materials for continuous glucose monitoring devices. Chem Rev. https://pubs.acs.org/doi/full/10.1021/cr300387j. Accessed 2025-04-30

Nikpour S, Mehrdad N, Sanjari M, Aalaa M, Heshmat R, Khabaz Mafinejad M, Larijani B, Nomali M, Najafi Ghezeljeh T (2022) Challenges of type 2 diabetes mellitus management from the perspective of patients: conventional content analysis. Interact J Med Res 11(2):e41933. https://doi.org/10.2196/41933

Oparaji NA (2024) Non-invasive blood glucose monitoring: a new approach for diabetic patients. Afr J Psychol Stud Soc Issues. https://ajpssi.org/index.php/ajpssi/article/view/618. Accessed 2025-04-23

Oser SM et al (2025) Assessing the iLet Bionic Pancreas deployed in primary care and via telehealth: a randomized clinical trial. Clin Diabetes. American Diabetes Association. https://diabetesjournals.org/clinical/article/doi/10.2337/cd24-0104/157928/Assessing-the-iLet-Bionic-Pancreas-Deployed-in. Accessed 2025-04-23

Padula F, Visioli A (2012) Optimal tuning rules for proportional-integral-derivative and fractional-order proportional-integral-derivative controllers for integral and unstable processes. IET Control Theory Appl 6(6):776–786. https://doi.org/10.1049/iet-cta.2011.0419

Palanisamy S, Cheemalapati S, Chen S-M (2014) Amperometric glucose biosensor based on glucose oxidase dispersed in multiwalled carbon nanotubes/graphene oxide hybrid biocomposite. Mater Sci Eng C 34:207–213. https://doi.org/10.1016/j.msec.2013.09.011

Palumbo PJ (2001) Glycemic control, mealtime glucose excursions, and diabetic complications in type 2 diabetes mellitus. Mayo Clin Proc 76(6):609–618. https://doi.org/10.4065/76.6.609

Panwar S, Sarkar P, Kasim DS, Anand R, Priya A, Prakash S, Jha SK (2024) Portable optical biosensor for point-of-care monitoring of salivary glucose using a paper-based microfluidic strip. Biosens Bioelectron X 17:100452. https://doi.org/10.1016/j.biosx.2024.100452

Park S, Boo H, Chung TD (2006) Electrochemical non-enzymatic glucose sensors. Anal Chim Acta 556(1):46–57. https://doi.org/10.1016/j.aca.2005.05.080

Perkins BA, Sherr JL, Mathieu C (2021) Type 1 diabetes glycemic management: insulin therapy, glucose monitoring, and automation. Science 373(6554):522–527. https://doi.org/10.1126/science.abg4502

Prieto-Avalos G, Cruz-Ramos NA, Alor-Hernández G, Sánchez-Cervantes JL, Rodríguez-Mazahua L, Guarneros-Nolasco LR (2022) Wearable devices for physical monitoring of heart: a review. Biosensors 12(5):292. https://doi.org/10.3390/bios12050292

Psoma SD, Kanthou C (2023) Wearable insulin biosensors for diabetes management: advances and challenges. Biosensors. https://www.mdpi.com/2079-6374/13/7/719. Accessed 2025-04-23

Puchalska P, Crawford PA (2021) Metabolic and signaling roles of Ketone bodies in health and disease. Ann Rev. https://www.annualreviews.org/content/journals/10.1146/annurev-nutr-111120-111518. Accessed 2025-04-29

Rachim VP, Chung W-Y (2019) Wearable-band type visible-near infrared optical biosensor for non-invasive blood glucose monitoring. Sens Actuators B Chem 286:173–180. https://doi.org/10.1016/j.snb.2019.01.121

Rahman MS, Hossain KS, Das S, Kundu S, Adegoke EO, Rahman MA, Hannan MA, Uddin MJ, Pang M-G (2021) Role of insulin in health and disease: an update. Int J Mol Sci 22(12):6403. https://doi.org/10.3390/ijms22126403

Rahul PK, Kafley CA, Sukumaran R, Kummari S, Kotagiri YG (2023) Electrochemical strip sensor based on a graphene-Au-PEDOT nanocomposite for detection and continuous monitoring of Dacarbazine. ACS Appl Nano Mater 6(18):16915–16926

Renard E (2008) Implantable continuous glucose sensors. Curr Diabetes Rev 4(3):169–174. https://doi.org/10.2174/157339908785294406

Ribet F, Stemme G, Roxhed N (2018) Real-time intradermal continuous glucose monitoring using a minimally invasive microneedle-based system. Biomed Microdevices 20(4):101. https://doi.org/10.1007/s10544-018-0349-6

Ridder FD, Braspenning R, Ordonez JS, Klarenbeek G, Lauwers P, Ledeganck KJ, Delbeke D, Block CD (2024) Early feasibility study with an implantable near-infrared spectroscopy sensor for glucose, ketones, lactate and ethanol. PLoS One 19(5):e0301041. https://doi.org/10.1371/journal.pone.0301041

Rizza RA, Gerich JE, Haymond MW, Westland RE, Hall LD, Clemens AH, Service, F. J (1980) Control of blood sugar in insulin-dependent diabetes: comparison of an artificial endocrine pancreas, continuous subcutaneous insulin infusion, and intensified conventional insulin therapy. N Engl J Med 303(23):1313–1318. https://doi.org/10.1056/NEJM198012043032301

Roep BO, Thomaidou S, van Tienhoven R, Zaldumbide A (2021) Type 1 diabetes mellitus as a disease of the β-cell (do not blame the immune system?). Nat Rev Endocrinol 17(3):150–161. https://doi.org/10.1038/s41574-020-00443-4

Russel SJ et al (2014) Outpatient glycemic control with a bionic pancreas in type 1 diabetes. N Engl J Med. https://www.nejm.org/doi/full/10.1056/NEJMoa1314474?casa_token=AFBq5VbgSmwAAAAA%3AYmsM1I_Gzx-2soLF0NbzvCHMaG0H02J_eYDxfX6EHW07xFfzuEPtIHyzYjLAFyxsLAJq_973HC9yQiU. Accessed 2025-04-23

Sabury S, Kazemi SH, Sharif F (2015) Graphene–gold nanoparticle composite: application as a good Scaffold for construction of glucose oxidase biosensor. Mater Sci Eng C 49:297–304. https://doi.org/10.1016/j.msec.2015.01.018

Sempionatto JR, Brazaca LC, García-Carmona L, Bolat G, Campbell AS, Martin A, Tang G, Shah R, Mishra RK, Kim J, Zucolotto V, Escarpa A, Wang J (2019) Eyeglasses-based tear biosensing system: non-invasive detection of alcohol, vitamins and glucose. Biosens Bioelectron 137:161–170. https://doi.org/10.1016/j.bios.2019.04.058

Sevil M, Rashid M, Hajizadeh I, Park M, Quinn L, Cinar A (2021) Physical activity and psychological stress detection and assessment of their effects on glucose concentration predictions in diabetes management. IEEE Trans Biomed Eng 68(7):2251–2260. https://doi.org/10.1109/TBME.2020.3049109

Shekher K, Sampath K, Vandini S, Satyanarayana M, Gobi KV (2023) Gold nanoparticle assimilated polymer layer on carbon nanotube matrices for sensitive detection of serotonin in presence of dopamine in-vitro. Inorganica Chim Acta 549:121399

Shen Y, Chen G, Chen Y, Gao Y, Hou C, Liao K, Ma B, Liu H (2024) Wearable microfluidic electrochemical sensor integrated with iontophoresis for non-invasive Sweat Ketone monitoring. Sens Actuators B Chem. 421:136518. https://doi.org/10.1016/j.snb.2024.136518

Shrestha SS et al (2019. American Diabetes Association. https://diabetesjournals.org/care/article/42/1/77/36332/Factors-Contributing-to-Increases-in-Diabetes. Accessed 2025-04-22) Factors contributing to increases in diabetes-related preventable hospitalization costs among U.S. adults during 2001–2014. Diabetes Care 42:77

Soni A, Jha SK (2015) A paper strip based non-invasive glucose biosensor for salivary analysis. Biosens Bioelectron 67:763–768. https://doi.org/10.1016/j.bios.2014.09.042

Soni A, Jha SK (2017) Smartphone based non-invasive salivary glucose biosensor. Anal Chim Acta 996:54–63. https://doi.org/10.1016/j.aca.2017.10.003

Sun H, Saeedi P, Karuranga S, Pinkepank M, Ogurtsova K, Duncan BB, Stein C, Basit A, Chan JCN, Mbanya JC, Pavkov ME, Ramachandaran A, Wild SH, James S, Herman WH, Zhang P, Bommer C, Kuo S, Boyko EJ, Magliano DJ (2022) IDF diabetes atlas: global, regional and country-level diabetes prevalence estimates for 2021 and projections for 2045. Diabetes Res Clin Pract 183:109119. https://doi.org/10.1016/j.diabres.2021.109119

Szadkowska A, Michalak A, Łosiewicz A, Kuśmierczyk H, Krawczyk-Rusiecka K, Chrzanowski J, Gawrecki A, Zozulińska-Ziółkiewicz D, Fendler W (2021) Impact of factory-calibrated Freestyle Libre system with new glucose algorithm measurement accuracy and clinical performance in children with type 1 diabetes during summer camp. Pediatr Diabetes 22(2):261–270. https://doi.org/10.1111/pedi.13135

Tai W, Hu Q, Dong X, Zhao F, Wu W, Wang Y, Yu L (2024) Hydrogel-integrated multimodal biosensor for the detection of glucose and carcinoembryonic antigen. Sens Actuators B Chem 412:135800. https://doi.org/10.1016/j.snb.2024.135800

Takeuchi K, Kim B (2018) Functionalized microneedles for continuous glucose monitoring. Nano Converg 5(1):28. https://doi.org/10.1186/s40580-018-0161-2

Teo E et al (2022) Effectiveness of continuous glucose monitoring in maintaining glycaemic control among people with type 1 diabetes mellitus: a systematic review of randomised controlled trials and meta-analysis. Diabetologia. https://link.springer.com/article/10.1007/s00125-021-05648-4. Accessed 2025-04-23

Teymourian H, Barfidokht A, Wang J (2020) Electrochemical glucose sensors in diabetes management: an updated review (2010–2020). Chem Soc Rev 49(21):7671–7709. https://doi.org/10.1039/D0CS00304B

Tong X, Jiang T, Yang J, Song Y, Ao Q, Tang J, Zhang L (2025) Continuous Glucose Monitoring (CGM) system based on protein hydrogel anti-biofouling coating for long-term accurate and point-of-care glucose monitoring. Biosens Bioelectron 277:117307. https://doi.org/10.1016/j.bios.2025.117307

Torrance T, Franklin V, Greene S (2003) Insulin Pumps. https://doi.org/10.1136/adc.88.11.949

Umanath K, Lewis JB (2018) Update on diabetic nephropathy: core curriculum 2018. Am J Kidney Dis 71(6):884–895. https://doi.org/10.1053/j.ajkd.2017.10.026

Valdes TI, Moussy F (2000) In vitro and in vivo degradation of glucose oxidase enzyme used for an implantable glucose biosensor. Diabetes Technol Ther. https://www.liebertpub.com/doi/abs/10.1089/15209150050194233?casa_token=prLp6EDWD3IAAAAA%3AIwEpovGG8_TZk2DQOIjwYFIsUMvSsvVIOnx-opmVpZos3bMPb8QELGxiQpVDstCM0A5fgqWdkOw. Accessed 2025-04-23

Vashist SK (2012) Non-invasive glucose monitoring technology in diabetes management: a review. Anal Chim Acta 750:16–27. https://doi.org/10.1016/j.aca.2012.03.043

Vettoretti M, Facchinetti A (2019) Combining continuous glucose monitoring and insulin pumps to automatically tune the basal insulin infusion in diabetes therapy: a review. Biomed Eng Online 18(1):37. https://doi.org/10.1186/s12938-019-0658-x

Vettoretti M, Cappon G, Facchinetti A, Sparacino G (2020) Advanced diabetes management using artificial intelligence and continuous glucose monitoring sensors. Sensors 20(14):3870. https://doi.org/10.3390/s20143870

Waataja JJ, Nihalani RK, Honda CN, Billington CJ (2022) Use of a bio-electronic device comprising of targeted dual neuromodulation of the hepatic and Celiac Vagal branches demonstrated enhanced glycemic control in a type 2 diabetic rat model as well as in an Alloxan treated Swine Model. Front Neurosci 16. https://doi.org/10.3389/fnins.2022.1005932

Wang G et al (2013) Non-enzymatic electrochemical sensing of glucose. Microchim Acta. https://link.springer.com/article/10.1007/s00604-012-0923-1. Accessed 2025-04-23

Wang X, Liu T, Liang R, Qin W (2024) Maintenance-free antifouling polymeric membrane potentiometric sensors based on self-polishing coatings. Analyst 149(10):2855–2863. https://doi.org/10.1039/D4AN00351A

Ward ZJ, Yeh JM, Reddy CL, Gomber A, Ross C, Rittiphairoj T, Manne-Goehler J, Abdalla AT, Abdullah MA, Ahmed A, Ankotche A, Azad K, Bahendeka S, Baldé N, Jain SM, Kalobu JC, Karekezi C, Kol H, Prasannakumar KM, Leik SK, Mbanya JC, Mbaye MN, Niang B, Paturi VR, Raghupathy P, Ramaiya K, Sethi B, Zabeen B, Atun R (2022) Estimating the total incidence of type 1 diabetes in children and adolescents aged 0–19 years from 1990 to 2050: a global simulation-based analysis. Lancet Diabetes Endocrinol 10(12):848–858. https://doi.org/10.1016/S2213-8587(22)00276-5

Weinstock RS et al (2020) The role of blood glucose monitoring in diabetes management | Compendia. American Diabetes Association. https://diabetesjournals.org/compendia/article/doi/10.2337/db2020-31/144940/The-Role-of-Blood-Glucose-Monitoring-in-Diabetes. Accessed 2025-04-23

Williams R, Karuranga S, Malanda B, Saeedi P, Basit A, Besançon S, Bommer C, Esteghamati A, Ogurtsova K, Zhang P, Colagiuri S (2020) Global and regional estimates and projections of diabetes-related health expenditure: results from the International Diabetes Federation Diabetes Atlas, 9th edition. Diabetes Res Clin Pract 162:108072. https://doi.org/10.1016/j.diabres.2020.108072

Wu C, Sun H, Li Y, Liu X, Du X, Wang X, Xu P (2015) Biosensor based on glucose oxidase-nanoporous gold co-catalysis for glucose detection. Biosens Bioelectron 66:350–355. https://doi.org/10.1016/j.bios.2014.11.037

Xu J et al (2021) Wearable biosensors for non-invasive sweat diagnostics. Biosensors. https://www.mdpi.com/2079-6374/11/8/245. Accessed 2025-04-23

Xu J et al (2022) Smartphone-based electrochemical systems for glucose monitoring in biofluids: a review. Sensors. https://www.mdpi.com/1424-8220/22/15/5670. Accessed 2025-04-23

Xu Y, De la Paz E, Paul A, Mahato K, Sempionatto JR, Tostado N, Lee M, Hota G, Lin M, Uppal A, Chen W, Dua S, Yin L, Wuerstle BL, Deiss S, Mercier P, Xu S, Wang J, Cauwenberghs G (2023) In-ear integrated sensor array for the continuous monitoring of brain activity and of lactate in sweat. Nat Biomed Eng 7(10):1307–1320. https://doi.org/10.1038/s41551-023-01095-1

Yamada S (2011) Historical achievements of self-monitoring of blood glucose technology development in Japan. J Diabetes Sci Technol 5(5):1300–1306

Yang Y, Gao W (2019) Wearable and flexible electronics for continuous molecular monitoring. Chem Soc Rev 48(6):1465–1491. https://doi.org/10.1039/C7CS00730B

Yang J, Gong X, Chen S, Zheng Y, Peng L, Liu B, Chen Z, Xie X, Yi C, Jiang L (2023) Development of smartphone-controlled and microneedle-based wearable continuous glucose monitoring system for home-care diabetes management. ACS Sens 8(3):1241–1251. https://doi.org/10.1021/acssensors.2c02635

Yao Y, Chen J, Guo Y, Lv T, Chen Z, Li N, Cao S, Chen B, Chen T (2021) Integration of interstitial fluid extraction and glucose detection in one device for wearable non-invasive blood glucose sensors. Biosens Bioelectron 179:113078. https://doi.org/10.1016/j.bios.2021.113078

Ye M, Chien P-J, Toma K, Arakawa T, Mitsubayashi K (2015) An acetone bio-sniffer (gas phase biosensor) enabling assessment of lipid metabolism from exhaled breath. Biosens Bioelectron 73:208–213. https://doi.org/10.1016/j.bios.2015.04.023

Yoon H, Xuan X, Jeong S, Park JY (2018) Wearable, robust, non-enzymatic continuous glucose monitoring system and its in vivo investigation. Biosens Bioelectron 117:267–275. https://doi.org/10.1016/j.bios.2018.06.008

Zafar H et al (2022) Comprehensive review on wearable sweat-glucose sensors for continuous glucose monitoring. Sensors. https://www.mdpi.com/1424-8220/22/2/638. Accessed 2025-04-30

Zhao J, Lin Y, Wu J, Nyein HYY, Bariya M, Tai L-C, Chao M, Ji W, Zhang G, Fan Z, Javey A (2019) A fully integrated and self-powered smartwatch for continuous sweat glucose monitoring. ACS Sens 4(7):1925–1933. https://doi.org/10.1021/acssensors.9b00891

Zheng Y, Ley SH, Hu FB (2018) Global aetiology and epidemiology of type 2 diabetes mellitus and its complications. Nat Rev Endocrinol 14(2):88–98. https://doi.org/10.1038/nrendo.2017.151

Zhou W, He Q, Ye H, Ye C, Wu X, Chu J (2021) Recent advances in flexible sweat glucose biosensors. J Phys Appl Phys 54(42):423001. https://doi.org/10.1088/1361-6463/ac14ef

Zhu H, Li L, Zhou W, Shao Z, Chen X (2016) Advances in non-enzymatic glucose sensors based on metal oxides. J Mater Chem B 4(46):7333–7349. https://doi.org/10.1039/C6TB02037B

Nano-Bioelectronics for Neurodegenerative Disease Monitoring and Management

9

Lakshmi R. Panicker, Kalyani Bukya, Naveen Kotagiri, and Yugender Kotagiri Goud

Abstract

Neurodegenerative diseases such as Alzheimer's, Parkinson's, and ALS present significant challenges in diagnosis and treatment due to their progressive nature and complex pathophysiology. In this chapter, we explore how nano-bioelectronics offers innovative solutions by integrating nanoscale materials with bioelectronic devices to monitor and intervene in disease progression at the cellular level. Biosensing-based approaches enable the early detection of critical biomarkers like amyloid-beta, tau proteins, and oxidative stress indicators, providing new avenues for timely diagnosis. In this chapter, we also discuss how nano-enabled biosensors allow real-time, continuous monitoring of disease states, enhancing the ability to track patient responses and adjust treatments accordingly. Therapeutically, nano-bioelectronics facilitates precise drug delivery to target neural cells and employs bioelectronic neuromodulation techniques to manage symptoms. Additionally, nanomaterials aid in neural regeneration and neuroplasticity, offering hope for long-term recovery and rehabilitation. In this chapter, we address the challenges of biocompatibility and ethical considerations, particularly in the use of implantable devices and data security. Despite these hurdles, the future of neurodegenerative disease management lies in the fusion of nano-bioelectronics with AI-driven analytics for personalized, precision healthcare. In this chapter, we highlight the potential of nano-bioelectronic systems to transform neurodegenerative disease management through early detection, continuous monitoring, and targeted therapy. The evolution of these technologies promises a more effective and tailored approach to tackling some of the most devastating neurological disorders.

L. R. Panicker · K. Bukya · N. Kotagiri · Y. K. Goud (✉)
Department of Chemistry, Indian Institute of Technology Palakkad, Palakkad, Kerala, India
e-mail: yugenderkotagiri@iitpkd.ac.in

K. Mahato, P. Chandra (eds.), *Nano-bioelectronics for Precision Health Monitoring*, https://doi.org/10.1007/978-981-95-2117-3_9

Keywords

Nano-bioelectronics · Neurodegenerative diseases · Biosensors · Personalized medicine · Brain · computer interfaces · Artificial intelligence

9.1 Introduction to Neurodegenerative Diseases

Neurodegenerative diseases (NDs) refer to a class of disorders that primarily affect the neurons in the human brain. Neurons are the core components of the nervous system, responsible for transmitting signals between different parts of the body. Unlike many other cells, neurons typically do not regenerate or replace themselves when damaged, making neurodegeneration a progressively worsening condition (Przedborski et al. 2003). The most common neurodegenerative diseases include Alzheimer's disease (AD), Parkinson's disease (PD), and amyotrophic lateral sclerosis (ALS), each of which presents distinct symptoms but shares underlying mechanisms related to neuronal loss and dysfunction (Kovacs 2016). Alzheimer's disease is the most prevalent form of dementia, marked by memory loss, cognitive decline, and behavioral changes (Scheltens et al. 2016). Alzheimer's typically affects older adults, though early-onset forms can strike individuals in their 40s or 50s. The exact cause of Alzheimer's is not fully understood, but it is characterized by the accumulation of amyloid-beta plaques and tau tangles in the brain, leading to synaptic dysfunction and neuronal death (Hardy and Selkoe 2002). Parkinson's disease primarily affects motor control, causing tremors, stiffness, and bradykinesia (Kalia and Lang 2015). It is caused by the degeneration of dopamine-producing neurons in the substantia nigra, a brain region that controls movement. As the disease progresses, non-motor symptoms such as cognitive decline, depression, and autonomic dysfunction may arise. Amyotrophic lateral sclerosis (ALS), also known as Lou Gehrig's disease, affects the neurons that control voluntary muscle movement. Patients with ALS experience muscle weakness, atrophy, and eventually paralysis as motor neurons deteriorate. While some cases of ALS are linked to genetic mutations, most cases occur sporadically without a clear cause (Hardiman et al. 2017). Although these diseases vary in their clinical manifestations, they share common pathophysiological features, including protein misfolding, mitochondrial dysfunction, oxidative stress, neuroinflammation, and excitotoxicity (Henstridge et al. 2019).

The pathophysiology of neurodegenerative diseases is complex and multifaceted, often involving a combination of genetic, environmental, and lifestyle factors. One hallmark of many NDs is the accumulation of misfolded proteins, such as amyloid-beta in Alzheimer's or alpha-synuclein in Parkinson's. These proteins aggregate to form toxic plaques or fibrils, which interfere with normal cellular function (Ross and Poirier 2004). Over time, this disruption leads to synaptic dysfunction, neuronal death, and widespread neurodegeneration. In Alzheimer's disease, amyloid-beta peptides form extracellular plaques, while tau proteins accumulate as neurofibrillary tangles inside neurons (Selkoe and Hardy 2016). These protein aggregates disrupt synaptic communication and trigger neuroinflammation. Microglial cells, the brain's resident immune cells, become activated in response to the toxic plaques, releasing pro-inflammatory cytokines that further damage neurons (Heneka et al. 2015). The result is widespread cortical atrophy, particularly in the hippocampus, which is crucial for memory formation. In Parkinson's disease, the misfolding and aggregation of alpha-synuclein lead to the formation of Lewy bodies within dopaminergic neurons. The loss of dopamine production in the substantia nigra causes the motor symptoms of the disease (Spillantini et al. 1998). Additionally, mitochondrial dysfunction and oxidative stress contribute to the degeneration of neurons in both the substantia nigra and other brain regions (Dias et al. 2013). ALS is characterized by the degeneration of both upper and lower motor neurons, leading to progressive muscle weakness. In some cases, ALS is linked to mutations in genes such as *SOD1*, which encodes the enzyme superoxide dismutase. Mutant forms of this enzyme may lead to increased oxidative stress and damage to motor neurons (Rosen et al. 1993). In sporadic ALS, environmental factors such as exposure to toxins may play a role, although the exact mechanisms are still under investigation (Chia et al. 2018).

Neurodegenerative diseases tend to follow a progressive course, with symptoms worsening over time as more neurons are lost. In the early stages, patients may experience subtle cognitive or motor deficits that can be difficult to diagnose. As the disease progresses, these deficits become more pronounced, eventually leading to severe disability and, in many cases, death. The rate of progression can vary widely among individuals, influenced by factors such as age, genetic predisposition, and overall health (Gao et al. 2011).

One of the greatest challenges in managing neurodegenerative diseases is the difficulty of early diagnosis. By the time symptoms become apparent, significant neuronal damage has often already occurred, limiting the effectiveness of interventions. For example, in Alzheimer's disease, amyloid-beta plaques may begin to accumulate in the brain decades before clinical symptoms of memory loss appear (Bateman et al. 2012). Similarly, in Parkinson's disease, motor symptoms do not emerge until approximately 60–80% of dopaminergic neurons have been lost (Fearnley and Lees 1991). Traditional diagnostic methods for neurodegenerative diseases rely on clinical assessments, neuroimaging, and biomarker analysis. However, these approaches are often not sensitive enough to detect the earliest stages of disease. Structural neuroimaging techniques such as magnetic resonance imaging (MRI) can identify brain atrophy, but only after significant neuronal loss has occurred. Functional imaging techniques such as positron emission tomography (PET) can detect abnormal protein deposits, but these methods are expensive and not widely available (Jack et al. 2018). The lack of early diagnostic tools also hampers the development of effective treatments. Most current therapies for neurodegenerative diseases are symptomatic rather than disease-modifying. For example, medications such as cholinesterase inhibitors and NMDA receptor antagonists are used to manage the cognitive symptoms of Alzheimer's, while dopamine replacement therapy is the mainstay of treatment for Parkinson's disease (Cummings et al. 2019). However, these treatments do not stop or reverse the underlying neurodegenerative process. A major obstacle in the development of disease-modifying therapies is the blood–brain barrier (BBB), a tightly regulated barrier that prevents most drugs from entering the brain. This makes it difficult to deliver therapeutic agents directly to affected brain regions (Pardridge 2012). Nanotechnology and bioelectronic systems hold promise for overcoming this challenge by enabling targeted drug delivery and real-time monitoring of drug efficacy (Bharadwaj et al. 2020).

Precision healthcare offers a new paradigm for managing neurodegenerative diseases by tailoring interventions to the individual patient based on their genetic, molecular, and clinical profile. Advances in genomics, proteomics, and biomarker discovery have paved the way for personalized approaches to diagnosis and treatment. In neurodegeneration, precision healthcare aims to identify at-risk individuals before symptoms appear, monitor disease progression in real time, and deliver targeted therapies to slow or halt neuronal loss. Nano-bioelectronics is a key enabling technology for precision healthcare in neurodegenerative diseases. By integrating nanoscale sensors with bioelectronic systems, it is possible to detect disease-specific biomarkers at extremely low concentrations and monitor neural activity with high temporal and spatial resolution. These capabilities are critical for early diagnosis, as well as for assessing the effectiveness of therapeutic interventions in real time. Biosensing technologies offer great potential for detecting early-stage neurodegenerative diseases. For example, electrochemical biosensors can detect biomarkers such as amyloid-beta or alpha-synuclein in cerebrospinal fluid or blood. These sensors are highly sensitive and can provide real-time data on biomarker levels, enabling earlier detection than traditional imaging or clinical assessments (Chung et al. 2019). In addition to diagnostics, nano-bioelectronics can also be used to deliver targeted therapies to specific brain regions. Nanoparticles can be engineered to cross the blood–brain barrier and release drugs in response to specific molecular triggers, such as the presence of misfolded proteins. Bioelectronic devices, such as deep brain stimulators, can modulate neural activity in real time to alleviate symptoms and potentially slow disease progression (Lo and Widge 2017).

Early diagnosis is critical to improving patient outcomes, yet current diagnostic tools are insufficient to detect these diseases in their earliest stages. Nano-bioelectronics offers a promising solution by enabling the early detection of biomarkers, real-time monitoring of disease progression, and precision delivery of therapeutic agents. By integrating biosensing technologies with bioelectronic systems, precision healthcare can transform the management of neurodegenerative diseases, offering new hope for patients and their families. The next sections of this chapter will delve deeper into the specific applications of nano-bioelectronics in diagnosing, monitoring, and treating neurodegenerative diseases, highlighting the potential of these technologies to revolutionize the field.

9.2 Nano-bioelectronics in Neurodegenerative Disease Management

Neurodegenerative diseases, such as Alzheimer's and Parkinson's, involve progressive changes in the brain's molecular and cellular environment, including the buildup of misfolded proteins, inflammation, and oxidative stress. Nano-bioelectronics offers advanced tools for monitoring these processes at an unprecedented level of detail, allowing for early diagnosis and ongoing tracking of disease progression One of the key applications of nano-bioelectronics in neurological monitoring is the development of biosensors that can detect biomarkers in bodily fluids, such as cerebrospinal fluid, blood, or even sweat (Fig. 9.1). For example, electrochemical biosensors based on nanomaterials like graphene or carbon nanotubes can detect the presence of amyloid-beta or tau proteins in very low concentrations, providing early indications of Alzheimer's disease before clinical symptoms arise (Kumar and Kim 2022). These sensors can also monitor other key indicators of neurodegeneration, such as oxidative stress markers (e.g., glutathione) and neuroinflammatory markers (e.g., cytokines) In addition to biomarker detection, nano-bioelectronics can be used to monitor neural activity. For instance, nanowire-based electrodes can be implanted in the brain to record electrical signals from neurons with high spatial and temporal resolution. This capability is particularly valuable for understanding the neural dysfunctions that underlie diseases like Parkinson's and epilepsy. By tracking changes in neural firing patterns, researchers can gain insights into disease progression and evaluate the effectiveness of therapeutic interventions.

Another significant development in nano-bioelectronics is the integration of these systems into wearable devices for continuous health monitoring. Wearable biosensors that utilize nanomaterials can provide real-time data on physiological parameters relevant to neurodegenerative diseases, such as changes in motor function, neural activity, or biomarker levels, for example, wearable electrochemical sensors can be used to monitor dopamine levels in the sweat of Parkinson's patients, providing insights into disease progression and medication effectiveness (Bandodkar and Wang 2014). These wearable devices can also be used in conjunction with mobile health platforms and artificial intelligence (AI) algorithms to track disease progression over time and deliver personalized treatment recommendations (Topol 2019).

9.3 Biosensing Approaches in Neurodegenerative Disease Diagnosis

Neurodegenerative diseases, including Alzheimer's, Parkinson's, and amyotrophic lateral sclerosis (ALS), present unique diagnostic challenges due to their progressive nature and complex pathophysiology. One of the key obstacles is the difficulty in diagnosing these diseases at an early stage when interventions may be most effective. Traditional diagnostic tools, such as neuroimaging and clinical assessments, often detect the disease only after significant neuronal damage has occurred, limiting the potential for early intervention and treatment (Jack et al. 2018). Biosensing technology, particularly in the context of nano-bioelectronics, offers a revolutionary approach to diagnosing neurodegenerative diseases by detecting disease-related biomarkers with high sensitivity and specificity. This section explores the various biosensing strategies that have emerged for early detection, continuous monitoring, and precise quantification of biomarkers associated with neurodegenerative

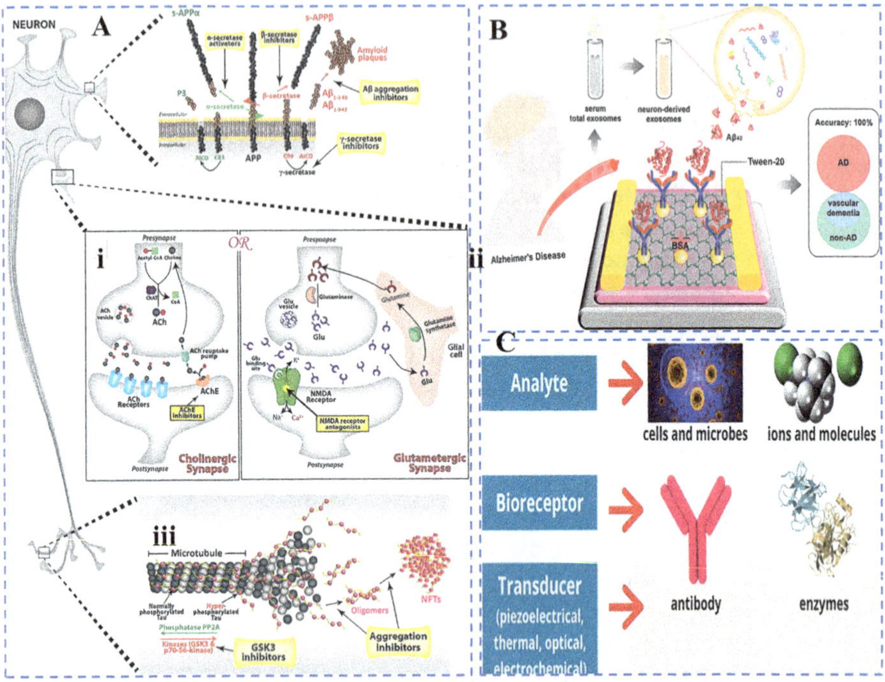

Fig. 9.1 (**a**) Summary of the main pathogenesis hypotheses for Alzheimer's disease, and treatment strategies targeting them. Amyloid cascade hypothesis, (i) cholinergic hypothesis, (ii) excitotoxicity hypothesis, (iii) tau hypothesis. ACh acetylcholine, AChE acetylcholinesterase, AICD amyloid precursor protein intracellular C-terminal domain, APP amyloid precursor protein, Aβ beta-amyloid protein, ChAT choline acetyltransferase, CoA coenzyme A, Glu glutamate, GSK3 glycogen synthase kinase 3, NFTs neurofibrillary tangles, NMDA *N*-methyl-d-aspartate (Wen et al. 2017). (**b**) Schematic representation of a graphene-based electrolyte-gated transistor (EGT) biosensor for the selective detection of neuron-derived exosomal $A\beta_{42}$ from serum. The platform enables accurate differentiation of Alzheimer's disease (AD), vascular dementia, and non-AD conditions, achieving 100% diagnostics (Li et al. 2023). (**c**) Examples of biosensors and their components on a graphene-based platform (Arshad et al. 2022). (**d**) Alpha-synuclein aggregation disrupts cellular function, leading to neuronal death and driving Parkinson's disease progression (Dhahi et al. 2024). (**e**) Schematic illustration of visible-light-driven hydrogen evolution using spherical nanoflower-like bimetallic (Mo,Ni) $(S,O)_{3-x}$ catalysts. The electron hopping mechanism facilitates charge transport across the catalyst surface, enhancing HER activity under $\lambda > 420$ nm irradiation. The presence of vacancies (V_o) and mixed anionic sites promotes water splitting and efficient electron transfer (Chen et al. 2021)

diseases. Biosensors are analytical devices that combine biological recognition elements, such as enzymes, antibodies, or nucleic acids, with a transducer that converts a biochemical signal into a measurable electrical, optical, or mechanical output. These devices are designed to detect specific biological molecules—referred to as analytes—in complex biological fluids, such as cerebrospinal fluid (CSF), blood, or saliva. The key to their utility in neurodegenerative disease management is their ability to detect disease biomarkers at very low concentrations, long before clinical symptoms appear (Kim et al. 2019). Nanotechnology plays a crucial role in enhancing the sensitivity and functionality of biosensors. By incorporating nanomaterials like nanoparticles, nanowires, and graphene into biosensing platforms, researchers have created devices capable of detecting minute quantities of biomarkers such as amyloid-beta, tau proteins, alpha-synuclein, and inflammatory molecules. These biosensors have the potential to revolutionize the field of neurodegenerative disease diagnostics, offering earlier diagnosis, better tracking of disease progression, and more accurate evaluation of therapeutic responses.

9.3.1 Early Detection of Biomarkers Using Nano-Enabled Biosensors

Early detection of neurodegenerative diseases is critical because it allows for interventions that may slow disease progression and improve patient outcomes. Biomarkers biological molecules that indicate disease presence or progression are key to early diagnosis. In the context of Alzheimer's disease, biomarkers such as amyloid-beta (Aβ) peptides and tau proteins are central to disease pathophysiology and are present in biological fluids long before cognitive symptoms manifest (Jack et al. 2018).

Amyloid-beta (Aβ) peptides are a major focus in Alzheimer's disease research because they aggregate to form plaques that disrupt neural communication and trigger neuroinflammation (Selkoe and Hardy 2016). Tau proteins, which normally stabilize microtubules in neurons, become hyperphosphorylated in Alzheimer's and aggregate into neurofibrillary tangles (Wang and Mandelkow 2016). Both Aβ and tau are found in cerebrospinal fluid (CSF) and blood, making them prime candidates for early diagnostic biomarkers. Nano-enabled biosensors have shown exceptional promise in detecting these biomarkers at very low concentrations, providing opportunities for early diagnosis. For instance, graphene-based field-effect transistors (GFETs) have been used to create highly sensitive biosensors for amyloid-beta detection. The high surface area and excellent electrical conductivity of graphene make it ideal for capturing low levels of Aβ in biological fluids. In one study, researchers functionalized graphene with antibodies specific to Aβ, achieving detection limits in the femtomolar range (10^{-15} M), which is much lower than conventional techniques like enzyme-linked immunosorbent assays (ELISA) Similarly, carbon nanotube-based biosensors have been developed for tau protein detection. By immobilizing anti-tau antibodies on the surface of carbon nanotubes, these biosensors can selectively bind to tau proteins and transduce the binding event into an electrical signal. This approach has been shown to detect tau at concentrations as low as picomolar levels, enabling earlier diagnosis than traditional methods.

Parkinson's disease is characterized by the accumulation of alpha-synuclein (α-syn), a protein that forms toxic aggregates in neurons. Detecting alpha-synuclein in biological fluids is essential for diagnosing Parkinson's at early stages and tracking disease progression. Similar to Aβ and tau in Alzheimer's, α-synuclein can be found in the blood, CSF, and other bodily fluids, making it a valuable biomarker for early detection (Atik et al. 2016). Researchers have developed biosensors that use nanomaterials such as gold nanoparticles, graphene, and quantum dots to detect α-synuclein with high sensitivity.

Gold nanoparticles, for instance, can be functionalized with aptamers or antibodies that specifically bind to α-synuclein. When the α-synuclein binds to the nanoparticles, the resulting change in optical properties or electrical signal can be measured One such biosensor achieved a detection limit in the nanomolar range, making it suitable for early-stage diagnosis. Graphene-based sensors have also been explored for detecting α-synuclein. By exploiting the high electrical conductivity of graphene and its ability to be functionalized with biomolecules, these sensors can detect very low concentrations of α-synuclein in CSF and blood samples. This is crucial for identifying Parkinson's patients before significant neuronal loss occurs, allowing for earlier intervention.

9.3.2 Real-Time Monitoring of Neurodegenerative Disease Progression

In addition to early detection, continuous monitoring of neurodegenerative diseases is essential for assessing disease progression and evaluating the effectiveness of therapeutic interventions (Khalil et al. 2018). Nano-enabled biosensors offer the ability to monitor disease-related biomarkers in real time, providing dynamic insights into the patient's condition.

Oxidative stress is a hallmark of many neurodegenerative diseases, including Alzheimer's, Parkinson's, and ALS (Butterfield and Halliwell 2019). It results from an imbalance between the production of reactive oxygen species (ROS) and the body's ability to detoxify these harmful molecules. ROS cause damage to lipids, proteins, and DNA, leading to neuronal dysfunction and cell death. Electrochemical biosensors based on nanomaterials such as carbon nanotubes, graphene, and metal nanoparticles have been developed to detect oxidative stress markers in biological fluids For example, sensors that detect glutathione (GSH), a key antioxidant, can provide information about the oxidative state of the brain. Carbon nanotube-based sensors functionalized with enzymes that react with ROS have been used to measure real-time changes in oxidative stress levels. By continuously monitoring oxidative stress, clinicians can track disease progression and adjust treatment strategies accordingly.

Neuroinflammation plays a critical role in the progression of neurodegenerative diseases, as microglial cells become activated in response to protein aggregates, releasing pro-inflammatory cytokines that contribute to neuronal damage (Heneka et al. 2015). Detecting and monitoring neuroinflammatory markers, such as interleukins (IL-6, IL-1β) and tumor necrosis factor-alpha (TNF-α), can provide insights into the inflammatory state of the brain and guide therapeutic interventions. Biosensors using nanomaterials such as gold nanoparticles and quantum dots have been developed to detect inflammatory cytokines with high sensitivity These sensors can be integrated into wearable devices or implantable systems to continuously monitor inflammation levels in real time. By tracking the inflammatory response, healthcare providers can adjust treatments, such as anti-inflammatory drugs or neuromodulation therapies, to slow disease progression and alleviate symptoms.

9.3.3 Wearable Biosensing Platforms for Neurodegenerative Disease Management

Wearable biosensing platforms represent a significant advancement in the field of neurodegenerative disease diagnostics, as they allow for continuous, noninvasive monitoring of biomarkers in real-world settings. These devices integrate nanomaterials with flexible electronics, enabling the detection of biomarkers in sweat, saliva, or interstitial fluid. Sweat contains a wealth of biological information, including biomarkers related to neurodegenerative diseases. Recent advances in wearable technology have led to the development of sweat-based biosensors that can detect key biomarkers in real time (Fig. 9.2) (Kim et al. 2019). For example, electrochemical sensors that detect dopamine, a neurotransmitter whose levels are dysregulated in Parkinson's disease, have been incorporated into wearable devices that can monitor patients during daily activities (Bandodkar et al. 2019). Graphene and carbon nanotube-based sensors are commonly used in these wearable platforms due to their flexibility, conductivity, and biocompatibility. By continuously monitoring dopamine levels in sweat, clinicians can gain insights into the progression of Parkinson's disease and adjust treatments, such as dopamine replacement therapy, based on real-time data. Saliva is another noninvasive biofluid that can be used for neurodegenerative disease diagnosis and monitoring. Biosensors that detect amyloid-beta, tau, or alpha-synuclein in saliva have been developed using nanomaterials such as gold nanoparticles and graphene. These wearable devices provide a noninvasive and convenient way to monitor biomarker levels without the need for invasive procedures like lumbar punctures. For example, a saliva-based biosensor functionalized with antibodies against amyloid-beta can detect changes in Aβ levels over time, providing real-time insights into the progression of Alzheimer's disease. The integration of such sensors into wearable devices allows for continuous monitoring, enabling more personalized and dynamic disease management.

9.3.4 Integration of Aptamer-Based and Antibody-Based Biosensors with Nanoelectronics

Aptamers are single-stranded DNA or RNA molecules that can fold into unique three-dimensional structures, allowing them to bind to specific target molecules with high affinity. Aptamers offer several advantages over antibodies, including greater stability, ease of synthesis, and the ability to recognize a wider range of targets. Aptamer-based biosensors have been developed for detecting key biomarkers in neurodegenerative diseases. For example, an aptamer specific to amyloid-beta can be immobilized on a graphene-based sensor to create a highly sensitive platform

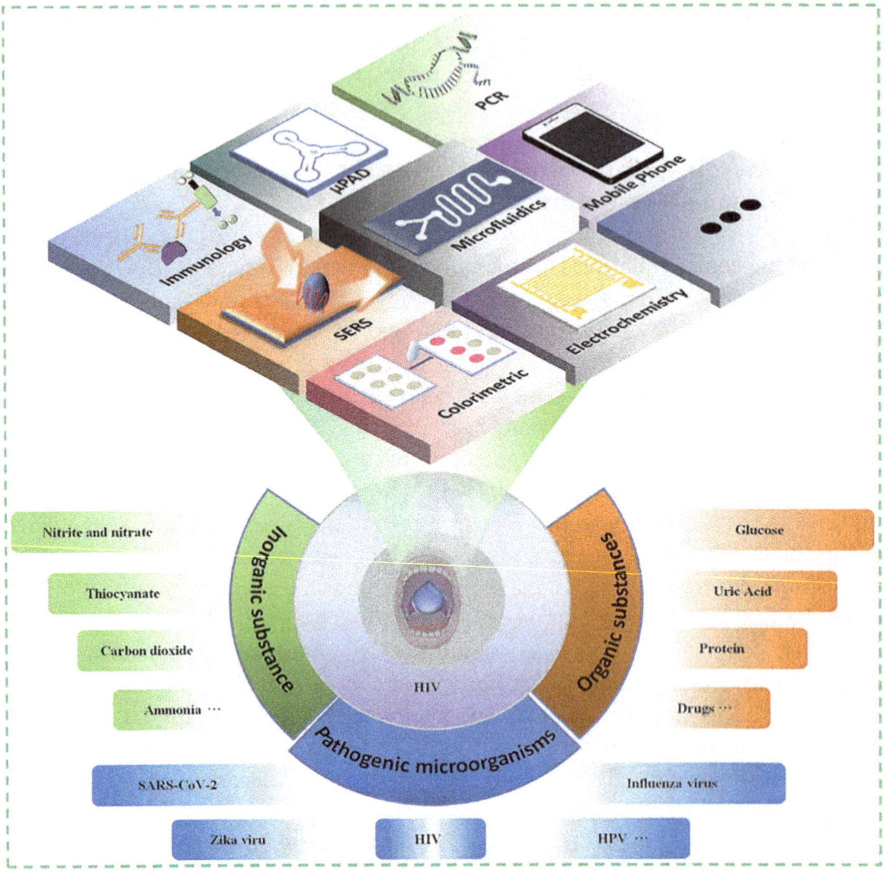

Fig. 9.2 (**a**) Device concept. (a) Schematic illustrating the exploded view of the complete hybrid battery-free system. PI polyimide, S.R. sweat rate. Close-up image of (b) microfluidic patch with embedded sensors and (c) battery-free NFC electronics. (d) Image illustrating the reversible magnetic attachment of the NFC electronics to the microfluidic patch. (e) Image of the complete system. (f) Image illustrating the device while sweating. (g) A phone interface that illustrates wireless communication and image acquisition (Bandodkar et al. 2019). (Photo credit: Philipp Gutruf, Northwestern University). (**b**) Biosensor components and the path of biosensor development for wearables (Kim et al. 2019). (a) Schematic representation of biosensor operation principles: target analyte detection by the corresponding receptor molecule followed by signal transduction method and output. (b) The concept of enzyme electrodes was proposed by Clark and Lyons in 1962. (**c**) Overview of microfluidic-based saliva analysis platforms integrating immunoassays, PCR, microfluidics, smartphones, electrochemistry, and genomics for the detection of a wide range of analytes (Liu et al. 2025). These include inorganic substances (e.g., nitrite, ammonia), organic substances (e.g., glucose, protein), and pathogenic microorganisms (e.g., HIV, SARS-CoV-2, HPV), highlighting the versatility of saliva as a diagnostic fluid

for early detection of Alzheimer's disease (Ohno et al. 2010). When the aptamer binds to Aβ, it induces a change in the electrical properties of the graphene, producing a measurable signal. Similarly, aptamer-based sensors have been developed for detecting alpha-synuclein in Parkinson's disease By integrating these sensors into wearable devices or implantable systems, real-time monitoring of biomarker levels can be achieved, providing continuous feedback on disease progression.

Antibodies are widely used in biosensing applications due to their high specificity for target antigens. In the context of neurodegenerative diseases, antibodies that recognize biomarkers such as amyloid-beta, tau, and alpha-synuclein can be immobilized on the surface of nanomaterials to create highly sensitive biosensors For example, antibody-functionalized gold nanoparticles have been used to detect tau protein in cerebrospinal fluid with high sensitivity (Rissin et al. 2011). When tau binds to the antibodies on the surface of the nanoparticles, it induces a change in the optical properties of the particles, which can be measured using techniques such as surface plasmon resonance (SPR) (Sharma et al. 2022). Antibody-based biosensors can also be integrated with nanoelectronics to create platforms capable of detecting multiple biomarkers simultaneously by functionalizing different areas of a graphene-based sensor with antibodies against amyloid-beta, tau, and alpha-synuclein; researchers have created multiplexed sensors that can provide comprehensive information about a patient's neurodegenerative disease status.

9.4 Challenges

Despite the significant advances in biosensing technology for neurodegenerative disease diagnosis, several challenges remain. One of the primary challenges is ensuring the specificity of biosensors in complex biological fluids, such as blood or saliva, which contain a wide range of proteins and other molecules that can interfere with the detection of target biomarkers. Additionally, biosensors must detect biomarkers at extremely low concentrations, often in the picomolar or femtomolar range, necessitating the use of highly sensitive materials and transducers. This requirement makes manufacturing at scale costly and complex. Another challenge is integrating biosensors into wearable or implantable devices for continuous, real-time biomarker monitoring. Such devices must be biocompatible, durable, and capable of maintaining sensitivity and accuracy over long periods. Finally, while biosensors offer promise for early diagnosis and monitoring, they are not yet widely used in clinical practice due to the need for large-scale clinical validation and standardized operational protocols. While nano-bioelectronics promises breakthroughs in neurodegenerative disease diagnosis, it is laden with technical, biological, and ethical hurdles. These span across material integration, biocompatibility, and regulatory and ethical frameworks. Ethical concerns around implantable neurodevices include issues of data privacy, autonomy, and potential misuse of neural data, emphasizing the need for proactive societal and legal consideration (Ienca and Andorno 2017; Yuste et al. 2017).

Fabricating nano-bioelectronic devices requires precise control at the nanoscale using advanced techniques such as electron-beam lithography, chemical vapor deposition, and self-assembly. Materials like graphene and carbon nanotubes must retain their electronic and mechanical properties during fabrication (Lee et al. 2019), and scalability for clinical use remains difficult. Nanosensors are highly sensitive and susceptible to false signals due to background noise from nonspecific interactions in biological fluids (Kaushik et al. 2014). Multimodal biosensors and machine learning algorithms are increasingly used to improve signal fidelity and support real-time clinical decisions, but they must be validated for reliability, transparency, and bias mitigation (Topol 2019; Esteva et al. 2019). Implantable nano-bioelectronic devices degrade under physiological conditions due to oxidation, biofouling, and immune responses. Long-term stability solutions involve surface coatings, biocompatible polymers, and soft materials matching tissue mechanics (Kozai et al. 2015; Green et al. 2013).

Nanomaterials can penetrate cells and cause oxidative stress, inflammation, or genetic damage, especially in the central nervous system (Nel et al. 2006; Functionalization (e.g., PEGylation) and use of biodegradable carriers like PLGA offer safer alternatives (Danhier et al. 2012). Implanted devices may induce foreign body responses such as glial scarring, thus limiting signal exchange between neurons and electrodes (Polikov et al. 2005). Strategies like anti-inflammatory coatings, flexible substrates, and miniaturization help mitigate immune responses (Kozai et al. 2015). Long-term stimulation devices like deep brain stimulators (DBS) can cause side effects including speech changes, cognitive decline, or emotional disturbances (Miocinovic et al. 2013). Adaptive DBS systems using AI for closed-loop feedback are being explored to reduce overstimulation and side effects.

9.5 Conclusions and Outlook

A future direction in neurodegenerative disease treatment is the combination of nanoparticle-based drug delivery systems with bioelectronic devices. For example, nanoparticles could be used to deliver drugs to specific brain regions, while bioelectronic devices provide electrical stimulation to enhance the therapeutic effects. In Parkinson's disease, combining deep brain stimulation with nanoparticle-based dopamine delivery could enhance the effectiveness of treatment. Nanoparticles loaded with dopamine or dopamine precursors could be delivered to the basal ganglia, while the bioelectronic device provides electrical stimulation to regulate neural activity. This integrated approach could improve motor function more effectively than either therapy alone. Future innovations may also include the use of magnetically guided nanoparticles for precise drug delivery in the brain. Magnetic nanoparticles can be directed to specific brain regions using external magnetic fields, delivering therapeutic agents exactly where they are needed. By combining this approach with bioelectronic monitoring, clinicians could track the patient's response to treatment in real time and adjust therapy as needed. As nano-bioelectronic devices generate vast amounts of data from biosensors and neural interfaces, the integration of AI and machine learning will be critical for analyzing this information and delivering personalized treatment strategies. AI-driven algorithms can analyze patterns in biomarker levels, neural activity, and patient symptoms to identify early signs of disease progression and recommend individualized interventions. AI has the potential to revolutionize the early diagnosis of neurodegenerative diseases by identifying subtle changes in biomarker levels or neural activity that may not be detectable by traditional methods. Machine learning algorithms can be trained to recognize patterns in biosensor data that are indicative of early-stage Alzheimer's, Parkinson's, or ALS, allowing for earlier intervention and more effective treatment.

AI-driven predictive models can also be used to forecast disease progression based on a patient's biomarker levels, neural activity, and other physiological data. These models can help clinicians anticipate changes in a patient's condition and adjust treatment plans accordingly. For example, in Alzheimer's disease, an AI model could predict when a patient is likely to experience cognitive decline based on their amyloid-beta levels and suggest preventive interventions. The integration of AI with nano-bioelectronics opens the door to personalized treatment strategies that adapt in real time to the patient's changing condition. Machine learning algorithms can analyze data from wearable and implantable devices to optimize neuromodulation and drug delivery based on the patient's individual needs. In closed-loop neuromodulation systems, AI algorithms can analyze real-time data from biosensors to adjust stimulation parameters dynamically. For example, in Parkinson's disease, an AI-driven closed-loop system could monitor the patient's motor symptoms and neural activity, automatically adjusting deep brain stimulation settings to optimize treatment. This personalized approach minimizes side effects and improves therapeutic outcomes. AI can also be integrated into nanoparticle-based drug delivery systems to optimize the timing and location of drug release. By analyzing patient data in real time, AI algorithms can trigger the release of drugs from nanoparticles when they are most needed, ensuring that therapeutic agents are delivered at the right time and in the right place.

References

Arshad F, Nabi F, Iqbal S, Khan RH (2022) Applications of graphene-based electrochemical and optical biosensors in early detection of cancer biomarkers. In: Colloids and surfaces B: biointerfaces, vol 212. Elsevier B.V. https://doi.org/10.1016/j.colsurfb.2022.112356

Atik A, Stewart T, Zhang J (2016) Alpha-synuclein as a biomarker for Parkinson's disease. Brain Pathol 26(3):410–418. https://doi.org/10.1111/bpa.12370

Bandodkar AJ, Wang J (2014) Non-invasive wearable electrochemical sensors: a review. TrAC Trends Anal Chem 76:173–185. https://doi.org/10.1016/j.trac.2015.07.001

Bandodkar AJ, Gutruf P, Choi J, Lee K, Sekine Y, Reeder JT, Jeang WJ, Aranyosi AJ, Lee SP, Model JB, Ghaffari R, Su C-J, Leshock JP, Ray T, Verrillo A, Thomas K, Krishnamurthi V, Han S, Kim J et al (2019) Battery-free, skin-interfaced microfluidic/electronic systems for simultaneous electrochemical, colorimetric, and volumetric analysis of sweat. Sci Adv. https://www.science.org

Bateman RJ et al (2012) Clinical and biomarker changes in dominantly inherited Alzheimer's disease. N Engl J Med 367(9):795–804

Bharadwaj LM et al (2020) Nanotechnology-based drug delivery systems and their potential for neurodegenerative diseases. J Control Release 328:892–910
Butterfield DA, Halliwell B (2019) Oxidative stress, dysfunctional glucose metabolism and Alzheimer disease. Nat Rev Neurosci 20(3):148–160. https://doi.org/10.1038/s41583-019-0132-6
Chen X, Sun H, Kuo DH, Abdeta AB, Zelekew OA, Guo Y, Zhang J, Yuan Z, Lin J (2021) Spherical nanoflower-like bimetallic (Mo,Ni)(S,O)3-x sulfo-oxide catalysts for efficient hydrogen evolution under visible light. Appl Catal B Environ 287. https://doi.org/10.1016/j.apcatb.2021.119992
Chia R, Chio A, Traynor BJ (2018) Novel genes associated with amyotrophic lateral sclerosis: diagnostic and clinical implications. Lancet Neurol 17(1):94–102
Chung S et al (2019) Biosensors for the detection of neurodegenerative diseases. Biosens Bioelectron 132:229–239
Cummings J, Lee G, Ritter A, Zhong K (2019) Alzheimer's disease drug development pipeline: 2019. Alzheimer's Dement 5:272–293
Danhier F et al (2012) PLGA-based nanoparticles: an overview of biomedical applications. J Control Release 161(2):505–522
Dhahi TS, Yousif Dafhalla AK, Al-Mufti AW, Elobaid ME, Adam T, Gopinath SCB (2024) Application of Nanobiosensor engineering in the diagnosis of neurodegenerative disorders. In: Results in engineering, vol 24. Elsevier B.V. https://doi.org/10.1016/j.rineng.2024.102790
Dias V, Junn E, Mouradian MM (2013) The role of oxidative stress in Parkinson's disease. J Parkinsons Dis 3(4):461–491
Esteva A et al (2019) A guide to deep learning in healthcare. Nat Med 25(1):24–29
Fearnley JM, Lees AJ (1991) Ageing and Parkinson's disease: substantia nigra regional selectivity. Brain 114(5):2283–2301
Gao HM et al (2011) Neuroinflammation and oxidative stress in Parkinson's disease and Alzheimer's disease. Adv Pharmacol 60:457–472
Green R et al (2013) Conducting polymers for neural interfaces. Biomaterials 34(29):7227–7233
Hardiman O et al (2017) Amyotrophic lateral sclerosis. Nat Rev Dis Prim 3:17071
Hardy J, Selkoe DJ (2002) The amyloid hypothesis of Alzheimer's disease: Progress and problems on the road to therapeutics. Science 297(5580):353–356
Heneka MT, Carson MJ, El Khoury J et al (2015) Neuroinflammation in Alzheimer's disease. Lancet Neurol 14(4):388–405. https://doi.org/10.1016/S1474-4422(15)70016-5
Henstridge CM, Hyman BT, Spires-Jones TL (2019) Beyond the neuron–cellular interactions early in Alzheimer disease pathogenesis. Nat Rev Neurosci 20:94–108
Ienca M, Andorno R (2017) Towards new human rights in the age of neuroscience and neurotechnology. Life Sci Soc Policy 13:5
Jack CR, Bennett DA, Blennow K et al (2018) NIA-AA Research Framework: toward a biological definition of Alzheimer's disease. Alzheimers Dement 14(4):535–562. https://doi.org/10.1016/j.jalz.2018.02.018
Kalia LV, Lang AE (2015) Parkinson's disease. Lancet 386(9996):896–912
Kaushik A et al (2014) Electrochemical biosensors for early detection of Alzheimer's disease. Anal Biochem 361(2):1–13
Khalil M, Teunissen CE, Otto M et al (2018) Neurofilaments as biomarkers in neurological disorders. Nat Rev Neurol 14(10):577–589. https://doi.org/10.1038/s41582-018-0058-z
Kim J, Campbell AS, de Ávila BE, Wang J (2019) Wearable biosensors for healthcare monitoring. Nat Biotechnol 37(4):389–406. https://doi.org/10.1038/s41587-019-0045-y
Kovacs GG (2016) Molecular pathology of neurodegenerative diseases: principles and practice. J Clin Pathol 69(8):643–658
Kozai TDY et al (2015) Reduction of neurovascular damage resulting from microelectrode insertion into the cerebral cortex using in vivo two-photon mapping. J Neural Eng 12(6):066019
Kumar S, Kim KH (2022) Electrochemical immunosensor for early detection of β-amyloid peptide in Alzheimer's disease. Biosensors 12(11):1059

Lee H et al (2019) Wearable/disposable sweat-based glucose monitoring device with multistage transdermal drug delivery module. Sci Adv 3(3):e1601314

Li J, Ni W, Jin D, Yu Y, Xiao MM, Zhang ZY, Zhang GJ (2023) Nanosensor-driven detection of neuron-derived Exosomal Aβ42 with graphene electrolyte-gated transistor for Alzheimer's disease diagnosis. Anal Chem 95(13):5719–5728. https://doi.org/10.1021/acs.analchem.2c05751

Liu X, Zhou X, Li X, Wei Y, Wang T, Liu S, Yang H, Sun X (2025) Saliva analysis based on microfluidics: focusing the wide spectrum of target analyte. In: Critical reviews in analytical chemistry, vol 55, Issue 2. Taylor and Francis Ltd., pp 330–352. https://doi.org/10.1080/10408347.2023.2287656

Lo MC, Widge AS (2017) Closed-loop neuromodulation systems: next-generation treatments for psychiatric illness. Int Rev Psychiatry 29(2):191–204

Miocinovic S et al (2013) History, applications, and mechanisms of deep brain stimulation. JAMA Neurol 70(2):163–171

Nel A et al (2006) Toxic potential of materials at the nanolevel. Science 311(5761):622–627

Ohno Y, Maehashi K, Yamashiro Y, Matsumoto K (2010) Label-free biosensors based on aptamer-modified graphene field-effect transistors. J Am Chem Soc 132(51):18012–18013. https://doi.org/10.1021/ja107440t

Pardridge WM (2012) Drug transport across the blood–brain barrier. J Cereb Blood Flow Metab 32(11):1959–1972. https://doi.org/10.1038/jcbfm.2012.126

Polikov VS et al (2005) Response of brain tissue to chronically implanted neural electrodes. J Neurosci Methods 148(1):1–18

Przedborski S, Vila M, Jackson-Lewis V (2003) Neurodegeneration: what is it and where are we? J Clin Invest 111(1):3–10

Rissin DM, Kan CW, Campbell TG, Howes SC, Fournier DR, Song L et al (2011) Single-molecule enzyme-linked immunosorbent assay detects serum proteins at subfemtomolar concentrations. Nat Biotechnol 28(6):595–599. https://doi.org/10.1038/nbt.1641

Rosen DR et al (1993) Mutations in Cu/Zn superoxide dismutase gene are associated with familial amyotrophic lateral sclerosis. Nature 362:59–62

Ross CA, Poirier MA (2004) Protein aggregation and neurodegenerative disease. Nat Med 10(7):S10–S17

Scheltens P et al (2016) Alzheimer's disease. Lancet 388(10043):505–517

Selkoe DJ, Hardy J (2016) The amyloid hypothesis of Alzheimer's disease at 25 years. EMBO Mol Med 8(6):595–608

Sharma S, Zapatero-Rodríguez J, Estrela P, O'Kennedy R (2022) Point-of-care diagnostics in low resource settings: present status and future role of microfluidics. Biosens Bioelectron 153:112035. https://doi.org/10.1016/j.bios.2020.112035

Spillantini MG et al (1998) Alpha-synuclein in Lewy bodies. Nature 388:839–840

Topol EJ (2019) High-performance medicine: the convergence of human and artificial intelligence. Nat Med 25(1):44–56

Wang Y, Mandelkow E (2016) Tau in physiology and pathology. Nat Rev Neurosci 17(1):22–35. https://doi.org/10.1038/nrn.2015.1

Wen MM, El-Salamouni NS, El-Refaie WM, Hazzah HA, Ali MM, Tosi G, Farid RM, Blanco-Prieto MJ, Billa N, Hanafy AS (2017) Nanotechnology-based drug delivery systems for Alzheimer's disease management: technical, industrial, and clinical challenges. J Control Release 245:95–107). Elsevier B.V.

Yuste R et al (2017) Four ethical priorities for neurotechnologies and AI. Nature 551:159–163

Multimodal Sensing Arrays for Comprehensive Health Monitoring and Disease Management

10

Kuldeep Mahato and Nirmal Roy

Abstract

Multimodal sensing arrays represent a groundbreaking advancement in healthcare technology, enabling the concurrent monitoring of diverse physiological parameters through the integration of biochemical, physical, optical, and thermal sensors. By capturing the dynamic interplay among biological systems, these platforms provide a comprehensive and real-time assessment of an individual's health status, paving the way for more precise, personalized, and proactive medical interventions. This chapter discusses the fundamental principles highlighting multimodal sensing technologies, examining their current applications and projecting future directions in the context of disease management and precision medicine. It introduces the conceptual framework of multimodal sensing and underscores its critical role in monitoring complex physiological phenomena, particularly in the management of chronic diseases, mental health conditions, and postoperative recovery. Emphasis is placed on the synergistic integration of biochemical markers with physical parameters including heart rate variability and biomechanical movement, to yield multidimensional health insights. The chapter further explores device design and fabrication strategies, emphasizing the use of biocompatible materials, flexible electronics, and low-power circuitry that support wearable and implantable form factors. Emerging innovations in data fusion and machine learning are also discussed, illustrating how artificial intelligence enables enhanced real-time monitoring, predictive modeling, and individualized therapeutic strategies. Key challenges, including regulatory com-

K. Mahato (✉)
Department of Biotechnology, Motilal Nehru National Institute of Technology Allahabad, Prayagraj, Uttar Pradesh, India

N. Roy
Electronics and Communication Engineering Department, Motilal Nehru National Institute of Technology Allahabad, Prayagraj, Uttar Pradesh, India

K. Mahato, P. Chandra (eds.), *Nano-bioelectronics for Precision Health Monitoring*, https://doi.org/10.1007/978-981-95-2117-3_10

pliance, data security, and system integration, are critically analyzed to illuminate the current limitations and future opportunities in this field. The chapter concludes with a forward-looking perspective, envisioning the convergence of multimodal sensing arrays with advances in nanotechnology, quantum sensing, and personalized medicine to transform healthcare delivery through early diagnosis, continuous monitoring, and tailored interventions.

Keywords

Multimodal sensing arrays · Precision healthcare · Wearable and implantable sensors · Data fusion and machine learning · Personalized medicine · Comprehensive health monitoring

10.1 Introduction to Multimodal Sensing in Healthcare

Healthcare has undergone significant transformations in the past decades due to the integration of advanced sensing technologies, particularly within the domain of personalized and precision medicine (Chandra and Mahato 2022; Mahato and Chandra 2024). The advent of bioelectronics, wearable devices, and implantable sensors has allowed healthcare professionals to monitor and assess physiological functions with greater accuracy and specificity. A critical development within this space is the emergence of multimodal sensing arrays, which offer a more comprehensive approach to health monitoring by integrating multiple sensor types into a unified system (Mahato et al. 2024). Multimodal sensing refers to the use of multiple sensor modalities to collect diverse data streams that reflect different aspects of physiological function. Unlike traditional single-sensor systems, which typically measure a specific parameter, multimodal sensing arrays combine various types of sensors to simultaneously monitor biochemical, electrical, and physical signals from the body. These signals provide a richer and more detailed profile of the patient's health, capturing interactions between different physiological systems (Xu et al. 2025). For instance, a multimodal sensor array designed for cardiovascular health monitoring might integrate electrochemical sensors for measuring blood lactate and glucose, physical sensors like accelerometers to track movement and activity levels, and optical sensors to monitor oxygen saturation and blood flow. This combination can offer a multidimensional view of cardiovascular function, enabling a more nuanced understanding of the patient's health. The core advantage of multimodal sensing is that it allows clinicians and researchers to go beyond isolated parameters. A single variable often fails to provide the whole picture, particularly in complex or dynamic conditions such as diabetes, where blood glucose is only one part of the metabolic system. By integrating multiple data streams, multimodal systems can provide a holistic picture, correlating biochemical changes with physical or electrical signals. This approach has great potential for identifying early signs of disease, monitoring disease progression, and optimizing personalized treatments based on real-time feedback.

Precision healthcare is a rapidly evolving approach that tailor's medical treatment to the individual characteristics of each patient. Rather than relying on general guidelines that apply broadly to populations, precision healthcare uses data from individual patients to guide decisions on prevention, diagnosis, and treatment (Johnson et al. 2021). Multimodal sensing arrays play a vital role in this paradigm, as they enable continuous and comprehensive health monitoring that can adapt to the unique needs of each patient. The central premise of conventional healthcare is that the "one-size-fits-all" approach in medicine is inherently limited (Mahato et al. 2024). Genetic variability, lifestyle factors, and environmental exposures all contribute to differences in how diseases manifest and how patients respond to treatment. By gathering real-time data from a variety of sensor types, multimodal sensing systems provide deeper insights into these differences, allowing for more personalized care (Van Calster et al. 2019). For instance, consider the management of chronic diseases, such as diabetes or hypertension. Multimodal sensing arrays can continuously monitor key indicators like blood glucose, blood pressure, heart rate variability, and physical activity, thereby providing real-time feedback to patients and clinicians (Saha and Del Caño 2023). This continuous flow of data allows for earlier detection of potential complications, enabling timely interventions. The rich data sets collected by multimodal sensors can also inform machine learning (ML) models, helping to predict disease flare-ups or responses to specific treatments (Khalifa and Albadawy 2024). Beyond chronic disease management, multimodal sensing has applications in preventive care. Sensors that track multiple biomarkers can help detect subtle physiological changes that precede disease onset, allowing for early intervention. For instance, a combination of heart rate variability and stress hormone levels measured by biochemical sensors could provide early warning signs of cardiovascular problems or mental health disorders (Smokovski et al. 2024). Furthermore, multimodal sensing arrays can improve postsurgical monitoring and rehabilitation. Patients recovering from surgery often require close monitoring to detect signs of infection, inflammation, or other complications. By integrating sensors that measure biomarkers related to wound healing, infection, and mobility, multimodal sensing arrays offer a more comprehensive tool for tracking recovery progress (Dong and Guo 2021). This information can be used to adjust postoperative care plans in real time, enhancing patient outcomes and reducing hospital readmission rates.

Technological advancements in materials science, microelectronics, and data analytics have catalyzed the development of sophisticated multimodal sensing arrays (Fig. 10.1). These systems typically consist of multiple sensors that are either integrated into a single device or distributed across different locations on or inside the body. Key trends shaping the evolution of multimodal sensing in healthcare include the miniaturization of sensors, the integration of flexible and wearable technologies, and the use of advanced data processing techniques such as artificial intelligence (AI) and machine learning (Shajari et al. 2023). The miniaturization of sensors has been crucial in enabling the development of multimodal systems, particularly for wearable and implantable devices. Advances in nanotechnology, microfabrication techniques, and low-power electronics have made it possible to create

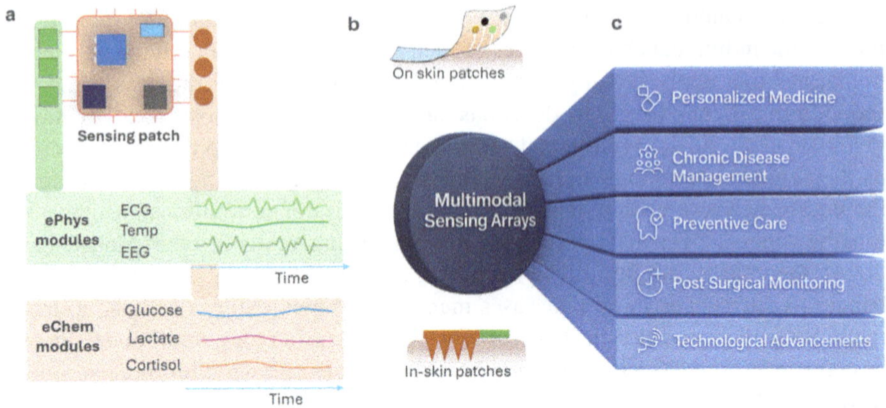

Fig. 10.1 Scheme showing (**a**) components of the multimodal sensing platform, the (**b**) wearable patches such as "on skin" and "in skin" biosensing arrays, (**c**) various applications of multimodal sensing arrays, including the personalized medicine, chronic disease management, preventive care, postsurgical monitoring, and other including the rehabilitation monitoring

sensors that are small enough to be integrated into minimally invasive devices, such as microneedles or smart contact lenses (Chandra and Mahato 2022). These miniaturized sensors are capable of continuously monitoring a range of physiological parameters without causing significant discomfort to the user. For instance, microneedle arrays, which penetrate only the outermost layer of the skin, can house multiple sensors capable of measuring glucose, lactate, and interstitial fluid metabolites (Teymourian and Tehrani 2021; Tehrani and Teymourian 2022). Such devices provide continuous biochemical monitoring without the need for invasive blood draws, enhancing patient comfort while delivering comprehensive health data. Wearable technologies such as smartwatches, patches, and textiles are also benefiting from sensor miniaturization. Many of these devices now integrate multiple sensors to monitor heart rate, blood oxygen levels, physical activity, and skin temperature simultaneously. Wearable patches, for example, can be designed to combine electrochemical sensors for sweat analysis with accelerometers and heart rate sensors, providing real-time insights into a user's hydration status, physical exertion, and cardiovascular health (Ates and Nguyen 2022).

10.2 Core Technologies and Sensors in Multimodal Arrays

The landscape of healthcare is continuously evolving, with emerging sensor technologies playing a critical role in advancing diagnostic and monitoring capabilities. Multimodal sensing arrays, which integrate multiple sensor types into a unified system, are among the most promising innovations. These arrays enable simultaneous monitoring of different physiological parameters, providing a comprehensive and holistic view of an individual's health status. This section will explore the core technologies that underpin multimodal sensing arrays, focusing on the distinct sensor types, electrochemical, biochemical, physical, optical, and thermal sensors, which make up these systems. Additionally, strategies for integrating these sensors into cohesive multimodal arrays will be discussed.

10.2.1 Biochemical Sensors

Biochemical sensors form the foundation of multimodal sensing arrays, especially for detecting and quantifying biological markers that provide insights into a person's biochemical state. These sensors operate by converting biochemical signals, such as the presence or concentration of a molecule, into electrical signals that can be easily measured and analyzed. Electrochemical sensors are particularly useful for monitoring analytes in bodily fluids like blood, sweat, saliva, and interstitial fluid, while biochemical sensors extend this capability to specific proteins, metabolites, and nucleic acids. Electrochemical sensors detect analytes by measuring the electrical signals generated from redox reactions, ion concentration changes, or other electrochemical processes. They are widely used in point-of-care devices, particularly for monitoring blood glucose levels, pH, electrolytes, and other small molecules (Zhu et al. 2015; Shanbhag et al. 2023). Electrochemical sensors are typically categorized into three main types: potentiometric, amperometric, and conductometric sensors, each operating on different electrochemical principles.

- Potentiometric sensors: Potentiometric sensors measure the voltage difference between a reference electrode and a working electrode in response to the presence of an ion or molecule of interest. These sensors are often used to measure pH, ion concentrations (e.g., sodium, potassium, calcium), and certain gases (e.g., carbon dioxide) in biological samples (Zdrachek and Bakker 2019; Sailapu et al. 2021). A common example of a potentiometric sensor is the ion-selective electrode, which has been employed in clinical settings to measure electrolyte imbalances in patients.
- Amperometric sensors: Amperometric sensors operate by measuring the current produced by the oxidation or reduction of an analyte at an electrode surface. These sensors are particularly effective for detecting glucose, lactate, and other metabolites, making them indispensable in the management of diabetes and sports medicine. In multimodal sensing arrays, amperometric sensors are often combined with other modalities to track the dynamic changes in biochemical markers alongside physiological signals such as heart rate or movement (Demkiv et al. 2024; Tajik et al. 2021).
- Conductometric sensors: Conductometric sensors measure changes in the electrical conductivity of a solution in response to the presence of an analyte. These sensors are typically less selective than potentiometric or amperometric sensors, but they are useful for measuring overall changes in ionic composition, which can be indicative of physiological stress or dehydration (Crispi et al. 2023; Crispi and Neri 2022). Conductometric sensors can be integrated into wearable devices that monitor sweat composition during physical exertion or heat stress.

These biochemical sensors detect specific biological molecules, such as proteins, nucleic acids, or metabolites, using recognition elements such as antibodies, aptamers, or enzymes. These sensors are highly specific, allowing for the detection of low concentrations of target analytes in complex biological fluids. Biochemical sensors are typically divided into several types, based on the recognition element and detection mechanism used.

- Enzymatic sensors: Enzymatic sensors are a type of biochemical sensor that utilize enzymes as the recognition element. The enzyme catalyzes a specific reaction with the target analyte, and the product of this reaction is then detected electrochemically or optically. One of the most widely known examples of enzymatic sensors is the glucose oxidase-based sensor used in continuous glucose monitoring (CGM) systems for diabetic patients (Zhang et al. 2023a). Enzymatic sensors are highly selective and are often integrated into multimodal arrays to track glucose levels in conjunction with other health parameters, such as heart rate variability or body temperature (Govindaraj et al. 2023).
- Aptamer-based sensors: Aptamers are short, single-stranded DNA or RNA molecules that can bind to specific target molecules with high affinity and specificity. Aptamer-based sensors have gained significant attention for their ability to detect a wide range of biomolecules, including small molecules, proteins, and even whole cells (Dauphin-Ducharme et al. 2019). In multimodal arrays, aptamer-based sensors are often used to detect disease-specific biomarkers, such as cancer antigens or viral proteins, in combination with other sensors that monitor physiological changes like inflammation or stress response (Zhou et al. 2023; Wang et al. 2020).
- Immunosensors: Immunosensors are biochemical sensors that use antibodies as the recognition element to detect specific antigens or proteins. These sensors are highly specific and can be used for detecting a wide range of biological markers, including hormones, cytokines, and pathogen-specific proteins (Zhu et al. 2019). Immunosensors are frequently integrated into multimodal arrays for applications in disease diagnosis and monitoring, where they can detect biomarkers related to infections, inflammation, or autoimmune disorders (Lee et al. 2022; Mehta et al. 2024).

10.2.2 Physical Sensors

Physical sensors monitor mechanical and physical parameters such as pressure, temperature, movement, and acceleration. These sensors are critical components of multimodal sensing arrays because they provide real-time data on the physical condition and activities of the individual being monitored. Physical sensors are particularly valuable in the context of wearable devices, where they can capture data related to movement, posture, and environmental conditions. Accelerometers and gyroscopes are commonly used in wearable devices to monitor movement, orientation, and balance. Accelerometers measure linear acceleration in one or more axes, while gyroscopes measure angular velocity. Together, these sensors provide a detailed picture of a person's physical activity, gait, and posture. In multimodal sensing arrays, accelerometers and gyroscopes are often combined with other sensors, such as heart rate monitors and biochemical sensors, to track the effects of physical activity on overall health (Wang et al. 2023; Qi et al. 2018; Jalal et al. 2020). For example, in the management of chronic conditions such as cardiovascular disease, accelerometers and gyroscopes can be used to monitor a patient's physical activity level and correlate it with biochemical markers of stress or inflammation. This integration allows for a more comprehensive assessment of the patient's condition and can help guide interventions such as lifestyle changes or medication adjustments (Ahmad et al. 2025; Bayoumy et al. 2021). Pressure sensors are used to measure mechanical pressure or force, and they have a wide range of applications in healthcare. In multimodal sensing arrays, pressure sensors can be used to monitor blood pressure, respiratory rate, or the pressure exerted on specific body parts, such as in bedsores or orthopedic rehabilitation. These sensors are often integrated into wearable or implantable devices to provide continuous, real-time monitoring of pressure-related health indicators (Manero et al. 2023). For instance, pressure sensors embedded in smart clothing can detect postural changes and body weight distribution, which is particularly useful for patients undergoing physical rehabilitation. In combination with biochemical sensors that track inflammation or muscle recovery, these systems can provide valuable insights into the patient's progress and optimize their rehabilitation program (Linh et al. 2025; Meena et al. 2023).

10.2.3 Optical and Thermal Sensors

Optical and thermal sensors have also become increasingly important in multimodal sensing arrays due to their ability to noninvasively monitor vital signs and other physiological parameters. Optical sensors detect light or changes in light intensity, while thermal sensors measure temperature and heat flux. Optical sensors are widely used in healthcare for applications such as pulse oximetry, blood flow monitoring, and glucose measurement (Kazanskiy et al. 2022). These sensors typically operate by emitting light at specific wavelengths and measuring the absorption, reflection, or transmission of light through tissue or fluids. Optical sensors are noninvasive and can provide continuous monitoring of key physiological parameters.

- Pulse oximeters: Pulse oximeters use optical sensors to measure the oxygen saturation of blood. They work by emitting light at two wavelengths (typically red and infrared) and measuring the absorption of light by oxygenated and deoxygenated hemoglobin. In multimodal sensing arrays, pulse oximeters can be combined with heart rate monitors and electrochemical sensors to track oxygen levels, heart function, and biochemical markers of stress or inflammation simultaneously (Chan et al. 2013).
- Photoplethysmography (PPG): PPG is an optical sensing technique that measures changes in blood volume in tissues, typically using infrared light. PPG

sensors are commonly integrated into wearable devices to monitor heart rate, blood flow, and vascular health. In multimodal arrays, PPG can be combined with other sensors, such as accelerometers or biochemical sensors, to provide a more comprehensive view of cardiovascular health and activity (Ahmad et al. 2025; Kim and Baek 2023).

Thermal sensors measure temperature and heat flux, which can provide valuable insights into a person's metabolic and inflammatory state. These sensors are often used to monitor body temperature, skin temperature, and localized inflammation. Thermal sensors are particularly useful in multimodal arrays for detecting fever, infection, or localized tissue damage. For example, a multimodal sensing array designed for wound monitoring might integrate thermal sensors to detect localized increases in temperature that could indicate infection, along with biochemical sensors to monitor inflammatory markers and pH levels. This combination of modalities allows for more accurate and timely detection of complications, enabling earlier interventions and improving patient outcomes (Lu et al. 2022; Mo et al. 2024; Tang et al. 2021).

10.2.4 Integration Strategies for Multimodal Sensing Arrays

The integration of multiple sensors into a unified multimodal array presents significant challenges, particularly in terms of data fusion, device design, and power management. However, several strategies have emerged to overcome these challenges and enable the development of cohesive, functional multimodal sensing systems (Wang et al. 2024a). Data fusion is a critical aspect of multimodal sensing arrays, as it involves combining data from multiple sensors into a single, coherent data stream. Effective data fusion requires sophisticated algorithms that can handle the diverse types of data generated by different sensors and extract meaningful insights (Puerta et al. 2021). Machine learning and artificial intelligence are increasingly being employed to analyze the vast amounts of data generated by multimodal arrays and identify patterns that may be indicative of disease or health changes. Designing a multimodal sensing array requires careful consideration of the physical layout and integration of multiple sensor types (Cai et al. 2019). Miniaturization of sensors and electronic components is essential for creating wearable or implantable devices that are comfortable, unobtrusive, and capable of long-term monitoring. Advances in flexible electronics, microfabrication techniques, and low-power circuitry have made it possible to integrate multiple sensors into compact devices that can be worn or implanted for extended periods (Liu et al. 2017). Power management is a critical challenge in multimodal sensing arrays, particularly for wearable and implantable devices that require continuous operation. Energy-efficient sensor designs and low-power communication protocols are essential for extending battery life and reducing the need for frequent recharging or replacement. Additionally, emerging energy-harvesting technologies, such as solar cells, thermoelectric generators, and piezoelectric materials, offer the potential to power multimodal arrays using energy from the environment or the body itself (Zeng et al. 2022).

10.3 Data Fusion and Processing in Multimodal Systems

Multimodal sensing arrays produce a large and diverse volume of data from various sensors, each monitoring distinct physiological parameters. To unlock the full potential of this data and translate it into actionable insights, advanced techniques for data fusion and processing are essential. The challenges associated with integrating data from multiple sensor modalities, such as biochemical, physical, optical, and thermal sensors, are significant. These challenges include dealing with heterogenous data types, managing the sheer volume of information, ensuring data accuracy, and making real-time decisions. This section will explore the core challenges of data integration, the methodologies and algorithms employed for data fusion, and the clinical relevance of multimodal data processing in healthcare.

10.3.1 Challenges in Data Integration

The primary challenge in multimodal sensing systems is that each sensor modality generates data in different forms, at different rates, and with varying degrees of reliability. Biochemical sensors, for example, may produce continuous streams of data on glucose levels, while physical sensors such as accelerometers may generate discrete data points related to motion. Optical sensors, on the other hand, may record waveforms of light absorption or reflection, which need to be translated into physiological parameters such as oxygen saturation. Combining these diverse data streams into a coherent and clinically meaningful dataset poses several challenges. One of the most significant obstacles in multimodal sensing arrays is the heterogeneity of data. Different sensors measure different types of signals, often in units that are not directly comparable. For example, electrochemical sensors might generate data in terms of voltage or current, while optical sensors measure changes in light intensity, and accelerometers produce data in units of acceleration (meters per second squared). This diversity makes direct comparisons or correlations between data streams difficult without a robust data processing framework (Casian et al. 2022; Bokade et al. 2021). Additionally, different sensor modalities operate at different sampling rates. Biochemical sensors may provide data continuously or at regular intervals (e.g., every few minutes), whereas accelerometers may capture data at high frequencies (e.g., hundreds of times per second). Synchronizing these data streams is necessary for any meaningful integration, as discrepancies in timing can lead to misleading correlations or gaps in the overall analysis. Another challenge in multimodal data integration is dealing with noise and ensuring data quality. Sensors are prone to errors due to environmental factors, mechanical issues, or biological variability. For example, physical sensors like accelerometers may pick up noise from external vibrations or movements unrelated to the physiological process being monitored. Similarly, biochemical sensors may experience signal drift due to enzyme degradation, contamination, or changes in the surrounding environment (e.g., pH or temperature). Dealing with noise requires sophisticated filtering and signal processing techniques. These techniques must be carefully tailored to each sensor modality to minimize the impact of irrelevant or erroneous data while preserving the integrity of the physiological signals of interest (John et al. 2022; Gashi et al. 2022). Furthermore, it is essential to account for sensor drift, calibration errors, and outliers in the data, particularly in long-term monitoring applications where sensor performance may degrade over time.

Many applications of multimodal sensing arrays, such as CGM or real-time cardiac health assessment, require real-time data processing. This introduces an additional layer of complexity, as data must be collected, integrated, and analyzed almost instantaneously to provide actionable feedback. Latency in data processing can render a system less effective, especially in critical care settings where timely interventions are necessary. Real-time processing demands high computational efficiency, particularly for wearable or implantable devices with limited processing power and energy resources. Low-power processors, on-device computation, and cloud-based solutions are increasingly being employed to handle the significant computational load associated with multimodal data fusion (Thilagavathi and Karthikeyan 2025). As with any system that collects personal health information, data privacy and security are major concerns in multimodal sensing arrays. Sensitive health data must be protected from unauthorized access, especially when it is being transmitted wirelessly between sensors, mobile devices, and cloud servers. Encrypting the data and ensuring secure communication protocols are essential to maintaining patient confidentiality and complying with regulatory standards such as the Health Insurance Portability and Accountability Act (HIPAA). In addition to privacy concerns, data integrity must be maintained throughout the data collection and fusion process. Any tampering or loss of data during transmission could compromise the accuracy of clinical insights and lead to incorrect diagnoses or treatment recommendations (Wang et al. 2022).

10.3.2 Algorithms for Data Fusion

Data fusion is the process of combining data from multiple sources to generate a more accurate and comprehensive picture of the system being monitored. In the context of multimodal sensing arrays, data fusion techniques are used to integrate the various data streams generated by different sensor modalities, enabling healthcare providers to make better-informed decisions. Several types of algorithms and methodologies are used for data fusion, ranging from simple statistical techniques to complex machine learning models.

10.3.2.1 Sensor-Level Fusion

Sensor-level fusion, also known as low-level fusion, involves combining raw data from multiple sensors before any preprocessing or analysis is performed. This method has the advantage of preserving all the information from each sensor, but it also presents significant challenges due to the high dimensionality and complexity of the data. Sensor-level fusion typically requires sophisticated signal processing techniques to align the data temporally and filter out noise. A common approach to sensor-level fusion is to use Kalman filtering, a mathematical algorithm that estimates the state of a dynamic system by minimizing the error between the predicted and observed data. Kalman filters are particularly useful in systems with noisy or incomplete data, as they can iteratively update the state estimates based on new sensor inputs. In multimodal sensing arrays, Kalman filters can be used to integrate data from multiple sensors, such as combining heart rate data from an optical sensor with accelerometer data to estimate overall cardiovascular health (Lian et al. 2022). Another technique for sensor-level fusion is Fourier transform-based analysis, which is used to combine frequency-domain data from different sensors (Zhao et al. 2023). For example, in systems that monitor both physical activity and cardiac signals, Fourier transforms can help identify overlapping frequencies between accelerometer data and heart rate variability, providing insights into how physical exertion affects cardiovascular function.

10.3.2.2 Feature-Level Fusion

Feature-level fusion involves extracting relevant features from each sensor's data stream before combining them into a single dataset for analysis. This approach reduces the dimensionality of the data, making it easier to process while still retaining important information. Feature extraction techniques, such as principal component analysis (PCA), independent component analysis (ICA), and wavelet transforms, are commonly used to identify the most relevant features from each sensor modality. Feature-level fusion is particularly useful in applications where data from different sensors are complementary. For instance, in a multimodal array designed for glucose monitoring, the features extracted from a biochemical sensor (e.g., glucose concentration) can be combined with physical activity data from an accelerometer to provide a more comprehensive understanding of how exercise influences glucose metabolism. Similarly, combining heart rate variability features with electrochemical markers of stress (e.g., cortisol levels) can provide insights into the physiological effects of psychological stress (Wan and Shao 2023; Tu et al. 2022; Albekairi et al. 2025). Deep learning algorithms are increasingly being used for feature-level fusion in multimodal sensing arrays. Convolutional neural networks and recurrent neural networks can automatically learn the most relevant features from raw sensor data and combine them in a way that enhances the overall predictive accuracy of the system. For example, a deep learning model might simultaneously analyze electrocardiogram (ECG) data, accelerometer data, and biochemical markers to predict the likelihood of a cardiac event in real time (Nweke et al. 2018).

10.3.2.3 Decision-Level Fusion

Decision-level fusion (also known as high-level fusion) involves analyzing each sensor's data independently and then combining the results at a higher decision-making level. This approach is commonly used in systems where the sensors measure fundamentally different parameters that do not need to be directly combined. For instance, a system monitoring both cardiovascular health and mental well-being might analyze heart rate and blood pressure data separately from cortisol levels or electroencephalogram (EEG) signals and then combine the results to assess overall health. Decision-level fusion is typically performed using voting-based algorithms or Bayesian inference models, which combine the outputs from different sensors to make a final decision. In healthcare, decision-level fusion might be used to integrate the outputs of various diagnostic tests, such as combining the results of blood tests, ECG readings, and physical activity data to generate a more comprehensive diagnosis (Valik et al. 2023; Haripriya et al. 2025). Fuzzy logic systems are another approach to decision-level fusion, particularly when dealing with uncertain or imprecise data. Fuzzy logic can be used to combine the outputs of different sensors that provide conflicting or uncertain information, generating a more nuanced and probabilistic assessment of the patient's health status (Chiang 2015).

10.3.2.4 Hybrid Fusion Techniques

Many multimodal sensing systems employ hybrid fusion techniques, which combine elements of sensor-level, feature-level, and decision-level fusion. For example, a hybrid approach might use sensor-level fusion to preprocess and align raw sensor data, feature-level fusion to extract the most relevant information from each sensor, and decision-level fusion to combine the results into a final diagnosis. Hybrid fusion is particularly useful in complex healthcare applications where data from multiple sensor modalities need to be integrated across different time scales. For instance, a multimodal system designed for diabetes management might use sensor-level fusion to integrate real-time glucose data with continuous physical activity monitoring, while using feature-level fusion to combine historical trends in glucose levels with periodic biochemical markers of insulin sensitivity. Finally, decision-level fusion could be employed to combine the results from different diagnostic tests to recommend a personalized treatment plan (Rajalingam et al. 2019).

10.3.3 Clinical Relevance of Multimodal Data Fusion

The primary motivation for employing multimodal data fusion in healthcare is to improve the accuracy, specificity, and sensitivity of diagnostic and monitoring systems. By integrating data from multiple sensors, healthcare providers can gain a more holistic understanding of a patient's health, leading to better outcomes in disease management, treatment planning, and preventive care. Below are several key areas where multimodal data fusion has demonstrated clinical relevance: Multimodal sensing arrays, combined with advanced data fusion techniques, offer the potential for early detection of diseases by capturing subtle changes in multiple physiological parameters. For example, integrating biochemical markers of inflammation with heart rate variability and physical activity data can help detect early signs of cardiovascular disease, long before symptoms become apparent. Similarly, combining CGM with insulin sensitivity markers and physical activity data can identify prediabetes in at-risk individuals (Sathya and Magesh 2025).

Machine learning models trained on multimodal datasets can also detect complex patterns in the data that are indicative of disease onset. These models can identify biomarkers, correlations, or trends that may not be immediately apparent to clinicians. In diseases like cancer, where early detection significantly improves outcomes, multimodal arrays can integrate tumor markers, imaging data, and metabolic signals to provide a more accurate and timely diagnosis (Kaczmarczyk et al. 2024; Mittermaier et al. 2023). Personalized treatment plans are one of the major benefits of multimodal sensing and data fusion. By integrating multiple streams of physiological data, healthcare providers can tailor treatments to the specific needs of each patient. For example, in managing chronic conditions like diabetes or hypertension, data from glucose sensors, blood pressure monitors, and physical activity trackers can be combined to adjust medication dosages, dietary recommendations, and exercise plans based on the patient's real-time condition (Yin et al. 2025). In precision medicine, multimodal data fusion allows for the customization of therapies based on individual genetic, biochemical, and lifestyle factors. Multimodal arrays that combine genetic data with biochemical and physical markers can be used to predict patient responses to medications or treatments, reducing the likelihood of adverse reactions and optimizing therapeutic outcomes. In critical care or high-risk patient populations, real-time monitoring using multimodal sensing arrays can provide early warnings of potential complications (Huang et al. 2025). For instance, in patients recovering from surgery, continuous monitoring of vital signs, biochemical markers of inflammation, and mobility data can help detect complications such as infection, deep vein thrombosis, or respiratory distress. This data can trigger automatic alerts, allowing clinicians to intervene before the patient's condition deteriorates. Wearable devices with multimodal sensing arrays, combined with real-time data fusion, have also been shown to be effective in monitoring patients with chronic diseases. For example, continuous monitoring of heart rate, respiratory rate, and blood oxygen levels in patients with chronic obstructive pulmonary disease (COPD) can help prevent exacerbations by identifying early signs of respiratory distress. As sensor technologies and data processing techniques continue to evolve, the future of multimodal data fusion in healthcare holds immense promise. Advances in machine learning, artificial intelligence, and big data analytics will enable even more sophisticated fusion techniques, capable of analyzing vast amounts of data in real time and providing actionable insights to clinicians. Wearable and implantable devices with multimodal sensing capabilities are expected to become even more compact, energy-efficient, and user-friendly, enabling continuous monitoring in a wide range of healthcare applications. The integration of 5G technology and cloud-based computing will further enhance the ability to process multimodal data in real time, enabling remote monitoring and telemedicine applications (Huang et al. 2018; Li et al. 2024).

10.4 Applications of Multimodal Sensing Arrays in Disease Management

The development of multimodal sensing arrays represents a significant leap in the field of healthcare, offering comprehensive and continuous monitoring of patients across multiple physiological domains. These integrated sensor systems combine various types of data, such as biochemical, electrical, and physical measurements, into a cohesive platform that provides real-time insights into an individual's health. This section explores the practical applications of multimodal sensing arrays in the management of diseases, focusing on how these systems are transforming the monitoring, diagnosis, and treatment of chronic diseases, mental health and neurological conditions, and postsurgical rehabilitation. Additionally, we will explore how multimodal sensing can enhance disease prediction and prevention.

10.4.1 Chronic Disease Management

Chronic diseases, including diabetes, cardiovascular diseases, and chronic respiratory conditions, are among the leading causes of mortality and morbidity worldwide. Effective management of these diseases requires continuous monitoring of various physiological parameters to prevent exacerbations, monitor treatment efficacy, and guide personalized interventions. Multimodal sensing arrays have emerged as a powerful tool for managing chronic diseases by providing comprehensive, real-time data on the patient's condition. Diabetes is a chronic condition that affects millions of people worldwide, and its management is highly dependent on the continuous monitoring of blood glucose levels. Traditional approaches to glucose monitoring involve intermittent fingerstick measurements, which are often insufficient for capturing the full picture of glucose fluctuations throughout the day (Templer 2022; Poletti et al. 2021). The detailed discussion on bioelectronics-based approaches for glucose monitoring is summarized in Chap. 8. In addition, CGM systems, which utilize electrochemical sensors to track glucose levels in real time, have already revolutionized diabetes care by providing continuous data streams that can guide insulin dosing and lifestyle adjustments. However, multimodal sensing arrays take this concept a step further by integrating additional sensors that monitor related physiological parameters, such as insulin levels, physical activity, heart rate variability, and even stress hormones like cortisol (Lucisano et al. 2017; Schubert-Olesen et al. 2022). These additional data streams provide deeper insights into the complex interactions between glucose metabolism, exercise, stress, and other factors that influence blood sugar levels. For example, a multimodal array that combines a glucose sensor with an accelerometer and heart rate monitor can provide more accurate predictions of how exercise affects blood sugar levels. Patients and clinicians can use this information to optimize insulin dosing before and after physical activity, reducing the risk of hypoglycemia or hyperglycemia. Similarly, integrating data on stress levels (e.g., via cortisol measurements from sweat or saliva) can help identify instances where psychological stress is driving glucose fluctuations, allowing for more targeted interventions. Machine learning algorithms can further enhance the capabilities of these multimodal systems by analyzing the combined data streams and predicting future glucose trends based on historical patterns. These predictive models can provide early warnings of impending hyperglycemia or hypoglycemia, enabling patients to take preventive actions before serious complications arise. Additionally, the ability to continuously monitor multiple physiological parameters enables clinicians to adjust treatment plans more precisely, improving long-term outcomes for diabetic patients.

10.4.1.1 Cardiovascular Disease Management

Cardiovascular diseases (CVDs) are another major area where multimodal sensing arrays have the potential to significantly improve patient outcomes. Conditions such as hypertension, heart failure, and coronary artery disease require continuous monitoring of various cardiovascular parameters to ensure that patients are receiving optimal care. Traditional methods of monitoring, such as periodic blood pressure checks or ECGs, provide only a snapshot of the patient's condition and may miss critical changes in cardiovascular function. Multimodal sensing arrays designed for cardiovascular disease management combine a range of sensors that monitor heart rate, blood pressure, respiratory rate, and blood oxygen levels in real time (Prasanna et al. 2022). These sensors can be integrated into wearable devices such as smartwatches, chest patches, or implantable sensors, allowing for continuous, noninvasive monitoring of cardiovascular health. One of the key advantages of multimodal sensing in cardiovascular care is the ability to monitor both hemodynamic parameters (e.g., blood pressure, heart rate variability) and biochemical markers (e.g., lactate, troponin, or NT-proBNP levels). Elevated levels of biochemical markers like troponin, which indicate myocardial injury, combined with changes in hemodynamic parameters, can provide early warning signs of a heart attack or heart failure exacerbation. By detecting these signs early, patients can seek medical attention before serious complications arise. In patients with heart failure, multimodal sensing arrays that monitor fluid status (via bioimpedance sensors), respiratory rate, and oxygen saturation can help prevent hospital readmissions by providing early warnings of fluid overload or respiratory distress (Sengupta et al. 2023). These systems allow for more timely interventions, such as adjusting diuretic doses or initiating oxygen therapy, thereby reducing the burden on healthcare systems and improving patient quality of life. Incorporating machine learning algorithms into multimodal sensing systems enhances their predictive power. By analyzing historical data and identifying patterns, these algorithms can provide personalized recommendations for medication adjustments, dietary changes, or physical activity levels based on each patient's unique physiological response. Such personalized care is a cornerstone of precision medicine and is critical in managing complex conditions like cardiovascular diseases.

10.4.1.2 Respiratory Disease Management

Chronic respiratory diseases, such as COPD and asthma, are characterized by recurrent episodes of airflow limitation and respiratory distress. Continuous monitoring of respiratory parameters is essential for preventing exacerbation, guiding treatment, and improving patient outcomes. Multimodal sensing arrays that monitor respiratory function, oxygen saturation, physical activity, and inflammatory markers offer a comprehensive solution for managing these diseases. Wearable devices with integrated optical and physical sensors can continuously monitor respiratory rate, oxygen saturation, and physical activity levels in real time. Combining this data with biochemical sensors that detect inflammatory markers (e.g., cytokines) in exhaled breath or sweat provides a more complete picture of the patient's respiratory status. For example, a rise in inflammatory markers coupled with an increase in respiratory rate and a drop in oxygen saturation may indicate an impending asthma attack or COPD exacerbation, allowing for early intervention with bronchodilators or corticosteroids (Cuperus et al. 2024). Multimodal sensing arrays also play a critical role in sleep apnea management, a common respiratory disorder characterized by repeated episodes of airway obstruction during sleep. Traditional sleep studies, which require patients to stay overnight in a sleep lab, are cumbersome and often fail to capture the full extent of the condition. Wearable multimodal systems that combine respiratory sensors, oxygen saturation monitors, and accelerometers can provide continuous, real-time data on the patient's breathing patterns and sleep quality in a more natural, home-based environment. This continuous monitoring allows for more accurate diagnoses and more personalized treatment plans, such as adjustments to continuous positive airway pressure therapy (Savito et al. 2023).

10.4.2 Mental Health and Neurological Conditions

Mental health disorders, including depression, anxiety, and posttraumatic stress disorder, as well as neurological conditions like Parkinson's disease and epilepsy, present unique challenges in terms of diagnosis and management. These conditions often manifest as a combination of psychological, biochemical, and physiological symptoms, making them particularly well-suited for monitoring with multimodal sensing arrays. By capturing data across multiple domains, such as heart rate variability, electrodermal activity, biochemical markers of stress, and movement patterns, multimodal sensing systems can provide a more comprehensive assessment of a patient's mental and neurological health. Stress and anxiety are highly prevalent conditions that can have significant impacts on physical health, contributing to cardiovascular disease, digestive disorders, and immune dysfunction. Current methods for assessing stress and anxiety often rely on self-reported questionnaires, which are subjective and can vary widely between patients. Multimodal sensing arrays offer an objective and continuous way to monitor stress and anxiety levels by integrating physiological, biochemical, and behavioral data. For example, heart rate variability (HRV) is a well-established marker of autonomic nervous system function and has been shown to correlate with stress and anxiety levels. Multimodal arrays that combine HRV monitoring with electrodermal activity, which measure skin conductance as an indicator of sympathetic nervous system activation, provide a more comprehensive assessment of the body's physiological response to stress. Biochemical sensors that detect stress hormones like cortisol in sweat or saliva can further enhance the accuracy of stress monitoring. By integrating data from these sensors with behavioral data (e.g., physical activity levels from accelerometers), multimodal systems can provide real-time feedback on how stress affects both physiological and behavioral responses. This information can be used to guide interventions such as mindfulness practices, breathing exercises, or pharmacological treatments. Machine learning algorithms can analyze the combined data streams to identify patterns that may indicate worsening anxiety or the onset of a panic attack. By detecting these patterns early, the system can provide personalized recommendations for coping strategies, helping patients manage their symptoms more effectively (Haque et al. 2024; Tsai and Chen 2022; Munch Nielsen et al. 2022).

Neurological conditions like Parkinson's disease, epilepsy, and multiple sclerosis are characterized by complex and often unpredictable symptom patterns. Multimodal sensing arrays can improve the management of these conditions by providing continuous monitoring of motor function, physiological parameters, and biochemical markers of disease activity. For example, in Parkinson's disease, tremor, bradykinesia, and gait abnormalities are hallmark symptoms that can fluctuate throughout the day. Traditional clinic-based assessments of these symptoms are often limited by the brief, episodic nature of the clinical encounter. Wearable multimodal sensing arrays that integrate accelerometers, gyroscopes, and EMG sensors can provide continuous, real-time data on motor function, allowing for more accurate assessments of disease progression and treatment efficacy. In epilepsy, multimodal sensing arrays can be used to detect early warning signs of seizures by monitoring a combination of EEG signals, heart rate variability, and physical activity. By integrating data from these sensors, the system can detect subtle changes in brain activity or autonomic function that preceded a seizure, allowing for timely interventions or the administration of rescue medications. For patients with multiple sclerosis, multimodal sensing arrays that combine physical sensors (e.g., accelerometers, gyroscopes) with biochemical markers of inflammation (e.g., cytokines) can provide a more comprehensive view of disease activity. These systems can help identify periods of disease flare-ups and guide adjustments to immunomodulatory therapies based on real-time data (Joseph et al. 2024; Choo et al. 2024).

10.4.3 Postsurgical and Rehabilitation Monitoring

The postsurgical period is a critical time for monitoring patient recovery and preventing complications such as infection, deep vein thrombosis, or poor wound healing. Traditional methods of monitoring, which often rely on intermittent check-ups and patient self-reporting, can miss early signs of complications. Multimodal sensing arrays offer a more continuous and comprehensive approach to postsurgical monitoring by integrating sensors that track vital signs, biochemical markers of infection or inflammation, and physical activity. Postsurgical infections and poor wound healing are common complications that can lead to prolonged hospital stays, increased healthcare costs, and, in severe cases, life-threatening conditions such as sepsis. Multimodal sensing arrays that monitor biochemical markers of infection (e.g., C-reactive protein, procalcitonin) and local wound temperature (via thermal sensors) can provide early warnings of infection before visible symptoms such as redness, swelling, or fever appear. Wearable patches or smart bandages with integrated optical and biochemical sensors can monitor the wound healing process by tracking changes in pH levels, oxygenation, and biochemical markers of inflammation. By providing continuous feedback on wound healing, these systems enable clinicians to intervene early, optimizing the use of antibiotics or other therapies and preventing further complications. Physical rehabilitation following surgery or injury often involves tracking the patient's range of motion, muscle activity, and pain levels to guide therapy and assess progress. Multimodal sensing arrays that integrate physical sensors (e.g., accelerometers, gyroscopes, EMG sensors) with biochemical markers of muscle damage (e.g., creatine kinase) and inflammation can provide a more comprehensive assessment of the patient's recovery process. For example, in patients undergoing rehabilitation following knee or hip replacement surgery, a multimodal sensing array that tracks joint movement, muscle activity, and pain biomarkers can help physical therapists tailor rehabilitation exercises to the patient's specific needs. The system can also provide real-time feedback to patients, encouraging them to perform exercises correctly and avoid movements that could lead to injury (Hossain et al. 2024; Wu et al. 2025a).

10.4.4 Disease Prediction and Prevention

One of the most exciting applications of multimodal sensing arrays is their potential to predict and prevent disease before symptoms become apparent. By continuously monitoring multiple physiological parameters, multimodal systems can identify early signs of disease and guide preventive interventions. For example, in cardiovascular disease, multimodal sensing arrays that monitor heart rate variability, blood pressure, and biochemical markers of inflammation can detect early signs of endothelial dysfunction or atherosclerosis, allowing for the implementation of lifestyle changes or pharmacological interventions to prevent heart attacks or strokes. Similarly, in metabolic syndrome, a condition characterized by a cluster of risk factors for diabetes and cardiovascular disease, multimodal arrays that monitor glucose levels, insulin sensitivity, lipid profiles, and physical activity can provide early warnings of metabolic dysfunction, guiding interventions such as dietary changes or exercise programs to prevent the onset of diabetes. In cancer prevention, multimodal sensing arrays that combine imaging data (e.g., from optical sensors) with biochemical markers (e.g., tumor antigens) can identify early-stage tumors before they become symptomatic, enabling earlier diagnosis and more effective treatment (Chaikijurajai et al. 2020; Wu et al. 2025b; Cai et al. 2023).

10.5 Design and Fabrication of Multimodal Sensing Devices

The design and fabrication of multimodal sensing devices require an interdisciplinary approach, combining knowledge from fields such as materials science, electrical engineering, biotechnology, and data science. The goal is to integrate multiple sensor types, electrochemical, biochemical, physical, optical, and thermal, into a single platform that can monitor a range of physiological parameters with high accuracy, sensitivity, and specificity. This section will explore the various considerations involved in the design and fabrication of multimodal sensing devices, including material selection, miniaturization, power management, biocompatibility, and interfacing with external systems for data transmission and analysis.

10.5.1 Material Selection

The choice of materials is a critical factor in the design of multimodal sensing devices, as it directly influences the device's performance, durability, biocompatibility, and overall user comfort. Multimodal sensing arrays often combine a variety of sensors, each with its own material requirements, depending on the nature of the signals they are designed to detect. Moreover, the materials must be suitable for the intended application, whether the device is wearable, implantable, or integrated into other medical systems. Key considerations in material selection include electrical conductivity, flexibility, mechanical strength, and biocompatibility. The core functionality of many sensors, especially electrochemical and biochemical sensors, depends on the use of conductive materials that can facilitate the transfer of electrical signals. Metals such as gold, platinum, and silver are commonly used in electrochemical sensors due to their high electrical conductivity and chemical stability. For instance, gold and platinum are often used as electrode materials in glucose sensors, where they facilitate the electron transfer processes that occur during the oxidation or reduction of glucose. However, these traditional metals are often rigid and not well-suited for flexible or wearable devices. As a result, researchers have explored the use of alternative conductive materials, such as graphene, carbon nanotubes, and conductive polymers (e.g., polypyrrole, polyaniline). These materials are highly conductive, flexible, and can be fabricated into thin films, making them ideal for wearable devices that require both conductivity and flexibility (Merchant et al. 2023; Roy et al. 2020; Liu et al. 2023). Graphene, in particular, has attracted significant attention due to its remarkable electrical, mechanical, and optical properties. It is a single layer of carbon atoms arranged in a two-dimensional honeycomb lattice, which gives it extraordinary electrical conductivity and mechanical flexibility. Graphene-based sensors have been used in a variety of applications, including glucose monitoring, pH sensing, and strain gauges for measuring physical deformations. Its biocompatibility, along with its ability to be functionalized with biomolecules (e.g., enzymes or aptamers), makes it a versatile material for multimodal sensing devices (Song et al. 2025; Kil et al. 2024).

For wearable devices, it is essential to use flexible and stretchable materials that can conform to the natural movements of the human body without losing their functionality. Silicone elastomers, such as polydimethylsiloxane (PDMS), are commonly used as substrates for flexible sensors because of their elasticity, biocompatibility, and ease of fabrication. PDMS can be molded into thin, flexible films that can accommodate various sensor types, making it an ideal material for skin-mounted or wearable devices. Polyimides are another class of flexible materials used in sensor fabrication. They have excellent thermal and chemical stability, making them suitable for use in harsh environments, such as inside the body or on the skin. Polyimides are frequently used in flexible printed circuits and microelectromechanical systems, where they serve as substrates for thin-film sensors and microfabricated components. The integration of stretchable electronics into multimodal sensing devices has opened up new possibilities for creating sensors that can stretch and deform along with the skin or other tissues. Stretchable materials, such as thermoplastic elastomers and liquid metals, can be embedded in flexible substrates to create sensors that maintain their performance even under mechanical strain. This is particularly important for applications that involve continuous movement, such as monitoring muscle activity or joint motion during physical rehabilitation (Gao et al. 2023).

For devices that come into direct contact with biological tissues, such as implantable sensors or skin-mounted wearables, biocompatibility is a critical consideration. The materials used must not provoke an immune response or cause irritation, and they must be able to withstand the harsh conditions of the body, such as exposure to bodily fluids, enzymes, and pH fluctuations. Hydrogels are a class of materials that are highly biocompatible and can be used in contact with biological tissues. Hydrogels are composed of crosslinked polymer networks that can absorb and retain large amounts of water, making them ideal for use in sensors that need to interact with biological fluids. For example, hydrogels can be functionalized with enzymes or antibodies to detect specific analytes in sweat, tears, or interstitial fluid. In implantable devices, the use of biodegradable materials is becoming increasingly popular. These materials, such as polylactic acid and polyglycolic acid, degrade over time into nontoxic byproducts that are naturally eliminated by the body. Biodegradable materials are particularly useful in temporary monitoring systems, where the sensor is designed to function for a limited period (e.g., during the healing process after surgery) and then degrade without the need for surgical removal (Chen et al. 2024; Zhang et al. 2023b).

10.5.2 Miniaturization and Power Management

As the demand for wearable and implantable devices increases, there is a growing need to miniaturize sensors while maintaining their performance and reliability. Miniaturization is essential for creating devices that are unobtrusive, comfortable, and capable of being integrated into everyday life. In addition, miniaturized devices must be designed with energy efficiency in mind, as power management is a critical challenge in portable and implantable systems. Miniaturization of multimodal sensors is achieved through the use of microfabrication techniques, which involve the precise construction of sensors and circuits on a microscopic scale. Microfabrication techniques, such as photolithography, thin-film deposition, and etching, are commonly used in the semiconductor industry and have been adapted for the fabrication of medical sensors. Photolithography is a process in which a light-sensitive material (photoresist) is applied to a substrate and then exposed to ultraviolet light through a patterned mask. The areas exposed to the photoresist are then developed, leaving behind a pattern that can be used to build sensor structures. Photolithography is used to create microelectrodes, transistors, and other components in multimodal sensing arrays. Thin-film deposition techniques, such as physical vapor deposition and chemical vapor deposition, are used to deposit layers of conductive, insulating, or semiconducting materials onto a substrate. These layers form the building blocks of the sensors, allowing for the creation of electrodes, conductive traces, and functional coatings. Thin-film depositions are particularly useful for creating highly sensitive sensors with small footprints, such as electrochemical or optical sensors. Etching processes, such as wet etching and dry etching, are used to remove material from specific areas of a substrate to create microstructures. These microstructures can serve as microfluidic channels, cavities for biological samples, or pathways for electrical signals. Microfluidic systems are often integrated into multimodal sensing arrays to guide the flow of biological fluids (e.g., blood, sweat) to the sensors for real-time analysis (Shim et al. 2022; Wei et al. 2020; Wang et al. 2018).

Power management is a significant challenge in multimodal sensing devices, particularly in wearable and implantable systems where long-term operation is required. Devices must be designed to consume as little power as possible to extend battery life or eliminate the need for frequent recharging. One approach to achieving this goal is the use of low-power circuitry, which minimizes energy consumption during data acquisition, signal processing, and wireless communication. In addition to low-power design, energy harvesting technologies are becoming an essential component of self-powered multimodal sensing devices. Energy harvesting involves capturing energy from the environment (e.g., body heat, motion, light) and converting it into electrical energy to power the sensors. This approach is particularly attractive for wearable devices, as it eliminates the need for bulky batteries and extends the operational life of the device. Thermoelectric generators (TEGs) are a type of energy harvesting technology that converts body heat into electrical energy. TEGs are based on the Seebeck effect, which generates a voltage when there is a temperature difference between two materials. In wearable devices, TEGs can be integrated into the device's design to capture heat from the skin and generate power for the sensors. Piezoelectric materials are another energy harvesting technology that can convert mechanical energy (e.g., body movements or vibrations) into electrical energy. Piezoelectric materials, such as lead zirconate titanate, generate an electrical charge when subjected to mechanical stress. These materials can be integrated into wearable devices that capture energy from walking, running, or other physical activities to power the sensors. Solar cells, particularly thin-film photovoltaics, can also be used to capture ambient light and convert it into electrical energy.

Thin-film solar cells are flexible and lightweight, making them suitable for integration into wearable devices that are exposed to natural or artificial light (Jin et al. 2024; Rao et al. 2025; Zhao and You 2014).

10.5.3 Biocompatibility and Sensor Integration

Biocompatibility is a critical consideration in the design of multimodal sensing devices, particularly for wearable and implantable systems that come into direct contact with the skin or internal tissues. Biocompatible materials must not provoke an immune response, cause irritation, or degrade over time in a way that could harm the patient. Additionally, sensors must be designed to function reliably in the complex and dynamic environment of the human body, where they may be exposed to sweat, blood, enzymes, and other biological fluids.

Wearable multimodal sensing devices are designed to be worn on the skin for extended periods, and they must be lightweight, flexible, and comfortable. The materials used for wearable devices must be nonirritating and allow for continuous contact with the skin without causing discomfort. In addition, wearable sensors must be designed to withstand the mechanical stresses associated with movement, such as bending, stretching, and twisting. Adhesive patches are a common form of wearable sensor, often used for continuous glucose monitoring, heart rate monitoring, or physical activity tracking. These patches are made from flexible materials, such as silicone or thermoplastic elastomers, and contain embedded sensors that can measure a range of physiological parameters. The adhesive used in these patches must be gentle on the skin and allow for long-term wear without causing irritation or skin damage. Textile-based sensors are another form of wearable device, where sensors are embedded into clothing or other textiles. These sensors are typically made from conductive threads or fabrics that can detect electrical signals, such as those generated by muscle contractions or heart activity. Textile-based sensors are often used in sports and fitness applications, where they can provide real-time data on physical performance (Ying and Liu 2021; Chen et al. 2022).

Implantable multimodal sensing devices are designed to be placed inside the body, where they can continuously monitor physiological parameters over long periods. These devices must be made from materials that are not only biocompatible but also able to withstand the harsh conditions of the internal environment. For example, implantable devices may be exposed to fluctuating pH levels, enzymatic degradation, and immune responses. Encapsulation materials, such as parylene or silicon nitride, are often used to protect implantable sensors from the body's internal environment. These materials form a protective barrier around the sensor, preventing the ingress of bodily fluids while allowing the sensor to function. Parylene, in particular, is widely used in medical implants due to its excellent barrier properties, chemical stability, and biocompatibility. Implantable devices must also be designed to minimize foreign body response, where the body's immune system reacts to the presence of the implant by forming scar tissue around it. One approach to mitigating this response is to use bioactive coatings that promote integration with the surrounding tissue. For example, coatings that release anti-inflammatory agents or promote the growth of endothelial cells can help reduce immune reactions and improve the longevity of the implant (Lecomte et al. 2017; Zhou et al. 2021; Miller et al. 2015).

10.5.4 Interfacing with External Systems

Multimodal sensing devices must interface with external systems to transmit the data they collect for processing, analysis, and decision-making. This typically involves wireless communication between the sensing device and an external platform, such as a smartphone, tablet, or cloud-based server. The design of the communication interface is critical for ensuring reliable data transmission, particularly in real-time monitoring applications where delays or data loss could have serious consequences.

Wireless communication is essential for wearable and implantable sensors, as it allows the devices to transmit data without the need for physical connections. The most commonly used wireless communication protocols in healthcare applications include Bluetooth Low Energy (BLE), Wi-Fi, and near-field communication. Each protocol has its advantages and limitations, depending on the specific application. BLE is widely used in wearable devices due to its low power consumption and short-range communication capabilities. BLE allows sensors to transmit data to nearby devices, such as smartphones or tablets, in real time. This makes it ideal for applications where continuous monitoring is required, such as glucose monitoring or heart rate tracking. BLE's low power consumption helps extend the battery life of wearable devices, making it suitable for long-term use. Wi-Fi offers higher data transfer rates and longer communication ranges than BLE, but it consumes more power. Wi-Fi is typically used in devices that require the transmission of large amounts of data or where long-range communication is necessary, such as in hospital-based monitoring systems or home healthcare devices. Wi-Fi-enabled sensors can transmit data directly to cloud-based servers for remote monitoring by healthcare providers. NFC is a short-range wireless technology that allows devices to communicate when they are in close proximity to each other. NFC is commonly used in implantable devices, where the sensor can transmit data to an external reader or smartphone when held near the implant site. NFC's low power requirements make it ideal for implantable sensors that need to minimize energy consumption. Once data is transmitted from the sensing device, it must be stored and processed to generate meaningful insights. Cloud-based platforms are increasingly being used to store and analyze data from multimodal sensing devices, allowing for real-time data processing and remote monitoring. Cloud platforms can aggregate data from multiple patients, enabling healthcare providers to monitor trends, detect anomalies, and make informed decisions about treatment. Edge computing is another emerging trend in the design of multimodal sensing systems. Edge computing involves processing data locally on the sensing device or nearby devices (e.g., smartphones), rather than transmitting all data to the cloud for processing. This approach reduces latency and allows for faster decision-making, which is particularly important in applications where real-time responses are critical, such as in cardiac monitoring or seizure detection (Anabtawi et al. 2016).

10.6 Future Directions

Multimodal sensing arrays are at the forefront of a technological revolution in healthcare, where the convergence of bioelectronics, data science, and personalized medicine is enabling unprecedented capabilities for disease monitoring, diagnosis, and treatment. While significant progress has been made in the development and deployment of multimodal sensors, there are numerous challenges that must be addressed to fully realize their potential. This section will discuss emerging trends in multimodal sensing, the challenges that lie ahead such as data privacy, regulatory issues, and system integration and the opportunities that these technologies present for the future of healthcare.

10.6.1 Personalized Medicine and Multimodal Sensors

One of the most transformative applications of multimodal sensing arrays lies in the realm of personalized medicine. Personalized medicine tailors healthcare strategies to the individual characteristics of each patient, taking into account genetic, environmental, and lifestyle factors. Multimodal sensing arrays, which provide continuous and comprehensive data on various physiological parameters, are ideally suited for enabling personalized care. As the integration of genomics and bioinformatics becomes increasingly prominent in healthcare, there is a growing interest in combining genetic information with real-time data from multimodal sensing arrays. By merging genetic data with physiological monitoring, clinicians can gain deeper insights into how a patient's unique genetic profile influences their response to medications, environmental factors, and disease progression (Lee et al. 2020). For example, patients with genetic predispositions to certain metabolic disorders (e.g., type 2 diabetes or familial hypercholesterolemia) can benefit from multimodal sensing arrays that continuously monitor glucose levels, lipid profiles, and physical activity. This integrated data allows for a more nuanced understanding of how genetic and environmental factors interact, enabling highly personalized treatment plans that optimize diet, medication, and exercise based on the individual's specific needs. In the future, we can envision multimodal sensing arrays that include genomic sensors, such as wearable or implantable devices that monitor specific RNA or protein expression levels in real time, enabling a more dynamic integration of genetic data with physiological signals. Such sensors could detect early signs of genetic mutations or epigenetic changes that might indicate the onset of diseases like cancer or neurodegenerative disorders, long before traditional symptoms appear (Chakraborty et al. 2022).

The field of pharmacogenomics, which studies how a person's genetic makeup influences their response to drugs, will likely be another significant beneficiary of advances in multimodal sensing technology. Currently, genetic testing for pharmacogenomic markers is performed once, usually before starting a medication. However, drug responses are often dynamic and can change based on a patient's overall physiological state, which can fluctuate due to factors such as diet, stress, or physical activity. In the future, multimodal sensing arrays could continuously monitor not only pharmacogenomic markers but also real-time indicators of drug metabolism (e.g., liver enzyme activity) and drug efficacy (e.g., changes in heart rate, blood pressure, or biochemical markers). By integrating these data streams, clinicians could adjust drug dosages in real time, minimizing adverse drug reactions and improving therapeutic outcomes. Such systems could prove invaluable in managing complex conditions like cancer, where chemotherapy drugs often have narrow therapeutic windows and require precise dosing (Leaning et al. 2024).

In addition to personalized treatment, multimodal sensing arrays hold great promise for continuous risk assessment and prevention of diseases. Wearable sensors that track heart rate variability, blood pressure, glucose levels, and other vital signs can provide a continuous stream of data that can be analyzed to identify patterns indicative of disease risk. This is particularly important for managing chronic diseases like cardiovascular disease or diabetes, where early intervention can prevent serious complications. Machine learning algorithms can be trained on multimodal datasets to predict disease onset by recognizing patterns that precede clinical symptoms (Huang et al. 2022; Persson et al. 2024). For instance, a combination of heart rate variability, inflammatory markers, and physical activity data might predict an impending heart attack or stroke in patients with cardiovascular disease, allowing for early intervention and prevention strategies. Similarly, patients at risk of developing diabetes could receive personalized recommendations for lifestyle modifications based on continuous monitoring of glucose levels and insulin sensitivity.

10.6.2 Next-Generation Sensor Technologies

As the field of multimodal sensing continues to evolve, emerging technologies in materials science, nanotechnology, and quantum sensing are expected to drive the development of next-generation sensor platforms. These new technologies will enable even more sophisticated and accurate monitoring of physiological parameters, expanding the range of applications for multimodal sensing arrays.

The development of flexible electronics and stretchable materials is already transforming the design of multimodal sensing devices, enabling the creation of wearables and implantable that conform to the human body's natural movements. In the future, advances in organic electronics, biocompatible polymers, and smart textiles will allow for even more comfortable and unobtrusive sensors that can be integrated into everyday clothing, skin patches, or even woven into the skin itself. For example, future smart tattoos, thin, flexible electronic circuits that can be applied to the skin, could monitor a range of physiological parameters, such as body temperature, hydration, and electrolyte levels, in real time (Joshi et al. 2024). These devices could be powered by body heat or motion, eliminating the need for external batteries. Similarly, implantable multimodal sensors made from biodegradable materials could provide temporary monitoring of critical parameters after surgery or during disease flare-ups, before safely degrading within the body.

One of the most exciting developments on the horizon is the use of quantum sensors in biomedical applications. Quantum sensing leverages the principles of quantum mechanics, such as superposition and entanglement, to detect extremely small changes in physical properties, making them far more sensitive than classical sensors (Das et al. 2024). For example, quantum sensors could be used to detect minute changes in magnetic fields, enabling the development of highly sensitive devices for monitoring brain activity (e.g., magnetoencephalography) or heart function (e.g., magnetocardiography). These sensors could provide unprecedented resolution and accuracy in monitoring neurological or cardiovascular conditions, opening up new possibilities for diagnosing and treating diseases like epilepsy, Parkinson's disease, or arrhythmias. Moreover, quantum sensors could enable the detection of single molecules or ions, allowing for real-time monitoring of molecular processes at the cellular level. This could revolutionize our understanding of disease mechanisms and enable the early detection of diseases like cancer at the molecular stage (Aslam et al. 2023).

Nanotechnology is expected to play an increasingly important role in the future of multimodal sensing. Nanosensors, which operate at the molecular or atomic level, offer unparalleled sensitivity and specificity for detecting small biomolecules, such as hormones, cytokines, or tumor markers. These sensors can be integrated into multimodal arrays to provide continuous monitoring of disease biomarkers with extremely high accuracy. For instance, nanowire-based biosensors could detect trace amounts of cancer biomarkers in blood or saliva, enabling early diagnosis and monitoring of cancer recurrence. Similarly, nanoparticle-based sensors could be used to monitor drug delivery and efficacy in real time, providing feedback on how well a treatment is working and allowing for dynamic adjustments to dosage or administration schedules (Wang et al. 2024b; Kaymaz et al. 2023).

10.6.3 Data Fusion and Artificial Intelligence

As multimodal sensing arrays generate increasingly complex and voluminous datasets, advanced data processing techniques will be essential for transforming this data into actionable insights. The integration of AI and machine learning into multimodal sensing systems is expected to play a central role in future developments, enabling more sophisticated data fusion and predictive analytics. Current data fusion techniques, while effective, are often limited by computational constraints and the need for real-time processing. As multimodal sensing arrays become more advanced, they will generate larger and more diverse datasets, making it increasingly difficult to process and interpret this data in real time. AI-driven data fusion algorithms will be critical for handling these complex datasets and extracting meaningful insights in real time (Mohammadi et al. 2025). Machine learning models, particularly deep learning algorithms, can be trained on multimodal datasets to automatically identify patterns and correlations between different physiological parameters. These models can then provide real-time feedback to clinicians or patients, offering personalized recommendations for treatment, lifestyle changes, or interventions based on the individual's unique physiological profile. For example, a deep learning model trained on data from glucose sensors, heart rate monitors, and accelerometers could predict the likelihood of hypoglycemia in a diabetic patient, alerting them to take preventive action before symptoms develop (Shaik et al. 2024). Similarly, AI-driven data fusion could combine data from wearable EEG sensors, heart rate monitors, and sleep trackers to provide early warnings of seizure activity in patients with epilepsy.

In addition to real-time monitoring, AI and machine learning will play a critical role in predictive analytics, enabling healthcare providers to identify future risks and intervene before a disease progresses. By analyzing historical data from multimodal sensing arrays, machine learning models can predict how a patient's condition is likely to evolve over time and recommend personalized treatment plans that minimize risks and optimize outcomes. Predictive analytics will be particularly valuable in the management of chronic diseases, where early intervention can prevent complications and improve long-term health (Krones et al. 2025; Singh et al. 2024). For instance, in patients with heart failure, AI-driven models could analyze data from cardiac sensors, respiratory monitors, and biochemical markers to predict worsening heart function, enabling clinicians to adjust medications or recommend lifestyle changes before a hospitalization is needed.

10.6.4 Regulatory and Ethical Challenges

Despite the many opportunities presented by multimodal sensing arrays, there are also significant regulatory and ethical challenges that must be addressed before these technologies can be widely adopted in clinical practice. Ensuring the safety, efficacy, and privacy of multimodal sensors is critical for gaining regulatory approval and ensuring patient trust. One of the biggest hurdles for multimodal sensing devices is navigating the complex regulatory landscape for medical devices. Regulatory agencies such as the US Food and Drug Administration (FDA) and the European Medicines Agency (EMA) require extensive testing and validation of new medical technologies to ensure their safety and efficacy. Multimodal sensing devices, which often integrate multiple sensor types and complex data processing algorithms, present unique challenges for regulators, as these systems may not fit neatly into existing categories of medical devices (Shaik et al. 2024). For example, ensuring the accuracy and reliability of multimodal sensing arrays can be challenging, as the integration of multiple sensor types increases the complexity of the device and the potential for errors. Regulators may require extensive clinical trials to demonstrate that these systems are both safe and effective in a wide range of patients and conditions. Moreover, regulatory agencies are increasingly scrutinizing the use of AI and machine learning in medical devices. AI-driven systems must be carefully validated to ensure that their predictions are accurate and that they do not introduce biases that could negatively impact patient care. As a result, the regulatory pathway for AI-enabled multimodal sensing arrays may be longer and more complex than for traditional medical devices.

Data privacy and security are major concerns for multimodal sensing devices, especially as these systems become more integrated with cloud-based platforms and Internet-of-Things (IoT) networks. Multimodal sensing arrays generate large amounts of sensitive health data, which must be protected from unauthorized access, tampering, or breaches. Ensuring compliance with data privacy regulations, such as the Health Insurance Portability and Accountability Act (HIPAA) in the United States or the General Data Protection Regulation (GDPR) in Europe, will be critical for the widespread adoption of multimodal sensing arrays. These regulations require healthcare providers and technology developers to implement robust data encryption, secure communication protocols, and other safeguards to protect patient information. Additionally, there are ethical concerns related to the use of AI in healthcare, particularly around issues of transparency and accountability. Patients and clinicians must understand how AI-driven algorithms make decisions and be able to trust that these decisions are based on accurate, unbiased data. Ensuring that AI systems are transparent and explainable will be crucial for building trust in these technologies (Ullah et al. 2025).

10.6.5 System Integration and Interoperability

Another challenge in the development of multimodal sensing arrays is system integration and interoperability. Multimodal sensing devices must be able to seamlessly integrate with existing healthcare infrastructure, including electronic health records (EHRs), hospital information systems, and remote monitoring platforms. Ensuring that multimodal data can be easily transmitted, stored, and accessed by healthcare providers is essential for delivering personalized, data-driven care. To achieve this goal, it will be necessary to develop standardized protocols for data exchange and interoperability between multimodal sensors and healthcare systems. Current standards, such as Health Level Seven (HL7) and Fast Healthcare Interoperability Resources (FHIR), provide frameworks for exchanging health data between systems, but these standards may need to be adapted or expanded to accommodate the unique challenges of multimodal sensing arrays (AlSaad et al. 2024; Chang et al. 2025). For example, multimodal sensing devices generate continuous streams of data that may need to be processed in real time. Ensuring that this data can be efficiently transmitted and integrated with EHRs and other healthcare systems will require advances in both hardware and software infrastructure. Additionally, healthcare providers will need user-friendly interfaces that allow them to visualize and interpret the complex data generated by multimodal sensors. As the use of telemedicine and remote patient monitoring continues to expand, there is a growing need for multimodal sensing arrays to integrate with remote monitoring platforms. This will enable clinicians to monitor patients in real time, regardless of their location, and make data-driven decisions about care. Ensuring seamless integration between multimodal sensing devices and remote monitoring platforms will be critical for managing chronic diseases, postsurgical recovery, and other conditions that require continuous monitoring.

10.7 Conclusions

Multimodal sensing arrays have emerged as a groundbreaking innovation in healthcare, enabling comprehensive, continuous, and real-time monitoring of diverse physiological and biochemical parameters. By integrating various sensor modalities, such as electrochemical, biochemical, physical, optical, and thermal sensors, these systems offer a multidimensional perspective on human health. This holistic approach is essential for capturing the complex interplay of biological processes that underline chronic diseases, mental health conditions, and postsurgical recovery. The convergence of advanced materials, flexible electronics, and miniaturization has facilitated the development of wearable and implantable devices that are not only minimally invasive but also capable of seamless integration with the human body, thereby enhancing patient comfort and compliance. Additionally, the application of data fusion techniques and AI significantly amplifies the utility of multimodal sensing arrays. AI-driven algorithms can process and integrate large volumes of heterogenous data from multiple sensors, uncovering patterns, and correlations that may not be evident through conventional analysis. This capability is pivotal for early disease detection, predictive modeling, and personalized therapeutic strategies tailored to an individual's unique physiological profile. For instance, real-time monitoring of metabolic, cardiovascular, and stress-related biomarkers can facilitate timely interventions, reduce hospital readmissions, and optimize long-term disease management. Despite these advances, several challenges need to be addressed to fully realize the potential of multimodal sensing systems. Data privacy, regulatory approval, energy efficiency, and system interoperability remain significant hurdles. Ensuring secure data transmission and compliance with healthcare standards is vital to protect patient information and maintain trust in these technologies. Furthermore, the integration of next-generation technologies, including nanotechnology, quantum sensors, and pharmacogenomics, holds the promise of elevating the precision and sensitivity of health monitoring to unprecedented levels. Looking forward, the synergy between multimodal sensing platforms, AI, and personalized medicine is poised to redefine healthcare paradigms. This integrated approach not only supports proactive and preventive care but also aligns with the vision of precision medicine, where interventions are customized based on comprehensive physiological and genetic insights. Eventually, multimodal sensing arrays could transform healthcare from a reactive model to a more anticipatory and individualized framework, significantly improving clinical outcomes and patient quality of life.

References

Ahmad RUS, Khan WU, Khan MS, Cheung P (2025) Emerging rapid detection methods for the monitoring of cardiovascular diseases: current trends and future perspectives. Mater Today Bio 32:101663

Albekairi M, Mohamed MVO, Kaaniche K, Abbas G, Alanazi MD, Alanazi TM, Emara A (2025) Multimodal medical image fusion combining saliency perception and generative adversarial network. Sci Rep 15(1):10609

AlSaad R, Abd-alrazaq A, Boughorbel S, Ahmed A, Renault M-A, Damseh R, Sheikh J (2024) Multimodal large language models in health care: applications, challenges, and future outlook. J Med Internet Res 26:e59505

Anabtawi N, Freeman S, Ferzli R (2016) A fully implantable, NFC enabled, continuous interstitial glucose monitor. In: IEEE-EMBS international conference on biomedical and health informatics. IEEE-EMBS international conference on biomedical and health informatics, pp 612–615

Aslam N, Zhou H, Urbach EK, Turner MJ, Walsworth RL, Lukin MD, Park H (2023) Quantum sensors for biomedical applications. Nat Rev Phys 5(3):157–169

Ates HC, Nguyen PQ (2022) End-to-end design of wearable sensors. Nat Rev Mater 7(11):887–907

Bayoumy K, Gaber M, Elshafeey A, Mhaimeed O, Dineen EH, Marvel FA, Martin SS, Muse ED, Turakhia MP, Tarakji KG, Elshazly MB (2021) Smart wearable devices in cardiovascular care: where we are and how to move forward. Nat Rev Cardiol 18(8):581–599

Bokade R, Navato A, Ouyang R, Jin X, Chou C-A, Ostadabbas S, Mueller AV (2021) A cross-disciplinary comparison of multimodal data fusion approaches and applications: accelerating learning through trans-disciplinary information sharing. Expert Syst Appl 165:113885

Cai Q, Wang H, Li Z, Liu X (2019) A survey on multimodal data-driven smart healthcare systems: approaches and applications. IEEE Access 7:133583–133599

Cai XJ, Tay JCK, Jiang Y, Yeo KK, Wong PEH, Ho KL, Chong DTT, Ti LK, Leong G, Wong K, Ching CK (2023) Non-invasive mid-term electrocardiogram patch monitoring is effective in detecting atrial fibrillation. J Electrocardiol 81:230–236

Casian T, Nagy B, Kovács B, Galata DL, Hirsch E, Farkas A (2022) Challenges and opportunities of implementing data fusion in process analytical technology—a review. Molecules (Basel, Switzerland) 27(15):4846

Chaikijurajai T, Laffin LJ, Tang WHW (2020) Artificial intelligence and hypertension: recent advances and future outlook. Am J Hypertens 33(11):967–974

Chakraborty J, Chaudhary AA, Khan S-U-D, Rudayni HA, Rahaman SM, Sarkar H (2022) CRISPR/Cas-based biosensor as a new age detection method for pathogenic bacteria. ACS Omega 7(44):39562–39573

Chan ED, Chan MM, Chan MM (2013) Pulse oximetry: understanding its basic principles facilitates appreciation of its limitations. Respir Med 107(6):789–799

Chandra P, Mahato K (2022) Miniaturized biosensing devices. Springer Singapore. https://doi.org/10.1007/978-981-16-9897-2

Chang JS, Kim H, Baek ES, Choi JE, Lim JS, Kim JS, Shin SJ (2025) Continuous multimodal data supply chain and expandable clinical decision support for oncology. npj Digit Med 8(1):128

Chen K, Hu Y, Wang F, Liu M, Liu P, Li C, Yu Y, Xiao X, Feng Q (2022) Ultra-stretchable, adhesive, and self-healing MXene/polyampholytes hydrogel as flexible and wearable epidermal sensors. Colloids Surf A Physicochem Eng Asp 645:128897

Chen R-S, Gao M, Chu D, Cheng W, Lu Y (2024) Self-powered hydrogel wearable bioelectronics. Nano Energy 128:109960

Chiang H-S (2015) ECG-based mental stress assessment using Fuzzy computing and Associative Petri Net. J Med Biol Eng 35(6):833–844

Choo M, Park D, Cho M, Bae S, Kim J, Han DH (2024) Exploring a multimodal approach for utilizing digital biomarkers for childhood mental health screening. Front Psych 15:1348319

Crispi S, Neri G (2022) Development of a Conductometric Sensor Based on Al,Ca-Doped ZnO for the Detection of Formaldehyde. Sensors 22(19):7465

Crispi S, Nocito G, Nastasi F, Condorelli G, Ricciardulli AG, Samorì P, Conoci S, Neri G (2023) Development of a novel C-dots conductometric sensor for NO sensing. Sensors Actuators B Chem 390:133957

Cuperus LJA, Bult L, van Zelst CM (2024) Wearable technology for detection of COPD exacerbations: feasibility of the Health Patch. ERJ Open Res 10(6):00396

Das S, Mazumdar H, Khondakar KR, Mishra YK, Kaushik A (2024) Review—quantum biosensors: principles and applications in medical diagnostics. ECS Sens Plus 3(2):025001

Dauphin-Ducharme P, Yang K, Arroyo-Currás N, Ploense KL, Zhang Y, Gerson J, Kurnik M, Kippin TE, Stojanovic MN, Plaxco KW (2019) Electrochemical Aptamer-based sensors for improved therapeutic drug monitoring and high-precision, feedback-controlled drug delivery. ACS Sens 4(10):2832–2837

Demkiv O, Nogala W, Stasyuk N, Klepach H, Danysh T, Gonchar M (2024) Highly sensitive amperometric sensors based on laccase-mimetic nanozymes for the detection of dopamine. RSC Adv 14(8):5472–5478

Dong R, Guo B (2021) Smart wound dressings for wound healing. Nano Today 41:101290

Gao F, Liu C, Zhang L, Liu T, Wang Z, Song Z, Cai H, Fang Z, Chen J, Wang J, Han M, Wang J, Lin K, Wang R, Li M, Mei Q, Ma X, Liang S, Gou G, Xue N (2023) Wearable and flexible electrochemical sensors for sweat analysis: a review. Microsyst Nanoeng 9(1):1

Gashi S, Min C, Montanari A, Santini S, Kawsar F (2022) A multidevice and multimodal dataset for human energy expenditure estimation using wearable devices. Sci Data 9(1):537

Govindaraj M, Srivastava A, Muthukumaran MK, Tsai P-C, Lin Y-C, Raja BK, Rajendran J, Ponnusamy VK, Arockia Selvi J (2023) Current advancements and prospects of enzymatic and non-enzymatic electrochemical glucose sensors. Int J Biol Macromol 253:126680

Haque Y, Zawad RS, Rony CSA, Al Banna H, Ghosh T, Kaiser MS, Mahmud M (2024) State-of-the-art of stress prediction from heart rate variability using artificial intelligence. Cogn Comput 16(2):455–481

Haripriya R, Khare N, Pandey M (2025) Privacy-preserving federated learning for collaborative medical data mining in multi-institutional settings. Sci Rep 15(1):12482

Hossain MI, Zahid MS, Chowdhury MA, Hossain MMM, Hossain N, Islam MA, Mobarak MH (2024) Smart bandage: a device for wound monitoring and targeted treatment. Results Chem 7:101292

Huang X, Wang D, Yuan Z, Xie W, Wu Y, Li R, Zhao Y, Luo D, Cen L, Chen B, Wu H, Xu H, Sheng X, Zhang M, Zhao L, Yin L (2018) A fully biodegradable battery for self-powered transient implants. Small 14(28):1800994

Huang W, Ying TW, Chin WLC, Baskaran L, Marcus OEH, Yeo KK, Kiong NS (2022) Application of ensemble machine learning algorithms on lifestyle factors and wearables for cardiovascular risk prediction. Sci Rep 12(1):1033

Huang W, Pang I, Bai J, Cui B, Qi X, Zhang S (2025) Artificial intelligence-enhanced, closed-loop wearable systems toward next-generation diabetes management. Adv Intell Syst 7:2400822

Jalal A, Quaid MAK, Tahir SBUD, Kim K (2020) A study of accelerometer and gyroscope measurements in physical life-log activities detection systems. Sensors 20(22):6670

Jin J, Hou Y, Li C, Liang X, Sun X, Ding Q, Zhang X, Xia Z, Yan S, Yang Y, Wang Z (2024) High-performance waterproof flexible thermoelectric generators for self-powered electronics. Nano Energy 132:110388

John A, Redmond SJ, Cardiff B, John D (2022) A multimodal data fusion technique for heartbeat detection in wearable IoT sensors. IEEE Internet Things J 9(3):2071–2082

Johnson KB, Wei WQ, Weeraratne D, Frisse ME, Misulis K, Rhee K, Zhao J, Snowdon JL (2021) Precision medicine, AI, and the future of personalized health care. Clin Transl Sci 14(1):86–93

Joseph AM, Kian A, Begg R (2024) Enhancing intelligent shoes with gait analysis: a review on the spatiotemporal estimation techniques. Sensors [Online] 24:7880

Joshi SR, Lee M, Lee HY, Lee M-E, Kim S (2024) Self-powered, multifunctional, and skin-compatible electronic-tattoos based on hybrid bio-nanomaterials as electrically functionalized skins. Chem Eng J 496:154160

Kaczmarczyk R, Wilhelm TI, Martin R, Roos J (2024) Evaluating multimodal AI in medical diagnostics. npj Digit Med 7(1):205
Kaymaz SV, Nobar HM, Sarıgül H, Soylukan C, Akyüz L, Yüce M (2023) Nanomaterial surface modification toolkit: principles, components, recipes, and applications. Adv Colloid Interf Sci 322:103035
Kazanskiy NL, Butt MA, Khonina SN (2022) Recent advances in wearable optical sensor automation powered by battery versus skin-like battery-free devices for personal healthcare—a review. Nano 12(3):334
Khalifa M, Albadawy M (2024) Artificial intelligence for clinical prediction: exploring key domains and essential functions. Comput Methods Prog Biomed Update 5:100148
Kil H-J, Kim JH, Lee K, Kang T-U, Yoo J-H, Lee Y-H, Park J-W (2024) A self-powered and supercapacitive microneedle continuous glucose monitoring system with a wide range of glucose detection capabilities. Biosens Bioelectron 257:116297
Kim KB, Baek HJ (2023) Photoplethysmography in wearable devices: a comprehensive review of technological advances, current challenges, and future directions. Electronics 12(13):2923
Krones F, Marikkar U, Parsons G, Szmul A, Mahdi A (2025) Review of multimodal machine learning approaches in healthcare. Inf Fusion 114:102690
Leaning IE, Ikani N, Savage HS, Leow A, Beckmann C, Ruhé HG, Marquand AF (2024) From smartphone data to clinically relevant predictions: a systematic review of digital phenotyping methods in depression. Neurosci Biobehav Rev 158:105541
Lecomte A, Degache A, Descamps E, Dahan L, Bergaud C (2017) In vitro and in vivo biostability assessment of chronically-implanted Parylene C neural sensors. Sensors Actuators B Chem 251:1001–1008
Lee KE, Yee J, Lee GY, Chung JE, Seong JM, Chang BC, Gwak HS (2020) Genotype-guided warfarin dosing may benefit patients with mechanical aortic valve replacements: randomized controlled study. Sci Rep 10(1):6988
Lee D, Bhardwaj J, Jang J (2022) Paper-based electrochemical immunosensor for label-free detection of multiple avian influenza virus antigens using flexible screen-printed carbon nanotube-polydimethylsiloxane electrodes. Sci Rep 12(1):2311
Li X, Chen W, Li H, Shen B, He J, Gao H, Bin F, Li H, Xiao D (2024) Temperature self-compensating intelligent wireless measuring contact lens for quantitative intraocular pressure monitoring. ACS Appl Mater Interfaces 16(17):22522–22531
Lian B, Lewis FL, Hewer GA, Estabridis K, Chai T (2022) Robustness analysis of distributed Kalman filter for estimation in sensor networks. IEEE Trans Cybernet 52(11):12479–12490
Linh VTN, Han S, Koh E, Kim S, Jung HS, Koo J (2025) Advances in wearable electronics for monitoring human organs: bridging external and internal health assessments. Biomaterials 314:122865
Liu Y, Pharr M, Salvatore GA (2017) Lab-on-skin: a review of flexible and stretchable electronics for wearable health monitoring. ACS Nano 11(10):9614–9635
Liu W, Xu W, Luan H-H, Li G, Liu J, Lu Z, Zhang F, Li H (2023) L-ribose specific recognition surface constructed by pillar[5]arene-based host–guest interaction. Biosens Bioelectron 241:115678
Lu S-H, Samandari M, Li C, Li H, Song D, Zhang Y, Tamayol A, Wang X (2022) Multimodal sensing and therapeutic systems for wound healing and management: a review. Sens Actuators Rep. 4:100075
Lucisano JY, Routh TL, Lin JT, Gough DA (2017) Glucose monitoring in individuals with diabetes using a long-term implanted sensor/telemetry system and model. IEEE Trans Biomed Eng 64(9):1982–1993
Mahato K, Chandra P (2024) Biosensors for personalized healthcare. Springer Singapore. https://doi.org/10.1007/978-981-97-5473-1
Mahato K, Saha T, Ding S, Sandhu SS, Chang A-Y, Wang J (2024) Hybrid multimodal wearable sensors for comprehensive health monitoring. Nat Electron 7(9):735–750

Manero A, Crawford KE, Prock-Gibbs H, Shah N, Gandhi D, Coathup MJ (2023) Improving disease prevention, diagnosis, and treatment using novel bionic technologies. Bioeng Transl Med 8(1):e10359

Meena JS, Choi SB, Jung S-B, Kim J-W (2023) Electronic textiles: new age of wearable technology for healthcare and fitness solutions. Mater Today Bio 19:100565

Mehta D, Gupta D, Kafle A, Kaur S, Nagaiah TC (2024) Advances and challenges in nanomaterial-based electrochemical immunosensors for small cell lung cancer biomarker neuron-specific enolase. ACS Omega 9(1):33–51

Merchant A, Batzner S, Schoenholz SS, Aykol M, Cheon G, Cubuk ED (2023) Scaling deep learning for materials discovery. Nature 624(7990):80–85

Miller MA, Neuzil P, Dukkipati SR, Reddy VY (2015) Leadless cardiac pacemakers: back to the future. J Am Coll Cardiol 66(10):1179–1189

Mittermaier M, Raza MM, Kvedar JC (2023) Bias in AI-based models for medical applications: challenges and mitigation strategies. npj Digit Med 6(1):113

Mo F, Zhou P, Lin S, Zhong J, Wang Y (2024) A review of conductive hydrogel-based wearable temperature sensors. Adv Healthc Mater 13(26):2401503

Mohammadi M, Tajik E, Martinez-Maldonado R, Sadiq S, Tomaszewski W, Khosravi H (2025) Artificial intelligence in multimodal learning analytics: a systematic literature review. Comput Educ Artif Intell 8:100426

Munch Nielsen J, Zibrandtsen IC, Masulli P, Lykke Sørensen T, Andersen TS, Wesenberg Kjær T (2022) Towards a wearable multi-modal seizure detection system in epilepsy: a pilot study. Clin Neurophysiol 136:40–48

Nweke HF, Teh YW, Al-garadi MA, Alo UR (2018) Deep learning algorithms for human activity recognition using mobile and wearable sensor networks: state of the art and research challenges. Expert Syst Appl 105:233–261

Persson I, Macura A, Becedas D, Sjövall F (2024) Early prediction of sepsis in intensive care patients using the machine learning algorithm NAVOY® Sepsis, a prospective randomized clinical validation study. J Crit Care 80:154400

Poletti F, Zanfrognini B, Favaretto L, Quintano V, Sun J, Treossi E, Melucci M, Palermo V, Zanardi C (2021) Continuous capillary-flow sensing of glucose and lactate in sweat with an electrochemical sensor based on functionalized graphene oxide. Sensors Actuators B Chem 344:130253

Prasanna R, Annaram K, Venkatalakshmi K (2022) Early detection of acute coronary syndrome through prothrombin time measurement using flexible UWB antenna for cardiac patient. Biomed Signal Proc Control 76:103636

Puerta JM, Aledo JA, Gámez JA, Laborda JD (2021) Efficient and accurate structural fusion of Bayesian networks. Inf Fusion 66:155–169

Qi J, Yang P, Waraich A, Deng Z, Zhao Y, Yang Y (2018) Examining sensor-based physical activity recognition and monitoring for healthcare using Internet of Things: a systematic review. J Biomed Inform 87:138–153

Rajalingam B, Priya R, Bhavani R (2019) Hybrid multimodal medical image fusion using combination of transform techniques for disease analysis. Proc Comput Sci 152:150–157

Rao Y, Bechtold T, Hohlfeld D (2025) Design of packaged thermoelectric generators for implantable medical devices: a comprehensive parameter study. Energy 320:134932

Roy N, Mitra S, Das NM, Mandal N, Bandyopadhyay D, Nemade HB, Mandal TK (2020) Paper based enzymatic Chemiresistor for POC detection of ethanol in human breath. IEEE Sensors J 20(5):2278–2286

Saha T, Del Caño R (2023) Wearable electrochemical glucose sensors in diabetes management: a comprehensive review. Chem Rev 123(12):7854–7889

Sailapu SK, Sabaté N, Bakker E (2021) Self-powered potentiometric sensors with memory. ACS Sens 6(10):3650–3656

Sathya K, Magesh G (2025) Multimodal deep learning for cardiovascular risk stratification: integrating retinal biomarkers and cardiovascular signals for enhanced heart attack prediction. IEEE Access 13:99672–99689

Savito L, Scarlata S, Bikov A, Carratù P, Carpagnano GE, Dragonieri S (2023) Exhaled volatile organic compounds for diagnosis and monitoring of asthma. World J Clin Cases 11(21):4996–5013

Schubert-Olesen O, Kröger J, Siegmund T, Thurm U, Halle M (2022) Continuous glucose monitoring and physical activity. Int J Environ Res Public Health 19(19):12296

Sengupta S, Biswal S, Titus J, Burman A, Reddy K, Fulwani MC, Khan A, Deshpande N, Shrivastava S, Yanamala N, Sengupta PP (2023) A novel breakthrough in wrist-worn transdermal troponin-I-sensor assessment for acute myocardial infarction. Eur Heart J Digit Health 4(3):145–154

Shaik T, Tao X, Li L, Xie H, Velásquez JD (2024) A survey of multimodal information fusion for smart healthcare: mapping the journey from data to wisdom. Inf Fusion 102:102040

Shajari S, Kuruvinashetti K, Komeili A (2023) The emergence of AI-based wearable sensors for digital health technology: a review. Sensors 23:23

Shanbhag MM, Manasa G, Mascarenhas RJ, Mondal K, Shetti NP (2023) Fundamentals of bioelectrochemical sensing. Chem Eng J Adv 16:100516

Shim H, Jang S, Yu C (2022) High-resolution patterning of organic semiconductors toward industrialization of flexible organic electronics. Matter 5(1):23–25

Singh M, Kumar A, Khanna NN, Laird JR, Nicolaides A, Faa G, Johri AM, Mantella LE, Fernandes JFE, Teji JS, Singh N, Fouda MM, Singh R, Sharma A, Kitas G, Rathore V, Singh IM, Tadepalli K, Al-Maini M, Isenovic ER, Chaturvedi S, Garg D, Paraskevas KI, Mikhailidis DP, Viswanathan V, Kalra MK, Ruzsa Z, Saba L, Laine AF, Bhatt DL, Suri JS (2024) Artificial intelligence for cardiovascular disease risk assessment in personalised framework: a scoping review. eClinicalMedicine 73:102660

Smokovski I, Steinle N, Behnke A, Bhaskar SMM, Grech G, Richter K, Niklewski G, Birkenbihl C, Parini P, Andrews RJ, Bauchner H, Golubnitschaja O (2024) Digital biomarkers: 3PM approach revolutionizing chronic disease management - EPMA 2024 position. EPMA J 15(2):149–162

Song J, Luo Y, Hao Z, Qu M, Huang C, Wang Z, Yang J, Liang Q, Jia Y, Song Q, Zhang Q, Luo S (2025) Graphene-based wearable biosensors for point-of-care diagnostics: from surface functionalization to biomarker detection. Mater Today Bio 32:101667

Tajik S, Dourandish Z, Jahani PM, Sheikhshoaie I, Beitollahi H, Shahedi Asl M, Jang HW, Shokouhimehr M (2021) Recent developments in voltammetric and amperometric sensors for cysteine detection. RSC Adv 11(10):5411–5425

Tang N, Zheng Y, Jiang X, Zhou C, Jin H, Jin K, Wu W, Haick H (2021) Wearable sensors and systems for wound healing-related pH and temperature detection. Micromachines (Basel) 12(4):430

Tehrani F, Teymourian H (2022) An integrated wearable microneedle array for the continuous monitoring of multiple biomarkers in interstitial fluid. Nat Biomed Eng 6(11):1214–1224

Templer S (2022) Closed-loop insulin delivery systems: past, present, and future directions. Front Endocrinol 13:919942

Teymourian H, Tehrani F (2021) Lab under the skin: microneedle based wearable devices. Adv Healthc Mater 10(17):e2002255

Thilagavathi G, Karthikeyan NK (2025) Edge computing for diabetic early disease prediction based on optimized ensemble boosting Fuzzified neural network using recursive support vector feature selection. Biomed Signal Proc Control 108:107913

Tsai CH, Chen PC (2022) Panic attack prediction using wearable devices and machine learning: development and Cohort Study. JMIR Med Inform 10(2):e33063

Tu Y, Lin S, Qiao J, Zhuang Y, Zhang P (2022) Alzheimer's disease diagnosis via multimodal feature fusion. Comput Biol Med 148:105901

Ullah I, Adhikari D, Su X, Palmieri F, Wu C, Choi C (2025) Integration of data science with the intelligent IoT (IIoT): current challenges and future perspectives. Digit Commun Netw 11(2):280–298

Valik JK, Ward L, Tanushi H, Johansson AF, Färnert A, Mogensen ML, Pickering BW, Herasevich V, Dalianis H, Henriksson A, Nauclér P (2023) Predicting sepsis onset using a machine learned causal probabilistic network algorithm based on electronic health records data. Sci Rep 13(1):11760

Van Calster B, Wynants L, Timmerman D, Steyerberg EW, Collins GS (2019) Predictive analytics in health care: how can we know it works? J Am Med Inform Assoc 26(12):1651–1654

Wan Z, Shao X (2023) Disease classification model based on multi-modal feature fusion. IEEE Access 11:27536–27545

Wang G, Liu T, Sun X-C, Li P, Xu Y-S, Hua J-G, Yu Y-H, Li S-X, Dai Y-Z, Song X-Y, Lv C, Xia H (2018) Flexible pressure sensor based on PVDF nanofiber. Sens Actuators, A 280:319–325

Wang Q-L, Cui H-F, Li C-L, Song X, Lv Q-Y, Li Z-Y (2020) A multimode aptasensor based on hollow gold nanoparticles and structure switching of aptamer: fast and sensitive detection of carcinoembryonic antigen. Sens Actuators Rep 2(1):100021

Wang K, Xie S, Rodrigues J (2022) Medical data security of wearable tele-rehabilitation under internet of things. Internet Things Cyber-Phys Syst 2:1–11

Wang X, Yu H, Kold S, Rahbek O, Bai S (2023) Wearable sensors for activity monitoring and motion control: a review. Biomimet Intell Robot 3(1):100089

Wang J, Luo Y, Loh XJ, Chen X (2024a) Integrated multimodal sensing for scent digitalization. Matter 7(7):2368–2381

Wang B, Hu S, Teng Y, Chen J, Wang H, Xu Y, Wang K, Xu J, Cheng Y, Gao X (2024b) Current advance of nanotechnology in diagnosis and treatment for malignant tumors. Signal Transduct Target Ther 9(1):200

Wei S, Hao Y, Ying Z, Xu C, Wei Q, Xue S, Cheng H-M, Ren W, Ma L-P, Zeng Y (2020) Transfer-free CVD graphene for highly sensitive glucose sensors. J Mater Sci Technol 37:71–76

Wu P, Jin L, Jiang W, Zhou Y, Lin L, Lin H, Chen H (2025a) Smart bandages for wound monitoring and treatment. Biosens Bioelectron 283:117522

Wu Y, Ehlert B, Metwally AA, Perelman D, Park H, Brooks AW, Abbasi F, Michael B, Celli A, Bejikian C, Ayhan E, Lu Y, Lancaster SM, Hornburg D, Ramirez L, Bogumil D, Pollock S, Wong F, Bradley D, Gutjahr G, Rangan ES, Wang T, McGuire L, Venkat Rangan P, Ræder H, Shipony Z, Lipson D, McLaughlin T, Snyder MP (2025b) Individual variations in glycemic responses to carbohydrates and underlying metabolic physiology. Nat Med 31:2232

Xu Y, Uppal A, Lee MS, Mahato K, Wuerstle BL, Lin M, Djassemi O, Chen T, Lin R, Paul A, Jain S, Chapotot F, Tasali E, Mercier P, Xu S, Wang J, Cauwenberghs G (2025) Earable multimodal sensing and stimulation: a prospective toward unobtrusive closed-loop biofeedback. IEEE Rev Biomed Eng 18:5–25

Yin Y, Zhao Z, Na J, Xue C, Qin K (2025) Physiologically inspired spatiotemporal adaptive multimodal fusion model for blood glucose prediction. Biomed Signal Proc Control 109:107998

Ying B, Liu X (2021) Skin-like hydrogel devices for wearable sensing, soft robotics and beyond. iScience 24(11):103174

Zdrachek E, Bakker E (2019) Potentiometric sensing. Anal Chem 91(1):2–26

Zeng X, Peng R, Fan Z, Lin Y (2022) Self-powered and wearable biosensors for healthcare. Mater Today Energy 23:100900

Zhang S, Zhao W, Zeng J, He Z, Wang X, Zhu Z, Hu R, Liu C, Wang Q (2023a) Wearable non-invasive glucose sensors based on metallic nanomaterials. Mater Today Bio 20:100638

Zhang Y, Xu Q, Wang F, Gao T, Wei T (2023b) Enzyme powered self-assembly of hydrogel biosensor for colorimetric detection of metabolites. Sensors Actuators B Chem 375:132942

Zhao J, You Z (2014) A shoe-embedded Piezoelectric energy harvester for wearable sensors. Sensors 14(7):12497–12510

Zhao L, Zhang H, Mbachu J (2023) Multi-sensor data fusion for 3D reconstruction of complex structures: a case study on a real high formwork project. Remote Sens 15(5):1264

Zhou H, Yang S, Wei D, Liang C, Yang Q, Yang H, Wang D, Li M, Yang L (2021) Development of hydrofluoric acid-cleaned silicon nitride implants for periprosthetic infection eradication and bone regeneration enhancement. Mater Sci Eng C 127:112241

Zhou L, Figueroa-Miranda G, Chen S, Neis M, Hu Z, Zhu R, Li Y, Prömpers M, Offenhäusser A, Mayer D (2023) Flexible multielectrode arrays based electrochemical aptasensor for glycated human serum albumin detection. Sensors Actuators B Chem 386:133730

Zhu C, Yang G, Li H, Du D, Lin Y (2015) Electrochemical sensors and biosensors based on nanomaterials and nanostructures. Anal Chem 87(1):230–249

Zhu G, Yin X, Jin D, Zhang B, Gu Y, An Y (2019) Paper-based immunosensors: current trends in the types and applied detection techniques. TrAC Trends Anal Chem 111:100–117

11 Smart Skin and Beyond: Unlocking the Full Potential of Wearable Health Sensors

Nuenghathai Chaiya, Tanin Tangkuaram, and Chochanon Moonla

Abstract

The integration of wearable sensors into healthcare has revolutionized health monitoring by enabling real-time, continuous tracking of physiological, biochemical, and behavioral parameters. This chapter explores the advances and challenges in wearable sensor technologies, focusing on their role in precision healthcare. It begins by discussing the evolution of wearables from single-sensor devices to sophisticated multimodal systems capable of capturing comprehensive health data. This chapter delves into key technological advancements, such as flexible materials, nanobiosensors, microfluidics, and artificial intelligence (AI)-driven analytics, which have enhanced the accuracy, reliability, and usability of wearable sensors. This chapter also examines the applications of wearable sensors in personalized medicine, emphasizing their potential to tailor treatments and preventive strategies to individual health profiles. It underscores the importance of hybrid systems that integrate physiological and biochemical sensors, enabling a holistic understanding of complex health conditions. Challenges such as data accuracy, privacy, and regulatory compliance are critically addressed, with strategies like sensor fusion, advanced calibration methods, and robust data

N. Chaiya
Chemistry program, Faculty of Science, Maejo University, Chiang Mai, Thailand

Department of Chemical and Nano Engineering, University of California San Diego, La Jolla, CA, USA

T. Tangkuaram
Chemistry program, Faculty of Science, Maejo University, Chiang Mai, Thailand

C. Moonla (✉)
Department of Chemical and Nano Engineering, University of California San Diego, La Jolla, CA, USA
e-mail: cmoonla@ucsd.edu

K. Mahato, P. Chandra (eds.), *Nano-bioelectronics for Precision Health Monitoring*, https://doi.org/10.1007/978-981-95-2117-3_11

security measures highlighted as solutions. Future directions are explored in this chapter, including next-generation sensors, the integration of wearables with AI and big data, and their impact on global health and preventive medicine. Ethical and societal implications, such as data ownership and health equity, are also considered. This chapter concludes by underscoring the transformative potential of wearable sensors to advance personalized, data-driven healthcare while addressing critical challenges to their widespread adoption.

Keywords

Wearable health sensors · Continuous health monitoring · Personalized medicine · Sensor fusion · Chronic disease monitoring

11.1 Introduction

Wearable health sensors, often called "wearables," represent a transformative technology in healthcare, offering noninvasive real-time continuous monitoring of vital health parameters. Over the past decade, these devices have transcended their original role in fitness tracking, evolving significantly into sophisticated tools for healthcare applications (Fig. 11.1) (Verma et al. 2023; Suman et al. 2024). This transition marks a pivotal step toward the implementation of precision medicine in healthcare, where continuous data acquisition facilitates personalized treatment strategies. Wearable technology for health monitoring fundamentally relies on the integration of compact, low-power sensors into devices worn on various body parts, such as the appendages, and abdomen, or even integrated directly into the skin. These devices

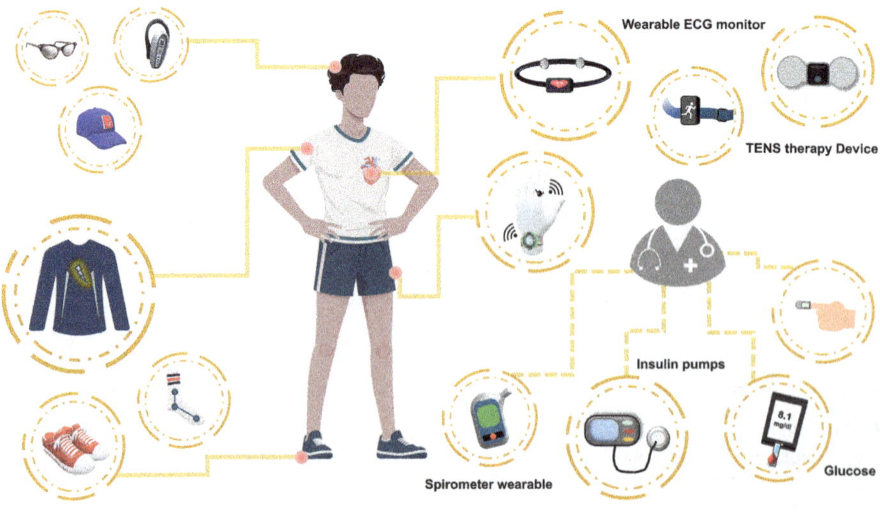

Fig. 11.1 Advancements in wearable health sensors technology: Transforming personal and clinical healthcare

are usually linked to or integrated with external platforms like smartphones or cloud-based systems, enabling real-time data transmission, storage, and analysis (Ates et al. 2022; Walter et al. 2024). The primary aim of these sensors is to allow for uninterrupted monitoring of key health markers without relying on intrusive or labor-intensive traditional medical practices, such as frequent blood sampling or hospital visits.

The growing interest in wearables is due to their potential to revolutionize healthcare delivery (Scataglini et al. 2023). Unlike traditional episodic methods, wearables provide continuous, longitudinal data that captures daily fluctuations in health metrics such as heart rate, oxygen saturation, and even biochemical markers like glucose or lactate in sweat. This continuous data stream can provide more nuanced insights into a patient's health, thus enabling the early detection of abnormalities and facilitating timely interventions (Engel et al. 2023; Lu et al. 2024). Moreover, wearables foster a patient-centric healthcare model by enabling individuals to actively manage their health (Mattison et al. 2022). One of the key features of wearable health sensors is their ability to connect the digital world with biological systems. Through the development of biocompatible materials and advanced electronics, these devices balance comfort, durability, and data accuracy. Recent advancements have led to flexible, skin-like devices that can closely conform to the body, offering a higher degree of comfort and improved sensor accuracy (Liu et al. 2017; Huang et al. 2022a). Wearable health sensors are part of the digital health ecosystem, combining sensors, data analytics, and telemedicine to provide personalized health interventions. These technologies are central to precision healthcare, which customizes medical treatments and recommendations based on the unique genetic, environmental, and lifestyle factors of each patient. By providing continuous real-time health data monitoring, wearable sensors enable a transition from reactive to proactive, preventive healthcare.

The use of wearable sensors has grown rapidly over the past decade, supported by advances in microelectronics, materials science, and data analytics (Kulkarni et al. 2024; Xiao et al. 2024; Xue et al. 2024). Early consumer devices, like activity trackers and smartwatches, primarily tracked fitness metrics such as steps, calories, and sleep. These devices, now widely available, have made health monitoring a common part of everyday life. As technology has advanced, wearable sensors have expanded beyond fitness tracking to monitor various clinical biomarkers (Xue et al. 2024). Modern devices can measure continuous heart rate, electrocardiograms (ECGs), blood pressure, and blood oxygen levels using pulse oximetry. Additionally, new wearable biochemical sensors are being developed to analyze biomarkers in bodily fluids like sweat, tears, and interstitial fluid. These sensors allow non-invasive tracking of important parameters such as glucose levels, electrolyte balance, and stress-related markers like cortisol. Wearable technology is shifting from fitness-focused devices to medical-grade applications, particularly for managing chronic diseases and supporting clinical decisions (Haghayegh et al. 2024). Continuous glucose monitors (CGMs) for diabetes management are a key example of this shift (Shaw et al. 2024). These medical devices undergo strict regulatory approval and thorough clinical testing to ensure they are accurate, reliable, and safe. The use of wearables in clinical practice is growing, driven by healthcare providers' interest in remote monitoring solutions. Wearable sensors play a key role in remote patient monitoring (RPM) programs, which are especially useful for managing chronic conditions like hypertension, diabetes, and heart disease. These programs provide real-time health data that support better decision-making. The COVID-19 pandemic accelerated the adoption of these technologies by enabling continuous care without in-person visits (Shaw et al. 2024). Hospitals and clinics are now integrating wearables into telemedicine platforms, creating a hybrid model of healthcare.

The progression from basic fitness tracking to clinical health monitoring in wearable sensors demonstrates their growing potential in healthcare (Xue et al. 2024; Szeto et al. 2024). Early fitness trackers target health-conscious users, offering features to monitor steps, distance, and calorie expenditure. While popular, these devices provided limited actionable health insights beyond general fitness data. The addition of heart rate sensors and other physiological monitors marked a major shift for wearables, expanding their capabilities. Heart rate monitoring enabled more comprehensive health tracking, including resting heart rate (RHR), heart rate variability (HRV), and other cardiovascular health markers (Xue et al. 2024). This development transformed wearables from simple fitness gadgets into versatile health tools. The addition of ECG functionality in wearable devices strengthened their role in clinical health. This feature enables users to perform on-demand ECGs to detect irregular heart rhythms, such as atrial fibrillation (AFib) (Ahluwalia et al. 2024). This advancement was pivotal, marking the shift of wearables from lifestyle gadgets to clinically relevant health tools. The development of continuous glucose monitors (CGMs) marked another major advancement in wearables for healthcare. Traditional glucose monitoring for diabetic patients requires frequent, painful fingerstick tests. CGMs revolutionized this process by enabling continuous, real-time glucose tracking with a small sensor placed under the skin. This innovation improved patient comfort and provided detailed data for better glycemic control. Furthermore, the inclusion of pulse oximetry sensors in wearable devices has enhanced users' ability to monitor respiratory health. This feature became particularly important during the COVID-19 pandemic, as oxygen saturation served as a key indicator for detecting respiratory distress in severe cases.

The wearable market is divided into consumer-grade and clinical-grade devices, which are evolving simultaneously. Consumer-grade wearables target the general population, emphasizing ease of use, affordability, and daily practicality. These accessible devices provide features for both fitness enthusiasts and those seeking basic health monitoring. However, while they have improved in offering clinically relevant features, their data is often less reliable and usually requires validation by medical devices before being used for clinical decision-making. Clinical-grade wearables are specifically designed for healthcare applications, prioritizing accuracy, reliability, and regulatory compliance. These devices undergo rigorous testing and validation to meet medical standards (Poh et al. 2023; Arslan et al. 2024; Hughes et al. 2023). For example, clinical-grade wearables used for long-term cardiac monitoring provide continuous ECG tracking over several days or weeks, enabling healthcare providers to detect arrhythmias that may not be captured during a short in-office ECG. The distinction between consumer and clinical wearables is becoming less clear, as more consumer-grade devices undergo clinical validation for medical applications. Manufacturers are increasingly pursuing regulatory approval and forming partnerships with healthcare providers (Hughes et al. 2023; Varma et al. 2024; Awad et al. 2023). For example, wearable devices have been used in studies on cardiovascular health, sleep apnea, and other conditions, with some companies collaborating with the FDA to develop devices specifically for clinical use.

11.2 Types of Wearable Sensors and Their Biomedical Applications

11.2.1 Physical Sensors for Vital Signs Monitoring

Wearable biosensors have become a key component of modern healthcare, providing a non-invasive, continuous method to track vital signs (Xue et al. 2024; Zhang et al. 2023). These sensors measure key metrics such as heart rate, ECG, respiratory rate, blood pressure, and body temperature. Real-time monitoring of these parameters offers healthcare professionals valuable insights into cardiovascular and respiratory health, supporting early diagnosis, ongoing monitoring, and preventive care.

Heart rate monitoring is one of the most widely used features in wearable sensors, found in devices from fitness trackers to clinical-grade wearables (Fioravanti et al. 2024). The primary technology for heart rate sensing, photoplethysmography (PPG), measures blood volume changes in tissue by detecting variations in light absorption. Over time, PPG-based monitoring has advanced from tracking basic fitness metrics to providing detailed cardiovascular health data, such as resting heart rate (RHR), heart rate variability (HRV), and real-time detection of irregular rhythms like atrial fibrillation (AFib). The addition of ECG sensors in smartwatches allows users to perform on-demand ECGs to detect AFib, empowering individuals and offering healthcare providers valuable diagnostic information for treatment planning (Ahluwalia et al. 2024; Pina Astete et al. 2024).

Electrocardiogram (ECG) monitoring is an important feature increasingly included in wearables. ECG sensors measure the heart's electrical activity, providing insights into its rhythm and function. Traditionally, ECG monitoring required clinical settings with bulky equipment, limiting accessibility and frequency. Wearable ECG sensors overcome this limitation by enabling remote and continuous ECG recording. Many devices now feature FDA-cleared ECG functionality, significantly enhancing their healthcare applications (Ahluwalia et al. 2024; Pina Astete et al. 2024). This advancement is especially beneficial for patients with cardiovascular conditions such as atrial fibrillation (AFib), where continuous monitoring can detect irregularities that may not be captured during standard in-clinic ECG tests.

Respiratory rate monitoring is an important feature of modern wearable biosensors. These devices use technologies such as bioimpedance, accelerometry, or PPG to track breathing patterns, providing vital information for managing conditions like asthma, chronic obstructive pulmonary disease (COPD), and respiratory infections (Shen et al. 2024; Yang et al. 2024a). Sensors capable of measuring thoracic bioimpedance offer real-time respiratory activity tracking, which is particularly valuable for identifying early signs of respiratory distress or dysfunction and supporting timely medical interventions.

Blood pressure monitoring in wearables is advancing, though progress is slower due to the challenge of achieving accurate, continuous measurements. Traditional methods use an inflatable cuff to measure systolic and diastolic pressure, which is impractical for wearables. Emerging techniques, such as optical and bioimpedance-based methods, estimate blood pressure non-invasively (Sel et al. 2023). Some devices still incorporate miniaturized cuffs for portability, but future wearable technology aims to eliminate cuffs, using methods like tonometry or pulse transit time (PTT) to achieve comparable accuracy (Zhao et al. 2023). Continuous blood pressure monitoring during daily activities offers significant benefits for hypertensive patients, allowing healthcare providers to track fluctuations and trends that might not be detected during occasional clinic visits.

Body temperature monitoring in wearables has gained increased attention during the COVID-19 pandemic, as temperature is a key indicator for early infection detection. Wearable temperature sensors enable continuous tracking of body temperature to identify fever or abnormal variations (Tao et al. 2023). This capability is particularly valuable for providing early warnings of infections or inflammatory processes and supporting timely medical intervention.

11.2.2 Biochemical Sensors for Metabolic Health

Biochemical wearable sensors mark a major advancement in health monitoring by delivering continuous, real-time data on metabolic markers. These sensors detect and measure analytes in body fluids such as sweat, saliva, tears, or interstitial fluid, providing a non-invasive alternative to blood tests. This technology shows great potential for managing conditions like diabetes, dehydration, and electrolyte imbalances, as well as tracking stress, fatigue, and other metabolic states.

Glucose monitoring is one of the most significant applications of biochemical sensors, especially for individuals with diabetes. Continuous glucose monitors (CGMs) have transformed diabetes management by allowing real-time, continuous glucose tracking without frequent finger pricks (Shaw et al. 2024; Mansour et al. 2024). These devices use small, minimally invasive sensors placed beneath the skin to measure glucose levels in interstitial fluid. CGMs provide detailed glucose data over time, supporting improved glycemic control, reducing the risk of hyperglycemia or hypoglycemia, and enhancing overall quality of life. Integration with mobile apps and insulin pumps has further optimized diabetes management through real-time alerts and data-driven recommendations.

Hydration status monitoring is an important application of wearable biochemical sensors, especially for individuals like athletes and military personnel, where dehydration can impair cognitive and physical performance (Kim et al. 2024a; Chiao et al. 2024). Wearable hydration sensors, often embedded in smart clothing or skin patches, track electrolyte concentrations to detect early signs of dehydration. This enables users to take preventive actions before dehydration becomes severe. These devices are particularly useful for monitoring electrolyte loss and maintaining hydration during physical activity.

Lactate monitoring has gained increasing attention in sports medicine and critical care. Lactate, a byproduct of anaerobic metabolism, is a key indicator of physical performance, fatigue, and metabolic stress (De Schrijver et al. 2024; Yang et al. 2024b). Athletes can use lactate monitoring to optimize training by identifying their lactate thresholds. In critical care, elevated lactate levels can signal tissue hypoxia and serve as an early warning for sepsis or shock. Wearable lactate monitors, often incorporated into sweat sensors, offer a non-invasive method for real-time tracking of this important metabolic marker.

Tear fluid and sweat analysis are emerging as promising avenues for non-invasive biochemical monitoring due to their shared biomarkers with blood and accessibility for wearable sensing (Chiao et al. 2024). Tear fluid offers potential applications in glucose monitoring and ocular disease management, such as diagnosing dry eye syndrome and glaucoma through biomarkers like osmolarity and intraocular pressure. However, challenges in sensor sensitivity and biocompatibility remain. Similarly, sweat analysis leverages biomarkers such as electrolytes (e.g., sodium, potassium), metabolites (e.g., lactate), and cortisol for monitoring hydration, metabolic fitness, and stress (Childs et al. 2024). Advances in microfluidics and flexible sensor technologies have enabled the development of epidermal patches for continuous sweat analysis, with applications ranging from cystic fibrosis diagnosis to stress management and athletic performance monitoring.

11.2.3 Motion Sensors for Activity and Sleep Tracking

Wearable motion sensors, such as accelerometers, gyroscopes, and magnetometers, are widely used in devices for monitoring physical activity and sleep. These sensors offer valuable data on activity levels, gait, posture, and sleep patterns, which are important indicators of overall health and well-being.

Accelerometers are among the most commonly used motion sensors in wearables. They measure changes in velocity and are widely used to track steps, activity intensity, and sedentary behavior (Olsen et al. 2024). Three-axis accelerometers enable precise movement detection in multiple directions, allowing differentiation between activities such as walking, running, and cycling. They also play a key role in estimating energy expenditure, providing calorie burn estimates commonly used in fitness tracking.

Gyroscopes complement accelerometers by measuring rotational motion and orientation, making them valuable for detailed movement analysis (Papafotis et al. 2022). They are particularly useful for applications like posture correction, fall detection, and gait analysis. Wearables that combine accelerometers and gyroscopes offer a more comprehensive understanding of movement patterns, supporting physical rehabilitation monitoring and fall risk detection in elderly populations.

Magnetometers, combined with accelerometers and gyroscopes, measure the Earth's magnetic field to provide an absolute reference for orientation. Together, these sensors form an inertial measurement unit (IMU), enabling precise tracking of body movements, balance, and spatial orientation (Shi et al. 2024; Gilbert et al. 2024). IMUs are particularly valuable for applications requiring accurate motion trackings, such as virtual reality (VR), sports performance analysis, and neurological assessments for conditions like Parkinson's disease.

Sleep Monitoring Poor sleep quality is linked to various health issues, including obesity, cardiovascular disease, and cognitive decline. Wearable devices use motion sensors to track sleep duration, stages (e.g., light, deep, REM), and disturbances (Zheng et al. 2024a). These devices estimate sleep stages by analyzing body movement patterns and heart rate variability, offering insights into sleep quality. Advanced algorithms combine data from accelerometers, gyroscopes, and heart rate sensors to provide more accurate assessments of sleep efficiency and recovery.

Gait analysis is increasingly used in clinical settings to evaluate motor function in patients with neurological or musculoskeletal disorders (Cen et al. 2024; Abuchi-Ogbonda et al. 2024; Voet et al. 2023). Wearable inertial measurement units (IMUs), placed on the body (e.g., ankles, hips, wrists), monitor gait parameters such as stride length, walking speed, and limb coordination. This data is valuable for tracking disease progression in conditions like Parkinson's, multiple sclerosis, or post-stroke recovery, as changes in gait can serve as early indicators of progression or improvement.

Fall detection is vital for elderly individuals and those with mobility impairments, as falls are a leading cause of injury and mortality among older adults (Newaz and Hanada 2023). Wearable devices use accelerometers and gyroscopes to identify sudden movement changes indicative of a fall. Many systems are equipped with emergency response features that automatically notify caregivers or medical personnel when a fall is detected, enabling timely intervention and improving outcomes.

11.2.4 Advanced Sensors for Stress, Fatigue, and Cognitive Monitoring

Wearable sensors are expanding beyond physical activity and vital signs to assess psychological and cognitive health. These devices measure physiological markers related to stress, fatigue, and cognitive function, providing new opportunities for managing mental health and improving occupational performance.

Stress monitoring is a promising application of wearable sensors in mental health. Chronic stress is associated with various health issues, including cardiovascular disease, metabolic disorders, anxiety, and depression. Wearable devices equipped with sensors for heart rate variability (HRV), electrodermal activity

(EDA), or cortisol enable continuous, real-time stress assessment (Kim et al. 2024b; Ding et al. 2024). HRV, a key marker of autonomic nervous system function, often decreases under higher stress. EDA sensors measure skin conductance, reflecting sympathetic nervous system activity and stress-induced sweating. Cortisol, a stress-related hormone, can be measured non-invasively via saliva or sweat sensors, providing an additional tool for stress evaluation.

Fatigue monitoring is a growing area of interest, particularly in occupational settings where long work hours or shift work can cause cognitive and physical fatigue. Wearable devices measuring heart rate, HRV, and motion patterns can identify signs of fatigue, such as reduced heart rate variability or decreased physical activity (Freitas et al. 2024). Cognitive fatigue, which impacts attention, memory, and decision-making, can also be assessed using wearable EEG headsets that track brain-wave patterns linked to alertness and fatigue (Fang et al. 2024).

Cognitive monitoring is still in its early stages but shows significant potential for applications in managing neurodegenerative diseases, cognitive behavioral therapy, and occupational health (Howie et al. 2024). Wearable EEG headsets, for example, can track brain activity related to attention, memory, and emotional states (Huang et al. 2024). This data can help monitor cognitive performance in individuals with conditions such as Alzheimer's disease, attention-deficit/hyperactivity disorder (ADHD), or traumatic brain injury. Additionally, advancements in neurofeedback technologies, which enable users to adjust their brain activity through real-time feedback, are being explored for integration into wearable cognitive monitoring devices.

11.3 Technological Advances in Wearable Sensor Design

Wearable sensors have improved greatly over the past decade thanks to advances in materials, electronics, and data analysis. These improvements have made sensors smaller, more efficient, and better at collecting detailed health information. This section will look at the main areas of progress in wearable sensors, including making them smaller and more flexible, improving energy use and power efficiency, and using artificial intelligence (AI) for better data processing. These innovations have made wearable sensors more useful and popular for health monitoring.

11.3.1 Miniaturization and Flexibility in Wearable Sensors

A key advancement in wearable sensor design is the miniaturization of components and the use of flexible, skin-friendly materials. Traditional sensors were bulky and rigid, making them impractical for continuous use. However, advancements in nanomaterials, flexible substrates, and stretchable electronics (Kundu et al. 2024; Duan et al. 2024; Khan et al. 2024) have led to sensors that are comfortable to wear on the skin for long periods without causing discomfort or irritation. Miniaturization involves reducing the size of electronic components while preserving or improving their functionality. In wearable sensors, this process has been essential for creating lightweight, discreet devices that do not interfere with daily activities. Microelectromechanical systems (MEMS) technology has been crucial in this advancement (Karimzadehkhouei et al. 2023). MEMS integrates tiny mechanical and electronic components onto a single chip, enabling sensing, actuation, and data processing. These microchips can be embedded in skin patches, smart clothing, or wristbands, enabling continuous health monitoring without the need for bulky external devices. An example of miniaturization in wearable sensor design is the creation of ultrathin epidermal electronics, or electronic skin (e-skin) (Sun et al. 2024a; Pradhan et al. 2024). These sensors are made from flexible, ultra-thin materials that adapt to the skin's contours, enabling the detection of physiological signals like temperature, hydration, and sweat biomarkers. Designed to mimic the flexibility of human skin, e-skin sensors can stretch, bend, and move with the body. This flexibility improves both user comfort and sensor accuracy by ensuring consistent contact with the skin.

Flexible and stretchable sensors are typically made from polymers, elastomers, and nanomaterials. For example, polyimide is a commonly used polymer in flexible electronics because of its high thermal stability and mechanical strength. Polydimethylsiloxane (PDMS) is another widely used material due to its flexibility and biocompatibility, making it suitable for skin-mounted devices (Zhang et al. 2024a). Nanomaterials like graphene and other two-dimensional (2D) materials are also promising for flexible sensors because of their exceptional electrical conductivity, mechanical strength, and flexibility (Sun et al. 2024b). Graphene-based sensors can be used in wearable devices to monitor physiological parameters such as heart rate, respiratory rate, and glucose levels. Printed electronics have significantly advanced the miniaturization and flexibility of wearable sensors. This technology uses conductive inks to print electronic circuits onto flexible materials like paper, fabric, or plastic (Campos-Arias et al. 2024; Hassanpoor et al. 2024). This method enables the production of thin, lightweight sensors that can be embedded in textiles or applied directly to the skin. For instance, inkjet printing is used to create wearable biosensors for sweat analysis. These sensors can be printed onto flexible substrates and worn as temporary tattoos, allowing continuous monitoring of biomarkers such as electrolytes, metabolites, and hydration levels. A major advancement in wearable technology is the development of flexible power sources. Traditional batteries are rigid and bulky, limiting the design and comfort of wearable devices. However, flexible batteries and energy storage materials now allow for power sources that can bend, stretch, or twist without losing functionality (Zhu et al. 2024). These flexible power sources can be seamlessly integrated into wearables, providing consistent power without adding bulk or discomfort.

The combination of miniaturization and flexibility has expanded the potential applications of wearable sensors in healthcare. For instance, smart textiles with embedded flexible sensors can track heart rate, respiration, and muscle activity in real-time, offering valuable data for athletes, patients in rehabilitation, and individuals managing chronic conditions (Xue et al. 2024; Peng et al. 2024). Additionally, skin-mounted sensors enable continuous, non-invasive monitoring of biochemical markers like glucose and lactate levels, paving the way for personalized diagnostics and health interventions.

11.3.2 Energy Harvesting and Low-Power Design

Power management is a key challenge in wearable sensor design. These devices often require continuous power to operate sensors, collect data, and transmit information. Traditional batteries have limited capacity and require regular recharging or replacement, which can be inconvenient and restrict device usability. Recent advances in energy harvesting technologies and low-power design strategies have addressed this issue by reducing the energy requirements of wearable sensors and enabling the creation of self-powered devices (Gao et al. 2024a). Energy harvesting is the process of capturing and converting ambient energy from the environment into electrical energy to power wearable devices. This approach removes the need for external power sources or frequent recharging, making it ideal for applications requiring continuous, long-term monitoring. Common energy harvesting methods for wearable sensors include solar, thermal, kinetic, and radio frequency (RF) energy (Yu et al. 2024). Solar energy harvesting, using photovoltaic cells to convert sunlight into electricity, is one of the most widely recognized methods (Wang et al. 2024a; Gao et al. 2024b). Advances in flexible and transparent solar cells have enabled their integration into wearable devices without adding bulk or weight. For instance, flexible solar panels can be embedded into smart clothing or worn as patches, providing a continuous power source for sensors monitoring physiological parameters like heart rate or temperature. These solar-powered devices can operate autonomously for long periods, making them ideal for remote or outdoor health monitoring applications. Thermal energy harvesting is a promising method, especially for wearables that make direct contact with the skin. The body naturally emits heat, which thermoelectric generators (TEGs) can capture and convert into electricity by exploiting the temperature difference between the skin and the surrounding environment (Jia et al. 2024). Flexible TEGs are designed to conform to the body, harvesting energy from body heat while ensuring user comfort. This method is particularly suitable for low-power devices like temperature sensors or electrocardiogram (ECG) monitors, which need only minimal energy to function. Kinetic energy harvesting uses the wearer's movements to generate power. It relies on piezoelectricity, where mechanical stress, such as bending or pressure, is converted into electrical energy. Wearable devices with piezoelectric materials can capture energy from activities like walking, running, or even subtle movements like breathing. For example, piezoelectric nanogenerators (PENGs) integrated into smart shoes or clothing can convert the mechanical energy from walking into electricity to power embedded sensors (Yu et al. 2024). This method is especially useful for sports and fitness wearables, where continuous movement is common.

Radio frequency (RF) energy harvesting captures ambient RF signals from sources like Wi-Fi routers, mobile phones, or broadcast towers and converts them into electrical energy (Yu et al. 2024; Lee et al. 2024). These systems can be integrated into wearable devices, providing a continuous power source independent of physical movement or environmental conditions. This method is especially effective in urban areas with abundant RF signals. However, the energy harvested is relatively small, making it suitable for low-power sensors or as a supplement to other energy harvesting methods. Low-power design strategies are a key focus in wearable sensor development. Advances in ultra-low-power microcontrollers, energy-efficient sensors, and low-power wireless communication protocols have significantly reduced power consumption in wearable devices. For instance, Bluetooth Low Energy (BLE) is now a standard in wearables, enabling data transmission with minimal energy use (Contoli et al. 2024). Compared to traditional Bluetooth, BLE uses far less power, allowing devices to continuously transmit data without quickly draining the battery. Sleep modes and duty cycling are effective strategies to reduce power consumption in wearable sensors. Since many sensors do not need to operate continuously, they can enter low-power sleep modes when not actively collecting data. Duty cycling alternates the sensor between active periods (data collection) and sleep periods (reduced power use), significantly extending battery life (Garcha et al. 2022). The integration of energy harvesting technologies and low-power design strategies has led to self-powered wearable sensors capable of operating autonomously for extended periods (Gao et al. 2024a). This is especially valuable in healthcare, where continuous, long-term monitoring is essential. Self-powered wearables reduce the need for frequent recharging, enhancing user convenience and enabling more effective health monitoring.

11.3.3 Improved Data Processing and AI Integration

As wearable sensors advance and collect larger volumes of health data, efficient data processing and analysis have become crucial. Traditional methods, where raw data is transmitted to an external device for analysis, are often constrained by bandwidth, latency, and power consumption. Recent advancements in on-device data processing and AI integration have overcome these limitations by enabling real-time data analysis directly on the wearable device (Fatima et al. 2024). This reduces the need for external data transmission and enhances the device's overall functionality. On-device data processing allows wearable devices to process and analyze sensor data locally, without depending on external devices like smartphones or cloud servers (Pankaj et al. 2024). This is achieved through edge computing, where data is processed directly on the device rather than in a centralized data center. Edge computing reduces latency and bandwidth needs, enabling real-time data analysis and decision-making.

A major innovation in on-device data processing is the development of machine learning (ML) algorithms that can run directly on wearable devices. These algorithms analyze physiological data, such as heart rate or respiration patterns, to make predictions or recommendations. For instance, a wearable ECG monitor could use ML to detect abnormal heart rhythms in real-time and alert the user or a healthcare provider (Wang et al. 2024b). By processing data locally, the device provides immediate feedback without needing to transmit large amounts of raw data to external servers. The integration of AI and deep learning into wearable devices has greatly enhanced real-time health monitoring and personalized healthcare. AI algorithms can analyze complex physiological signals, identify subtle patterns, and deliver personalized health insights. For instance, AI-powered wearables can detect early signs of disease, monitor chronic conditions, and even predict health events like heart attacks or seizures (Fatima et al. 2024). By continuously analyzing data from multiple sensors, AI provides more accurate and comprehensive health assessments compared to traditional methods. A promising application of AI in wearable sensors is predictive analytics, where AI models forecast future health outcomes using historical data. For instance, a wearable device could predict a patient's risk of a hypertensive crisis by analyzing trends in blood pressure, heart rate, and activity levels (Xu and Xu 2024). Such predictions enable early interventions, improving patient outcomes and reducing the need for emergency medical care.

Natural language processing (NLP) is increasingly used in wearable devices, especially for mental health monitoring (McElroy et al. 2024). NLP algorithms can analyze speech patterns, tone, and language use to identify changes in mood or cognitive function. This technology offers the potential to detect early signs of conditions like depression or anxiety, enabling timely intervention and treatment. AI in wearable sensors goes beyond health monitoring to support personalized healthcare interventions. Wearable devices with AI algorithms can offer tailored recommendations for lifestyle changes, exercise routines, or medication adjustments based on individual health data (Fatima et al. 2024; Huang et al. 2022b). For instance, an AI-powered fitness tracker could analyze a user's heart rate, sleep patterns, and activity levels to suggest a customized workout plan that aligns with their fitness goals and health needs. The combination of advanced data processing and AI integration has significantly improved wearable sensors, enabling accurate real-time health monitoring and personalized healthcare interventions. These advancements have the potential to transform healthcare by shifting the focus from reactive care to proactive, data-driven health management.

11.4 Multimodal Sensor Integration for Comprehensive Monitoring

Multimodal sensor integration is a major advancement in wearable health monitoring, enabling the simultaneous capture and analysis of multiple physiological and biochemical signals. This approach provides a more comprehensive view of an individual's health compared to single-mode sensors that monitor only one parameter. By combining data from various sensors, multimodal systems offer deeper insights into complex biological processes, disease conditions, and overall wellness. This section examines how hybrid systems integrate physiological and biochemical data, explores sensor fusion and data correlation strategies, and highlights the role of multimodal wearables in managing chronic diseases.

11.4.1 Hybrid Systems Combining Physiological and Biochemical Data

Hybrid wearable systems combining physiological and biochemical sensors have become powerful tools for comprehensive health monitoring (Xue et al. 2024). Physiological sensors measure parameters like heart rate, respiration, blood pressure, and temperature, while biochemical sensors detect biomarkers such as glucose, lactate, cortisol, and other metabolites in bodily fluids like sweat, saliva, or interstitial fluid. Integrating these sensors enables simultaneous monitoring of vital signs and metabolic processes, providing a more complete assessment of an individual's health. Hybrid systems are necessary because physiological parameters alone cannot fully capture the complexity of human health (Zhang et al. 2024b). For example, heart rate and respiration provide useful information about cardiovascular and respiratory health but do not reflect metabolic changes like glucose metabolism or electrolyte balance. Similarly, biochemical sensors provide detailed metabolic data but lack context about overall physiological function. By integrating both types of sensors, hybrid systems offer more comprehensive insights and a deeper understanding of how different biological systems interact. A notable example of a hybrid system is a continuous glucose monitor (CGM) combined with a heart rate monitor (Thomas et al. 2024). This system tracks glucose levels and cardiovascular activity simultaneously, offering valuable insights for individuals with diabetes. By monitoring both parameters, users and healthcare providers can better understand how factors like physical activity, diet, and stress affect glucose levels. This enables more precise glycemic control and improved disease management. Such integration is especially useful for identifying periods of hypoglycemia or hyperglycemia triggered by physiological stress, such as during exercise or illness. Hybrid systems are also valuable in exercise physiology and sports medicine, where tracking both physiological and biochemical markers provides detailed insights into athletic performance and recovery (Zhang et al. 2024b; Ucar et al. 2024). For instance, a wearable device monitoring lactate levels, muscle oxygenation, and heart rate variability (HRV) can assess an athlete's anaerobic threshold, recovery time, and cardiovascular health. This information is crucial for optimizing training programs and preventing overtraining or injury. By correlating lactate levels with heart rate and oxygen saturation, athletes and coaches can create personalized training plans that enhance performance while reducing the risk of fatigue or injury. Hybrid systems also show great promise in managing chronic diseases that require continuous monitoring of multiple parameters. For example, a system combining ECG, blood pressure, and sweat biomarker sensors can monitor patients with cardiovascular diseases. By tracking cardiac function alongside biochemical markers like sodium, potassium, or cortisol in sweat, healthcare providers can identify early signs of electrolyte imbalances, dehydration, or stress—common precursors to serious cardiovascular events (Arslan et al. 2024; Zhang et al. 2024b; Hannan et al. 2024). This comprehensive monitoring enables early interventions, reducing the risk of complications and hospitalizations.

11.4.2 Sensor Fusion and Data Correlation

A major challenge in developing multimodal wearable systems is integrating data from different sensors, a process known as sensor fusion (Arslan et al. 2024; Zhang et al. 2024b). Sensor fusion combines data from multiple sensors to create a unified and coherent picture of an individual's health. This approach harnesses the strengths of each sensor type, enabling more accurate and meaningful data interpretation. While individual sensors provide valuable information, integrating and correlating their data offers deeper insights into complex physiological processes and disease states. A key advantage of sensor fusion is its ability to correlate different types of data, revealing patterns that may not be evident when analyzing each parameter separately. For instance, combining accelerometer data (tracking physical activity) with heart rate and blood glucose data provides a deeper understanding of how exercise affects glucose metabolism (Nemat et al. 2024). Analyzing these data streams together can uncover trends, such as how specific activities impact blood sugar levels or the time it takes for glucose to normalize after exercise. This insight is especially beneficial for individuals with diabetes, enabling more personalized and effective management strategies. Sensor fusion in wearable devices relies on advanced algorithms to process and analyze data from multiple sources. Many of these algorithms use machine learning (ML) techniques to identify patterns and correlations between different data streams (Xiao et al. 2024; Kim et al. 2024b; Ding et al. 2024). For example, a wearable monitoring sleep, heart rate, and cortisol levels might use ML to predict stress-related sleep disturbances or detect signs of chronic stress. By linking changes in cortisol levels with heart rate and sleep quality, the system provides accurate insights into stress levels and offers recommendations to improve sleep and overall health. In healthcare, multivariate data analysis is a key approach in sensor fusion, analyzing multiple variables simultaneously to uncover relationships between them. For instance, a wearable device for cardiovascular health may combine ECG, blood pressure, and accelerometer data to detect arrhythmias or hypertensive episodes (Wang et al. 2024b; Spatz et al. 2024). By analyzing these variables together, the system can identify subtle changes that signal an increased risk of a cardiovascular event, enabling early intervention.

Data synchronization is a crucial aspect of sensor fusion in multimodal wearables. Different sensors often collect data at varying rates and frequencies (Zhang et al. 2024b; Celik and Godfrey 2023)—for example, ECG sensors might sample at high frequencies (e.g., 1000 Hz), while accelerometers sample at lower frequencies (e.g., 50 Hz). To accurately correlate data from these sensors, their data streams must be aligned in time. This is achieved using time-stamping techniques or synchronization protocols, ensuring that data from different sensors can be accurately combined and analyzed. Interpreting multimodal data is a significant challenge in sensor fusion due to the large volume of data generated by multiple sensors. Extracting meaningful insights requires advanced algorithms to filter noise, identify relevant patterns, and make predictions or recommendations. Machine learning (ML) and artificial intelligence (AI) are essential in this process, allowing wearables to analyze large datasets in real-time and provide actionable insights for users and healthcare providers (Xiao et al. 2024). A successful example of sensor fusion is the development of multimodal wearable systems for monitoring sleep disorders. These devices combine data from actigraphy, heart rate variability (HRV), and respiratory sensors to evaluate sleep quality, detect sleep apnea, and identify circadian rhythm disturbances (Tang et al. 2024; Bello et al. 2024). By integrating multiple data streams, these systems provide a more detailed understanding of sleep patterns and potential causes of disruptions that single sensors might miss. For instance, a correlation between low HRV and restlessness could indicate stress or anxiety, while respiratory data might point to sleep apnea (Tomasi et al. 2022). This combined analysis enables more accurate and personalized recommendations for improving sleep quality.

11.4.3 Wearables in Chronic Disease Management

The use of multimodal wearable systems in chronic disease management is a highly promising area of healthcare innovation. Conditions like diabetes, hypertension, cardiovascular disease, and respiratory disorders require continuous monitoring of multiple physiological and biochemical parameters to manage symptoms, prevent complications, and improve outcomes. Multimodal wearables provide a non-invasive, real-time solution for monitoring these parameters, delivering actionable data to patients and healthcare providers to guide treatment decisions and lifestyle changes. A well-established use case for multimodal wearables in chronic disease management is diabetes monitoring. Devices that combine continuous glucose monitoring (CGM) with heart rate and physical activity sensors provide valuable insights into how factors like exercise, diet, and stress affect glucose levels (Thomas et al. 2024; Di Molfetta et al. 2024; Gómez-Peralta et al. 2024). By correlating these variables, individuals with diabetes can receive personalized recommendations to optimize glucose control. For instance, a multimodal system might identify that a specific type of exercise consistently causes hypoglycemia and suggest adjustments to the user's exercise routine or dietary intake. These insights help individuals avoid dangerous glucose fluctuations, improve glycemic control, and reduce the risk of long-term complications such as neuropathy, retinopathy, or cardiovascular disease. Multimodal wearables are also highly valuable in managing cardiovascular diseases. Patients with conditions like hypertension, heart failure, or arrhythmias often need continuous monitoring of parameters such as blood pressure, ECG, heart rate variability (HRV), and respiratory rate. By combining these sensors, multimodal devices provide a more comprehensive view of cardiovascular health than single-mode devices (Lu et al. 2024; Abbas et al. 2024; Wang et al. 2022). For instance, correlating blood pressure readings with ECG data can help healthcare providers detect early signs of hypertension-induced arrhythmias and initiate timely treatment. In heart failure patients, tracking both HRV and respiratory rate can reveal signs of fluid overload or respiratory distress, common indicators of worsening heart failure. Early detection enables prompt intervention, potentially preventing hospitalizations and improving outcomes. Multimodal wearables are also valuable in managing chronic respiratory conditions like chronic obstructive pulmonary disease (COPD) and asthma (Shah et al. 2023; Oliveira et al. 2024). These devices combine sensors for respiratory rate, oxygen saturation (SpO2), and activity levels to monitor lung function and detect early signs of exacerbations. By integrating data from these sensors, they provide insights into how factors like physical activity, environmental conditions, and medication adherence impact respiratory health. For instance, a wearable monitoring of SpO2, respiratory rate, and air quality might detect that a patient's respiratory function worsens during exposure to specific triggers, such as pollen or pollution, and offer personalized recommendations to avoid them. In COPD patients, early detection of exacerbations through multimodal monitoring enables timely interventions, reducing hospitalizations and improving quality of life. Multimodal wearables are increasingly being used to manage mental health conditions such as anxiety, depression, and post-traumatic stress disorder (PTSD) (Ding et al. 2024; Ahmed et al. 2023; Zhang et al. 2024c; Dias et al. 2024). These devices combine sensors for heart rate variability (HRV), electrodermal activity (EDA), and cortisol levels to assess stress and emotional states. By analyzing patterns in these physiological markers, they provide insights into how stress, sleep quality, and physical activity affect mental health. For instance, a wearable detecting high cortisol levels, low HRV, and poor sleep quality may indicate chronic stress or anxiety. It could then offer personalized recommendations for stress management, such as breathing exercises, physical activity, or improved sleep hygiene. Additionally, these insights can be shared with healthcare providers to guide treatment plans, including adjustments to therapy or medication. The use of multimodal wearables in chronic disease management extends beyond individual patients to remote patient monitoring (RPM) programs. These programs collect and analyze real-time data from multiple patients, allowing healthcare providers to continuously monitor chronic conditions, detect early signs of deterioration, and intervene promptly (Tan et al. 2024; Frerichs et al. 2024; Pauly et al. 2024). For example, in an RPM program for heart failure patients, clinicians can track changes in heart rate, blood pressure, and respiratory rate to identify signs of fluid overload or worsening heart function. Early intervention helps prevent hospitalizations, lower healthcare costs, and improve patient outcomes.

11.5 Challenges and Opportunities in Wearable Sensor Implementation

The integration of wearable sensors into healthcare is transforming how health is monitored, diagnosed, and managed. These sensors offer real-time, continuous data on various physiological and biochemical markers, enabling personalized and proactive healthcare. However, to fully unlock their potential in clinical and consumer settings, several challenges must be addressed. This section examines key challenges and opportunities, including data accuracy, reliability, and calibration; privacy and security concerns; and regulatory and ethical considerations.

11.5.1 Data Accuracy, Reliability, and Calibration

A major challenge in implementing wearable sensors is ensuring data accuracy and reliability. For these devices to be effective in clinical and consumer health applications, they must consistently deliver accurate, dependable data. Inaccurate or unreliable data can lead to misinterpretations, incorrect diagnoses, and inappropriate treatments, potentially endangering patient safety.

Wearable sensors often function in dynamic and unpredictable environments where factors like motion artifacts, sweat, temperature changes, and skin tone can impact measurement accuracy. For example, photoplethysmography (PPG) sensors, used for heart rate monitoring, are prone to motion artifacts during physical activity, which can distort optical signals and produce inaccurate readings, especially during vigorous exercise (Hao et al. 2024; Rehman et al. 2024). Similarly, electrocardiogram (ECG) sensors, which measure the heart's electrical activity, may yield inaccurate results due to electrode movement on the skin or poor contact during physical activity (Umar et al. 2021). The biocompatibility of wearable sensors directly impacts data accuracy (Moudgil et al. 2023). Skin irritation or allergic reactions to sensor materials can disrupt sensor-skin contact, reducing data quality. To mitigate this, developers are utilizing biocompatible materials like flexible polymers, silicone, and textiles, which ensure stable contact while minimizing irritation. The reliability of sensor data is crucial, especially in clinical applications requiring long-term monitoring. Wearable sensors must deliver consistent accuracy over weeks or months, but issues like sensor drift, battery depletion, and wear can affect performance. For example, electrochemical sensors in continuous glucose monitoring (CGM) may face enzyme degradation or membrane fouling, leading to inaccurate glucose readings. Maintaining long-term accuracy is essential for successful clinical use (Lee et al. 2022; Sharma et al. 2023; Lielpetere et al. 2023).

To enhance data accuracy and reliability, wearable sensor systems often use calibration protocols. Calibration adjusts the sensor's output to align with a known reference, ensuring accurate readings. For instance, continuous glucose monitors require periodic calibration with a finger-stick blood glucose measurement. However, frequent calibration can be inconvenient, especially for sensors designed for continuous, long-term use. This has driven the development of self-calibrating sensors, which use advanced algorithms to automatically adjust for environmental factors, motion artifacts, and sensor drift, maintaining accuracy without user intervention (Hanna et al. 2022; Sun et al. 2024c). Recent advancements in sensor fusion have significantly improved data accuracy and reliability. By integrating data from multiple sensors, wearable systems can cross-check and validate readings, reducing errors in individual sensors. For example, combining accelerometer data with PPG readings helps correct motion artifacts in heart rate monitoring. Similarly, integrating ECG and blood pressure sensors enhances cardiovascular monitoring by comparing their data streams (Celik and Godfrey 2023; Zhang et al. 2024d; Duro 2024). While sensor fusion is highly effective in improving accuracy, it requires advanced algorithms and robust data processing to integrate and interpret the data efficiently.

Artificial intelligence (AI) and machine learning (ML) algorithms offer a promising solution to improving data accuracy in wearable sensors. These algorithms can identify and correct anomalies, such as those caused by motion artifacts or environmental factors, ensuring reliable sensor readings (Xiao et al. 2024; Marvasti et al. 2024; Li and Zhang 2024). They are especially useful for wearable devices that generate continuous data, as they automatically detect patterns and trends, delivering more accurate and insightful results. However, developing and validating these algorithms requires large, high-quality datasets, and ensuring the accuracy of AI-driven insights remains a key challenge.

11.5.2 Privacy and Security Concerns

Wearable sensors generate large amounts of personal health data, including metrics like heart rate, sleep patterns, glucose levels, and stress markers. While this data offers valuable insights into health and well-being, it also raises significant privacy and security concerns, which must be addressed to build trust and promote the widespread adoption of wearable technologies. One major concern is the potential for data breaches or unauthorized access to personal health information. Wearable devices often transmit data wirelessly to external platforms for storage and analysis, creating vulnerabilities during transmission (Jian et al. 2024). For instance, if a device's wireless connection is inadequately secured, malicious actors could intercept sensitive information such as heart rate or glucose levels. In healthcare contexts, such breaches could compromise patient confidentiality and violate data protection regulations. Data aggregation poses another risk, as continuous health data generated by wearable sensors is often analyzed by companies or third parties (Jian et al. 2024; Mughal et al. 2024; Gelinas et al. 2023). While such analysis can lead to personalized health insights and improved healthcare services, it also raises the potential for misuse. For example, health data could be sold to advertisers or insurers, leading to concerns about targeting or discrimination based on perceived health risks. These practices highlight ethical issues around privacy, consent, and data control.

Anonymizing health data by removing personally identifiable information (PII) can help protect users' privacy. However, even anonymized datasets can sometimes be re-identified through advanced data mining techniques if they contain unique patterns, such as specific combinations of heart rate and activity data (Gadotti et al. 2024). This makes true anonymization a complex challenge. Robust encryption methods are critical for protecting data during transmission and storage (Rehman and Shafqiue 2024; Agrawal et al. 2012). For example, end-to-end encryption ensures that data remains secure as it moves between devices, smartphones, and cloud platforms, allowing only authorized parties to access it. Multi-factor authentication (MFA) further enhances security by requiring users to verify their identity through multiple methods, such as passwords and biometric data. Clarifying data ownership is essential. While users may assume they own the data collected by their devices, manufacturers or third-party providers often retain control. Transparent guidelines on data ownership, user consent, and sharing practices are necessary to ensure users maintain control over their health information. Consent mechanisms should be simple and explicit, enabling users to decide how their data is used and shared. Decentralized approaches to health data management, where users store and control data locally on their devices, offer additional privacy benefits. By reducing reliance on centralized cloud servers, this approach minimizes the risk of data breaches and enhances user control. Blockchain technology has emerged as a potential solution, providing secure, tamper-proof data storage (Xu et al. 2024). Users could use private keys to control access, granting permissions on a case-by-case basis. However, blockchain-based systems for wearable devices remain in early development and face challenges related to scalability and technical implementation.

11.5.3 Regulatory and Ethical Considerations

The adoption of wearable sensors in healthcare relies on both technological progress and adherence to regulatory standards, along with addressing ethical concerns. To integrate into clinical practice, wearable sensors must comply with strict regulations to ensure safety, accuracy, and reliability. Moreover, ethical considerations, including privacy, autonomy, and equity, must be addressed when using these devices to monitor personal health data. In the United States, wearable sensors intended for medical use must comply with US Food and Drug Administration (FDA) regulations (McGrath and Scanaill 2013). The FDA categorizes medical devices into three classes based on risk: Class I (low risk), Class II (moderate risk), and Class III (high risk). Wearable sensors that monitor vital signs, such as heart rate or glucose levels, are typically classified as Class II devices. These require premarket notification (510(k)) to prove their safety and effectiveness. This process includes rigorous testing, such as clinical trials, to ensure the device meets standards for accuracy, reliability, and safety. The FDA's Digital Health Innovation Action Plan provides a framework for regulating digital health technologies, including wearable sensors (Zinchenko et al. 2022; Yu et al. 2023). This plan aims to streamline the approval process and foster innovation in the field. Additionally, the Software as a Medical Device (SaMD) framework clarifies regulatory requirements for software-based wearables, such as AI-driven diagnostic algorithms integrated into sensors. In Europe, wearable sensors must adhere to the Medical Device Regulation (MDR), implemented in 2021 (Bretthauer et al. 2023; Jurczak et al. 2024). Similar to the FDA system, the MDR classifies medical devices based on their intended use and risk level. Wearable sensors providing clinical data, such as ECG or glucose monitoring, must pass conformity assessments and obtain CE marking to confirm compliance with European safety and performance standards. The MDR also emphasizes post-market surveillance, requiring manufacturers to continuously monitor the safety and effectiveness of their devices after market release.

A key regulatory challenge for wearable sensors is distinguishing between consumer-grade and medical-grade devices. Consumer-grade wearables, like fitness trackers, are not subject to strict medical device regulations. However, the line is blurring as devices like smartwatches with ECG functionality or continuous glucose monitors are marketed to consumers but are also used in clinical settings. Ensuring these devices meet regulatory standards for medical use is essential for their integration into healthcare (Wang et al. 2024b; Linh et al. 2025). Wearable sensors also present ethical concerns, particularly around privacy, autonomy, and equity. A major issue is the potential for surveillance and loss of privacy, as these devices continuously monitor health data. Questions arise about how this data is used and who can access it. For instance, employers or insurers might use wearable data to track employees' health or adjust insurance premiums based on perceived risks (Eversdijk et al. 2024; Dobson et al. 2023). This highlights the importance of informed consent and balancing the benefits of wearable sensors with the risks to individual privacy. Wearable sensors also raise concerns about autonomy and the role of technology in health management. While these devices empower individuals to monitor and manage their health, they may lead to data dependency and over-reliance on technology (Varma et al. 2024; Kang and Exworthy 2022; Wall et al. 2023). For example, users might rely too heavily on wearable devices for health decisions, which could reduce their ability to make informed choices independently. Ensuring users maintain control over their health data and are free from external pressures is a key ethical consideration. Health equity is another important concern with wearable sensors. While these devices can democratize healthcare by providing real-time health data, they may also worsen health disparities (Wilson et al. 2024; Swahn et al. 2024; Cruz et al. 2024). High costs and the need for smartphones or digital devices make wearable sensors inaccessible to low-income populations or those in underserved areas. Ensuring these technologies are affordable and accessible to everyone, regardless of socioeconomic status, is essential for promoting health equity.

11.5.4 Opportunities for Future Growth

Despite challenges in implementing wearable sensors, the field offers significant opportunities for growth. Wearable sensors have the potential to transform healthcare by enabling continuous, real-time health monitoring, improving chronic disease management, and empowering individuals to take control of their health. Wearable sensors are key to advancing personalized medicine by providing real-time data on physiological and biochemical markers. This data allows healthcare providers to tailor treatment plans to individual needs. For example, continuous glucose monitors (CGMs) provide insights into how glucose levels respond to food, activity, and medications, enabling precise insulin dosing and dietary adjustments (Mansour et al. 2024). Similarly, sensors that track heart rate variability (HRV) and stress markers offer personalized insights into stress management, recommending interventions such as meditation or exercise (Xue et al. 2024). Wearable sensors enhance remote patient monitoring by enabling real-time tracking of patients with chronic conditions like diabetes, hypertension, or heart failure (Xue et al. 2024). RPM systems that integrate wearable sensors allow healthcare providers to detect complications early and intervene promptly, reducing hospitalizations, improving outcomes, and lowering healthcare costs (Kulkarni et al. 2024; Adashi et al. 2024). In preventive care, wearable sensors can detect early signs of diseases, such as elevated blood pressure, abnormal heart rhythms, or metabolic changes, before symptoms appear. Early detection enables timely intervention and improved long-term health outcomes. For instance, cardiac sensors can identify early signs of atrial fibrillation (AFib), allowing for prompt treatment to reduce stroke risk (Xue et al. 2024). The integration of AI and machine learning into wearable systems opens new possibilities. AI algorithms can process large volumes of sensor data in real-time, providing personalized insights, predictive analytics, and automated recommendations. For example, AI-powered devices could predict the likelihood of a hypertensive crisis based on blood pressure and heart rate trends or identify subtle patterns, like changes in gait indicating early Parkinson's disease or glucose fluctuations signaling prediabetes (Xiao et al. 2024; Marvasti et al. 2024). Wearable sensors can advance global health by providing real-time health data in underserved and remote areas. In regions with limited healthcare access, these devices enable remote monitoring and telemedicine, allowing providers to track patient health and deliver care from a distance. Sensors monitoring vital signs, respiratory function, and metabolic markers are especially valuable in managing infectious diseases, such as COVID-19, by enabling continuous symptom tracking to prevent disease spread.

11.6 Future Directions and Potential of Wearable Sensors in Precision Healthcare

The integration of wearable sensors into healthcare has already transformed how individuals monitor their health and how care is delivered. The future promises even greater advancements as technology, data analytics, and artificial intelligence (AI) converge to enhance their capabilities (Fig. 11.2). As these devices become more sophisticated and seamlessly integrated into daily life, they are poised to play a pivotal role in precision healthcare, tailoring treatments and interventions to each individual's unique biological, behavioral, and environmental characteristics. This section examines the future potential of wearable sensors in healthcare, focusing on their contributions to personalized medicine, next-generation biosensing technologies, and their impact on global health and preventive medicine. Additionally, it explores the broader implications for society and healthcare systems, highlighting how wearable sensors could shape the future of medicine.

11.6.1 Wearables for Personalized Medicine

Wearable sensors hold significant potential to advance personalized medicine, which tailors medical treatments to individual characteristics such as genetics,

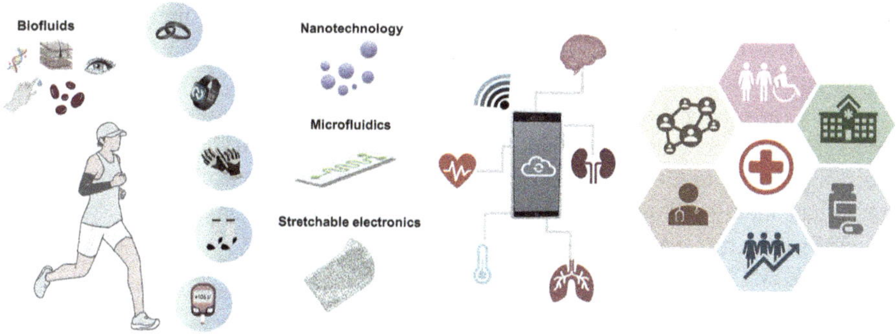

Fig. 11.2 Future directions global health and preventive care

lifestyle, and environmental factors. By providing continuous, real-time data on physiological, biochemical, and behavioral health metrics, wearable sensors are uniquely positioned to support this approach. Currently, most healthcare practices rely on generalized guidelines based on population averages, such as standard blood pressure targets or medication dosages. These approaches often fail to account for individual differences in treatment response. Wearable sensors can address this gap by delivering detailed, real-time data, enabling healthcare providers to create personalized interventions tailored to an individual's specific health needs. Future wearable sensors are expected to capture a broader range of health metrics (Yang and Gao 2019; Mallin et al. 2023; Singar et al. 2024). For example, genetic sensors could monitor how a person's genetic makeup influences their responses to medications or exercise, while devices tracking epigenetic markers—chemical modifications that regulate gene expression—could reveal how factors like diet, stress, or environmental exposures affect long-term health. These insights could inform precise therapeutic strategies, including personalized drug regimens and preventive measures targeting specific risks based on genetic and lifestyle profiles. Wearable sensors could play a transformative role in pharmacogenomics, the study of how genetic factors affect drug responses. For instance, sensors that track liver enzymes involved in drug metabolism could provide real-time data on how quickly a patient processes a medication, allowing providers to adjust dosages to minimize side effects and optimize treatment (Yang et al. 2024c). In oncology, wearable devices could monitor tumor markers, offering personalized feedback on treatment effectiveness and enabling real-time adjustments rather than waiting for periodic lab results (Chow et al. 2024). Wearable sensors have the potential to revolutionize chronic disease management and preventive medicine by providing personalized and data-driven insights. For diabetes, continuous glucose monitors integrated with advanced analytics can assess not only glucose levels but also factors like sleep, exercise, and stress, offering tailored recommendations for diet and insulin dosing to minimize complications (Hermanns et al. 2022; Frank et al. 2024). Similarly, for cardiovascular conditions, wearables that track heart rate variability, blood pressure, and physical activity can generate individualized exercise and medication plans based on real-time data (Sousa et al. 2023). In preventive medicine, sensors monitoring stress biomarkers such as cortisol or heart rate variability can offer early warnings for mental health conditions like anxiety or depression, enabling timely interventions such as therapy or lifestyle changes (Ding et al. 2024). Additionally, wearables tracking metabolic health markers, including glucose and insulin sensitivity, can identify individuals at risk for type 2 diabetes long before symptoms appear, supporting early prevention strategies like diet and exercise adjustments (Lee et al. 2023).

11.6.2 Next-Generation Biosensing Technologies

The future of wearable sensors lies in next-generation biosensing technologies, which will expand the range of measurable biomarkers and enhance the accuracy and reliability of devices. Advances in nanotechnology, biophotonics, microfluidics, and flexible materials will enable sensors to monitor complex physiological and biochemical processes in real-time, with improved sensitivity and miniaturization. One promising innovation is nano biosensors, which use nanoscale materials to detect molecular changes. For example, graphene-based sensors offer exceptional sensitivity for detecting biomolecules such as proteins, nucleic acids, or metabolites in non-invasive fluids like sweat, saliva, or interstitial fluid (Goodrum et al. 2024; Zhao et al. 2024). These sensors hold the potential for monitoring health conditions such as cancer, infections, or metabolic disorders. Biophotonics, which explores interactions between light and biological tissues, is another exciting area. Wearable biophotonic sensors can use light-based techniques, such as fluorescence or Raman spectroscopy, to monitor markers like blood oxygen levels, glucose, or lactate (Amjad and Xian 2025). These non-invasive sensors provide continuous monitoring of critical health indicators, with applications ranging from diabetes management to detecting sepsis. Microfluidic devices, which analyze small fluid volumes in microscale channels, are also advancing wearable sensors (Childs et al. 2024; Saha et al. 2024). Integrated into patches or textiles, these devices can analyze bodily fluids like sweat, saliva, or tears in real-time. Sweat-based sensors, for instance, could monitor electrolytes, glucose, or pH, providing personalized feedback for athletes or patients with metabolic disorders. Flexible and stretchable electronics are revolutionizing the design of wearable biosensors by improving comfort and adaptability (Xing et al. 2024; Liu et al. 2024). Made from materials like conductive polymers or flexible silicone, these sensors conform to the body's shape and movement, enhancing usability and accuracy. Applications already include ECG monitoring, glucose sensing, and drug delivery, with future possibilities for continuous hormone or inflammation tracking. Implantable and ingestible sensors represent another frontier, allowing internal monitoring of processes inaccessible from the skin's surface. Implantable sensors could track blood glucose continuously, eliminating the need for finger-stick tests, while ingestible sensors might monitor gastrointestinal health by analyzing pH, microbiota composition, or nutrient absorption.

These technologies could provide unprecedented precision and personalization for managing conditions such as diabetes, gastrointestinal disorders, or metabolic diseases (Chimene et al. 2024; Zheng et al. 2024b).

11.6.3 Impact of Wearables on Global Health and Preventive Medicine

Wearable sensors have the potential to significantly impact global health, particularly in low-resource settings where healthcare access is limited. By enabling continuous, remote health monitoring, these devices can bridge healthcare gaps, providing individuals with real-time health insights and empowering proactive disease prevention. Wearables can facilitate telemedicine and remote patient monitoring (RPM), addressing geographic and infrastructure barriers and mitigating healthcare workforce shortages. For instance, patients with chronic conditions such as hypertension or diabetes can use devices to monitor blood pressure or glucose levels, transmitting data to healthcare providers via mobile networks (Spatz et al. 2024; Wall et al. 2023; Izu et al. 2024). Providers can adjust treatments remotely, reducing the need for in-person visits and improving outcomes. Wearables play a critical role in managing infectious diseases, particularly in regions prone to outbreaks of malaria, tuberculosis, or HIV. Sensors monitoring vital signs like body temperature, heart rate, and respiratory rate can provide early warnings of infections, enabling timely interventions and reducing disease spread. For example, wearable devices can detect early symptoms of malaria or dengue fever, such as fever or abnormal respiratory patterns, and monitor treatment effectiveness, minimizing drug resistance risks (Xue et al. 2024; Haghayegh et al. 2024; Wu et al. 2024). During pandemics like COVID-19, wearable sensors have been invaluable for monitoring symptoms, detecting early signs of infection, and tracking disease progression (Nkengue et al. 2024; Botonis et al. 2024). Devices with pulse oximeters, for example, monitored blood oxygen levels in patients, helping prioritize care for high-risk individuals (Alboksmaty et al. 2022). Future wearables equipped with advanced sensors could detect viral or bacterial biomarkers, enabling earlier detection and containment of infectious diseases (Haghayegh et al. 2024; Quer et al. 2024). Wearable sensors can promote preventive healthcare by providing real-time data and personalized recommendations. Devices monitoring stress biomarkers, such as cortisol or heart rate variability, can help manage chronic stress, reducing risks for cardiovascular disease, metabolic disorders, and mental health conditions (Ding et al. 2024). Similarly, wearables tracking physical activity, sleep, and diet can offer tailored guidance to improve lifestyle habits and prevent chronic diseases like obesity, diabetes, or hypertension (Sempionatto et al. 2021; Del-Valle-Soto et al. 2024). Wearable devices can also contribute to public health by collecting data from large populations. This information can reveal trends in disease spread, environmental health impacts, or chronic disease prevalence. Governments and public health organizations can use these insights to design targeted interventions, allocate resources effectively, and reduce health disparities in underserved communities.

11.6.4 Integration of Wearable Sensors with AI and Big Data

The integration of wearable sensors with artificial intelligence (AI) and big data analytics is key to maximizing their potential in precision healthcare. Wearables generate continuous, real-time data that can be difficult to process using traditional methods. AI, particularly through machine learning (ML) models, excels at analyzing complex datasets, identifying patterns, and providing actionable insights. AI-powered wearables are already transforming healthcare. For instance, AI algorithms can analyze ECG data in real time to detect arrhythmias like atrial fibrillation, alerting users or healthcare providers for timely intervention and reducing stroke risk (Fatima et al. 2024). Similarly, AI can process sleep data to identify conditions such as sleep apnea or insomnia and offer personalized recommendations to improve sleep quality (Haghayegh et al. 2024; Fatima et al. 2024). In the future, AI-driven wearables are expected to provide even more accurate and nuanced health predictions. By analyzing physiological, biochemical, and behavioral data, AI-enabled devices could predict risks for chronic conditions such as type 2 diabetes or cardiovascular disease before symptoms appear (Haghayegh et al. 2024; Fatima et al. 2024; Marvasti et al. 2024). These insights could support personalized preventive measures, including tailored exercise or dietary plans, helping to reduce the burden of chronic diseases on healthcare systems. AI can also enhance real-time feedback and personalized recommendations (Campanella et al. 2024). For example, wearables monitoring glucose levels, physical activity, and sleep patterns could use AI to offer customized advice on insulin dosing, exercise, or meal timing, empowering individuals to take proactive control of their health and make informed decisions. The combination of wearables with big data analytics presents significant opportunities for public health. Aggregated data from millions of devices can reveal population health trends, track disease outbreaks, or assess the impact of environmental factors on health (Corponi et al. 2024; Roos and Slavich 2023; Alruwaili et al. 2024). For example, wearable data could monitor the spread of infectious diseases, evaluate the effects of air pollution on respiratory health, or detect early signs of mental health crises in specific populations. These insights could guide public health policies, improve resource allocation, and enable targeted interventions for at-risk groups.

References

Abbas S, Avelino Sampedro G, Krichen M, Alamro MA, Mihoub A, Kulhanek R (2024) Effective hypertension detection using predictive feature engineering and deep learning. IEEE Access 12:89055–89068. https://doi.org/10.1109/ACCESS.2024.3418553

Abuchi-Ogbonda S, Ahmad M, Davies AH (2024) 1103 Gait analysis using wearables and AI in predicting peripheral arterial disease: a systematic review. Br J Surg 111(znae163):198. https://doi.org/10.1093/bjs/znae163.198

Adashi EY, O'Mahony DP, Cohen IG (2024) Remote patient monitoring: a leading anchor of the "Hospital-at-Home" paradigm. Am J Med 137:81–82. https://doi.org/10.1016/J.AMJMED.2023.10.018

Agrawal D, El Abbadi A, Wang S (2012) Secure and privacy-preserving data services in the cloud. Proc VLDB Endowment 5:2028–2029. https://doi.org/10.14778/2367502.2367569
Ahluwalia N, Saengkrajang G, Maclean E, Abbass H, Hussain A, Butcher C, Schilling RJ (2024) Validation of diagnostic wearable ECG data from high-risk patients and integration into the atrial fibrillation catheter ablation pathway. Eur Heart J 45(ehae666):3421. https://doi.org/10.1093/eurheartj/ehae666.3421
Ahmed A, Ramesh J, Ganguly S, Aburukba R, Sagahyroon A, Aloul F (2023) Evaluating multimodal wearable sensors for quantifying affective states and depression with neural networks. IEEE Sensors J 23:22788–22802. https://doi.org/10.1109/JSEN.2023.3303436
Alboksmaty A, Beaney T, Elkin S, Clarke JM, Darzi A, Aylin P, Neves AL (2022) Effectiveness and safety of pulse oximetry in remote patient monitoring of patients with COVID-19: a systematic review. Lancet Digit Health 4:e279–e289. https://doi.org/10.1016/S2589-7500(21)00276-4
Alruwaili O, Yousef A, Armghan A (2024) Monitoring the transmission of data from wearable sensors using probabilistic transfer learning. IEEE Access 12:97460–97475. https://doi.org/10.1109/ACCESS.2024.3428444
Amjad A, Xian X (2025) Optical sensors for transdermal biomarker detection: a review. Biosens Bioelectron 267:116844. https://doi.org/10.1016/J.BIOS.2024.116844
Arslan MM, Yang X, Zhang Z, Rahman SU, Ullah M, Abbasi QH (2024) Advancing healthcare monitoring: integrating machine learning with innovative wearable and wireless systems for comprehensive patient care. IEEE Sensors J 24:29199–29210. https://doi.org/10.1109/JSEN.2024.3434409
Ates HC, Nguyen PQ, Gonzalez-Macia L, Morales-Narváez E, Güder F, Collins JJ, Dincer C (2022) End-to-end design of wearable sensors. Nat Rev Mater 7:887–907. https://doi.org/10.1038/s41578-022-00460-x
Awad S, Aljuburi L, Lumsden RS, Mpandzou M, Marinus R (2023) Connected health in US, EU, and China: opportunities to accelerate regulation of connected health technologies to optimize their role in medicines development. Front Med (Lausanne) 10:1248912. https://doi.org/10.3389/fmed.2023.1248912
Bello H, Lukowicz P, Zhou B, Suh S (2024) Unimodal and multimodal sensor fusion for wearable activity recognition. In: 2024 IEEE international conference on pervasive computing and communications workshops and other affiliated events (PerCom workshops), pp 364–365. https://doi.org/10.1109/PerComWorkshops59983.2024.10502797
Botonis OK, Mendley J, Aalla S, Veit NC, Fanton M, Lee JY, Tripathi V, Pandi V, Khobragade A, Chaudhary S, Chaudhuri A, Narayanan V, Xu S, Jeong H, Rogers JA, Jayaraman A (2024) Feasibility of snapshot testing using wearable sensors to detect cardiorespiratory illness (COVID infection in India). npj Digit Med 7:1–12. https://doi.org/10.1038/s41746-024-01287-2
Bretthauer M, Gerke S, Hassan C, Ahmad OF, Mori Y (2023) The new European Medical Device Regulation: balancing innovation and patient safety. Ann Intern Med 176:844–848. https://doi.org/10.7326/M23-0454
Campanella S, Paragliola G, Cherubini V, Pierleoni P, Palma L (2024) Towards personalized AI-based diabetes therapy: a review. IEEE J Biomed Health Inform. https://doi.org/10.1109/JBHI.2024.3443137
Campos-Arias L, Peřinka N, Costa P, Vilas-Vilela JL, Lanceros-Méndez S (2024) Stretchable conductive inks with carbon-based fillers for conformable printed electronics. Adv Eng Mater 26:2400354. https://doi.org/10.1002/ADEM.202400354
Celik Y, Godfrey A (2023) Bringing it all together: wearable data fusion. npj Digit Med 6:1–3. https://doi.org/10.1038/s41746-023-00897-6
Cen S, Zhang H, Li Y, Gu Z, Yuan Y, Ruan Z, Cai Y, Chhetri JK, Liu S, Mao W, Chan P (2024) Gait analysis with wearable sensors in isolated REM sleep behavior disorder associated with Phenoconversion: An explorative study. J Parkinsons Dis 14:1027–1037. https://doi.org/10.3233/JPD-230397
Chiao JC, Bing S, Chawang K, Crowe B (2024) Thirsty for a noninvasive wearable to detect dehydration: a review. IEEE Antennas Propag Mag 66:66–76. https://doi.org/10.1109/MAP.2024.3442823

Childs A, Mayol B, Lasalde-Ramírez JA, Song Y, Sempionatto JR, Gao W (2024) Diving into sweat: advances, challenges, and future directions in wearable sweat sensing. ACS Nano 18:24605–24616. https://doi.org/10.1021/acsnano.4c10344

Chimene D, Queener KMK, Ko BS, McShane M, Daniele M (2024) Insertable biosensors: combining implanted sensing materials with wearable monitors. Annu Rev Biomed Eng 26:197–221. https://doi.org/10.1146/annurev-bioeng-110222-101045

Chow R, Drkulec H, Im JHB, Tsai J, Nafees A, Kumar S, Hou T, Fazelzad R, Leighl NB, Krzyzanowska M, Wong P, Raman S (2024) The use of wearable devices in oncology patients: a systematic review. Oncologist 29:e419–e430. https://doi.org/10.1093/ONCOLO/OYAD305

Contoli C, Freschi V, Lattanzi E (2024) Energy-aware human activity recognition for wearable devices: a comprehensive review. Pervasive Mob Comput 104:101976. https://doi.org/10.1016/J.PMCJ.2024.101976

Corponi F, Li BM, Anmella G, Valenzuela-Pascual C, Mas A, Pacchiarotti I, Valentí M, Grande I, Benabarre A, Garriga M, Vieta E, Young AH, Lawrie SM, Whalley HC, Hidalgo-Mazzei D, Vergari A (2024) Wearable data from subjects playing Super Mario, taking university exams, or performing physical exercise help detect acute mood disorder episodes via self-supervised learning: prospective, exploratory, observational study. JMIR Mhealth Uhealth 12:e55094. https://doi.org/10.2196/55094

Cruz S, Lu C, Ulloa M, Redding A, Hester J, Jacobs M (2024) Perceptions of wearable health tools post the COVID-19 emergency in low-income Latin Communities: qualitative study. JMIR Mhealth Uhealth 12:e50826. https://doi.org/10.2196/50826

De Schrijver L, Saeidi A, Sheibani S, Locca D, Ionescu A (2024) Scalable enzymatic lactate sensor for continuous monitoring in interstitial fluid. Eur Heart J 45(ehae666):3742. https://doi.org/10.1093/eurheartj/ehae666.3742

Del-Valle-Soto C, Briseño RA, Valdivia LJ, Nolazco-Flores JA (2024) Unveiling wearables: exploring the global landscape of biometric applications and vital signs and behavioral impact. BioData Min 17:1–25. https://doi.org/10.1186/S13040-024-00368-Y/TABLES/9

Di Molfetta S, Rossi A, Boscari F, Irace C, Laviola L, Bruttomesso D (2024) Criteria for personalised choice of a continuous glucose monitoring system: An expert opinion. Diabetes Ther 15:2263–2278. https://doi.org/10.1007/s13300-024-01666-8

Dias SB, Jelinek HF, Hadjileontiadis LJ (2024) Wearable neurofeedback acceptance model for students' stress and anxiety management in academic settings. PLoS One 19:e0304932. https://doi.org/10.1371/JOURNAL.PONE.0304932

Ding Y, Tan K, Sheng L, Ren H, Su Z, Yang H, Zhang X, Li J, Hu PA (2024) Integrated mental stress smartwatch based on sweat cortisol and HRV sensors. Biosens Bioelectron 265:116691. https://doi.org/10.1016/J.BIOS.2024.116691

Dobson R, Stowell M, Warren J, Tane T, Ni L, Gu Y, Mccool J, Whittaker R (2023) Use of consumer wearables in health research: issues and considerations. J Med Internet Res 25:e52444. https://doi.org/10.2196/52444

Duan H, Zhang Y, Zhang Y, Zhu P, Mao Y (2024) Recent advances of stretchable nanomaterial-based hydrogels for wearable sensors and electrophysiological signals monitoring. Nano 14:1398. https://doi.org/10.3390/nano14171398

Duro N (2024) Sensor data fusion analysis for broad applications. Sensors 24:3725. https://doi.org/10.3390/S24123725

Engel E, Hill AC, Granchelli A, Lockhart GC, Lyng G, Kaye L, Lyson H, Bowler RP (2023) Wearable sensors for early detection of COPD exacerbations. Chest 164:A5034. https://doi.org/10.1016/j.chest.2023.07.3260

Eversdijk M, Habibović M, Willems DL, Kop WJ, Ploem MC, Dekker LRC, Tan HL, Vullings R, Bak MAR (2024) Ethics of wearable-based out-of-hospital cardiac arrest detection. Circ Arrhythm Electrophysiol 17:e012913. https://doi.org/10.1161/CIRCEP.124.012913

Fang X, Li H, Ma J, Xing X, Fu Z, Antwi-Afari MF, Umer W (2024) Assessment of construction workers' spontaneous mental fatigue based on non-invasive and multimodal in-ear EEG sensors. Buildings 14:2793. https://doi.org/10.3390/buildings14092793

Fatima M, Sahebkar A, Kesharwani P (2024) From data to diagnosis: how AI-enabled wearable sensors are leading the charge. Microchem J 205:111397. https://doi.org/10.1016/J.MICROC.2024.111397

Fioravanti VBO, Freitas PG, Rodrigues PG, de Lima RG, Lucafo GD, Cabello FC, Seidel I, Penatti OAB (2024) Machine learning framework for Inter-Beat Interval estimation using wearable Photoplethysmography sensors. Biomed Signal Proc Control 88:105689. https://doi.org/10.1016/J.BSPC.2023.105689

Frank J, Son D, Szafranski K, Poon Y (2024) Continuous glucose monitoring for self-management of diabetes in people living with type 2 diabetes mellitus on basal insulin therapy: a microsimulation model and cost-effectiveness analysis from a US perspective with relevance to Medicaid. J Manag Care Spec Pharm 30:917–928. https://doi.org/10.18553/jmcp.2024.24025

Freitas A, Almeida R, Gonçalves H, Conceição G, Freitas A (2024) Monitoring fatigue and drowsiness in motor vehicle occupants using electrocardiogram and heart rate – a systematic review. Transp Res Part F Traffic Psychol Behav 103:586–607. https://doi.org/10.1016/J.TRF.2024.05.008

Frerichs M, Li H, Andersson A, Andelid K, Crona M, Vanfleteren LEGW (2024) Remote monitoring of patients with chronic obstructive pulmonary disease using a tablet system: a randomised cross over study of quality of life measurements. ERJ Open Res 00532–02024. https://doi.org/10.1183/23120541.00532-2024

Gadotti A, Rocher L, Houssiau F, Crețu AM, de Montjoye YA (2024) Anonymization: the imperfect science of using data while preserving privacy. Sci Adv 10:eadn7053. https://doi.org/10.1126/SCIADV.ADN7053

Gao N, Xu G, Chang G, Wu Y (2024a) From lab to life: self-powered sweat sensors and their future in personal health monitoring. Adv Sci 2409178. https://doi.org/10.1002/ADVS.202409178

Gao Z, Zhou Y, Zhang J, Foroughi J, Peng S, Baughman RH, Wang ZL, Wang CH (2024b) Advanced energy harvesters and energy storage for powering wearable and implantable medical devices. Adv Mater 36:2404492. https://doi.org/10.1002/ADMA.202404492

Garcha P, Schaffer V, Haroun B, Ramaswamy S, Wieser J, Lang J, Chandraksan A (2022) A duty-cycled integrated-fluxgate magnetometer for current sensing. IEEE J Solid State Circuits 57:2741–2751. https://doi.org/10.1109/JSSC.2022.3156572

Gelinas L, Morrell W, Bierer BE (2023) Terms and conditions apply: an ethical analysis of mobile health user agreements in research. J Law Biosci 10:lsad021. https://doi.org/10.1093/jlb/lsad021

Gilbert TJ, Lin Z, Day S, de Hamilton AFC, Ward JA (2024) A magnetometer-based method for in-situ syncing of wearable inertial measurement units. Front Comput Sci 6:1385392. https://doi.org/10.3389/fcomp.2024.1385392

Gómez-Peralta F, Leiva-Gea I, Duque N, Artime E, Rubio de Santos M (2024) Impact of continuous glucose monitoring and its glucometrics in clinical practice in Spain and future perspectives: a narrative review. Adv Ther 41:3471–3488. https://doi.org/10.1007/s12325-024-02943-5

Goodrum R, Weldekidan H, Li H, Mohanty AK, Misra M (2024) Graphene-based nanostructures from green processes and their applications in biomedical sensors. Adv Ind Eng Polym Res 7:37–53. https://doi.org/10.1016/J.AIEPR.2023.03.001

Haghayegh F, Norouziazad A, Haghani E, Feygin AA, Rahimi RH, Ghavamabadi HA, Sadighbayan D, Madhoun F, Papagelis M, Felfeli T, Salahandish R (2024) Revolutionary point-of-care wearable diagnostics for early disease detection and biomarker discovery through intelligent technologies. Adv Sci 11:2400595. https://doi.org/10.1002/advs.202400595

Hanna J, Tawk Y, Azar S, Ramadan AH, Dia B, Shamieh E, Zoghbi S, Kanj R, Costantine J, Eid AA (2022) Wearable flexible body matched electromagnetic sensors for personalized non-invasive glucose monitoring. Sci Rep 12:1–12. https://doi.org/10.1038/s41598-022-19251-z

Hannan A, Cheema SM, Pires IM (2024) Machine learning-based smart wearable system for cardiac arrest monitoring using hybrid computing. Biomed Signal Proc Control 87:105519. https://doi.org/10.1016/J.BSPC.2023.105519

Hao Z, Wang J, Zhang G, Gao L, Zhang X, Liu J, Zhang X, Yang X, Lai Z (2024) PPG heart rate extraction algorithm based on the motion artifact intensity classification and removal framework. Biomed Signal Proc Control 94:106287. https://doi.org/10.1016/J.BSPC.2024.106287

Hassanpoor S, Uddin MA, Strange SA, Shearod K, Kim T (2024) Wearable tactile sensors customized with hierarchical microdomes via direct ink writing. Adv Eng Mater 2401390. https://doi.org/10.1002/ADEM.202401390

Hermanns N, Ehrmann D, Shapira A, Kulzer B, Schmitt A, Laffel L (2022) Coordination of glucose monitoring, self-care behaviour and mental health: achieving precision monitoring in diabetes. Diabetologia 65:1883–1894. https://doi.org/10.1007/s00125-022-05685-7

Howie EE, Harari R, Dias RD, Wigmore SJ, Skipworth RJE, Yule S (2024) Feasibility of wearable sensors to assess cognitive load during clinical performance: lessons learned and blueprint for success. J Surg Res 302:222–231. https://doi.org/10.1016/j.jss.2024.07.009

Huang Z, Xu Y, Cheng Y, Xue M, Deng M, Jaffrezic-Renault N, Guo Z (2022a) Recent advances in skin-like wearable sensors: sensor design, health monitoring, and intelligent auxiliary. Sens Diagn 1:686–708. https://doi.org/10.1039/D2SD00037G

Huang J, Kaewunruen S, Ning J (2022b) AI-based quantification of fitness activities using smartphones. Sustainability 14:690. https://doi.org/10.3390/su14020690

Huang H, Chen J, Xiao J, Chen D, Zhang J, Pan J, Li Y (2024) Real-time attention regulation and cognitive monitoring using a wearable EEG-based BCI. IEEE Trans Biomed Eng. https://doi.org/10.1109/TBME.2024.3468351

Hughes A, Shandhi MMH, Master H, Dunn J, Brittain E (2023) Wearable devices in cardiovascular medicine. Circ Res 132:652–670. https://doi.org/10.1161/CIRCRESAHA.122.322389

Izu L, Scholtz B, Fashoro I (2024) Wearables and their potential to transform health management: a step towards sustainable development goal 3. Sustainability 16:1850. https://doi.org/10.3390/SU16051850

Jia Y, Zhang S, Li J, Han Z, Zhang D, Qu X, Wu Z, Wang H, Chen S (2024) Wearable device with high thermoelectric performance and long-lasting usability based on gel-thermocells for body heat harvesting. Small 2401427. https://doi.org/10.1002/SMLL.202401427

Jian W, Tabassum A, Li JP (2024) Systematic survey on data security in wireless body area networks in IoT healthcare system. Front Med (Lausanne) 11:1422911. https://doi.org/10.3389/FMED.2024.1422911/BIBTEX

Jurczak KM, van der Boon TAB, Devia-Rodriguez R, Schuurmann RCL, Sjollema J, van Huizen L, De Vries JPPM, van Rijn P (2024) Recent regulatory developments in EU Medical Device Regulation and their impact on biomaterials translation. Bioeng Transl Med:e10721. https://doi.org/10.1002/BTM2.10721

Kang HS, Exworthy M (2022) Wearing the future-wearables to empower users to take greater responsibility for their health and care: scoping review. JMIR Mhealth Uhealth 10:e35684. https://doi.org/10.2196/35684

Karimzadehkhouei M, Ali B, Jedari Ghourichaei M, Alaca BE (2023) Silicon nanowires driving miniaturization of microelectromechanical systems physical sensors: a review. Adv Eng Mater 25:2300007. https://doi.org/10.1002/ADEM.202300007

Khan MA, Attique S, Ali N, Shehzad K, Gong N, Zhou N, Chen X, Li Z, Gao Y, Yan M, Qiu J, Ma Z, Xu B (2024) Development of a highly sensitive and stretchable charge-transfer fiber strain sensor for wearable applications. ACS Appl Mater Interfaces 54520–54528. https://doi.org/10.1021/acsami.4c07698

Kim KR, Kang TW, Kim H, Lee YJ, Lee SH, Yi H, Kim HS, Kim H, Min J, Ready J, Millard-Stafford M, Yeo WH (2024a) All-in-one, wireless, multi-sensor integrated athlete health monitor for real-time continuous detection of dehydration and physiological stress. Adv Sci 11:2403238. https://doi.org/10.1002/ADVS.202403238

Kim N, Lee S, Kim J, Choi SY, Park SM (2024b) Shuffled ECA-Net for stress detection from multimodal wearable sensor data. Comput Biol Med 183:109217. https://doi.org/10.1016/J.COMPBIOMED.2024.109217

Kulkarni MB, Rajagopal S, Prieto-Simón B, Pogue BW (2024) Recent advances in smart wearable sensors for continuous human health monitoring. Talanta 272:125817. https://doi.org/10.1016/j.talanta.2024.125817

Kundu A, Arief I, Mandal S, Meena KK, Krause B, Staudinger U, Mondal T, Wießner S, Das A (2024) Elastomeric sensor-triboelectric nanogenerator coupled system for multimodal strain sensing and organic vapor detection. ACS Appl Mater Interfaces 16:53083–53097. https://doi.org/10.1021/acsami.4c14011

Lee I, Wakako T, Ikebukuro K, Sode K (2022) In vitro continuous 3 months operation of direct electron transfer type open circuit potential based glucose sensor: heralding the next CGM sensor. J Diabetes Sci Technol 16:1107–1113. https://doi.org/10.1177/19322968221092449

Lee MA, Song MK, Bessette H, Roberts Davis M, Tyner TE, Reid A (2023) Use of wearables for monitoring cardiometabolic health: a systematic review. Int J Med Inform 179:105218. https://doi.org/10.1016/J.IJMEDINF.2023.105218

Lee S, Baek CM, Kim GH, Pattipaka S, Song H, Jang J, Hwang GT, Ryu J (2024) Driving Wi-Fi IoT sensors by a hybrid magneto-mechano-electric energy generator extracting a power of over 50 mW. Adv Sci 11:2405526. https://doi.org/10.1002/ADVS.202405526

Li Z, Zhang X (2024) An enhanced autoencoder-based anomaly detection model for time series data from wearable medical devices. IEEE J Biomed Health Inform. https://doi.org/10.1109/JBHI.2024.3434420

Lielpetere A, Jayakumar K, Leech D, Schuhmann W (2023) Cross-linkable polymer-based multilayers for protecting electrochemical glucose biosensors against uric acid, ascorbic acid, and biofouling interferences. ACS Sens 8:1756–1765. https://doi.org/10.1021/acssensors.3c00050

Linh VTN, Han S, Koh E, Kim S, Jung HS, Koo J (2025) Advances in wearable electronics for monitoring human organs: bridging external and internal health assessments. Biomaterials 314:122865. https://doi.org/10.1016/J.BIOMATERIALS.2024.122865

Liu Y, Pharr M, Salvatore GA (2017) Lab-on-skin: a review of flexible and stretchable electronics for wearable health monitoring. ACS Nano 11:9614–9635. https://doi.org/10.1021/acsnano.7b04898

Liu L, Dou Y, Wang J, Zhao Y, Kong W, Ma C, He D, Wang H, Zhang H, Chang A, Zhao P (2024) Recent advances in flexible temperature sensors: materials, mechanism, fabrication, and applications. Adv Sci 11:2405003. https://doi.org/10.1002/ADVS.202405003

Lu HY, Feng XM, Zhang J (2024) Early detection of cardiorespiratory complications and training monitoring using wearable ECG sensors and CNN. BMC Med Inform Decis Mak 24:1–21. https://doi.org/10.1186/s12911-024-02599-9

Mallin M, Hall J, Herlihy M, Gelman EJ, Stone MB (2023) A pilot retrospective study of a physician-directed and genomics-based model for precision lifestyle medicine. Front Med (Lausanne) 10:1239737. https://doi.org/10.3389/FMED.2023.1239737/BIBTEX

Mansour M, Saeed Darweesh M, Soltan A (2024) Wearable devices for glucose monitoring: a review of state-of-the-art technologies and emerging trends. Alex Eng J 89:224–243. https://doi.org/10.1016/J.AEJ.2024.01.021

Marvasti TB, Gao Y, Murray KR, Hershman S, McIntosh C, Moayedi Y (2024) Unlocking tomorrow's health care: expanding the clinical scope of wearables by applying artificial intelligence. Can J Cardiol 40:1934–1945. https://doi.org/10.1016/J.CJCA.2024.07.009

Mattison G, Canfell O, Forrester D, Dobbins C, Smith D, Töyräs J, Sullivan C (2022) The influence of wearables on health care outcomes in chronic disease: systematic review. J Med Internet Res 24:e36690. https://doi.org/10.2196/36690

McElroy E, Wood T, Bond R, Mulvenna M, Shevlin M, Ploubidis GB, Hoffmann MS, Moltrecht B (2024) Using natural language processing to facilitate the harmonisation of mental health questionnaires: a validation study using real-world data. BMC Psychiatry 24:1–9. https://doi.org/10.1186/s12888-024-05954-2

McGrath MJ, Scanaill CN (2013) Regulations and standards: considerations for sensor technologies. Sens Technol:115–135. https://doi.org/10.1007/978-1-4302-6014-1_6

Moudgil A, Hou K, Li T, Leong WL (2023) Biocompatible solid-state ion-sensitive organic electrochemical transistor for physiological multi-ions sensing. Adv Mater Technol 8:2300605. https://doi.org/10.1002/ADMT.202300605

Mughal MA, Ullah A, Yu X, He W, Jhanjhi NZ, Ray SK (2024) A secure and privacy preserved data aggregation scheme in IoMT. Heliyon 10:e27177. https://doi.org/10.1016/J.HELIYON.2024.E27177

Nemat H, Khadem H, Elliott J, Benaissa M (2024) Physical activity integration in blood glucose level prediction: different levels of data fusion. IEEE J Biomed Health Inform. https://doi.org/10.1109/JBHI.2024.3483999

Newaz NT, Hanada E (2023) The methods of fall detection: a literature review. Sensors 23:5212. https://doi.org/10.3390/s23115212

Nkengue MJ, Zeng X, Koehl L, Tao X, Dassonville F, Dumont N, Ye-Lehmann S, Akwa Y, Ye H (2024) An intelligent garment for long COVID-19 real-time monitoring. Comput Biol Med 181:109067. https://doi.org/10.1016/J.COMPBIOMED.2024.109067

Oliveira TRA, Fernandes AT, do NSF, Santino TA, da Silva Menescal FEP, de Miranda Silva Nogueira PA (2024) Effects of using wearable devices to monitoring physical activity in pulmonary rehabilitation programs for chronic respiratory diseases: a systematic review protocol. PLoS One 19:e0308109. https://doi.org/10.1371/journal.pone.0308109

Olsen RJ, Hasan SS, Woo JJ, Nawabi DH, Ramkumar PN (2024) The fundamentals and applications of wearable sensor devices in sports medicine: a scoping review. Arthrosc - J Arthrosc Relat Surg. https://doi.org/10.1016/j.arthro.2024.01.042

Pankaj KA, Kumar M, Komaragiri R (2024) Edge-based computation of super-resolution Superlet spectrograms for real-time estimation of heart rate using an IoMT-based reference-signal-less PPG sensor. IEEE Internet Things J 11:8647–8657. https://doi.org/10.1109/JIOT.2023.3322947

Papafotis K, Nikitas D, Sotiriadis PP (2022) Improving gyroscope's noise performance using multiple accelerometers in a closed-loop configuration. IEEE Sensors J 22:22062–22068. https://doi.org/10.1109/JSEN.2022.3207110

Pauly T, Lüscher J, Wilhelm LO, Amrein MA, Boateng G, Kowatsch T, Fleisch E, Bodenmann G, Scholz U (2024) Using wearables to study biopsychosocial dynamics in couples who cope with a chronic health condition: Ambulatory Assessment Study. JMIR Mhealth Uhealth 12:e49576. https://doi.org/10.2196/49576

Peng Y, Sun F, Jing J, Zhao J, Zhang N, Zhang P, Pan R (2024) Switchable pseudo-triaxial structure enabled Mechanosensory textiles with ultra-wide detection range for flexible E-wearables. Adv Funct Mater 2411177. https://doi.org/10.1002/ADFM.202411177

Pina Astete L, Buelga Suarez ML, Tur Sainz A, Munoz Gomez MI, Rodriguez Sanchez P, Penaranda Romero N, Garcia Montalvo J, Rebolleda Sanchez A, Alonso Salinas GL (2024) Smartwatch a new tool in the assesment of ischaemic heart disease. Eur J Cardiovasc Nurs 23(zvae098):128. https://doi.org/10.1093/eurjcn/zvae098.128

Poh MZ, Battisti AJ, Cheng LF, Lin J, Patwardhan A, Venkataraman GS, Athill CA, Patel NS, Patel CP, Machado CE, Ellis JT, Crosson LA, Tamura Y, Plowman RS, Turakhia MP, Ghanbari H (2023) Validation of a deep learning algorithm for continuous, real-time detection of atrial fibrillation using a Wrist-Worn Device in an ambulatory environment. J Am Heart Assoc 12:e030543. https://doi.org/10.1161/JAHA.123.03054

Pradhan GB, Jeong S, Sharma S, Lim SJ, Shrestha K, Lee YY, Park JY (2024) A breathable and strain-insensitive multi-layered E-skin patch for digital healthcare wearables. Adv Funct Mater 34:2407978. https://doi.org/10.1002/adfm.202407978

Quer G, Coughlin E, Villacian J, Delgado F, Harris K, Verrant J, Gadaleta M, Hung TY, Ter Meer J, Radin JM, Ramos E, Adams M, Kim L, Chien JW, Baca-Motes K, Pandit JA, Talantov D, Steinhubl SR (2024) Feasibility of wearable sensor signals and self-reported symptoms to prompt at-home testing for acute respiratory viruses in the USA (DETECT-AHEAD): a decentralised, randomised controlled trial. Lancet Digit Health 6:e546–e554. https://doi.org/10.1016/S2589-7500(24)00096-7

Rehman MU, Shafqiue A (2024) Robust encryption framework for IoT devices based on bit-plane extraction, chaotic sine models, and quantum operations. IoT 27:101241. https://doi.org/10.1016/J.IOT.2024.101241

Rehman RZU, Chatterjee M, Manyakov NV, Daans M, Jackson A, O'Brisky A, Telesky T, Smets S, Berghmans PJ, Yang D, Reynoso E, Lucas MV, Huo Y, Thirugnanam VT, Mansi T, Morris M (2024) Assessment of physiological signals from photoplethysmography sensors compared to an electrocardiogram sensor: a validation study in daily life. Sensors 24:6826. https://doi.org/10.3390/S24216826

Roos LG, Slavich GM (2023) Wearable technologies for health research: opportunities, limitations, and practical and conceptual considerations. Brain Behav Immun 113:444–452. https://doi.org/10.1016/J.BBI.2023.08.008

Saha T, Khan MI, Sandhu SS, Yin L, Earney S, Zhang C, Djassemi O, Wang Z, Han J, Abdal A, Srivatsa S, Ding S, Wang J (2024) A passive perspiration inspired wearable platform for continuous glucose monitoring. Adv Sci 11:2405518. https://doi.org/10.1002/ADVS.202405518

Scataglini S, Kandappu T, Ghose A, Sinha A (2023) Editorial: wearable sensors role in promoting health and wellness via reliable and longitudinal monitoring. Front Physiol 14:1201847. https://doi.org/10.3389/fphys.2023.1201847

Sel K, Osman D, Huerta N, Edgar A, Pettigrew RI, Jafari R (2023) Continuous cuffless blood pressure monitoring with a wearable ring bioimpedance device. npj Digit Med 6:1–11. https://doi.org/10.1038/s41746-023-00796-w

Sempionatto JR, Montiel VRV, Vargas E, Teymourian H, Wang J (2021) Wearable and Mobile sensors for personalized nutrition. ACS Sens 6:1745–1760. https://doi.org/10.1021/acssensors.1c00553

Shah AJ, Althobiani MA, Saigal A, Ogbonnaya CE, Hurst JR, Mandal S (2023) Wearable technology interventions in patients with chronic obstructive pulmonary disease: a systematic review and meta-analysis. npj Digit Med 6:1–17. https://doi.org/10.1038/s41746-023-00962-0

Sharma H, Kalita D, Naskar U, Mishra BK, Kumar P, Mirza KB (2023) Prediction of glucose sensor sensitivity in the presence of biofouling using machine learning and electrochemical impedance spectroscopy. IEEE Sensors J 23:18785–18797. https://doi.org/10.1109/JSEN.2023.3289619

Shaw JLV, Bannuru RR, Beach L, ElSayed NA, Freckmann G, Füzéry AK, Fung AWS, Gilbert J, Huang Y, Korpi-Steiner N, Logan S, Longo R, MacKay D, Maks L, Pleus S, Rogers K, Seley JJ, Taxin Z, Thompson-Hutchison F, Tolan NV, Tran NK, Umpierrez GE, Venner AA (2024) Consensus considerations and good practice points for use of continuous glucose monitoring systems in hospital settings. Diabetes Care 47:2062–2075. https://doi.org/10.2337/dci24-0073

Shen S, Zhou Q, Chen G, Fang Y, Kurilova O, Liu Z, Li S, Chen J (2024) Advances in wearable respiration sensors. Mater Today 72:140–162. https://doi.org/10.1016/J.MATTOD.2023.12.003

Shi H, Tang R, Wang Q, Song T (2024) Performance evaluation of a bioinspired geomagnetic sensor and its application for geomagnetic navigation in simulated environment. Sensors 24:6477. https://doi.org/10.3390/S24196477

Singar S, Nagpal R, Arjmandi BH, Akhavan NS (2024) Personalized nutrition: tailoring dietary recommendations through genetic insights. Nutrients 16:2673. https://doi.org/10.3390/NU16162673

Sousa P, Ferreira P, Ventura F, Dixe M, Martinho R, Dias S, Morais J, Goncalves L (2023) Using m-health and wearable technologies to promote self-management and treatment adherence among cardiovascular disease patients: a clinical trial protocol. Eur J Cardiovasc Nurs 22(zvad064):015. https://doi.org/10.1093/EURJCN/ZVAD064.015

Spatz ES, Ginsburg GS, Rumsfeld JS, Turakhia MP (2024) Wearable digital health technologies for monitoring in cardiovascular medicine. N Engl J Med 390:346–356. https://doi.org/10.1056/NEJMra2301903

Suman S, Choudhary VK, Pacharaney U, Kumar A, Kumar D, Raushan R (2024) A review on implementation of wearable devices in healthcare. In: 7th international conference on inventive computation technologies, ICICT 2024, pp 1660–1665. https://doi.org/10.1109/ICICT60155.2024.10544749

Sun X, Guo X, Gao J, Wu J, Huang F, Zhang JH, Huang F, Lu X, Shi Y, Pan L (2024a) E-skin and its advanced applications in ubiquitous health monitoring. Biomedicine 12:2307. https://doi.org/10.3390/biomedicines12102307

Sun M, Wang S, Liang Y, Wang C, Zhang Y, Liu H, Zhang Y, Han L (2024b) Flexible graphene field-effect transistors and their application in flexible biomedical sensing. Nano-Micro Lett 17:1–62. https://doi.org/10.1007/s40820-024-01534-x

Sun T, Liu J, Chen CJ (2024c) Calibration algorithms for continuous glucose monitoring systems based on interstitial fluid sensing. Biosens Bioelectron 260:116450. https://doi.org/10.1016/J.BIOS.2024.116450

Swahn MH, Gittner KB, Lyons MJ, Nielsen K, Mobley K, Culbreth R, Palmier J, Johnson NE, Matte M, Nabulya A (2024) Advancing mHealth research in low-resource settings: young women's insights and implementation challenges with wearable smartwatch devices in Uganda. Sensors 24:5591. https://doi.org/10.3390/S24175591

Szeto K, Arnold J, Maher C (2024) The Wearable Activity Tracker Checklist for Healthcare (WATCH): a 12-point guide for the implementation of wearable activity trackers in healthcare. Int J Behav Nutr Phys Act 21:1–10. https://doi.org/10.1186/s12966-024-01567-w

Tan SY, Sumner J, Wang Y, Wenjun Yip A (2024) A systematic review of the impacts of remote patient monitoring (RPM) interventions on safety, adherence, quality-of-life and cost-related outcomes. npj Digit Med 7:1–16. https://doi.org/10.1038/s41746-024-01182-w

Tang C, Yi W, Xu M, Jin Y, Zhang Z, Chen X, Liao C, Smielewski P, Occhipinti LG (2024) A deep learning-enabled smart garment for accurate and versatile sleep conditions monitoring in daily life. arXiv preprint arXiv 2408.00753. https://doi.org/10.48550/arXiv.2408.00753

Tao Q, Liu S, Zhang J, Jiang J, Jin Z, Huang Y, Liu X, Lin S, Zeng X, Li X, Tao G, Chen H (2023) Clinical applications of smart wearable sensors. iScience 26:107485. https://doi.org/10.1016/j.isci.2023.107485

Thomas A, Haak T, Tombek A, Kulzer B, Ehrmann D, Kordonouri O, Kröger J, Schubert-Olesen O, Kolassa R, Siegmund T, Haller N, Heinemann L (2024) How to use continuous glucose monitoring efficiently in diabetes management: opinions and recommendations by German experts on the status and open questions. J Diabetes Sci Technol. https://doi.org/10.1177/19322968241267776

Tomasi J, Lisoway A, Zai C, Zai G, Kennedy JL, Tiwari A (2022) P118. Preliminary study of heart rate variability association with target anxiety genes across Europeans and East Asians. Biol Psychiatry 91:S135. https://doi.org/10.1016/j.biopsych.2022.02.352

Ucar MHB, Adjevi A, Aktaş F, Solak S (2024) Utilizing IoMT-based smart gloves for continuous vital sign monitoring to safeguard athlete health and optimize training protocols. Sensors 24:6500. https://doi.org/10.3390/s24206500

Umar AH, Othman MA, Harun FKC, Yusof Y (2021) Dielectrics for non-contact ECG bioelectrodes: a review. IEEE Sensors J 21:18353–18367. https://doi.org/10.1109/JSEN.2021.3092233

Varma N, Han JK, Passman R, Rosman LA, Ghanbari H, Noseworthy P, Avari Silva JN, Deshmukh A, Sanders P, Hindricks G, Lip G, Sridhar AR (2024) Promises and perils of consumer mobile technologies in cardiovascular care: JACC scientific statement. J Am Coll Cardiol 83:611–631. https://doi.org/10.1016/J.JACC.2023.11.024

Verma K, Preity P, Ranjan R (2023) An insight into wearable devices for smart healthcare technologies. In: Proceedings of the 13th international conference on cloud computing, data science and engineering, confluence 2023, pp 426–431. https://doi.org/10.1109/Confluence56041.2023.10048809

Voet N, Altmann V, Saris C (2023) P106 Gait analysis for support in diagnostics of neuromuscular diseases. Neuromuscul Disord 33:S171. https://doi.org/10.1016/j.nmd.2023.07.413

Wall C, Hetherington V, Godfrey A (2023) Beyond the clinic: the rise of wearables and smartphones in decentralising healthcare. npj Digit Med 6:1–3. https://doi.org/10.1038/s41746-023-00971-z

Walter JR, Xu S, Rogers JA (2024) From lab to life: how wearable devices can improve health equity. Nat Commun 15:1–4. https://doi.org/10.1038/s41467-023-44634-9

Wang Y, Miao F, An Q, Liu Z, Chen C, Li Y (2022) Wearable multimodal vital sign monitoring sensor with fully integrated analog front end. IEEE Sensors J 22:13462–13471. https://doi.org/10.1109/JSEN.2022.3177205

Wang Z, Wang Y, Zhang X, Yang D, Ma D, Ramakrishna S, Yuan W, Ye T (2024a) Flexible photovoltaic micro-power system enabled with a customized MPPT. Appl Energy 367:123425. https://doi.org/10.1016/J.APENERGY.2024.123425

Wang R, Veera SCM, Asan O, Liao T (2024b) A systematic review on the use of consumer-based ECG wearables on cardiac health monitoring. IEEE J Biomed Health Inform. https://doi.org/10.1109/JBHI.2024.3456028

Wilson S, Tolley C, Mc Ardle R, Lawson L, Beswick E, Hassan N, Slight R, Slight S (2024) Recommendations to advance digital health equity: a systematic review of qualitative studies. npj Digit Med 7:1–9. https://doi.org/10.1038/s41746-024-01177-7

Wu J, Lin X, Yang C, Yang S, Liu C, Cao Y (2024) Wearable sensors based on miniaturized high-performance hybrid nanogenerator for medical health monitoring. Biosensors (Basel) 14:361. https://doi.org/10.3390/BIOS14080361/S1

Xiao X, Yin J, Xu J, Tat T, Chen J (2024) Advances in machine learning for wearable sensors. ACS Nano 18:22734–22751. https://doi.org/10.1021/acsnano.4c05851

Xing Y, Wang J, Li J (2024) Design and manufacturing of soft electronics for in situ biochemical sensing. Int J Extrem Manuf 6:062005. https://doi.org/10.1088/2631-7990/AD65A0

Xu D, Xu Z (2024) Machine learning applications in preventive healthcare: a systematic literature review on predictive analytics of disease comorbidity from multiple perspectives. Artif Intell Med 156:102950. https://doi.org/10.1016/J.ARTMED.2024.102950

Xu Z, Zheng E, Han H, Dong X, Dang X, Wang Z (2024) A secure healthcare data sharing scheme based on two-dimensional chaotic mapping and blockchain. Sci Rep 14:1–19. https://doi.org/10.1038/s41598-024-73554-x

Xue Z, Gai YS, Wu Y, Liu Z, Li Z (2024) Wearable mechanical and electrochemical sensors for real-time health monitoring. Commun Mater 5:1–7. https://doi.org/10.1038/s43246-024-00658-2

Yang Y, Gao W (2019) Wearable and flexible electronics for continuous molecular monitoring. Chem Soc Rev 48:1465–1491. https://doi.org/10.1039/C7CS00730B

Yang G, Kang Y, Charlton PH, Kyriacou PA, Kim KK, Li L, Park C (2024a) Energy-efficient PPG-based respiratory rate estimation using spiking neural networks. Sensors 24:3980. https://doi.org/10.3390/s24123980

Yang G, Hong J, Park SB (2024b) Wearable device for continuous sweat lactate monitoring in sports: a narrative review. Front Physiol 15:1376801. https://doi.org/10.3389/fphys.2024.1376801

Yang JW, Khorsandi D, Trabucco L, Ahmed M, Khademhosseini A, Dokmeci MR, Ye JY, Jucaud V (2024c) Liver-on-a-Chip integrated with label-free optical biosensors for rapid and continuous monitoring of drug-induced toxicity. Small 20:2403560. https://doi.org/10.1002/SMLL.202403560

Yu J, Zhang J, Sengoku S (2023) Innovation process and industrial system of us food and drug administration approved software as a medical device: review and content analysis. J Med Internet Res 25:e47505. https://doi.org/10.2196/47505

Yu R, Feng S, Sun Q, Xu H, Jiang Q, Guo J, Dai B, Cui D, Wang K (2024) Ambient energy harvesters in wearable electronics: fundamentals, methodologies, and applications. J Nanobiotechnology 22:1–26. https://doi.org/10.1186/S12951-024-02774-0

Zhang W, Sun F, Lu Z, Fan S, Huang Z, Hao Y, Pan Z, Chen L, Lou Y, Liu J (2023) A wearable medical sensors system for in-situ monitoring vital signs of patients with hemorrhagic shock in big disaster scenes. Sens Actuators B Chem 395:134448. https://doi.org/10.1016/j.snb.2023.134448

Zhang H, Yan Z, Zhang T, Wang J, Wang X, Chen Y, Zhu S, Li Z, Chen Y, Hong W, Zhao Y, Chen S, Hong Q, Xu Y, Guo X (2024a) Bioinspired high-linearity, wide-sensing-range flexible stretchable bioelectronics based on MWCNTs/GR/Nd2Fe14B/PDMS nanocomposites for human-computer interaction and biomechanics detection. ACS Sens 9:3947–3957. acssensors.4c00664

Zhang Y, Zheng XT, Zhang X, Pan J, Thean AVY (2024b) Hybrid integration of wearable devices for physiological monitoring. Chem Rev 124:10386–10434. https://doi.org/10.1021/acs.chemrev.3c00471

Zhang Y, Folarin A, Dobson R (2024c) Exploring digital biomarkers for depression using mobile technology. Int J Popul Data Sci 9. https://doi.org/10.23889/IJPDS.V9I4.2422

Zhang H, Zhang P, Lin F, Chao L, Wang Z, Ma F, Li Q (2024d) Co-learning-assisted progressive dense fusion network for cardiovascular disease detection using ECG and PCG signals. Expert Syst Appl 238:122144. https://doi.org/10.1016/J.ESWA.2023.122144

Zhao L, Liang C, Huang Y, Zhou G, Xiao Y, Ji N, Zhang YT, Zhao N (2023) Emerging sensing and modeling technologies for wearable and cuffless blood pressure monitoring. npj Digit Med 6:1–15. https://doi.org/10.1038/s41746-023-00835-6

Zhao W, Zhang W, Chen J, Li H, Han L, Li X, Wang J, Song W, Xu C, Cai X, Wang L (2024) Sensitivity-enhancing strategies of graphene field-effect transistor biosensors for biomarker detection. ACS Sens 9:2705–2727. https://doi.org/10.1021/acssensors.4c00322

Zheng NS, Annis J, Master H, Han L, Gleichauf K, Ching JH, Nasser M, Coleman P, Desine S, Ruderfer DM, Hernandez J, Schneider LD, Brittain EL (2024a) Sleep patterns and risk of chronic disease as measured by long-term monitoring with commercial wearable devices in the All of Us Research Program. Nat Med 30:2648–2656. https://doi.org/10.1038/s41591-024-03155-8

Zheng Z, Zhu R, Peng I, Xu Z, Jiang Y (2024b) Wearable and implantable biosensors: mechanisms and applications in closed-loop therapeutic systems. J Mater Chem B 12:8577–8604. https://doi.org/10.1039/D4TB00782D

Zhu X, Zhang H, Huang Y, He E, Shen Y, Huang G, Yuan S, Dong X, Zhang Y, Chen R, Zhang X, Wang Y (2024) Recent progress of flexible rechargeable batteries. Sci Bull (Beijing) 69:3730–3755. https://doi.org/10.1016/J.SCIB.2024.09.032

Zinchenko V, Chetverikov S, Akhmad E, Arzamasov K, Vladzymyrskyy A, Andreychenko A, Morozov S (2022) Changes in software as a medical device based on artificial intelligence technologies. Int J Comput Assist Radiol Surg 17:1969–1977. https://doi.org/10.1007/s11548-022-02669-1

Integrated Bioelectronics for Precision Health Monitoring

12

Yuchen Xu, Min Suk Lee, Adyant Balaji, Abhinav Uppal, Omeed Djassemi, Yuhong Zhang, Hubin Zhao, and Gert Cauwenberghs

Abstract

Precision health monitoring systems represent a transformative approach to collecting and analyzing physiological and biochemical data in real time. By leveraging integrated bioelectronics encompassing advanced sensors, low-power analog front ends, on-chip computational units, and embedded wireless communication interfaces, these systems enable continuous, non-invasive monitoring in compact, user-friendly devices. This chapter provides a comprehensive overview of the fundamental building blocks of precision health monitoring integrated bioelectronics, highlighting the unique challenges and circuit-level solutions required for high-fidelity electrophysiological and electrochemical signal acquisition. Key considerations for signal integrity, noise reduction, and energy efficiency are detailed, with discussions on front-end amplification techniques for weak biopotential signals, robust potentiostat designs for electrochemical sensing, and the selection and implementation of energy-efficient analog-to-digital

Y. Xu (✉)
Institute for Neural Computation, University of California, San Diego, CA, USA

Aizip inc., Cupertino, CA, USA
e-mail: yux013@ucsd.edu

M. S. Lee · A. Balaji · A. Uppal · O. Djassemi · Y. Zhang
Shu Chien-Gene Lay Department of Bioengineering, University of California, San Diego, CA, USA

H. Zhao
Division of Surgery and Interventional Science, University College London, London, UK

G. Cauwenberghs
Institute for Neural Computation, University of California, San Diego, CA, USA

Shu Chien-Gene Lay Department of Bioengineering, University of California, San Diego, CA, USA

K. Mahato, P. Chandra (eds.), *Nano-bioelectronics for Precision Health Monitoring*, https://doi.org/10.1007/978-981-95-2117-3_12

converter architectures. The chapter also examines the integration of emerging technologies, such as edge artificial intelligence and multi-modal sensor fusion, which further enhance system performance and functionality. These innovations, combined with advanced packaging and form factors, pave the way for the next generation of mobile, wearable, and precision health monitoring systems for personalized and preventive care.

Keywords

Precision health monitoring · Integrated bioelectronics · Electrophysiological and electrochemical sensing

Abbreviations

2DSM	Second-order delta-sigma modulator
AAF	Anti-aliasing filter
AC	Alternate current
ADC	Analog to digital converter
AFE	Analog front end
Ag/AgCl	Silver/silver chloride
BLE	Bluetooth Low Energy
CA	Control amplifier
CDS	Correlated double sampling
CE	Counter electrode
CGM	Continuous glucose monitor
CM	Current mirror
CMOS	Complementary metal-oxide-semiconductor
CMRR	Common-mode rejection ratio
C-TIA	Capacitive feedback transimpedance amplifier
CV	Cyclic voltammetry
DC	Direct current
DPV	Differential pulse voltammetry
DRL	Driven right leg
DSP	Digital signal processor
EEG	Electroencephalography
EIS	Electrochemical impedance spectroscopy
EMI	Electromagnetic interference
EOV	Electrode offset voltage
ESD	Electrostatic discharge
ExG	Electrophysiological signals
IA	Instrumental amplifier
IC	Integrated circuit

IMU	Inertia measurement unit
LSB	Least significant bit
LSL	Lab Streaming Layer
MCU	Micro-controller unit
MOSFET	Metal oxide semiconductor field-effect transistor
MSB	Most significant bit
NEF	Noise efficiency factor
NFC	Near field communication
NPU	Neural processing unit
OECT	Organic electrochemical transistors
PDA	Predictive digital auto-ranging
PEF	Power efficiency factor
PGA	Programmable gain amplifier
PMIC	Power management integrated circuits
PMU	Power management unit
RE	Reference electrode
R-TIA	Resistive feedback transimpedance amplifier
SAR	Successive approximation register
SGS	Silvered glass silicone
SNR	Signal-to-noise ratio
SoC	System on chip
TIA	Transimpedance amplifier
TinyML	Tiny machine learning
VF	Voltage follower
WE	Working electrodes
WIFI	IEEE 802.11x standards
ΔΣ	Delta-sigma

12.1 Architectural Overview of Interfacing Electronics for Wearable Precision Health Monitoring Systems

Precision health monitoring systems offer non-invasive, continuous tracking of physiological and biochemical parameters in everyday settings, often through wearable form factors. These integrated systems incorporate a range of components to achieve efficient signal acquisition, data processing, and secure transmission, ultimately enabling real-time health assessments. As illustrated in Fig. 12.1, the architecture involves multiple subsystems designed for specific sensing modalities, all while maintaining a compact, user-friendly profile. At the core of such a precision health monitoring device are several key subsystems: a low-power computational unit, diverse sensors, carefully engineered analog front-ends (AFEs), a wireless communication interface, and a power management unit (PMU) (Sharma 2022).

The computational unit manages system control, online preprocessing of sensor data, and communication with external devices (Biglari and Tang 2023). The choice of computational platform depends on the system's complexity and power

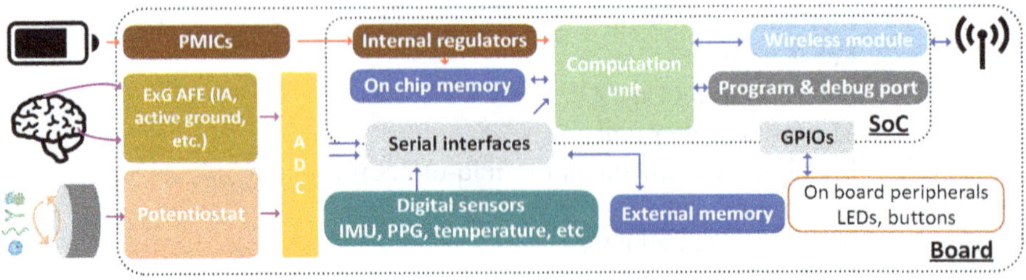

Fig. 12.1 System architecture for precision health monitoring devices, highlighting the integration of electrophysiological and electrochemical sensor pathways. The architecture features power management integrated circuits (PMICs) for energy regulation, with separate analog front ends (AFEs) tailored for electrophysiological signals (e.g., EEG, ECG) and electrochemical signals (e.g., glucose, lactate). The electrophysiological AFE includes key components such as instrumentation amplifiers (IAs) and active grounding, while the potentiostat in the electrochemical pathway enables accurate current and voltage monitoring. Both analog signals converge at a high-resolution ADC for digitization before entering the computational unit. The computational unit, embedded in a system-on-chip (SoC), handles data processing, internal memory management, and communication commonly via serial interfaces. Such wearable systems usually support external memory and interfaces with digital sensors for additional health related modalities like PPG, temperature, and IMU signals. Wireless communication modules ensure seamless data transmission to external devices, while GPIOs provide connections to onboard peripherals such as LEDs and buttons. Such modular and scalable design underpins wearable systems' ability to perform real-time, multimodal physiological monitoring in compact and power-efficient platforms

requirements. The prevalent choices of computational units among wearable precision health monitoring systems are microcontrollers (MCUs), which are suitable for low-power wearable applications where basic data processing and control are required. Additionally, digital signal processors (DSPs) are ideal for computation-intensive tasks, such as audio signal processing or complex filtering of physiological signals. Field-programmable gate arrays (FPGAs), though less commonly used in wearable health monitoring systems due to cost and power consumption, provide customizable hardware acceleration for real-time processing of multiple data streams. Recently, advances in semiconductor manufacturing and the rising demand for edge artificial intelligence (AI) have led to the integration of neural processing units (NPUs) in system-on-chip (SoC) designs. NPUs are specialized cores designed to accelerate machine learning workloads, such as deep learning inference, while maintaining energy efficiency. This trend aligns with the broader adoption of tiny machine learning (TinyML) (Lin et al. 2020; Nguyen et al. 2023), a framework for deploying AI models on resource-constrained devices like wearables. By enabling real-time AI inference directly on the device, NPUs reduce latency and dependence on cloud processing, paving the way for more sophisticated applications such as predictive health analytics and anomaly detection in wearable systems. Heterogeneous SoC architecture combining MCUs, DSPs, FPGAs, and NPUs is increasingly common to balance flexibility, power efficiency, and processing capabilities (Ito 2023). Functionally, the computational unit preprocesses signals through filtering, noise reduction, or feature extraction. For advanced applications, it can also implement algorithms for machine learning-based anomaly detection or trend analysis, leveraging TinyML frameworks to process data efficiently on the edge.

Precision health monitoring systems commonly utilize a range of sensors to detect physiological and biochemical signals. These sensors include the two main types of analog sensors: electrophysiological sensors, which detect bioelectric signals such as ECG, EEG, EMG, and EOG. Electrophysiological sensors require tight coupling with the skin through electrodes and rely on analog front-end circuits for amplification and digitization. And then electrochemical sensors measure the chemical composition of biofluids like sweat, saliva, and tears. These sensors often include potentiostats to monitor current and voltage responses during chemical reactions (Heikenfeld et al. 2018). Modern wearable health systems also have established digital sensor solutions to expand sensing modalities. Devices such as inertia measurement unit (IMU), photoplethysmography (PPG), and temperature sensors provide direct digital outputs through serial interfaces, enabling direct digital readout. These sensors typically include built-in signal processing capabilities, reducing the computational burden on the main processing unit (Röddiger et al. 2022).

For analog signal sensors among the sensors introduced, the direct circuit interfacing component is the analog front end (Bretschneider and De Weille 2019; Webster 2009). The AFE plays a crucial role in signal acquisition, particularly for analog signals such as those generated by electrophysiological and electrochemical sensors. It ensures the integrity of weak signals before further processing. AFEs commonly perform the tasks of amplification, filtering, and digitization of signals (Fig. 12.2) (Petkos et al. 2019). For amplification, AFEs boost weak signals to a level suitable for digitization. For instance, signals like EEG (~1–100 μV) and ECG (~10 μV–4 mV) require high-gain amplification with low noise. For filtering, AFEs implement hardware-based filtering that removes noise and artifacts using high-pass, low-pass, or notch filters in an efficient manner, which is essential for mitigating baseline drift, motion artifacts, and power line interference. Lastly, the preconditioned analog signals are converted into digital form using high-resolution analog-to-digital converters (ADCs) (e.g., 12-bit, 16-bit, or 24-bit, depending on applications) for accurate downstream processing in computational units (Bretschneider and De Weille 2019). Additionally, AFEs are also customized toward different applications. For example, in electrochemical sensing, the AFE includes potentiostats to control and measure the electrochemical interactions at the sensor interface. These circuits monitor current and voltage responses and ensure accurate real-time biochemical analysis.

Lastly, the wireless communication module and the power management unit are essential for the functionality of wearable precision health monitoring platforms given their wearability requirements (Gao et al. 2023). Wireless communication modules enable data transmission to external devices like smartphones, cloud servers, or host machines for further processing. Wireless modules nowadays can be a part of the core SoC itself and often integrate antenna designs tailored for wearable form factors, ensuring robust connectivity in diverse environments. Common wireless protocols include Bluetooth Low Energy (BLE) for low-power, short-range communication, wireless network protocols based on the IEEE 802.11 family of

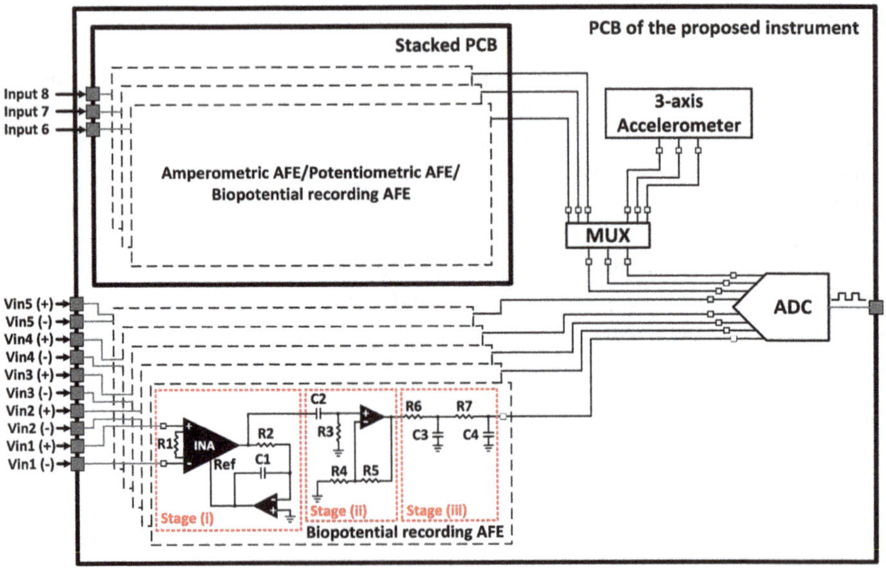

Fig. 12.2 Example implementation of an interfacing electronics comprising of AFEs and ADC for electrophysiological and electrochemical sensing instrumentations. (From Petkos et al. (2019), reprinted with permission under Creative Commons Attribution license)

standards (WIFI) for higher throughput transmissions, and Near-Field Communication (NFC) for short-range, low-data-rate applications. Built upon these protocols, previous research has investigated integrating different body sensors to form a body area network (BAN), capable of aggregating various biosignal streams over multiple worn devices simultaneously. For example, the WIFI-based BAN enables scalable, flexible configurations for simultaneous monitoring of EEG, EOG, and EMG signals across different body locations, facilitating comprehensive health and activity monitoring in naturalistic settings (Paul et al. 2022a). The PMU ensures efficient energy usage and commonly includes voltage regulators, which regulate input voltage to levels required by different chips on board, battery management circuits, which handle charging, discharging, and battery protection, and lastly, energy harvesting modules that harvest different forms of energy sources from the body to extend the power budget. Examples include thermoelectric or piezoelectric generators.

Integrating all these core subsystems, a typical wearable precision health monitoring system processes signals through distinct stages (Moin et al. 2021; Chung et al. 2019). Initially, raw physiological or biochemical signals are captured and conditioned through the AFE or received from digital sensors. Subsequently, the computational unit performs preprocessing tasks like artifact removal or feature extraction. Finally, the processed data is transmitted wirelessly for storage, visualization, or advanced analytics. Multi-modal sensing capabilities are increasingly in demand; hence, robust data synchronization techniques, such as those provided by the Lab Streaming Layer (LSL) library, are imperative. Such synchronization ensures that diverse data streams—ranging from ECG and PPG to EEG and audio—are time-aligned for integrative analyses and advanced health insights (Dasenbrock et al. 2022).

With the understanding of the underlying architecture and system components, the subsequent sections will delve deeper into the specialized sensing modalities and the intricacies of their interfacing electronics, mainly the AFE designs, which play a critical role in achieving reliable, high-quality data acquisition in precision health monitoring systems. Electrophysiological and electrochemical sensing modalities.

As discussed in the last section, wearable sensors have become integral to modern precision health monitoring systems, providing continuous, real-time, and non-invasive access to a wide array of physiological and biochemical information. Among the diverse sensing approaches, electrophysiological and electrochemical modalities are particularly prominent (Yeung et al. 2021; Ullah et al. 2022; Mohan et al. 2020). Electrophysiological sensors detect biopotential signals originating from the electrical activity of excitable cells, such as neurons in the brain (EEG), ocular muscles (EOG), cardiac muscle (ECG), and skeletal muscles (EMG). In contrast, electrochemical sensors measure analytes within biofluids (e.g., sweat, saliva, interstitial fluid), translating chemical reactions into electrical signals that reflect underlying metabolic processes. Leveraging these complementary data streams can provide a more comprehensive picture of an individual's health status. Indeed, recent efforts have explored multi-modal sensor integration, effectively combining electrophysiological and electrochemical data to enhance diagnostics, personalized health assessments, and disease management (Imani et al. 2016; Xu et al. 2023).

12.1.1 Electrophysiological Signals

Electrophysiological signals arise from the electrical potential generated by ionic movements across cellular membranes. These potentials, caused by the depolarization and repolarization of excitable cells such as neurons, cardiac cells, and muscle fibers, are detected as weak bioelectric signals. The main types of electrophysiological signals (ExG) monitored by wearable sensors include electrocardiography (ECG), electroencephalography (EEG), electromyography (EMG), and electrooculography (EOG) (Fig. 12.3a–d) (Das et al. 2019). ECG signal represents the electrical activity of the heart, generated by the sequential depolarization and repolarization of cardiac cells. The vector sum of these activities produces the characteristic ECG waveform, which is used to diagnose arrhythmias, coronary artery disease, and other heart conditions (Randazzo et al. 2019). EEG signals originate from neuronal activities in the brain and are recorded from the scalp, forehead (Cannard et al. 2021), or recently at the ear (Kaveh et al. 2024). These signals are used for assessing brain function and diagnosing neurological disorders such as epilepsy and sleep disorders. EMG signals are produced by the electrical activity of muscle fibers during contraction, enabling applications in neuromuscular

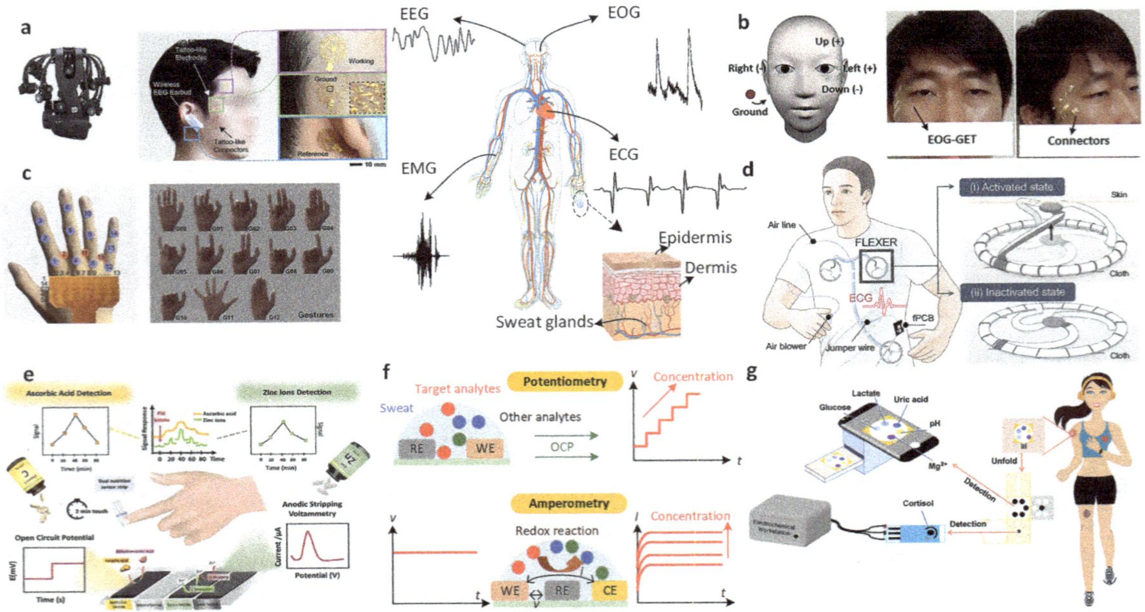

Fig. 12.3 Examples of wearable electrophysiological and electrochemical sensors. (**a**) A multi-channel wireless and wearable EEG headset design (Cognionics, USA) (from Niso et al. (2023)). Elsevier, reprinted with permission) and a network of EEG electrodes with one earbud and three open-mesh, structured, tattoo-like electrodes (from Shin et al. (2022)), reprinted with permission under Creative Commons Attribution license). (**b**) Graphene electronic tattoos (GET) for EOG based human–robot interface (from Ameri et al. (2018), reprinted with permission under Creative Commons Attribution license). (**c**) High-density EMG sensor arrays tailored for decoding fine finger movements (from Hu et al. (2022), reprinted with permission under Creative Commons Attribution license). (**d**) A fabric-based controllable, lamina emergent MXene-based electrode for ECG monitoring (Lee et al. 2024), reprinted with permission under Creative Commons Attribution license). (**e**) Touch-based disposable zinc (Zn(II))/ascorbic acid (AA) multisensory platform for parallel measurements micronutrients (from Seker et al. (2024)). Elsevier, reprinted with permission). (**f**) Schematic illustrations of three electrochemical sensing techniques: Potentiometric sensing utilizes an ion-selective membrane as the recognition element, generating a potential difference (V_{cell}) across the WE and RE proportional to ion concentration. Amperometric sensing operates in a three-electrode configuration where electroactive species undergo enzymatic or non-enzymatic oxidation, producing a current proportional to analyte concentration under a constant applied potential. (**g**) Wearable sweat sensor based on a 3D paper structure for simultaneous analysis of multiple biomarkers (Cheng et al. 2023), reprinted with permission under Creative Commons Attribution license)

diagnostics and rehabilitation (Maragliulo et al. 2019). EOG sensors capture the corneal-retinal potential generated by eyeball movements and eye blinks, offering applications in fatigue monitoring and virtual reality (Kołodziej et al. 2020).

A common form of electrophysiological sensors are on-body electrodes, commonly in flexible formats in recent days for tight interfacing with the body. Depending on their interfacing mechanisms, it is usually characterized as wet, dry or non-contact electrodes, with silver/silver chloride (Ag/AgCl) electrodes as a common selection (Chi et al. 2010). These electrodes must exhibit low impedance, biocompatibility, and mechanical compatibility with the skin or tissue. Materials such as gold nanowires, graphene, and conductive polymers like PEDOT:PSS have been widely employed to enhance signal quality and user comfort (Zhu et al. 2021; Li et al. 2024). These electrophysiological electrodes have been integrated into various wearable devices (Fig. 12.3a–d) to monitor physiological states non-invasively. Smartwatches and Fitness Bands: devices like the Apple Watch and Fitbit incorporate ECG sensors to monitor heart rate and detect arrhythmias (Thomson et al. 2019; Wang et al. 2022). These wearables provide continuous cardiac monitoring in real-life settings. Products like the Muse headband track brain activity for stress management and sleep monitoring (Niso et al. 2023; Kappel et al. 2019; Hunkin et al. 2021). Flexible electrodes ensure consistent signal acquisition even during motion. Devices such as the Myo armband capture muscle activity for gesture recognition and prosthetic control. These wearables enable applications in assistive technologies and physical rehabilitation. Smart eyewear integrates EOG sensors to track eye movement for applications in virtual reality interfaces and fatigue detection (Lin et al. 2021). The impedance between the electrode and the skin is commonly referred to as the main performance factor for electrophysiological electrodes, it can vary significantly due to factors like sweat, movement, and electrode material. This variability can distort the signal and reduce the signal-to-noise ratio (SNR) (Kappel and Kidmose 2015).

Electrophysiological signals exhibit unique properties that pose significant challenges and requirements for the design of the interfacing systems in wearable devices (Yazicioglu et al. 2008; Ha et al. 2014; Ha et al. 2021). Electrophysiological signals are typically low in amplitude, making them susceptible to noise and interference. For example, ECG signals range from 10 μV to 5 mV, with frequencies between 0.05 and 150 Hz (Satija et al. 2015). EEG signals are even weaker, ranging from 1 to 100 μV, with frequencies under 100 Hz (Siuly et al. 2016). EMG signals are typically stronger, with amplitudes of ~50 μV to a few millivolts and frequencies between 10 and 500 Hz. EOG signals fall within a range of 0.05 to 35 Hz, with amplitudes typically in the tens of microvolt. The AFE must have a high dynamic range and sensitivity to capture these weak signals without distortion or clipping. This necessitates low-noise amplifiers with high input impedance to prevent signal attenuation and distortion. Active impedance matching circuits or impedance measurement features are needed to adapt to changing electrode-skin impedance dynamically. The use of conductive gels or advanced electrode materials (e.g., graphene (Shao et al. 2019) or PEDOT:PSS (Park et al. 2020)) shows promises to lower impedance. Electrophysiological signals are also highly prone to contamination from various noise sources, including motion artifacts resulted from electrode movement relative to the skin, power line interference typically at 50 or 60 Hz, and electromagnetic interference (EMI) caused by surrounding electronic devices. The signal conditioning circuits, or signal processing pipelines must incorporate filtering techniques to reject noise. High-pass filters can remove baseline wander, notch filters can mitigate power line interference, and shielding is essential to reduce EMI. Additionally, electrophysiological signals are often superimposed on a high common-mode voltage generated by the body, which can be several orders of magnitude higher than the signal of interest. Differential amplifiers with high common-mode rejection ratio (CMRR) are critical to isolate the small differential signal from the large common-mode noise. Driven right leg (DRL) circuits are commonly used in ExG systems to minimize common-mode noise further.

12.1.2 Electrochemical Signals

Electrochemical sensors measure the chemical composition of biofluids such as sweat, interstitial fluid, saliva, and tears. The primary sensing mechanisms include potentiometry and amperometry (Fig. 12.3f) (Mahato and Wang 2021). Potentiometry sensors measure the potential difference between a working electrode (WE) and a reference electrode (RE) that is proportional to the ion concentration. Potentiometric sensors are widely used for monitoring electrolytes like sodium, potassium, and chloride ions. Ion-selective membranes and nanostructured materials are critical components for achieving high sensitivity and selectivity. Amperometric sensors detect analytes by monitoring the current produced during a redox reaction. For instance, glucose levels are measured using enzymatic sensors where glucose oxidase catalyzes the reaction, producing an electrical current proportional to glucose concentration (Sempionatto et al. 2021a). Amperometric sensors commonly employ a constant potential across the working and reference electrodes, sensors that use non-constant potentials are referred to as voltametric sensing. Such examples include differential pulse voltammetry (DPV), which employs a staircase potential waveform for detecting trace levels of specific analytes (e.g., certain metabolites or drugs) (Crespi 2020), and cyclic voltammetry (CV), which employs a linear potential waveform to study the oxidation and reduction of molecular species (Alatraktchi et al. 2016).

Electrochemical sensors often incorporate innovative materials such as molecularly imprinted polymers and nanostructures to improve their performance. These materials enhance ion-to-electron transduction, stability, and biocompatibility, making electrochemical sensors suitable for wearable applications (Sempionatto et al. 2022). Electrochemical sensing modalities have expanded the capabilities of wearable sensors to include biochemical monitoring. Several successful electrochemical sensors have been available these days, examples include the continuous glucose monitors (CGMs) produced by Dexcom G6 and Abbott FreeStyle Libre, which measure interstitial glucose levels using minimally invasive sensors in a continuous and real-time manner (Price et al. 2023). Wearable patches have been developed to analyze sweat for metabolites like lactate, electrolytes such as sodium and potassium, and stress-related biomarkers like cortisol (Fig. 12.3e, g). These devices can be used in athletic performance monitoring and hydration assessment (Sempionatto et al. 2021b; Jia et al. 2013; Yu et al. 2020; Yin et al. 2022).

Electrochemical signals are determined by the chemical interactions at the electrode interface. For example, potentiometric sensors measure voltage changes, while amperometric sensors detect currents. Both signals are directly related to analyte concentration. For the interfacing electronic design, AFEs must be capable of either clamping the current or voltage, and then measuring both low voltages or currents, respectively, with high precision, requiring high input impedance for potentiometric sensing and low-noise current sensing (Ma et al. 2023; Tehrani et al. 2022). Electrochemical signals vary over several orders of magnitude, depending on analyte concentration. For instance, sodium concentrations in sweat can range from millimolar to micromolar levels, demanding wide dynamic range and high resolution (Bandodkar et al. 2014). The selectivity and sensitivity of these electrochemical sensors are achieved through the use of biorecognition elements like enzymes or ion-selective membranes. Signal stability and long-term usability are the key challenges for measuring electrochemical signals, which are affected by electrode fouling and material degradation. Thus, for the interfacing electronics, the AFE may need to compensate for drift or calibration errors over time. Periodic recalibration, self-calibration routines, temperature compensation, and real-time drift corrections are common techniques to maintain long-term accuracy (Aymerich Gubern 2021). Another key metric for electrochemical measurement system is the measurement latency, the response time of electrochemical sensors depends on the diffusion of analytes to the electrode surface and the reaction kinetics of the electrode-analyte interaction. Faster kinetics enable real-time monitoring but may introduce noise. Thus, the AFE must balance signal acquisition speed with noise filtering to ensure accurate real-time measurements.

In summary, both electrophysiological and electrochemical sensing modalities provide critical insights into physiological and biochemical processes, forming the foundation of modern wearable and precision health monitoring systems. Electrophysiological signals, characterized by their low amplitude, susceptibility to noise, and variability in electrode-skin impedance, require analog front-end designs that prioritize high dynamic range, low-noise amplification, and advanced filtering techniques. Similarly, electrochemical signals, driven by chemical interactions at the electrode interface, present challenges such as wide dynamic range, sensitivity to drift, and varying response times. The seamless integration of sensing and electronics underpins the functionality and accuracy of wearable biosensing platforms.

12.2 Circuit Techniques for Electrophysiological Sensing

To accurately measure electrophysiological signals such as EEG, EMG, ECG, and EOG, interfacing circuits must address the unique challenges posed by low-amplitude signals, high levels of noise, and variability in electrode-skin impedance

(ExG signals vs noise range shown in Fig. 12.4b). Designing interfacing circuits for precision electrophysiological sensing devices necessitates a careful balance between high precision, low power consumption, robust noise rejection, and the constraints of size and portability inherent in wearable form factors. This section details the key techniques and methodologies employed in the design of biopotential acquisition circuits for precision health systems, highlighting recent advancements in low-power design and state-of-the-art instrumentation that cater to the specific needs of wearable electrophysiological devices.

A common architecture for electrophysiological AFE design features three stages, initial front-end filtering to suppress DC offsets and high-frequency noise, followed by a low-noise instrumentation amplifier for differential amplification, and a secondary stage providing additional filtering before the signal is sent to the ADC input (Fig. 12.4a) (Hasan et al. 2024). Several key challenges arise in designing circuits capable of accurately capturing electrophysiological signals as introduced in the last section. First, the low signal amplitudes mean that amplifiers must detect and process signals that are often buried in noise levels similar to or greater than the signal itself. This necessitates extremely low-noise circuit designs to achieve acceptable SNRs. Second, high levels of noise from EMI—such as power line interference at 50/60 Hz, electronic devices, and muscle activity (EMG artifacts)—can corrupt the desired signals. Effective noise rejection techniques are essential to mitigate these interferences. Third, variability in electrode-skin impedance poses a significant challenge. The impedance (lumped-element circuit model as shown in Fig. 12.4c) at the electrode-skin interface can vary widely due to factors like movement, sweat, skin condition, and electrode placement, affecting signal quality and consistency (Taji et al. 2018). Fourth, power consumption is a critical consideration in wearable devices, which require low-power operation to prolong battery life while maintaining high-performance signal acquisition. Lastly, size and portability constraints mean that the circuits must be compact and lightweight to be integrated into wearable form factors without compromising user comfort or device aesthetics. To address these challenges, circuits designed for biopotential acquisition must exhibit several common and important characteristics (Ha et al. 2021).

High input impedance is essential to prevent loading of the electrode signal and to ensure accurate voltage measurements across the electrode-skin interface. Amplifiers with input impedances exceeding 10 GΩ are typically required to minimize the loading effect, significantly higher than the electrode-body interface impedance (Chi et al. 2011). This high input impedance is often achieved using complementary metal-oxide-semiconductor (CMOS) designs with techniques such as capacitive coupling, which isolates the direct current (DC) offset voltage while allowing the alternating current (AC) biopotential signal to pass, and active feedback mechanisms that maintain high input impedance and stability.

Minimizing intrinsic noise is also crucial for detecting weak biopotential signals. Target input-referred noise levels are typically between 1 and 5 μV_{RMS}, particularly important for applications like EEG where signal amplitudes are minimal. Techniques to reduce noise include the use of chopper-stabilized amplifiers (Xu et al. 2013), which modulate the input signal to a higher frequency band where

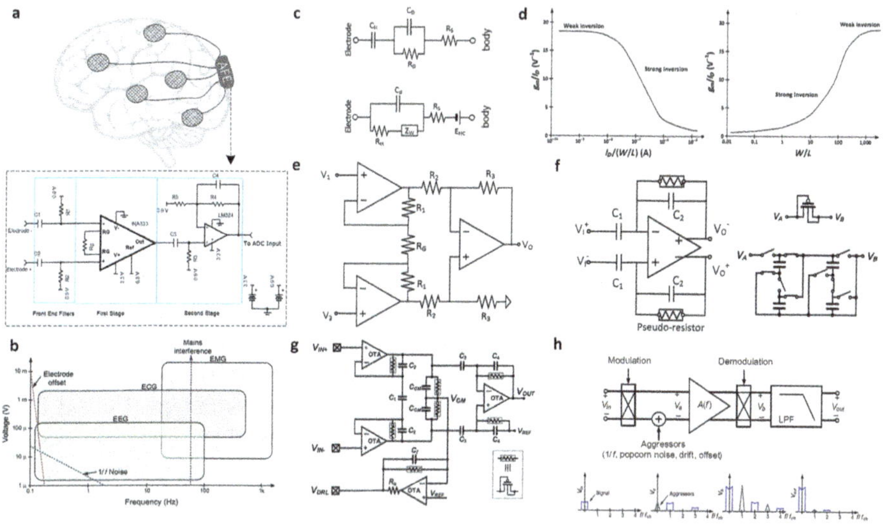

Fig. 12.4 Circuit techniques for electrophysiological sensing. (**a**) Common circuit architecture of a multi-channel electrophysiological sensing AFE (bottom dashed box shows one channel of differential electrophysiological signal input (from Hasan et al. (2024)), Elsevier, reprinted with permission). The AFE processes biopotential signals acquired through electrodes placed on the scalp for applications like EEG monitoring. The architecture includes three key stages: *front-end filters*, a *first-stage amplifier*, and a *second-stage amplifier*. The front-end filters remove high-frequency noise and ensure signal integrity before amplification. The first stage employs an instrumentation amplifier (e.g., INA333) with high input impedance to amplify low-amplitude biopotential signals while rejecting common-mode noise. The second stage, implemented with an operational amplifier (e.g., LM324), provides additional gain and low-pass filtering to suppress high-frequency noise. The output is conditioned for digitization by an ADC, ensuring accurate acquisition of electrophysiological signals for further processing. (**b**) Characteristics of ExG signals (e.g., EEG, ECG, EMG) illustrating their frequency and amplitude ranges relative to mains interference and noise sources, including *1/f* noise (Webster 2009; Yazicioglu et al. 2008; Ha et al. 2021). (**c**) Lumped-element circuit model of the electrode-body interface (Akinin et al. 2020): (top) Polarizable, non-faradaic electrode and (bottom) non-polarizable, faradaic electrode. (**d**) Subthreshold weak inversion operation of MOS transistors, showing: (left) transconductance efficiency as a function of bias current; (right) transconductance efficiency as a function of transistor size (Ha et al. 2021). (**e, f**) Common architectures for differential and instrumentation amplifiers (Akinin et al. 2020). (**e**) 3-opamp instrumentation amplifier and (**f**) fully differential capacitively coupled amplifier, the pseudo-resistors implementations are further illustrated on the right with (top) single MOS-bipolar pseudoresistor and (bottom) switched-capacitor based pseudoresistor (Ha et al. 2021). (**g**) General architecture of an instrumentation amplifier, incorporating a driven right leg circuit for active noise cancellation and improved common-mode rejection (Ha et al. 2021) (**h**) Block diagram and frequency-domain representation of the chopper stabilization technique, illustrating modulation and demodulation for reducing low-frequency noise, such as *1/f* noise (Ha et al. 2021). **d, f, g, h** are from Ha et al. (2021), Elsevier, reprinted with permission. **c, e, f** are from Akinin et al (2020), Springer International Publishing, reprinted with permission

flicker noise (*1/f* noise) is less significant, and operating transistors in the weak inversion (subthreshold) region (Vittoz 2006), where CMOS transistors offer high transconductance efficiency, reducing thermal noise. Additionally, the circuits must operate with low power consumption, typically less than 100 μW, to be suitable for wearable applications with limited power budgets.

The amplifiers must provide application-dependent signal gain, ranging from 1 to 1000, with variable gain settings in some applications to accommodate different signal amplitudes. This flexibility supports a medium-to-high signal dynamic range of 40–80 dB, essential for capturing both low-amplitude signals like EEG and higher amplitude signals like EMG. The circuits must support a low-to-medium frequency range, typically from 0.1 to 10 kHz, with configurable bandwidth in some applications to focus on the relevant frequency components of the biopotential signals. DC blocking is also critical to prevent electrode offset voltages (EOVs) from saturating the amplifiers (Bagheri et al. 2017). Alternatively, if DC coupling is used, the circuits must be designed to tolerate large EOVs without affecting the amplification of the desired signal. Motion artifact rejection techniques are necessary to mitigate disturbances caused by movement, which can introduce significant low-frequency noise and affect signal integrity.

A high common-mode rejection ratio, typically exceeding 80 dB, is necessary to reject common-mode noise, including power line interference and environmental electromagnetic interference. Achieving high CMRR involves differential amplification, which amplifies the difference between two input signals while rejecting signals common to both, active shielding to reduce capacitive coupling and interference from external noise sources and driven right leg circuits that actively cancel common-mode voltages by injecting a compensating current into the body (Zhu et al. 2024). Similarly, a high power supply rejection ratio (PSRR), also exceeding 80 dB, ensures that fluctuations in the supply voltage do not affect the signal integrity. Proper grounding techniques are essential to minimize ground loops and further reduce noise.

Lastly, rejection of main power interference at 50/60 Hz is critical, often achieved through notch filters or differential amplification, to mitigate power line noise, which is a common interference in biopotential measurements. Analog filtering is vital for eliminating unwanted noise and artifacts before digitization. High-pass filters remove baseline drift and low-frequency artifacts, often caused by motion or electrode polarization. Low-pass filters suppress high-frequency noise, such as EMI or muscle artifacts in EEG recordings.

To facilitate integration into a compact or even wearable format, the circuits must occupy a small silicon area, typically less than 10 mm^2, and minimize the use of off-chip components. Having no or few off-chip components not only reduces physical size but also lowers manufacturing costs and improves reliability. Energy-efficient analog-to-digital converter (ADC) architectures, such as successive approximation register (SAR) (Khayatzadeh et al. 2012; Tang et al. 2020) and delta-sigma (ΔΣ) ADCs (Custódio et al. 2013), are commonly employed in wearable devices to meet these requirements while providing adequate sampling rates and resolutions. The ADC must have sampling rates at least twice as high as the highest frequency component of the signal, typically requiring 500 Hz to 1 kHz for EEG and ECG signals. Resolutions of 12–16 bits are necessary to discern the small voltage variations in biopotential signals. Detailed ADC designs will be introduced in the following sections.

While these essential characteristics establish the foundational requirements for biopotential acquisition circuits, the following sections delve into specialized methodologies that elevate the performance of biopotential sensing interfaces, focusing on innovative amplification strategies, noise reduction mechanisms, and low-power design approaches tailored for precision electrophysiological applications.

12.2.1 Circuit Techniques to Enhance Electrophysiological Sensing

Building on the foundational requirements for electrophysiological signal acquisition, this subsection explores advanced circuit techniques designed to overcome key challenges, such as ultra-low signal amplitudes, high susceptibility to noise, and the need for reliable performance in dynamic, real-world conditions.

12.2.1.1 Enhancing Common-Mode Rejection and Signal Integrity

Differential Amplifiers and Instrumentation Amplifiers (IAs)
Differential amplifiers serve as the cornerstone of biopotential acquisition systems, primarily for their ability to amplify the small voltage differences between two electrodes while rejecting common-mode signals, such as noise from power lines or motion artifacts. Among differential amplifiers, IAs are particularly valuable for biopotential sensing due to their low input-referred noise, high CMRR, and precise gain control. Most IAs are implemented using a three-opamp topology (Fig. 12.4e), which offers several key benefits: the ability to provide adjustable gain settings, high input impedance to prevent signal attenuation from electrode-skin impedance, and low distortion for accurate amplification of low-amplitude signals such as EEG and ECG (Kouhalvandi et al. 2023). The transfer function of an IA is helpful to understanding its behavior in biopotential amplification. For the classical three-opamp IA, the gain is determined by the following (parameters from Fig. 12.4e):

$$V_{\text{out}} = (v_1 - v_2)\left(1 + \frac{2R_1}{R_G}\right)\left(\frac{R_3}{R_2}\right)$$

This classical three-opamp approach often requires large passive components, and may consume more power and silicon area, factors that can be limiting in ultra-low-power, miniature wearable systems. In response to these constraints, more integrated and power-efficient IA topologies have emerged, which leverage capacitive coupling, pseudoresistors, and switched-capacitor techniques to realize both high performance and compact form factors. One of the improvement is to use capacitor coupled differential amplifiers (Fig. 12.4f), the gain and frequency response of the

IA are set primarily by capacitor ratios rather than by large discrete resistors. This approach offers several advantages: integrated CMOS processes can achieve excellent capacitor matching, enabling precise control of gain and filtering characteristics, and reducing sensitivity to process variations. A large resistor R_f, crucial for setting the high-pass cutoff frequency, is often implemented using a pseudoresistor or a switched-capacitor circuit, both of which can achieve ultra-high effective resistances on-chip without occupying excessive area. For a typical single-ended IA design utilizing capacitive coupling, the gain depends on the ratio of the coupling and feedback capacitors (C_f and C_C), as well as the choice of feedback resistor R_f and the operational transconductance amplifier (OTA) parameters. A representative transfer function for such an IA is given by Ha et al. (2014):

$$A_v(s) = -\frac{C_c}{C_f}\frac{1}{1+\frac{1}{sR_fC_f}}\frac{1}{1+s\frac{C_CC_L\left(\frac{1}{C_L}+\frac{1}{C_C}+\frac{1}{C_f}\right)}{g_m}}\left(1-\frac{sC_f}{g_m}\right)$$

Here g_m is the transconductance of the OTA, R_f is the feedback resistor, C_C and C_f are the coupling and feedback capacitors, C_L is the load capacitance contributing to the pole locations. Capacitive coupling is also particularly effective in handling EOVs, which can arise from polarization at the electrode-skin interface. Large DC offsets, if not addressed, can saturate the amplifier and obscure the low-amplitude biopotential signals of interest. By incorporating input coupling capacitors, the IA can block these DC components, effectively acting as a high-pass filter that isolates the AC components of the biopotential signal. The high-pass cutoff frequency, f_{HP}, for the amplifier is determined by the feedback resistor R_f and coupling capacitor C_f by:

$$f_{HP} = \frac{1}{2\pi R_fC_f}$$

This cutoff frequency is typically set below 1 Hz to ensure the preservation of low-frequency signals, such as those found in EEG and ECG.

The IA design aims to achieve low noise performance. One of the primary challenges in biopotential amplifiers is minimizing noise, particularly input-referred noise, which can obscure the small physiological signals. The input-referred noise of an IA has two dominant contributors: Noise from the OTA ($V^2_{ni,\mathrm{amp}}$ time-averaged mean square noise voltage), determined by its transconductance (g_m), and thermal noise (V^2_{ni,R_f}) from the feedback resistor (R_f). The two main noise sources are modeled as Ha et al. (2021):

$$\sqrt{\overline{V^2_{ni,\mathrm{amp}}}} = \sqrt{\overline{V^2_{ni,\mathrm{system}}}}\,\frac{C_f + C_{in} + C_C}{C_C}$$

$$\sqrt{V_{ni,R_f}^2} = \sqrt{\frac{4KT}{R_f}\frac{1}{2\pi fC_{\text{in}}}}$$

Here k is the Boltzmann constant and T is the absolute temperature, C_{in} refers to parasitic capacitance at the OTA input nodes. To minimize the thermal noise, R_f must be carefully chosen to balance low noise with the desired high-pass cutoff frequency. The noise efficiency factor (*NEF*) is often used to compare the performance of different amplifier designs, combining noise and power consumption metrics (Steyaert and Sansen 1987):

$$\text{NEF} = \frac{V_{\text{n,rms}}}{\sqrt{\frac{2\pi V_t BWkT}{I_{\text{D}}}}}$$

where $V_{\text{n, rms}}$ is the input-referred noise voltage, I_{D} is the total supply current, BW is the bandwidth V_t is the thermal voltage. In practice, differential IAs with input differential pairs incur twice the supply current for the same transconductance, with *NEF* values commonly greater than 2. The power efficiency factor *(PEF)* is an extension of *NEF* that accounts for supply voltage, providing a more comprehensive efficiency measure (Muller et al. 2012):

$$\text{PEF} = \text{NEF}^2 \times V_{\text{DD}}$$

Driven Right Leg Circuit

The driven right leg circuit is another common technique used to reduce common-mode voltage in biopotential measurements. The DRL circuit actively drives the common-mode voltage of the body toward a reference potential (usually ground) by injecting feedback current into the body through a reference electrode (Fig. 12.4g shows an IA design with DRL) (Fay et al. 2009). This feedback loop enhances the CMRR of the system by reducing the common-mode voltage at the amplifier inputs. The DRL circuit compensates for electrode impedance mismatches and environmental noise sources, effectively rejecting common-mode interference such as power line noise. Designing an effective DRL circuit involves ensuring stability in the feedback loop and protecting against overcurrent conditions that could affect patient safety.

12.2.1.2 Noise Reduction Techniques

Chopper Stabilization

Chopper stabilization is a highly effective technique for reducing low-frequency noise, including flicker noise and baseline drift, in amplifiers used for biopotential signal acquisition (Enz et al. 1987). Flicker noise, also known as *1/f* noise, is particularly significant at low frequencies and arises due to carrier trapping and detrapping at the Si/SiO_2 interface. The power spectral density ($i_{n,f}^2$) of flicker noise is inversely proportional to frequency, making it a critical limitation in applications like EEG and ECG recording. The chopper stabilization process mitigates flicker noise by modulating the input signal with a square wave carrier, effectively shifting it to a higher frequency band where the amplifier's flicker noise is negligible. After amplification in this higher frequency band, the signal is demodulated back to its original band. This approach reduces the contribution of *1/f* noise to the overall noise spectrum and enhances the signal-to-noise ratio for microvolt-level signals.

$$\overline{i_{n,\mathrm{f}}^2} = \frac{g_{\mathrm{m}}^2 K}{C_{\mathrm{ox}} WL} \frac{1}{f} \Delta f$$

$$f_{\mathrm{c}} = \frac{K}{C_{\mathrm{ox}} WL} \frac{1}{2nKT}$$

Here K is a process-dependent constant that varies with fabrication technology, A higher K generally indicates worse flicker noise performance. g_{m} is transconductance of the transistor. C_{ox} is the gate oxide capacitance per unit area, representing the ability of the MOSFET gate to control the channel and f represents the frequency at which the noise power is measured. n is the subthreshold slope factor, a dimensionless parameter that quantifies how the MOSFET's current changes with gate-to-source voltage V_{GS} in the subthreshold (weak inversion) region, a smaller n implies a steeper subthreshold slope and thus a stronger modulation of the channel current by V_{GS}. As shown in the equation (i _ (n,f) ^ 2), Flicker noise decreases with increasing frequency, hence the *1/f* dependency. Flicker noise reduction can also be characterized by the *1/f* noise corner frequency (f_{c}), which defines the point where thermal noise begins to dominate over flicker noise.

Chopper stabilization ensures the chopping frequency is set significantly higher than f_{c}, often in the range of 100 Hz to 10 kHz for biomedical applications, thereby effectively suppressing flicker noise. Chopper amplifiers have demonstrated exceptionally low input-referred noise densities, sometimes below 50 nV/$\sqrt{\mathrm{Hz}}$, making them suitable for high-precision biopotential measurements in wearable and implantable systems. However, challenges such as ripple and spikes introduced by switching must be addressed through careful filter design, ensuring minimal distortion and high fidelity in the amplified signals (Wu et al. 2009). This combination of noise suppression and high precision makes chopper stabilization indispensable in low-noise amplifier design for biopotential applications.

Correlated Double Sampling (CDS)

Correlated double sampling is a technique used to reduce low-frequency noise including flicker noise (*1/f* noise), offset voltages, and drift in amplifiers. By sampling the input signal at two closely spaced times and subtracting the samples, CDS effectively cancels out noise components and offsets that are correlated over the sampling interval (Paul et al. 2022b; Enz and Temes 1996).

$$V_{out}(t) = A\left[V_{in}(t) + V_{OS} + n(t)\right]$$

Here $V_{in}(t)$ is the input signal, V_{OS} is the offset voltage, $n(t)$ is the low-frequency noise component and A is the amplifier gain, with CDS $n(t)$ cancel out because they are common to both samples.

$$\Delta V_{out} = V_{out}(t2) - V_{out}(t1) = A\Delta V_{in}$$

In other words, CDS preserves the instantaneous change in the input signal while nulling any slowly varying offset or noise component common to both measurement points. This approach is particularly effective in low-frequency applications such as electrophysiological signal acquisition or electrochemical sensing, where flicker noise and offset voltages can mask the true signal of interest. By carefully choosing the sampling interval to be short relative to the noise variation timescale, CDS ensures that the correlated noise and offsets are eliminated, improving accuracy and stability without necessitating complex additional circuitry. The result is a cleaner, more robust signal that supports the low-noise, low-power operation crucial for precision and wearable sensing platforms.

12.2.1.3 Improving Electrode-Skin Interface and Mitigate Interference

Lead-Off Detection

Maintaining good electrode-skin contact is crucial for reliable electrophysiological signal acquisition. Lead-off detection techniques monitor contact impedance to ensure signal integrity (Li et al. 2018). Alternating current injection methods inject a small AC current between electrodes, and the resulting voltage is measured to estimate impedance. Square wave injection methods inject a known square wave signal and monitor the attenuation of harmonics to assess contact quality. These methods enable the system to detect when an electrode has become detached or if contact impedance has increased beyond acceptable levels, allowing for corrective actions. Some systems incorporate automatic gain control or adaptive filtering based on lead-off detection to maintain signal quality.

Active Electrodes

Active electrodes integrate amplification directly at the electrode site, offering several advantages (Xu et al. 2011; Mathewson et al. 2017). By minimizing the length of conductors carrying weak signals, the susceptibility to noise and interference is reduced. This is particularly important in wearable systems where long cable runs cannot be entirely avoided. Active electrodes reduce motion artifacts by providing buffering and initial amplification close to the signal source, improving signal quality in dry or capacitive electrode setups where contact impedance can fluctuate. They also simplify cabling by reducing the number of necessary cables, facilitating compact designs suitable for wearable systems. Implementing active electrodes requires miniaturized, low-power amplifiers capable of operating in close proximity to the body. Thermal management and isolation have become critical considerations to ensure user safety and comfort.

Active Shielding
Active shielding is essential in wearable systems where electrode cables are prone to EMI, especially when not fully integrated. Active shielding involves driving a shield surrounding the signal-carrying conductors with a voltage that closely follows the signal line. By matching the shield voltage to the signal, the capacitance between them is minimized, and external noise sources have less influence on the signal path. This technique enhances signal quality, particularly during movement or in electrically noisy environments, by reducing capacitive coupling between the signal and external noise sources. Implementing active shielding requires additional circuitry to generate the shield voltage and maintain stability, but the benefits in noise reduction often justify the complexity.

12.2.2 Low-Power Analog Circuit Design Techniques

As precision health monitoring turns to wearable formats recently, they must operate within strict power budgets, often relying on small batteries. The following low-power design techniques have been reported for electrophysiological sensing circuits in achieving the required performance without compromising energy efficiency (Sarpeshkar 2010).

12.2.2.1 Subthreshold Operation in CMOS Transistors

Subthreshold weak inversion operation leverages the exponential relationship between the gate-to-source voltage (V_{gs}) and the and the drain current (I_D) for low-power circuit design, making it ideal for biomedical applications. Operating CMOS transistors in the subthreshold region leverages diffusion currents, providing high transconductance efficiency (Fig. 12.4d) (Sarpeshkar 2010; Sharroush et al. 2009).

$$I_{\mathrm{D}} = I_{\mathrm{DS0}} \frac{W}{L} e^{\frac{V_{\mathrm{gs}}}{nV_{\mathrm{th}}}} \left(1 - e^{-\frac{V_{\mathrm{ds}}}{V_{\mathrm{th}}}} \right)$$

$$g_{\mathrm{m}} = \frac{I_{\mathrm{D}}}{nV_{\mathrm{th}}}$$

Here I_{DS0} is the current at threshold, V_{th} is the threshold voltage, n is the subthreshold slope factor. In the subthreshold region, the drain current exhibits an exponential dependence on the gate-source voltage, allowing for significant current levels at low gate voltages. Subthreshold operation allows achieving a certain transconductance at lower current, improving noise efficiency ($g_{\mathrm{m}}/I_{\mathrm{D}}$), which is beneficial for low-power amplification. The noise performance is optimized with thermal noise proportional to I_{D}, and flicker noise reduced by increasing the transistor dimensions (W and L). Techniques like chopper stabilization and correlated double sampling which we discuss in later sections further mitigate noise for high SNR. While subthreshold operation minimizes power consumption, it requires careful management of limited bandwidth and process variations to ensure robust performance in wearable systems.

12.2.2.2 Pseudoresistors

Pseudoresistors are essential components in biomedical signal processing, particularly for implementing ultra-high resistance in AFE circuits designed for electrophysiological sensing. These elements enable the design of high-pass filters with sub-Hz cutoff frequencies, a critical requirement for capturing slow physiological signals like those found in ExG. Unlike bulky discrete resistors, pseudoresistors are compact and compatible with standard CMOS processes, making them a preferred solution for miniaturized biomedical systems.

A common approach to implement the pseudoresistor is the PMOS-based MOS-bipolar pseudoresistors (Fig. 12.4f right). The most basic topology among these is a PMOS whose gate and body terminals are connected. The gate-connected drain is recommended to be placed on a more leakage-sensitive side such as the floating inputs of an OTA, where the maximum resistance (r_{pr}, incremental resistance of the pseudoresistor) is achieved when the source and drain voltages are equal (Harrison and Charles 2003; Tajalli and Leblebici 2010).

$$r_{pr}\Big|_{V_{DS}=0} = \frac{1}{\left.\dfrac{\partial I_{\mathrm{D}}}{\partial V_{DS}}\right|_{V_{DS}=0}} = \frac{U_{\mathrm{T}}}{I_{\mathrm{D0}}}$$

Here U_{T} is the thermodynamic voltage and I_{D0} is the residual charge current given by:

$$I_{\mathrm{D0}} = 2n\mu C_{\mathrm{ox}}\frac{W}{L}$$

Here n is the subthreshold slope factor, accounting for the MOSFET's deviation from the ideal subthreshold behavior, μ is the carrier mobility, which is a material property determining how easily charge carriers (electrons or holes) move through the MOSFET channel under an electric field. C_{ox} is the gate oxide capacitance per unit area, a parameter that depends on the MOSFET's fabrication process and determines how strongly the gate voltage influences the channel. W/L is the width-to-length ratio of the MOSFET. A larger ratio increases the current (I_{D0}) and decreases the pseudoresistor resistance. This behavior, however, introduces challenges such as signal-dependent distortion when the voltage moves away from zero. To address these limitations, more advanced pseudoresistor designs have been developed. For example, PMOS-based MOS-bipolar pseudoresistors configured with symmetrical or outwardly connected gates can achieve higher resistance values and a greater voltage dynamic range (Ha et al. 2021). Configurations with inwardly connected gates further enhance resistance tunability (Xu et al. 2011). Another notable design employs a bias voltage applied to the PMOS gate, enabling controlled resistance tuning (Han et al. 2013). This design widens the linear operating range, minimizing signal-dependent distortion and improving performance in applications requiring high linearity. Balanced tunable pseudoresistors, which incorporate an auxiliary amplifier for precise control, offer even broader dynamic ranges and more consistent resistance characteristics (Zou et al. 2009). Despite their advantages, MOS-bipolar pseudoresistors are sensitive to process, voltage, and temperature variations, as well as electromagnetic interference. These sensitivities can lead to variability in the resistance values, impacting the cutoff frequency of the associated filters. To overcome these challenges, switched-capacitor pseudoresistors have been proposed as an alternative (Verma et al. 2010; Centurelli et al. 2021) (Fig. 12.4d bottom). In these designs, a capacitor and switching frequency f_s are used to emulate high resistance, with the equivalent resistance given by $R = 1/(f_s C)$. While this method provides better process variation insensitivity, it introduces constraints on the capacitor size and switching frequency. For example, achieving resistance values in the GΩ range requires extremely small capacitances (e.g., 1 fF) and high switching frequencies (e.g., 1 kHz), which may not be practical or compatible with biomedical signal requirements.

In summary, the combination of advanced circuit techniques, such as differential and instrumentation amplifiers, chopper stabilization, correlated double sampling, and active shielding, enables robust, low-power acquisition of electrophysiological signals. These methods effectively suppress noise, improve common-mode rejection, and preserve signal integrity, ensuring accurate detection of low-amplitude biopotential signals in real-world precision monitoring scenarios. With these enhanced analog front-end solutions in place, the next critical step involves digitizing the conditioned signals using appropriate analog-to-digital converter architectures, which will be explored in subsequent sections.

12.3 Circuit Techniques for Electrochemical Sensing

The emergence of wearable electrochemical biosensors has opened new horizons in health monitoring and diagnostics, enabling real-time detection of a wide array of biochemical markers directly from bodily fluids such as sweat, saliva, and interstitial fluid. These sensors operate by transducing biochemical reactions into measurable electrical signals through electrochemical processes. To fully exploit their potential, specialized interfacing circuits are essential to ensure accurate signal acquisition, low power consumption, and seamless integration with wearable platforms. This section explores the foundational techniques and innovations in the design of readout circuits for electrochemical biosensors, focusing on the challenges and solutions inherent in creating reliable and efficient systems tailored for wearable applications.

Electrochemical sensors often rely on redox reactions occurring at the working electrode (WE) to transduce chemical interactions into measurable electrical signals. Operating within a three-electrode system—which includes a working electrode (WE), a counter electrode (CE), and a reference electrode (RE)—allows precise control of the potential and current during measurements. This configuration is crucial for accurate detection and quantification of analytes in various applications. Redox sensing techniques are broadly classified into two main categories: amperometric and potentiometric methods (Fig. 12.3f). Each technique utilizes different measurement principles and requires specific circuit designs to optimize performance and reliability (Ma et al. 2023).

In a typical electrochemical sensing system (Lu et al. 2021a) (amperometric sensing example shown in Fig. 12.5a), the readout circuitry includes the potentiostat circuit, which controls and measures the electrochemical reactions, a signal conditioning block that amplifies and filters the measured signal to improve the signal-to-noise ratio, a processing unit, often a microcontroller or digital signal processor, that interprets the conditioned signal; and a communication module that facilitates data transmission to external devices for analysis and display. Integrating these components into a compact, low-power system suitable for wearable devices presents significant design challenges, requiring innovative circuit techniques and careful consideration of the electrochemical principles involved. As the core component of an electrochemical sensing system, a typical potentiostat circuit architecture comprises three key components (Fig. 12.5b) (Snizhko et al. 2023): the control amplifier (CA), the voltage follower (VF), and the transimpedance amplifier (TIA). Detailed potentiostat circuit design methodologies will be discussed further in the following sections.

12.3.1 Redox-Based Electrochemical Interface

12.3.1.1 Potentiometric Sensor Interface

Potentiometric sensors (Fig. 12.3f) measure the open-circuit potential difference between the WE and RE without drawing significant current (Snizhko et al. 2023; Grieshaber et al. 2008). This potential difference is related to the ion concentration of the analyte via the Nernst equation (Bard et al. 2022):

$$E = E^0 + \frac{RT}{nF} \ln \frac{[\mathrm{Red}]}{[\mathrm{Ox}]}$$

Here, E is the electrode potential, E^0 is the standard electrode potential, R is the universal gas constant, T is the absolute temperature in Kelvin, n is the number of electrons transferred, F is Faraday's constant, and [Red] and [Ox] are the concentrations of the reduced and oxidized forms of the analyte.

To preserve the open-circuit condition that ensures accurate potentiometric measurements, the interfacing circuitry must maintain extremely high input impedance so that virtually no current flows between the WE and RE. Drawing any noticeable current would disturb the equilibrium at the electrode interface, introducing polarization effects that shift the measured potential. As a result, no potentiostat function—enforcing a specific voltage or current—is needed. Instead, the system relies

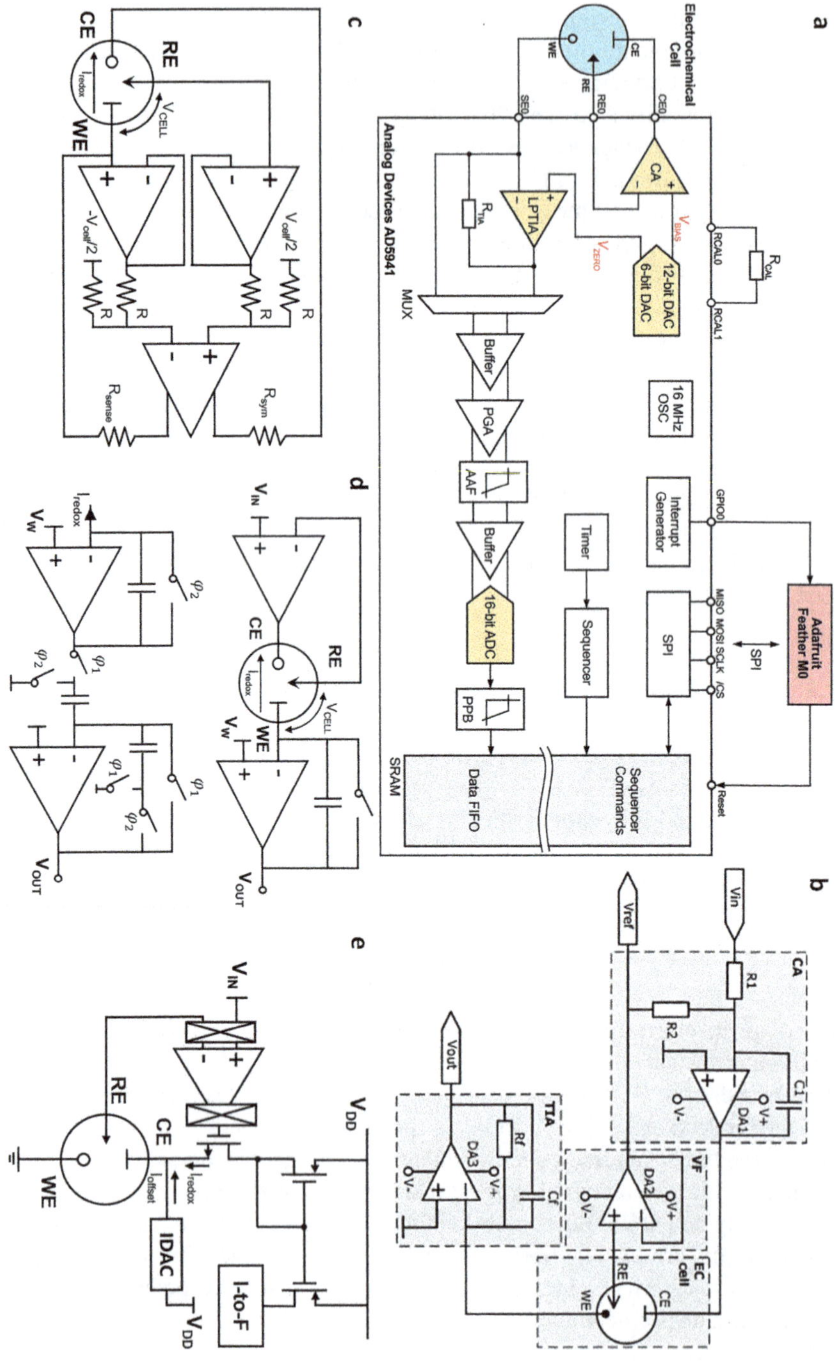
a
Electrochemical Cell
CE
RE
WE
CE0
RE0
SE0
CA
LPTIA
R_{TIA}
V_{BIAS}
V_{ZERO}
12-bit DAC
6-bit DAC
R_{CAL}
RCAL0
RCAL1
MUX
Buffer
PGA
AAF
Buffer
16-bit ADC
PPB
16 MHz OSC
Interrupt Generator
GPIO0
Timer
Sequencer
SPI
MISO MOSI SCLK /CS
SPI
Adafruit Feather M0
Reset
Sequencer Commands
Data FIFO
SRAM
Analog Devices AD5941
b
Vin
Vref
CA
R1
R2
C1
DA1
V+
V-
VF
DA2
EC cell
TIA
Rf
Cf
DA3
Vout
c
RE
CE
WE
I_{redox}
V_{CELL}
$V_{cell}/2$
$-V_{cell}/2$
R
R_{sym}
R_{sense}
d
V_{IN}
V_{CELL}
V_W
V_{OUT}
φ_1
φ_2
e
V_{IN}
V_{DD}
I_{redox}
I_{offset}
IDAC
I-to-F

Fig. 12.5 Designs of potentiostat circuits and sensing configurations for electrochemical sensing systems. (**a**) The architecture of the AD5941 potentiostat chipset (Analog Devices, USA) for electrochemical biosensors. The system integrates a control amplifier to maintain a fixed potential across the working and reference electrodes, a low-power transimpedance amplifier for current-to-voltage conversion, and a signal processing chain comprising a programmable gain amplifier (PGA), anti-aliasing filter (AAF), and 16-bit ADC. A sequencer and buffer facilitate digital command execution and data transfer to an external microcontroller via an SPI interface, enabling multiplexed sensor readings and low-power operation (from Bill et al. (2023)), reprinted with permission under Creative Commons Attribution license). (**b**) Schematic diagram of a common potentiostat circuit architecture for electrochemical sensing. The circuit consists of three main units: the control amplifier, voltage follower, and transimpedance amplifier. The control amplifier polarizes the working electrode by maintaining the desired potential difference between the working electrode and the reference electrode. The voltage follower ensures accurate measurement of the reference electrode's equilibrium potential while isolating it from the rest of the circuit. The transimpedance amplifier measures the redox current generated at the working electrode by converting it into a proportional voltage signal for analytical applications (from Snizhko et al. (2023), Elsevier, reprinted with permission). (**c**–**e**) Potentiostat circuit implementations (from Ma et al. (2023)), reprinted with permission under Creative Commons Attribution license): (**c**) Fully differential potentiostat designs, employing symmetrical feedback to stabilize V_{cell} while minimizing offset and noise for high-precision electrochemical measurements. (**d**) Capacitive feedback potentiostat with correlated double sampling technique, ideal for noise suppression and drift correction in long-term, low-current sensing applications. (**e**) Current-mirror-based readout circuits, which convert redox currents (I_{redox}) into frequency or other measurable parameters, providing compact and energy-efficient solutions for wearable electrochemical biosensors

on a high-impedance instrumentation amplifier or buffer stage that passively measures the electrode potential. Input impedances on the order of GΩ to TΩ prevent loading effects and preserve the integrity of the open-circuit measurement. Differential amplification techniques are often employed to reject common-mode noise and improve signal fidelity, enabling the detection of subtle potential differences even in low-concentration analyte scenarios. Additional design considerations include minimizing offset voltages and noise, both of which can obscure small signals, as well as ensuring temperature stability (Wang et al. 2011). Because the Nernst equation depends on temperature, fluctuations can affect the measured potential, making components with low temperature coefficients and potential temperature compensation strategies essential.

12.3.1.2 Amperometric Sensor Interface

Amperometric sensors (Fig. 12.3f) measure the current produced by the oxidation or reduction of an analyte at the WE (Grieshaber et al. 2008; Polat et al. 2022). The redox current is directly proportional to the analyte concentration and is governed by the Cottrell equation (Gmucová 2012):

$$I_{\text{redox}} = nFAc\sqrt{\frac{D}{\pi t}}$$

Here, in this equation, I_{redox} represents the redox current, n is the number of electrons transferred in the redox reaction, F is Faraday's constant, A is the area of the WE, c is the analyte concentration, D is the diffusion coefficient of the analyte, and t is the time since the application of the potential.

Voltage clamping, also known as potentiostatic control, maintains a fixed potential difference ($V_{\text{WE}} - V_{\text{RE}}$) between the WE and RE (Akinin et al. 2020). By controlling this potential, the electrochemical environment at the WE is precisely regulated, ensuring that the redox reaction occurs efficiently and that the resulting current is accurately proportional to the analyte concentration. In designing circuits for amperometric sensors, several circuit components are essential. The control amplifier adjusts the voltage at the CE to maintain stable potential at the WE relative to the RE. It operates in a feedback loop, continuously sensing the potential difference and compensating for any deviations by modulating the CE voltage. The transimpedance amplifier converts the redox current from the WE into a measurable voltage signal, providing the necessary gain and bandwidth to accurately capture current changes corresponding to varying analyte concentrations. A precise voltage reference is also critical, providing a stable baseline potential against which the CA sets the desired $V_{\text{WE}} - V_{\text{RE}}$ potential difference. This ensures the accuracy and stability of the measurement. Design considerations for amperometric sensor circuits include minimizing noise, ensuring control loop stability, and accommodating a wide dynamic range. Thermal and electronic noise can significantly affect the accuracy of current measurements, so employing low-noise amplifiers and proper filtering techniques is essential. The feedback loop must be carefully designed to prevent oscillations and ensure a rapid response to changes in the electrochemical environment. Additionally, the circuit must handle a wide range of currents, from picoamperes to microamperes, requiring components that maintain linearity across this spectrum.

12.3.2 Potentiostat Circuit Designs

Potentiostats are integral for precise control and measurement within three-electrode electrochemical systems. A common potentiostat architecture comprises three principal functional blocks, a voltage follower, a control amplifier, and a transimpedance amplifier, each serving distinct roles to achieve stable and accurate electrochemical measurements (Fig. 12.5b) (Snizhko et al. 2023). The voltage follower, often implemented with a high-impedance operational amplifier, is connected to the reference electrode to measure its equilibrium potential with minimal loading. By offering ultra-high input impedance and low input bias current, the VF preserves the RE's open-circuit condition, ensuring that the measured potential is a true representation of the solution's electrochemical state. The control amplifier then uses the VF output to adjust the CE potential according to the desired polarization program while maintaining the WE at a precise, stable potential relative to the RE. This closed-loop control relies on negative feedback to prevent the CA from becoming an uncontrolled source, ensuring stable operation across varying load conditions. The TIA amplifier measures the current flowing through the WE by converting it into a voltage signal proportional to analyte concentration. Operating the WE at a virtual ground, the TIA ensures accurate current measurement over a wide dynamic range without disturbing the electrochemical equilibrium. These three units often share similar power supply levels and must tolerate the compliance voltage range of the potentiostat. By carefully selecting op-amps with low noise, low offset, high gain, and suitable slew rates, and by incorporating stabilization elements (such as small feedback capacitors) as needed, designers create a robust and versatile potentiostat platform. This platform can handle various electrochemical cells, from three-electrode configurations to simplified two-electrode setups, and can be adapted with buffering stages to improve load driving capability. The closed-loop control of the potentiostat relies on feedback, which is important to accurately regulate the potential or current at the working electrode relative to a stable reference electrode. Previous potentiostat designs have employed resistive, capacitive, or transistor-based approaches to address specific performance needs, including sensitivity, noise reduction, power efficiency, and suitability for integration into precision electrochemical sensing devices (Snizhko et al. 2023).

12.3.2.1 Resistive Feedback Potentiostats

Resistive feedback transimpedance amplifiers (R-TIAs) are commonly used in amperometric sensing. In an R-TIA, the redox current from the WE is converted into a voltage signal through a feedback resistor (R_f) in the TIA (Rajabzadeh et al. 2018), following the relationship:

$$V_{\text{out}} = -I_{\text{redox}} \times R_{\text{f}}$$

The performance of an R-TIA hinges on balancing sensitivity and noise. Increasing the feedback resistance R_f amplifies small currents into measurable voltage levels, thereby enhancing the sensor's sensitivity. However, a higher R_f also increases thermal noise, calculated as $4kTR_f\Delta f$, where k is Boltzmann's constant, T is the absolute temperature, and Δf is the bandwidth. This thermal noise can obscure low-level signals, making it crucial to strike an optimal balance between sensitivity and noise.

To mitigate the trade-off between sensitivity and noise, PGAs or adjustable feedback networks are often incorporated. These configurations allow dynamic tuning of the gain to accommodate varying current levels without causing saturation. Additionally, limiting the sensor's bandwidth suppresses noise, enhancing the stability and clarity of the output signal. High values of R_{f}, while improving the detection of low-current signals, can lead to output saturation when measuring higher redox currents. To extend the dynamic range while preserving linearity, multi-stage amplification is utilized, distributing gain across multiple stages to prevent early saturation. Implementing fully differential R-TIAs offers several advantages (Ghodsevali et al. 2015; Martin et al. 2009). Differential configurations inherently reject common-mode noise, enhancing signal integrity (Fig. 12.5c). The output voltage swing is effectively doubled compared to single-ended designs, providing additional headroom for large signals. Furthermore, better matching of the signal paths reduces distortion, leading to more accurate measurements. Hybrid feedback schemes that combine resistive and capacitive feedback elements can also adjust the frequency response and extend the range of measurable currents. For measuring picoampere-level currents, ultra-high resistances are required, which are impractical with conventional resistors. Pseudoresistors, which has been introduced in earlier low-power design section, implemented using MOS transistors biased in the subthreshold (weak inversion) region, serve as effective alternatives, achieving resistance values in the gigaohm to teraohm range. Configuring back-to-back diode-connected MOSFETs minimize leakage currents and maintains linearity over a wide input range. Using matched pairs of MOSFETs improves process, voltage, and temperature stability, reducing resistance variability due to fabrication inconsistencies or environmental changes (Djekic et al. 2018).

12.3.2.2 Capacitive Feedback Potentiostats

Capacitive feedback transimpedance amplifiers (C-TIAs) integrate the redox current over time to produce a voltage signal (Fig. 12.5d top) (Rajabzadeh et al. 2018; Dai et al. 2016), according to the equation:

$$V_{\text{out}}(t) = -\frac{1}{C_{\text{f}}}\int_{t}^{0} I_{\text{redox}}(t')\,\text{d}t'$$

Here, C_{f} is the feedback capacitor. C-TIAs offer advantages in terms of low noise, as capacitors do not generate thermal noise, enhancing the signal-to-noise ratio. The integration process amplifies low-level currents over time, improving the detection of minute analyte concentrations.

A primary challenge with C-TIAs is preventing integrator saturation due to the accumulation of charge on the feedback capacitor, which can drive the output voltage beyond the amplifier's dynamic range and lead to signal distortion. To address this, various techniques are employed to reset or stabilize the integrator. One common method is the periodic reset, where a switch parallel to C_f periodically discharges the capacitor. This requires precise timing control to avoid introducing offset errors or discontinuities in the output signal. Another technique is correlated double sampling which is shared between electrophysiological and electrochemical sensing (Fig. 12.5d bottom) (Yang et al. 2009). CDS samples the integrator output at two closely spaced intervals and computes their difference. This effectively cancels offset voltages and low-frequency noise components, including flicker noise, thereby preventing saturation and improving stability. However, CDS may become less effective at higher frequencies where noise components are less correlated. Design considerations for C-TIAs also include minimizing leakage currents through C_f by selecting high-quality capacitors with low dielectric absorption. Additionally, the operational amplifier used must have low input-referred voltage and current noise to maintain a high SNR. Ensuring the stability of the feedback loop is crucial, requiring careful design of the frequency response and phase margin to prevent oscillations.

Capacitive feedback designs are particularly suited for applications requiring high sensitivity and low noise, such as detecting nanoampere or picoampere-level currents in electrochemical sensors. Their compatibility with CMOS fabrication processes facilitates integration into compact, low-power wearable and implantable devices. By effectively addressing challenges related to saturation and noise through advanced reset mechanisms and feedback stabilization techniques, C-TIAs enable precise and reliable current measurements in modern electrochemical potentiostats, enhancing the performance and reliability of sensors in biomedical and environmental monitoring applications.

12.3.2.3 Current Mirror Potentiostats

Current mirror (CM) potentiostats utilize transistor-based circuits to replicate the redox current and convert it into a usable signal (Fig. 12.5e). The redox current is directed into the input of a current mirror circuit, typically consisting of matched transistors. The output current is a scaled version of the input current, which can be used to generate a voltage across a load resistor or to charge a capacitor, producing a voltage or frequency signal proportional to the analyte concentration. CM potentiostats are advantageous in terms of compactness and power efficiency. They can be highly integrated using standard CMOS processes, making them suitable for multi-channel applications in wearable devices (Ahmadi and Jullien 2009).

However, CM potentiostats have limitations in bandwidth and sensitivity. The frequency response is limited by transistor characteristics and parasitic capacitances, which may affect the ability to track rapid changes in analyte concentration. Achieving linear response over a broad current range requires careful matching of transistors and consideration of non-ideal effects such as channel length modulation. Temperature effects on transistor parameters can also impact accuracy, necessitating temperature compensation techniques or circuit designs that minimize temperature sensitivity.

12.3.3 Circuit Techniques to Enhance Electrochemical Sensing

To further improve the performance of electrochemical sensors, additional circuit techniques are employed. Several circuit techniques are common among both electrophysiological and electrochemical sensing. Differential measurement configurations are fundamental in enhancing noise rejection for both sensing modalities. By amplifying the difference between two input signals while rejecting common-mode signals, differential amplifiers significantly improve the signal-to-noise ratio. Instrumentation amplifiers with high input impedance and precise gain settings are essential for matching electrode characteristics and ensuring accurate measurements (Martin et al. 2009). Physical symmetry in circuit layout and meticulous component matching further enhance common-mode rejection, reducing susceptibility to environmental noise and interference. This technique is crucial for minimizing the impact of electromagnetic interference and power line noise, which are common in both electrochemical and electrophysiological measures. Besides, guarding and shielding techniques are employed to minimize leakage currents and capacitive coupling that can degrade signal integrity, especially in low-current or low-voltage measurements typical of both sensor types (Shi et al. 2008). Guard rings can be placed around sensitive nodes with traces driven at the same potential, effectively reducing leakage paths and preventing unwanted current flow. Shielded cables and enclosures protect signal lines from EMI and electrostatic discharge. Proper grounding establishes a low-impedance ground plane, reducing noise coupling and ensuring stable reference potentials. Implementing these physical design practices is critical in achieving the low-noise performance required for accurate sensing in wearable devices. Lastly, chopper stabilization is a technique widely used in both electrochemical and electrophysiological sensing circuits to reduce low-frequency noise and offset errors in amplifiers, particularly flicker noise (1/f noise) that is prominent in CMOS devices (Shi et al. 2008; Lu et al. 2021b). In this method, the input signal is modulated to a higher frequency by a chopper circuit before amplification. Since flicker noise diminishes at higher frequencies, the amplifier operates in a range where this noise is less significant. After amplification, the signal is demodulated back to its original frequency, effectively filtering out the low-frequency noise components. Implementing chopper stabilization requires careful selection of the chopping frequency to balance noise reduction without interfering with the signal bandwidth.

While the aforementioned techniques are common to both electrochemical and electrophysiological sensing, electrochemical sensors have unique challenges that necessitate specialized solutions. Auto-zeroing and offset calibration (Enz and Temes 1996; Kern and Chen 2013) are particularly important in electrochemical sensing due to the need for high DC accuracy and long-term stability. Electrochemical measurements often involve detecting small changes in analyte concentrations over extended periods, making them susceptible to drift and offset errors caused by temperature fluctuations, component aging, and other factors. For example, in continuous lactate monitoring in sweat, offset drift and calibration stability are critical concerns. Lactate levels are often monitored over extended periods, ranging from several hours during endurance sports to full days in clinical settings, requiring the electrochemical sensor interface to maintain accuracy despite changing environmental conditions, electrode degradation, and component aging. Periodic auto-zeroing and background calibration techniques address these challenges by actively measuring and compensating for offset drift. In auto-zeroing, the system periodically disconnects the input signal and measures the inherent offset of the readout circuitry (Ghoreishizadeh et al. 2017). This offset value is then stored and subtracted from subsequent measurements, effectively canceling out slow drifts that accumulate over time. Offset calibration further refines accuracy by incorporating a controlled reference or known baseline condition against which the system can adjust its output. For example, in a lactate monitoring scenario, the device might periodically measure the response of the sensor in a neutral environment (where lactate concentration should be negligible) or during a brief pause in measurement (Hanitra et al. 2018). Using this known reference, the circuit's parameters can be fine-tuned to ensure that the reported lactate concentration remains accurate, compensating for electrode fouling, component drift, or changes in temperature or humidity.

The design of potentiostats for redox-based electrochemical interfaces is critical for the accurate and reliable operation of wearable biosensors. By carefully selecting and optimizing circuit components whether using resistive or capacitive feedback, current mirrors, or advanced noise reduction techniques. Specific requirements of the sensing application guides the choice of potentiostat design, ensuring that the device meets the necessary performance criteria while adhering to the constraints of wearable technology, such as size, power consumption, and user comfort.

12.4 Analog-to-Digital Converter Techniques for Biopotential Signal Acquisition

Optimized analog front-end designs for electrophysiological and electrochemical sensing play a critical role in preparing signals for accurate digitization by ADCs. The low amplitudes and susceptibility to noise in electrophysiological signals, as well as the current or voltage characteristics of electrochemical signals, necessitate careful amplification, filtering, and noise suppression before digitization. Electrophysiological circuits ensure high signal integrity by addressing challenges such as common-mode rejection and motion artifacts, while electrochemical circuits stabilize redox current and voltage measurements in potentiostat designs. These preparatory steps enhance the quality and dynamic range of the input signals, allowing ADCs to operate efficiently within their resolution and bandwidth constraints. Consequently, ADCs bridge the gap between the analog and digital domains, leveraging the refined outputs of electrophysiological and electrochemical circuits to ensure precise digital representations of biosignals for downstream processing and analysis. The selection and design of ADCs must complement the preceding circuit techniques to ensure overall system performance meets the stringent requirements of wearable biopotential monitoring. The ADC must efficiently translate analog signals into digital data while operating within power and size constraints. Two predominant ADC architectures used in low-power biopotential systems are successive approximation register ADCs, delta-sigma ADCs and recently the ADC-direct architecture (Ha et al. 2014, 2019).

12.4.1 Successive Approximation Register (SAR) ADCs

SAR ADCs are popular for medium-resolution (8–12 bits) applications due to their simplicity and energy efficiency (Fig. 12.6a). The typical SAR ADC consists of SAR logic that implements the binary search algorithm for successive approximation, a comparator that compares the input voltage with the reference voltage, and a capacitor digital-to-analog converter (DAC) array that adjusts the reference voltage for each bit decision. The operation involves a binary search algorithm that approximates the signal input over successive clock cycles, refining from the most significant bit (MSB) to the least significant bit (LSB). On each cycle, the SAR logic sets a bit in the DAC, and the comparator determines whether the input voltage is higher or lower than the DAC voltage, adjusting the bits accordingly (Min et al. 2013; Khayatzadeh et al. 2012). To enhance the energy efficiency of SAR ADCs, several architectural innovations are employed. Split capacitor DACs reduce the total capacitance and switch energy by splitting the capacitor array into smaller segments (Agnes et al. 2008). Charge recycling techniques minimize the energy required for charging and discharging capacitors during the conversion process (Ginsburg and Chandrakasan 2005).

While SAR ADCs are energy-efficient, they often require significant analog front-end amplification to bring biopotential signals within the input range of the ADC. Biopotential signals are typically in the microvolt to millivolt range, while SAR ADCs have input ranges of several hundred millivolts to a few volts. Additionally, anti-aliasing filters are necessary to prevent high-frequency noise from folding into the signal band, as SAR ADCs have limited inherent anti-aliasing properties.

12.4.2 Delta-Sigma (ΔΣ) ADCs

ΔΣ ADCs are well-suited for high-resolution (12–16 bits or more) biopotential applications due to their inherent noise shaping and oversampling capabilities. The architecture consists of integrators that accumulate the input signal differences over time, a quantizer that converts the integrated signal into a digital output, and a

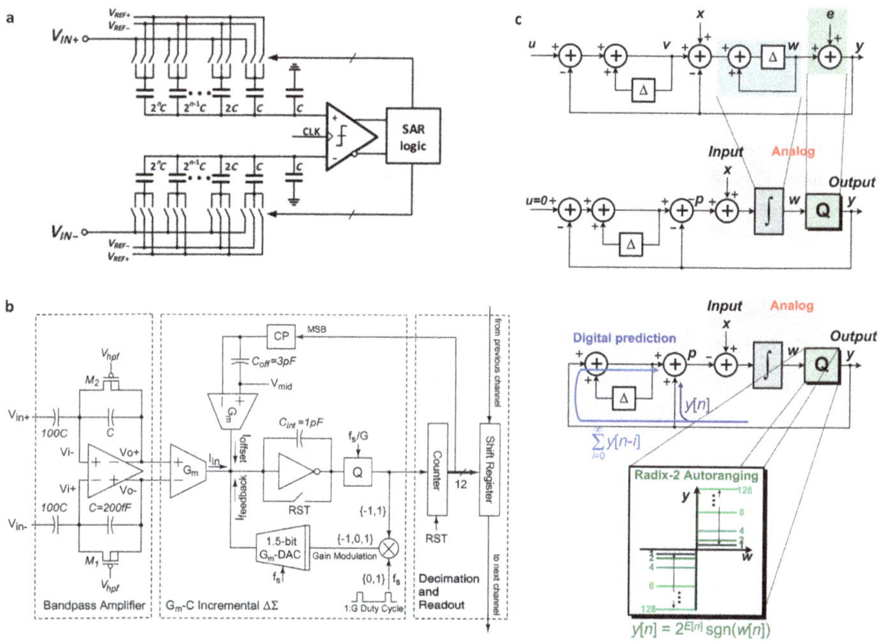

Fig. 12.6 ADC architectures for biopotential acquisition (from Ha et al. (2019)), Elsevier, reprinted with permission). (**a**) General architecture of a differential successive approximation register (SAR) ADC designed for low-power consumption and high-resolution digitization of biopotential signals, ensuring efficient energy use in physiological monitoring systems. (**b**) Implementation of a g_m-C incremental ΔΣ ADC, optimized for high-resolution and low-noise operation in biopotential recording systems. (**c**) ADC-direct frontend with predictive digital autoranging (PDA): (A) Generalized architecture of a second-order delta-sigma modulator (2DSM) for noise shaping and high dynamic range. (B) Continuous-time analog implementation of the second integrator, which directly couples to the input signal, ensuring stable operation and enhanced signal fidelity. (C) Digital accumulator implementation of the first integrator, with the auxiliary input $u = 0$, generating a digital prediction p for improved quantization accuracy. (D) PDA functionality through radix-2 dynamic expansion and compression of the quantization y, effectively adapting the quantizer range to accommodate large transients and maintain high-resolution signal integrity

feedback DAC that subtracts the quantized signal from the input to form a closed-loop system (Somappa and Baghini 2021; Marateb et al. 2009).

The ΔΣ modulator oversamples the input signal at a rate much higher than the Nyquist rate, typically 64 to 256 times the signal bandwidth. Noise shaping decreases the quantization noise by a factor of this oversampling rate, which can then be filtered out in the digital domain. Techniques such as chopping can also be employed to push the signal band of interest away from the effects of low-frequency noise. This design offers high resolution and accuracy due to oversampling and noise shaping. Next, they have inherent anti-aliasing properties because the modulator acts as a low-pass filter, reducing the need for complex analog anti-aliasing filters. Third, they offer flexibility, as the trade-off between resolution and bandwidth can be adjusted by modifying the oversampling ratio. Adding to the existing architecture, continuous-time ΔΣ ADCs offer additional benefits over discrete-time (switched-capacitor) implementations. They eliminate the need for sample-and-hold circuits, simplifying the analog front-end. Continuous-time operation can be more power-efficient because it avoids the power-hungry sampling switches and associated clocking circuitry. Furthermore, continuous time ΔΣ ADCs have enhanced linearity and reduced sensitivity to clock jitters and other non-idealities due to their inherent filtering properties.

Modern ΔΣ ADCs incorporate advanced features to improve performance in wearable biopotential applications. G_m-C integrators use transconductance amplifiers for integration (Fig. 12.6b), offering high linearity and low noise (Zhang et al. 2019; Mollazadeh et al. 2009). A transconductance cell converts the differential input voltage signal to a current, approximately linear over the voltage range of typical biopotentials. The difference between this current and a feedback current is integrated, and the resulting voltage is compared for three-level quantization of the feedback current implementing a continuous-time first-order modulator. A continuous-time oversampling ADC avoids the need for anti-aliasing filter and sample-and-hold circuits preceding the ADC. In addition, direct digital control over duty cycle in the feedback offers precise digital gain programmability from 1 to 4096. These integrators can be designed to operate in the subthreshold region for low-power applications.

12.4.3 ADC-Direct Frontend Design

The ADC-direct frontend represents a significant advancement in integrating analog-to-digital conversion for physiological signal acquisition. By combining high dynamic range, noise-shaping, and predictive techniques, this approach minimizes analog complexity while providing accurate biopotential sensing. The ADC-direct frontend leverages a hybrid analog-digital architecture, prominently featuring a second-order delta-sigma modulator (2nd-order ΔΣ modulator) with predictive digital auto-ranging (PDA). At the core of the ADC-direct frontend is the second-order delta-sigma modulator (2DSM) (Pavan et al. 2017), which employs two integrators to shape quantization noise and maximize resolution. The dynamics of the modulator are described by the following equations governing the integrators and the quantizer (Fig. 12.6c) (Kim et al. 2018).

First Integrator Output: $v[n] = v[n-1] + u[n] - y[n-1]$.

Second Integrator Output: $w[n] = w[n-1] + v[n] - y[n-1] + x[n]$

Quantizer Output: $y[n] = \text{Quantize}(w[n])$, for a simple 1-bit quantizer, the output is $+1$ for $w[n] \geq 0$ and -1 for $w[n] < 0$

Here $u[n]$ is the input to the first integrator (often zero in biopotential acquisition), $x[n]$ is the biopotential signal input, coupled directly to the second integrator, $v[n]$ is the output of the first integrator, $w[n]$ is the output of the second integrator, $y[n]$ is the output of the quantizer (ADC output), and $y[n-1]$ is the previous quantizer output, fed back for noise shaping. By directly coupling the biopotential signal $x[n]$ to the second integrator, the design reduces analog complexity and avoids saturation due to large electrode offset voltages, allowing seamless integration with biopotential signals and high conversion gain while maintaining a dynamic range suitable for biopotential measurements.

The key enabling techniques for the ADC-direct frontend are PDA and 2DSM. PDA enables adaptive quantization ranges for efficient signal tracking and dynamically adjusts the quantizer's range based on the transient behavior of the input signal, particularly during large excursions caused by artifacts (e.g., motion artifacts, electrode discharges). The 2DSM inherently shapes quantization noise by pushing it to higher frequencies, away from the signal band of interest. This is achieved through the modulator's feedback structure, where the quantization error is effectively integrated, resulting in a higher-order noise transfer function. By oversampling, the shaped quantization noise is spread over a wider frequency range, and the in-band noise (within the signal bandwidth) is significantly reduced. This enhances the SNR for low amplitude biopotential signals.

By minimizing analog complexity and leveraging advanced digital techniques, the ADC-direct frontend approach offers a compelling solution for precision and wearable biopotential monitoring. The combination of predictive digital auto-ranging and second-order delta-sigma modulation ensures high resolution, reduced noise, and adaptive handling of large transients. As a result, these designs deliver accurate and stable digital representations of physiological signals while operating within the stringent power and size constraints demanded by next-generation wearable healthcare devices.

12.5 Conclusion

This chapter presents a comprehensive exploration of the core building blocks and design considerations central to the integrated bioelectronics of precision health monitoring systems. Beginning with an architectural overview, we examined how emerging devices integrate diverse sensors, analog front ends, computational units, wireless communication interfaces, and power management circuits into compact, energy-efficient platforms. By combining electrophysiological and electrochemical sensing modalities, these systems capture a broad spectrum of physiological and biochemical signals, enriching our understanding of human health in real-world contexts.

Key challenges in developing such wearable devices, including low signal amplitudes, susceptibility to noise, variable electrode-skin impedance, and the need for low-power operation, necessitate specialized solutions at the circuit and system levels. For electrophysiological sensing, advanced instrumentation amplifiers, driven right leg circuits, chopper stabilization, correlated double sampling, and active shielding techniques effectively reduce noise and improve common-mode rejection. Similarly, electrochemical sensing strategies leverage potentiostats, pseudoresistors, and capacitive or current mirror feedback loops to achieve accurate, stable, and long-term monitoring of analyte concentrations in biofluids. Throughout the chapter, we also highlight the importance of careful analog front end design, including the selection of transistors operating in subthreshold for improved noise efficiency, and the use of on-chip pseudoresistors for ultra-high resistances at sub-Hz cutoff frequencies. These low-power techniques are paramount to ensuring that wearable and implantable devices remain unobtrusive and comfortable, operating for extended periods on minimal energy reserves.

The chapter also underscored the critical role of data conversion through energy-efficient ADC architectures—such as SAR and ΔΣ topologies—and introduced new concepts like the ADC-direct frontend with predictive digital auto-ranging. By aligning the ADC choice and configuration with upstream circuit techniques, designers can realize systems that dynamically adapt to changing signal conditions, achieve high resolution, and handle transient artifacts without sacrificing signal fidelity.

Looking ahead, the integration of edge intelligence, multi-modal sensor fusion, and tight coupling between front-end analog circuits and digital processing units will continue to enhance the capabilities of precision health monitoring platforms. The synergistic adoption of machine learning at the sensor node, improved materials for electrodes and sensing interfaces, and advanced packaging solutions will pave the way for next-generation devices that are not only accurate and reliable but also seamlessly blend into users' daily lives.

References

Agnes A, Bonizzoni E, Malcovati P, Maloberti F (2008) A 9.4-ENOB 1V 3.8μW 100kS/s SAR ADC with time-domain comparator. In: 2008 IEEE International Solid-State Circuits Conference - Digest of Technical Papers, pp 246–610. https://doi.org/10.1109/ISSCC.2008.4523149

Ahmadi MM, Jullien GA (2009) Current-Mirror-based potentiostats for three-electrode amperometric electrochemical sensors. IEEE Trans Circuits Syst Regul Pap 56:1339–1348

Akinin A, Paul A, Wang J, Buccino A, Cauwenberghs G (2020) Biopotential measurements and electrodes. In: He B (ed) Neural engineering. Springer International Publishing, Cham, pp 65–96. https://doi.org/10.1007/978-3-030-43395-6_2

Alatraktchi FA, Breum Andersen S, Krogh Johansen H, Molin S, Svendsen WE (2016) Fast selective detection of pyocyanin using cyclic voltammetry. Sensors 16:408

Ameri SK et al (2018) Imperceptible electrooculography graphene sensor system for human–robot interface. Npj 2D Mater Theory Appl 2:1–7

Aymerich Gubern J (2021) Low-power read-out ICs for smart electrochemical sensors. TDX (Tesis Doctorals en Xarxa) Universitat Politècnica de Catalunya. https://doi.org/10.5821/dissertation-2117-347911

Bagheri A, Salam MT, Perez Velazquez JL, Genov R (2017) Low-frequency noise and offset rejection in DC-coupled neural amplifiers: a review and digitally-assisted design tutorial. IEEE Trans Biomed Circuits Syst 11:161–176

Bandodkar AJ et al (2014) Epidermal tattoo potentiometric sodium sensors with wireless signal transduction for continuous non-invasive sweat monitoring. Biosens Bioelectron 54:603–609

Bard AJ, Faulkner LR, White HS (2022) Electrochemical methods: fundamentals and applications. John Wiley & Sons

Biglari A, Tang W (2023) A review of embedded machine learning based on hardware, application, and sensing scheme. Sensors 23:2131

Bill D et al (2023) Electrochemical methods in the cloud: FreiStat, an IoT-enabled embedded Potentiostat. Anal Chem 95:13003–13009
Bretschneider F, De Weille J (2019) Electronic and electrophysiological instrumentation. In: Introduction to electrophysiological methods and instrumentation, Elsevier, pp 77–121. https://doi.org/10.1016/B978-0-12-814210-3.00004-9
Cannard C, Wahbeh H, Delorme A (2021) Validating the wearable MUSE headset for EEG spectral analysis and Frontal Alpha Asymmetry. In: 2021 IEEE International Conference on Bioinformatics and Biomedicine (BIBM), pp 3603–3610. https://doi.org/10.1109/BIBM52615.2021.9669778
Centurelli F, Fava A, Scotti G, Trifiletti A (2021) A detailed model of the switched-resistor technique. IEEE Open J Circuits Syst 2:497–507
Cheng Y et al (2023) Dual-signal readout paper-based wearable biosensor with a 3D origami structure for multiplexed analyte detection in sweat. Microsyst Nanoeng 9:1–10
Chi YM, Jung T-P, Cauwenberghs G (2010) Dry-contact and noncontact biopotential electrodes: methodological review. IEEE Rev Biomed Eng 3:106–119
Chi YM, Maier C, Cauwenberghs G (2011) Ultra-high input impedance, low noise integrated amplifier for noncontact biopotential sensing. IEEE J Emerg Sel Top Circuits Syst 1:526–535
Chung HU et al (2019) Binodal, wireless epidermal electronic systems with in-sensor analytics for neonatal intensive care. Science 363:eaau0780
Crespi F (2020) Differential pulse voltammetry: evolution of an in vivo methodology and new chemical entries, a short review. J New Dev Chem 2:20–28
Custódio JR, Goes J, Paulino N, Oliveira JP, Bruun E (2013) A 1.2-V 165–/spl mu/W 0.29-mm2 multibit sigma-delta ADC for hearing aids using nonlinear DACs and with over 91 dB dynamic-range. IEEE Trans Biomed Circuits Syst 7:376–385
Dai S, Perera RT, Yang Z, Rosenstein JK (2016) A 155-dB dynamic range current measurement front end for electrochemical biosensing. IEEE Trans Biomed Circuits Syst 10:935–944
Das DM et al (2019) Design and development of an Internet-of-Things enabled wearable ExG measuring system with a novel signal processing algorithm for electrocardiogram. IET Circuits Devices Syst 13:903–907
Dasenbrock S et al (2022) Synchronization of ear-EEG and audio streams in a portable research hearing device. Front Neurosci 16:13
Djekic D, Fantner G, Lips K, Ortmanns M, Anders J (2018) A 0.1% THD, 1-M Ømega to 1-G Ømega tunable, temperature-compensated transimpedance amplifier using a multi-element pseudo-resistor. IEEE J Solid State Circuits 53:1913–1923
Enz CC, Temes GC (1996) Circuit techniques for reducing the effects of op-amp imperfections: autozeroing, correlated double sampling, and chopper stabilization. Proc IEEE 84:1584–1614
Enz CC, Vittoz EA, Krummenacher F (1987) A CMOS chopper amplifier. IEEE J Solid State Circuits 22:335–342
Fay L, Misra V, Sarpeshkar R (2009) A micropower electrocardiogram amplifier. IEEE Trans Biomed Circuits Syst 3:312–320
Gao M et al (2023) Wearable power management system enables uninterrupted battery-free data-intensive sensing and transmission. Nano Energy 107:108107
Ghodsevali E, Gosselin B, Boukadoum M, Miled A (2015) High accuracy and sensitivity differential potentiostat with amplifier-based error cancellation feedback loop. In: 2015 IEEE Biomedical Circuits and Systems Conference (BioCAS), pp 1–4. https://doi.org/10.1109/BioCAS.2015.7348424
Ghoreishizadeh SS, Taurino I, De Micheli G, Carrara S, Georgiou P (2017) A differential electrochemical readout ASIC with heterogeneous integration of bio-nano sensors for amperometric sensing. IEEE Trans Biomed Circuits Syst 11:1148–1159
Ginsburg BP, Chandrakasan AP (2005) An energy-efficient charge recycling approach for a SAR converter with capacitive DAC. In: 2005 IEEE International Symposium on Circuits and Systems, vol 1, pp 184–187. https://doi.org/10.1109/ISCAS.2005.1464555

Gmucová K (2012) A review of non-Cottrellian diffusion towards micro-and nano-structured electrodes. In: Electrochemical cells - New advances in fundamental researches and applications, pp 3–16

Grieshaber D, MacKenzie R, Vörös J, Reimhult E (2008) Electrochemical biosensors - sensor principles and architectures. Sensors 8:1400–1458

Ha S et al (2014) Integrated circuits and electrode interfaces for noninvasive physiological monitoring. IEEE Trans Biomed Eng 61:1522–1537

Ha S, Kim C, Mercier PP, Cauwenberghs G (2019) ECoG signal coding and digitization. In: High-density integrated electrocortical neural interfaces. Elsevier, pp 51–71. https://doi.org/10.1016/B978-0-12-815115-0.00010-6

Ha S et al (2021) Low-power integrated circuits for wearable electrophysiology. In: Wearable sensors. Elsevier, pp 163–199. https://doi.org/10.1016/B978-0-12-819246-7.00006-1

Han D, Zheng Y, Rajkumar R, Dawe GS, Je M (2013) A 0.45 V 100-channel neural-recording IC With sub-\mu \rm W/channel consumption in 0.18 \mu\rm m CMOS. IEEE Trans Biomed Circuits Syst 7:735–746

Hanitra IN et al (2018) A flexible front-end for wearable electrochemical sensing. In: 2018 IEEE International Symposium on Medical Measurements and Applications (MeMeA), pp 1–6. https://doi.org/10.1109/MeMeA.2018.8438787

Harrison RR, Charles C (2003) A low-power low-noise CMOS amplifier for neural recording applications. IEEE J Solid State Circuits 38:958–965

Hasan S, Pantha T, Arafat MA (2024) Design and development of a cost-effective portable IoT enabled multi-channel physiological signal monitoring system. Biomed Eng Adv 7:100124

Heikenfeld J et al (2018) Wearable sensors: modalities, challenges, and prospects. Lab Chip 18:217–248

Hu X, Song A, Wang J, Zeng H, Wei W (2022) Finger movement recognition via high-density electromyography of intrinsic and extrinsic hand muscles. Sci Data 9:373

Hunkin H, King DL, Zajac IT (2021) Evaluating the feasibility of a consumer-grade wearable EEG headband to aid assessment of state and trait mindfulness. J Clin Psychol 77:2559–2575

Imani S et al (2016) A wearable chemical–electrophysiological hybrid biosensing system for real-time health and fitness monitoring. Nat Commun 7:1–7

Ito M (2023) Architecture challenges for heterogeneous processors in embedded SoCs. In: 2023 IEEE Asian Solid-State Circuits Conference (A-SSCC), pp 1–2. https://doi.org/10.1109/A-SSCC58667.2023.10348007

Jia W et al (2013) Electrochemical tattoo biosensors for real-time noninvasive lactate monitoring in human perspiration. Anal Chem 85:6553–6560

Kappel SL, Kidmose P (2015) Study of impedance spectra for dry and wet EarEEG electrodes. In: 2015 37th Annual International Conference of the Ieee Engineering in Medicine and Biology Society, pp 3161–3164

Kappel SL, Rank ML, Toft HO, Andersen M, Kidmose P (2019) Dry-contact electrode ear-EEG. IEEE Trans Biomed Eng 66:150–158

Kaveh R, Schwendeman C, Pu L, Arias AC, Muller R (2024) Wireless ear EEG to monitor drowsiness. Nat Commun 15:6520

Kern T, Chen T (2013) A low-power, offset-corrected potentiostat for chemical imaging applications. In: 2013 IEEE 4th Latin American Symposium on Circuits and Systems (LASCAS), pp 1–4. https://doi.org/10.1109/LASCAS.2013.6519041

Khayatzadeh M, Zhang X, Tan J, Liew W-S, Lian Y (2012) A 0.7-V 17.4-μW 3-lead wireless ECG SoC. In: 2012 IEEE Biomedical Circuits and Systems Conference (BioCAS), pp 344–347. https://doi.org/10.1109/BioCAS.2012.6418426

Kim C et al (2018) Sub- μ Vrms-noise sub- μ W/channel ADC-direct neural recording with 200-mV/ms transient recovery through predictive digital autoranging. IEEE J Solid State Circuits 53:3101–3110

Kołodziej M et al (2020) Fatigue detection caused by office work with the use of EOG signal. IEEE Sensors J 20:15213–15223

Kouhalvandi L, Matekovits L, Peter I (2023) Amplifiers in biomedical engineering: a review from application perspectives. Sensors 23:2277

Lee S et al (2024) Fabric-based lamina emergent MXene-based electrode for electrophysiological monitoring. Nat Commun 15:5974

Li S, Wang Z, Li C (2018) A wearable EEG real-time measure and analysis platform for home applications. In: 2018 IEEE International Conference on Intelligence and Safety for Robotics (ISR), pp 504–509. https://doi.org/10.1109/IISR.2018.8535959

Li Y et al (2024) Advancing EEG-based brain-computer interface technology via PEDOT:PSS electrodes. Matter 7:2859–2895

Lin J et al (2020) MCUNet: Tiny deep learning on IoT devices. In: Advances in neural information processing systems, vol 33. Curran Associates, Inc., pp 11711–11722

Lin C-T, Jiang W-L, Chen S-F, Huang K-C, Liao L-D (2021) Design of a Wearable eye-Movement Detection System Based on electrooculography signals and its experimental validation. Biosensors 11:343

Lu S-Y et al (2021a) A review of CMOS electrochemical readout Interface designs for biomedical assays. IEEE Sensors J 21:12469–12483

Lu S-Y et al (2021b) 18.4 A wireless multimodality system-on-a-chip with time-based resolution scaling technique for chronic wound monitoring. In: 2021 IEEE International Solid-State Circuits Conference (ISSCC), vol 64, pp 282–284

Ma Y et al (2023) A review of electrochemical electrodes and readout Interface designs for biosensors. IEEE Open J Solid-State Circuits Soc 3:76–88

Mahato K, Wang J (2021) Electrochemical sensors: from the bench to the skin. Sens Actuators B Chem 344:130178

Maragliulo S, Lopes PFA, Osório LB, De Almeida AT, Tavakoli M (2019) Foot gesture recognition through dual channel wearable EMG system. IEEE Sensors J 19:10187–10197

Marateb HR, Introzzi R, Merletti R, Cescon C (2009) A low-noise front-end amplifier for biopotential recordings. In: Dössel O, Schlegel WC (eds) World Congress on Medical Physics and Biomedical Engineering, September 7–12, 2009, Munich, Germany. Springer, Berlin, Heidelberg, pp 608–611. https://doi.org/10.1007/978-3-642-03885-3_169

Martin SM, Gebara FH, Strong TD, Brown RB (2009) A fully differential Potentiostat. IEEE Sensors J 9:135–142

Mathewson KE, Harrison TJL, Kizuk SAD (2017) High and dry? Comparing active dry EEG electrodes to active and passive wet electrodes. Psychophysiology 54:74–82

Min Y-J et al (2013) Design of wavelet-based ECG detector for implantable cardiac pacemakers. IEEE Trans Biomed Circuits Syst 7:426–436

Mohan AMV, Rajendran V, Mishra RK, Jayaraman M (2020) Recent advances and perspectives in sweat based wearable electrochemical sensors. TrAC Trends Anal Chem 131:116024

Moin A et al (2021) A wearable biosensing system with in-sensor adaptive machine learning for hand gesture recognition. Nat Electron 4:54–63

Mollazadeh M, Murari K, Cauwenberghs G, Thakor N (2009) Micropower CMOS integrated low-noise amplification, filtering, and digitization of multimodal neuropotentials. IEEE Trans Biomed Circuits Syst 3:1–10

Muller R, Gambini S, Rabaey JM (2012) A 0.013 \hbox mm^2, 5 \mu\hboxW, DC-coupled neural signal acquisition IC With 0.5 V supply. IEEE J Solid State Circuits 47:232–243

Nguyen H-T, Mai N-D, Lee BG, Chung W-Y (2023) Behind-the-ear EEG-based wearable driver drowsiness detection system using embedded tiny neural networks. IEEE Sensors J 23:23875

Niso G, Romero E, Moreau JT, Araujo A, Krol LR (2023) Wireless EEG: a survey of systems and studies. NeuroImage 269:119774

Park J, Sun F, Xie Y, Xiong Z, Xu G (2020) Low-impedance low-artifact PEDOT: PSS-coated graphene electrodes towards high density optogenetic electrophysiology. IEEE Electron Device Lett 41:1261–1264

Paul A, Lee M, Xu Y, Deiss S, Cauwenberghs G (2022a) A versatile in-ear biosensing system for continuous brain and health monitoring. IEEE. https://doi.org/10.1109/iscas48785.2022.9937994

Paul A et al (2022b) Neural recording analog front-end noise reduction with digital correlated double sampling. IEEE. https://doi.org/10.1109/biocas54905.2022.9948618

Pavan S, Schreier R, Temes GC (2017) Understanding delta-sigma data converters. John Wiley & Sons

Petkos K et al (2019) A high-performance 8 nV/$\sqrt{}$Hz 8-channel wearable and wireless system for real-time monitoring of bioelectrical signals. J NeuroEng Rehabil 16:156

Polat EO et al (2022) Transducer technologies for biosensors and their wearable applications. Biosensors 12:385

Price CE et al (2023) Feasibility of intraoperative continuous glucose monitoring: an observational study in general surgery patients. J Clin Anesth 87:111090

Rajabzadeh M et al (2018) Comparison study of integrated potentiostats: Resistive-TIA, capacitive-TIA, CT ΣΔ modulator. In: 2018 IEEE International Symposium on Circuits and Systems (ISCAS), pp 1–5. https://doi.org/10.1109/ISCAS.2018.8351029

Randazzo V, Ferretti J, Pasero E (2019) ECG WATCH: a real time wireless wearable ECG. In: 2019 IEEE International Symposium on Medical Measurements and Applications (MeMeA), pp 1–6. https://doi.org/10.1109/MeMeA.2019.8802210

Röddiger T et al (2022) Sensing with Earables: a systematic literature review and taxonomy of phenomena. Proc ACM Interact Mob Wearable Ubiquitous Technol 6:1–57

Sarpeshkar R (2010) Ultra low power bioelectronics: fundamentals, biomedical applications, and bio-inspired systems. Cambridge University Press, Cambridge. https://doi.org/10.1017/CBO9780511841446

Satija U, Ramkumar B, Manikandan MS (2015) A simple method for detection and classification of ECG noises for wearable ECG monitoring devices. In: 2015 2nd International Conference on Signal Processing and Integrated Networks (SPIN), pp 164–169. https://doi.org/10.1109/SPIN.2015.7095425

Seker S, Soulis D, Moon J-M, Economou A, Wang J (2024) Decentralized touch-based micronutrition sensor towards personalized nutrition: parallel detection of sweat ascorbic acid and zinc ions. Sens Actuators B Chem 408:135525

Sempionatto JR et al (2021a) An epidermal patch for the simultaneous monitoring of haemodynamic and metabolic biomarkers. Nat Biomed Eng 5:737–748

Sempionatto JR, Moon J-M, Wang J (2021b) Touch-based fingertip blood-free reliable glucose monitoring: personalized data processing for predicting blood glucose concentrations. ACS Sens 6:1875–1883

Sempionatto JR, Lasalde-Ramírez JA, Mahato K, Wang J, Gao W (2022) Wearable chemical sensors for biomarker discovery in the omics era. Nat Rev Chem 6:899–915

Shao L, Guo Y, Liu W, Sun T, Wei D (2019) A flexible dry electroencephalogram electrode based on graphene materials. Mater Res Express 6:085619

Sharma M (2022) Hardware architecture of IoT and wearable devices. In: Internet of Things. CRC Press

Sharroush SM, Abdalla YS, Dessouki AA, El-Badawy E-SA (2009) Subthreshold MOSFET transistor amplifier operation. In: 2009 4th International Design and Test Workshop (IDT), pp 1–6. https://doi.org/10.1109/IDT.2009.5404144

Shi Y, Dou H, Zhou A, Chen Y (2008) Design and fabrication of a miniaturized electrochemical instrument and its preliminary evaluation. Sens Actuators B Chem 131:516–524

Shin JH et al (2022) Wearable EEG electronics for a brain–AI closed-loop system to enhance autonomous machine decision-making. Npj Flex Electron 6:1–12

Siuly S, Li Y, Zhang Y (2016) EEG signal analysis and classification. Springer International Publishing, Cham. https://doi.org/10.1007/978-3-319-47653-7

Snizhko D, Zholudov Y, Kukoba A, Xu G (2023) Potentiostat design keys for analytical applications. J Electroanal Chem 936:117380

Somappa L, Baghini MS (2021) A 400 mV 160 nW/Ch compact energy efficient \Delta \Sigma modulator for multichannel biopotential signal acquisition system. IEEE Trans Biomed Circuits Syst 15:765–776

Steyaert MSJ, Sansen WMC (1987) A micropower low-noise monolithic instrumentation amplifier for medical purposes. IEEE J Solid State Circuits 22:1163–1168

Tajalli A, Leblebici Y (2010) Subthreshold source-coupled logic. In: Tajalli A, Leblebici Y (eds) Extreme low-power mixed signal IC design: subthreshold source-coupled circuits. Springer, New York, pp 61–97. https://doi.org/10.1007/978-1-4419-6478-6_3
Taji B, Chan ADC, Shirmohammadi S (2018) Effect of pressure on skin-electrode impedance in wearable biomedical measurement devices. IEEE Trans Instrum Meas 67:1900–1912
Tang T, Goh WL, Yao L, Cheong JH, Gao Y (2020) An integrated multi-channel biopotential recording analog front-end IC with area-efficient driven-right-leg circuit. IEEE Trans Biomed Circuits Syst 14:297–304
Tehrani F et al (2022) An integrated wearable microneedle array for the continuous monitoring of multiple biomarkers in interstitial fluid. Nat Biomed Eng 6:1214–1224
Thomson EA et al (2019) Heart rate measures from the Apple Watch, Fitbit Charge HR 2, and electrocardiogram across different exercise intensities. J Sports Sci 37:1411–1419
Ullah H et al (2022) Recent advances in stretchable and wearable capacitive electrophysiological sensors for long-term health monitoring. Biosensors 12:630
Verma N et al (2010) A micro-power EEG acquisition SoC with integrated feature extraction processor for a chronic seizure detection system. IEEE J Solid State Circuits 45:804–816
Vittoz EA (2006) Origins of weak inversion (or sub-threshold) circuit design. In: Wang A, Calhoun BH, Chandrakasan AP (eds) Sub-threshold design for ultra low-power systems. Springer US, Boston, pp 7–s. https://doi.org/10.1007/978-0-387-34501-7_2
Wang W-S et al (2011) Real-time telemetry system for amperometric and potentiometric electrochemical sensors. Sensors 11:8593–8610
Wang H et al (2022) Robust tattoo electrode prepared by paper-assisted water transfer printing for wearable health monitoring. IEEE Sensors J 22:3817–3827
Webster JG (2009) Medical instrumentation: application and design. John Wiley & Sons
Wu R, Makinwa KAA, Huijsing JH (2009) A chopper current-feedback instrumentation amplifier with a 1 mHz 1/f noise corner and an AC-coupled ripple reduction loop. IEEE J Solid State Circuits 44:3232–3243
Xu J et al (2011) A 160 \mu \rm W 8-channel active electrode system for EEG monitoring. IEEE Trans Biomed Circuits Syst 5:555–567
Xu J et al (2013) Measurement and analysis of current noise in chopper amplifiers. IEEE J Solid State Circuits 48:1575–1584
Xu Y et al (2023) In-ear integrated sensor array for the continuous monitoring of brain activity and of lactate in sweat. Nat Biomed Eng 7:1307–1320
Yang C, Huang Y, Hassler BL, Worden RM, Mason AJ (2009) Amperometric electrochemical microsystem for a miniaturized protein biosensor array. IEEE Trans Biomed Circuits Syst 3:160–168
Yazicioglu RF, van Hoof C, Puers R (2008) Biopotential readout circuits for portable acquisition systems. Springer Science & Business Media
Yeung KK et al (2021) Recent advances in electrochemical sensors for wearable sweat monitoring: a review. IEEE Sensors J 21:14522–14539
Yin L et al (2022) A stretchable epidermal sweat sensing platform with an integrated printed battery and electrochromic display. Nat Electron 5:694–705
Yu Y et al (2020) Biofuel-powered soft electronic skin with multiplexed and wireless sensing for human-machine interfaces. Sci Robot 5:eaaz7946
Zhang Y, Basak D, Pun K-P (2019) A highly linear multi-level SC DAC in a power-efficient Gm-C continuous-time delta-sigma modulator. IEEE Trans Circuits Syst Regul Pap 66:4592–4605
Zhu M et al (2021) Flexible electrodes for in vivo and in vitro electrophysiological signal recording. Adv Healthc Mater 10:2100646
Zhu L et al (2024) A high CMRR differential difference amplifier employing combined input pairs for neural signal recordings. IEEE Trans Biomed Circuits Syst 18:100–110
Zou X, Xu X, Yao L, Lian Y (2009) A 1-V 450-nW fully integrated programmable biomedical sensor Interface Chip. IEEE J Solid State Circuits 44:1067–1077

GPSR Compliance
The European Union's (EU) General Product Safety Regulation (GPSR) is a set of rules that requires consumer products to be safe and our obligations to ensure this.

If you have any concerns about our products, you can contact us on

ProductSafety@springernature.com

In case Publisher is established outside the EU, the EU authorized representative is:

Springer Nature Customer Service Center GmbH
Europaplatz 3
69115 Heidelberg, Germany

www.ingramcontent.com/pod-product-compliance
Lightning Source LLC
Chambersburg PA
CBHW051424031125
34911CB00001B/26
9789819521166